Particle Technology Forum

Presentations at the 2009 AIChE Annual Meeting

Nashville, Tennessee, USA
8-13 November 2009

ISBN: 978-1-61567-927-0

Printed from e-media with permission by:

Curran Associates, Inc.
57 Morehouse Lane
Red Hook, NY 12571

Some format issues inherent in the e-media version may also appear in this print version.

Printed by Curran Associates, Inc. (2010)

For permission requests, please contact AIChE
at the address below.

AIChE
3 Park Avenue
New York, NY 10016-5991

Phone: (203) 702-7660
Fax: (203) 775-5177

www.aiche.org

Additional copies of this publication are available from:

Curran Associates, Inc.
57 Morehouse Lane
Red Hook, NY 12571 USA
Phone: 845-758-0400
Fax: 845-758-2634
Email: curran@proceedings.com
Web: www.proceedings.com

TABLE OF CONTENTS

An Experimental Investigation of Nucleation Phenomenon in a Static Powder Bed

Andrew C.S. Lee, School of Mechanical Engineering, Purdue University, West Lafayette, IN

Paul E. Sojka, School of Mechanical Engineering, Purdue University, West Lafayette, IN

Wet granulation is a particle agglomeration process where fine particles agglomerate together with a granulating liquid to form larger aggregates. Despite its wide application in industries such as pharmaceuticals, chemical plants, food processing, detergent production, and agriculture, the scientific understanding of the process is very limited. One of the primary reasons is that wet granulation is a complex process that involves several of tightly coupled sub-processes, which exist in both dry and wet states. Over the years, researchers have established qualitative descriptions of the sub-processes that control the wet granulation behavior. However, the quantitative understanding, or the ability to predict the outcome of wet granulation, is far from complete. Consequences are high manufacturing cost, high recycle ratio, dependency on empirical correlations, inability to predict the effects of scale up on the final products, and inability to analyze or understand reasons for manufacturing failures [1], which ultimately translates into high product cost. This warrants a continued endeavor of both theoretical and experimental investigations of wet granulation.

One of the least understood areas in wet granulation is nucleation. It is generally described as the process where the granulating liquid fist comes in contact with the powder bed and forms highly saturated initial agglomerates called nuclei. Although nucleation is thought to occur immediately upon the initial liquid-powder bed interaction, only few studies have investigated nucleation at such time regime [2, 3]. We do not know how nucleation kinetic or nucleus size is influenced by various impact conditions that may exist in wet granulation.

In this study, influences of the liquid drop impact velocity and liquid's physical properties on nucleation kinetics and nucleus size were investigated. Drops of water and glycerin/water mixtures were released from various heights onto a bed of glass beads. The moment of impact and formation of nucleus were recorded using a high speed video camera with a frame rate of up to 5000 frames per second. The high speed video reveals that, when a liquid drop impacts the glass beads, a nucleus is formed as the liquid drop initially spreads over the glass beads and retracts into a spherical shape while the glass beads are pulled into the liquid. The surface of nucleus appears to be highly saturated with liquid and the shape continues to oscillate even after the initial formation, which only takes a few milliseconds to occur. Once the oscillation stabilizes, the degree of surface saturation changes as the liquid in the nucleus is drained through the bottom of the nucleus via capillary action until equilibrium is reached. The shape and the size of nuclei, as well as the time it takes to reach the equilibrium, are found to be highly dependent on the impact velocity and liquid viscosity.

A time dependent nucleation model has been developed based on the capillary penetration of liquid into a porous bed. It correctly predicts the effect of liquid viscosity on the nucleus size. In order for the model to be completely analytical, we need a mathematical model that can predict the spreading profile of an impacting liquid drop on an unconsolidated powder bed surface. Such model is not available in the literature. Using experimental data for the spreading profile, the nucleation model predicts the nucleus size within a reasonable accuracy.

References

1. Iveson, S.M., Litster, J.D., Hapgood, K.P., and Ennis, B.J., "Nucleation, growth and breakage phenomena in agitated wet granulation processes: a review," Powder Technology, Vol. 117, pp. 3-39, 2001.

2. Agland, S. and Iveson, S.M., "The impact of liquid drops on powder bed surfaces," CHEMECA 99, New

Castle, Australia, 1999.

3. Hapgood, K.P., PhD Thesis, University of Queensland, Brisbane, Australia, 2000.

See more of this Session: Agglomeration and Granulation Processes
See more of this Group/Topical: Particle Technology Forum

Design of Regime Separated Granulation

Defne Kayrak-Talay, School of Chemical Engineering, Purdue University, West Lafayette, IN

Heather N. Emady, School of Chemical Engineering, Purdue University, West Lafayette, IN

William C. Schwerin, UOP, a Honeywell Company, Des Plaines, IL

James D. Litster, School of Chemical Engineering; Department of Industrial and Physical Pharmacy, Purdue University, West Lafayette, IN

Currently, many of the granulation rate processes occur simultaneously in industrial applications, making it difficult to control product properties, such as size and density. These three rate processes include wetting and nucleation, consolidation and growth, and breakage and attrition. The aim of this research is to physically separate the different rate processes to get better control of the product granule attributes.

A new design was developed that incorporates the wetting and nucleation, and also the consolidation and growth mechanisms in two separate pieces of process equipment. The first piece of equipment is a conveyor belt apparatus that achieves monosized nuclei granules from monosized drops on a bed of powder. Following nucleation, the granules and fines are transferred to a tumbling drum, where they densify and grow by layering of the fines. By separating the two rate processes, extremely tight control of the granule size and density is achieved. Results for the granulation of two different model powders using this new regime separated granulation process will be presented.

See more of this Session: Agglomeration and Granulation Processes
See more of this Group/Topical: Particle Technology Forum

A Quantitative Assessment of the Effect of Primary Particle Size Distribution On Granule Inhomogeneity: Modelling and Experiments

Rohit Ramachandran, Department of Chemical Engineering, Massachusetts Institute of Technology, Cambridge, MA

Mansoor A. Ansari, Departiment of Chemical Engineering, Imperial College London, London, United Kingdom

Frantisek Stepanek, Department of Chemical Engineering, Institute of Chemical Technology, Prague, Prague 6, Czech Republic

Inter-granular heterogeneity within a product batch is one of the major issues in granulation processes [1]. Inhomogeneous (or undesired) distribution of binder, active ingredients and primary particles in different size fractions of granule may influence the uniformity of the final dosage form and hence the application performance [2-5]. The distribution of all the above mentioned components among different size class of granules is profoundly affected by both process/design parameters and operating conditions. For instance, the impeller speed in high shear mixers, droplet size of liquid binder in fluid bed process, binder viscosity and its solidification/drying rate and the corresponding process temperature all play important role in uniform (or otherwise) distribution of the binder in the final granulated particles. Similarly the method of addition of active ingredient and its relative size are among the many factors that control its radial arrangement in the final product [6]. .

The phenomenon of preferential growth of smaller particles and their segregation towards the larger granules has been observed by several workers [1,2]. . Their work provided useful insights into inter-granule heterogeneity and suggested possible ways to control it. Their studies were mostly empirical and involved experiments with particles distributions with different median sizes. Furthermore, these studies were empirically based and do not account for the decoupled effect of particle size distribution on granule inhomogeneity, which can be quantified by fixing the mean diameter and varying the size distribution. In this work we present a systematic approach combining both experimental and computational tools to quantitatively assess the effect of size distribution width on granule inhomogeneity among different size classes. Results obtained are encouraging toward the use of model-based feedback control of granule properties (e.g. porosity, composition) using primary particle size distribution as a control handle.

References

1. Schæfer T., Johnsen D., Johansen A. 2004. Effects of powder particle size and binder viscosity on intergranular and intragranular particle size heterogeneity during high shear granulation. European Journal of Pharmaceutical Sciences 21, 525-531.

2. Scott A.C., Hounslow M.J., Instone T. 2000. Direct evidence of heterogeneity during high-shear granulation. Powder Technology 113, 205-213.

3. Iveson S.M., Litster J.D., Hapgood K., Ennis B.J. 2001. Nucleation, growth and breakage phenomena in agitated wet granulation processes: a review. Powder Technology 117, 3-39.

4. Knight P.C. 2001. Structuring agglomerated products for improved performance. Powder Technology 119, 14-25.

5. Stepanek F. 2004. Computer-aided product design: granule dissolution. Chemical Engineering Research and Design, 82(A11), 1458-1466.

6. Ansari M.A., Stepanek F. 2006. Formation of hollow core granules by fluid bed in situ melt granulation: Modelling and experiments. International Journal of Pharmaceuticals, 321, 108-116.

See more of this Session: Agglomeration and Granulation Processes
See more of this Group/Topical: Particle Technology Forum

Population Balance Modeling of Granulation of Excipient with Continuous Addition of Binder

Carl L. Marshall Jr., Chemical Engineering, The Pennsylvania State University, University Park, PA

Themis Matsoukas, Chemical Engineering, The Pennsylvania State University, University Park, PA

The simulation of granulation based on population balance (PB) models has increased substantially through a combination of advances in the speed of desktop computing and increased contact between the field of granulation and that of the population balance community. Nonetheless, granulation is treated as involving a single component, the solid powder of interest. In reality, granulation involves at least two components, excipient and binder; simulations that treat the process as one that involves a single pseudo-component account for the effects of the binder indirectly via the granulation kernel. This lumped approach treats the kernel in an empirical fashion and, while this may be both acceptable and practical in many situations, it is also of limited use when results in one system are used to extrapolate behavior in other systems. In this paper, we present simulations of granulation in which the binder is treated explicitly. Specifically, we conduct simulations of an idealized fluidized bed granulator in semi-batch mode with the excipient being loaded at time zero and the binder introduced continuously during granulation. This requires a two-component PB model in which each granule is characterized by the amount of excipient and the amount of binder. The agglomeration kernel is modeled based on the collision rate obtained by the kinetic theory of granular flow (KTGF) combined with a geometric model for the sticking probability of granules based on the fractional coverage of the granule surface by the binder. Three numerical methodologies are used to establish the accuracy of the numerical solution: (i) a discrete system model in which a granule with i units of excipient and j units of binder is represented as an explicit size class whose evolution is followed in time; (ii) PB model based on the quadrature method of moments; and (iii) a Monte Carlo simulation of the bicomponent problem. The goal of the study is two-fold: to compare the results of the simulation to actual granulation experiments with a binder and a single excipient, and to establish the accuracy of the DQMOM method against Monte Carlo and the rigorous discrete method, both of which are more accurate but computationally more demanding that DQMOM.

See more of this Session: Agglomeration and Granulation Processes
See more of this Group/Topical: Particle Technology Forum

Estimation of Optimal Coating Performance Via Optimal Mixing Conditions in a Pancoater

Ekneet Sahni, Department of Pharmaceutical Sciences, University of Connecticut, Storrs, CT

Bodhisattwa Chaudhuri, Department of Pharmaceutical Sciences, University of Connecticut, Storrs, CT

Coating of particles (tablets) is done to mask the unpleasant taste or odor of the drug, to control the bioavailability of the API, and to increase shelf-life. The coating solution is sprayed in specific locations of the granular bed and uniformity of coating is achieved by interparticle collisions and overall mixing behavior in the coater. Thus, the knowledge of particle flow and mixing in a pan coater is critical to optimize the design and operation of coating equipment. Lactose made placebos of two colors are used to investigate axial and radial mixing in the pancoater. Video-imaging and discrete pocket sampler is used to quantify mixing and to finally estimate the optimal operating conditions for coating in a pan coater. DEM (Discrete element method) based numerical model is also developed and fine tuned with the experimental findings. The optimal mixing condition is estimated and used as the starting point of coating studies.

Optimal mixing conditions comprise of a particular orientation of the coater, a specific vessel speed and a fill volume. Keeping these conditions same, the coating fluid (aqueous solution of Opadry II) is sprayed intermittently at different flow rates and concentration. Vernier Caliper is used to measure the change in diameter of the particles. Moreover, DEM based modeling of spray coating is also achieved for same operational parameter set and spray characteristic (centers and the radii of the spray zones) used in the experiments. DEM Simulation provides the residence time distribution of all the particles passing through the spray zone. The coating variability in experiments is estimated at different spary fluid concentration and flow rate. The optimal values of spray fluid concentration and flow rate is estimated for higher and uniform coating. Coating mass distribution is quantified in the numerical model by the total number of particles passing through the spray zone and also the frequency distribution of the residence time of the coated particles. From, the simulations the coating variability decreases with increase in speed of the pancoater vessel. It was observed that the simulations over predicted the effect of speed when compared with the experiments but the trends are in good agreement for the effect of tilt of the vessel.

See more of this Session: Agglomeration and Granulation Processes
See more of this Group/Topical: Particle Technology Forum

Intra-Particle Coating Variation: Monte-Carlo Coating of a Single Particle

Ben Freireich, School of Mechanical Engineering, Purdue University, West Lafayette, IN

Carl R. Wassgren, Purdue University, West Lafayette, IN

In this presentation, a simple analytical model is presented which describes the intra-particle coating variability of a single particle by a uniform spray. For random orientation distributions, the expected coefficient of variation of film coat thickness asymptotes to zero as the number of coating trials increases. However, if the spray zone visits have preferential orientation, the limiting value of the coefficient of variation is non-zero. The analytical model is also compared to Monte Carlo simulations of a single particle subject to a coating spray.

See more of this Session: Agglomeration and Granulation Processes
See more of this Group/Topical: Particle Technology Forum

Modeling and Validation of Flow Fields in a Mixer-Granulator

Hong Sing Tan, Procter & Gamble Ltd., Newcastle, United Kingdom

Paul Mort, Procter & Gamble, Cincinnati, OH

The use of Discrete Element Method (DEM) modeling is rapidly expanding as a modelling tool to analyse particulate processes and to address technological needs in various aspects of pharmaceutical, agricultural, food and consumer product industries. Examples include estimation of parameters that are difficult to measure or quantify experimentally (e.g. internal flow fields and mixing patterns) and facilitating the scale-up of particulate processes from laboratory to pilot plant. This paper compares DEM simulations of particle flow in a paddle mixer to experimental measurements using Positron Emission Particle Tracking (PEPT).

See more of this Session: Agglomeration and Granulation Processes
See more of this Group/Topical: Particle Technology Forum

An Innovative Method Using Magnetic Particle Tracking to Measure Solids Circulation Rates in a Spouted Bed

J. S. Halow, Chemistry, Waynesburg University, Waynesburg, PA

Emily E. Patterson, Chemistry, Waynesburg University, Waynesburg, PA

C. Stuart Daw, Fuels, Engines, and Emissions Research Center, Oak Ridge National Laboratory, Knoxville, TN

Charles E.A. Finney, Fuels, Engines, and Emissions Research Center, Oak Ridge National Laboratory, Knoxville, TN

Introduction

Spouted fluidized beds are used extensively for many types of solid-fluid reaction applications where uniform inter-phase contacting is important. While various commercial applications have been developed, the basic hydrodynamics and mixing patterns are not well characterized. Most past studies of spouted bed dynamics have relied on visual observation and pressure measurements to determine hydrodynamic parameters such as minimum spouting velocity and spout height. Studies of bulk solids circulation, however, have been very limited because of the difficulty of directly observing solids motion in these beds. In three-dimensional beds, it is usually impossible to track the motion of individual particles due to shielding by surrounding neighbors. For this reason, two-dimensional transparent beds and beds with semicircular cross-sections (with a flat viewing window) have been used to enhance optical access. While providing some information, visual methods typically introduce wall effects and anomalous flow patterns that do not match the behavior in the three-dimensional beds. Other non-visual techniques have been used to measure solids motion, including piezoelectric probes and computer aided radioactive particle tracking (CARPT). These techniques, however, also have limitations. In-bed probes can disturb solids and gas flows from their normal condition leading to uncertainty in the measurements. Radioactive particle tracking is less intrusive but requires expensive specialized equipment and access to radioactive isotopes that involve significant safety issues.

Experimental Materials and Methods

The particle tracking method we describe here is suited for use in three-dimensional beds, does not involve the introduction of flow-disturbing probes, and uses safe, inexpensive magnetic tracers thus avoiding the issues associated with past methods. Our particle tracking method utilizes neodymium magnets embedded in plastic spherical particles and externally positioned Hall-effect magnetic field detectors to continuously locate the position of a tracer particle over time. The method takes advantage of the tendency of the tracer particles to align their magnetic axes with the earth's magnetic field like a compass needle. Proper positioning of the probes can take advantage of this tendency, which simplifies the data analysis.

We studied the solids flow behavior of various spoutable materials in a 39-mm-diameter spouted bed with a 45-degree conical bottom. Four externally mounted magnetic field detectors located at the level of the bed surface tracked our tracer particles. We have developed semi-empirical relationships to relate tracer particle position with the magnetic field measurements. The vertical position of the tracer versus time is calculated and circulation frequencies and solids circulation rates determined from this parameter. A range of particle properties, bed depths, and superficial velocities were studied in our 39-mm bed.

Results and Discussion

Our experiments show that magnetic particle tracking in spouted beds is indeed feasible and can provide quantitative measurements of bulk particle circulation rates. We have correlated circulation rates for the limited range of parameters studied so far. With further testing and improvements in the technique, we expect to develop a more general correlation of solids circulation in spouted beds and provide useful data to develop and validate spouted bed models.

See more of this Session: Fundamentals of Fluidization - I
See more of this Group/Topical: Particle Technology Forum

Gas / Solids Mixing IN Rotating Fluidized Beds IN A Static Fluidization Chamber

Nicolas Staudt, Imap, UCL, Louvain-la-Neuve, Belgium

Mourad Abouahi, Imap, UCL, Louvain-la-Neuve, Belgium

Juray De Wilde, Materials and Process Engineering, UCL, Louvain-la-Neuve, Belgium

The gas and solids mixing patterns in a rotating fluidized bed in a static geometry are experimentally investigated. To quantify the solids mixing, a colored particles step response technique and mixtures of different types of particles are used. The gas phase mixing pattern is visualized by means of smoke. The influence of the fluidization gas flow rate and the solids loading is studied.

1. INTRODUCTION

The concept of a rotating fluidized bed in a static fluidization chamber with a static or rotating chimney was recently presented and experimentally proven [de Broqueville, 2004; de Broqueville and De Wilde, 2007; De Wilde and de Broqueville, 2007, 2008, 2008b]. The particle bed rotational motion is introduced by injecting the fluidization gas tangentially in the fluidization chamber, via multiple slots in the fluidization chamber outer wall. The rotating chimney can be used to reduce the rate of solids losses via the chimney at a given solids loading, but its direct impact, that is, on the particle bed rotational speed, is limited to the vicinity of the chimney.

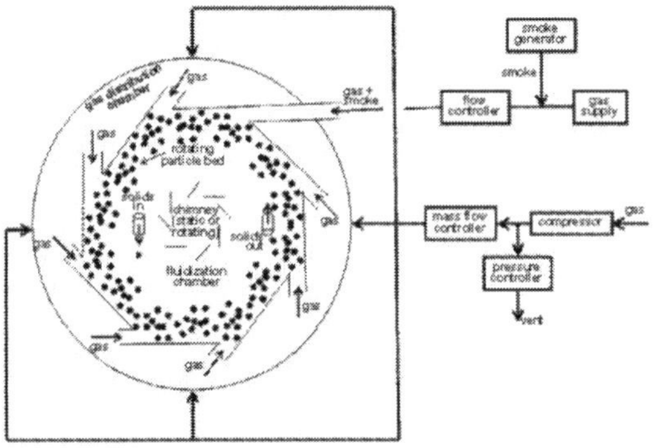

Fig. 1: Schematic representation of a rotating fluidized bed in a static fluidization chamber with a static or rotating chimney. Application of smoke visualization technique.

Whereas previous work mainly focused on proof of concept aspects [De Wilde and de Broqueville, 2007, 2008b] and stability considerations [De Wilde and de Broqueville, 2008, 2008b], the present work addresses the gas and solids mixing patterns and how these are affected by the fluidization gas flow rate and the solids loading in the fluidization chamber. An improved understanding and the quantification of the gas and solids mixing patterns must facilitate the development of simple engineering models for reactor design and scale-up.

2. PARTICLE BED MIXING AND SEGREGATION BEHAVIOR

Details on the geometry of the fluidization chamber and chimney can be found in De Wilde and de Broqueville [2007, 2008, 2008b]. The fluidization chamber is 24 cm in diameter and 11.5 cm long and contains 24, 2.3 mm wide gas inlet slots. The chimney is 16 cm in diameter and consists of 32 blades. The chimney can be rotated fast around its axis of symmetry by means of a motor. Chimney rotational speeds between 0 and 2800 rpm (rotations per minute) can be set. The fluidization gas flow rate is controlled with a mass flow controller and can be varied between 650 and 900 Nm^3/h. The solids are fed from a container using a screw feed controller and a rotating sealing valve.

First, the particle bed mixing was studied experimentally using colored particles of the same type and applying a step response technique. The latter is schematically represented in Fig. 2. The rotating fluidized bed is operated in a continuous solids feeding mode with the solids outlet closed. This implies that the rate of solids losses via the chimney equals the solids feeding rate. Initially, white particles are fed to the fluidization chamber and the flow pattern is fully developed. At time t_0, the solids feeding is switched to colored particles. The particles are fed via a tube through one of the fluidization chamber end plates. Radially, the solids feeding tube is positioned in the middle of the particle bed. Hence, both longitudinal and radial mixing and segregation can be studied. For comparison reasons, the response to the step change in the colored particle concentration (from 0 to 100 wt%) is measured, both inside the particle bed and in the chimney outlet. To quantify the behavior in the particle bed, a fast digital camera is used which faces the fluidization chamber plexi-glass end plate opposite the solids inlet. The concentration of colored particles in the chimney outlet is measured by sampling (Fig. 3).

Next, the particle bed mixing and segregation behavior with particles of different density and / or size was investigated.

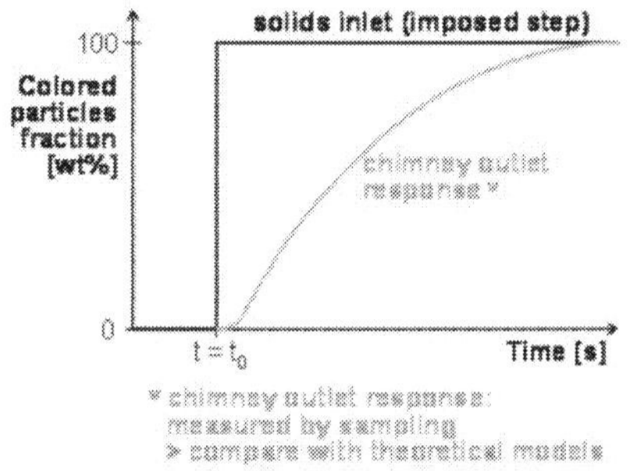

Fig. 2: Colored particles step response technique.

(a) (b)

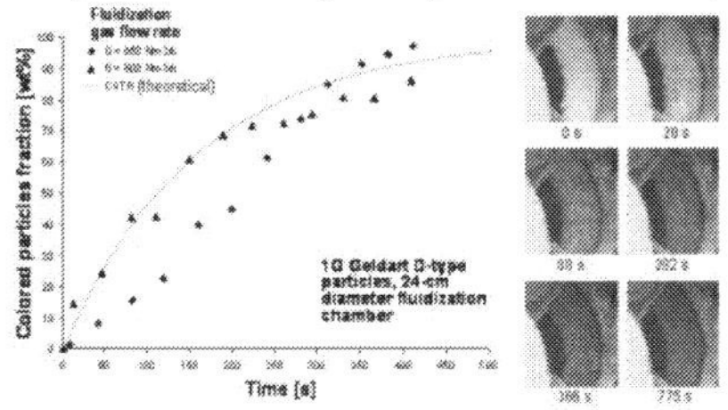

Fig. 3: Response of the concentration of colored, but otherwise identical particles to a step change in the solids inlet colored particle concentration at time $t_0 = 0$ s from 0 to 100 wt%. Particles: 2-5 mm polymer pellets. (a) In the chimney outlet for fluidization gas flow rates G of 650 and 800 Nm^3/h; (b) At the end plate of the fluidization chamber opposite the solids inlet for a fluidization gas flow rate G of 800 Nm^3/h. Theoretical ideal CSTR behavior of the particle bed shown in (a) for comparison.

3. FLUIDIZATION GAS MIXING BEHAVIOR

To visualize the gas phase flow path and mixing behavior in the rotating fluidized bed, smoke has been used as a tracer (Fig. 1). The smoke is injected through a section of one of the gas inlet slots of the reactor, which was compartimented from the gas inlet slots (Fig. 1) and the gas distribution chamber and which was positioned close to one of the end plates of the fluidization chamber to ensure visibility. The smoke injection compartment surface area has been chosen such that (i) with the smoke flow rate generated, the smoke is sufficiently concentrated to be visible; (ii) smoke injection velocities comparable with the gas injection velocities in the gas inlet slots could be obtained. To avoid gas bypassing smoke rich zones and a maldistribution of the gas resulting in poor smoke injection and an incorrect visualization of the gas flow pattern, the smoke requires a separate flow rate control system, supplied by a separate compressor or gas bottles. Equal smoke and main gas injection velocities have to be ensured. The smoke flow path is captured by means of different rapid cameras (Fig. 4). Because smoke detection within the particle bed is not possible at this stage, observations focused on the particle bed freeboard where the smoke leaves the particle bed.

A comparison with Computational Fluid Dynamics (CFD) simulations is made for an improved understanding of the gas and solids flow patterns.

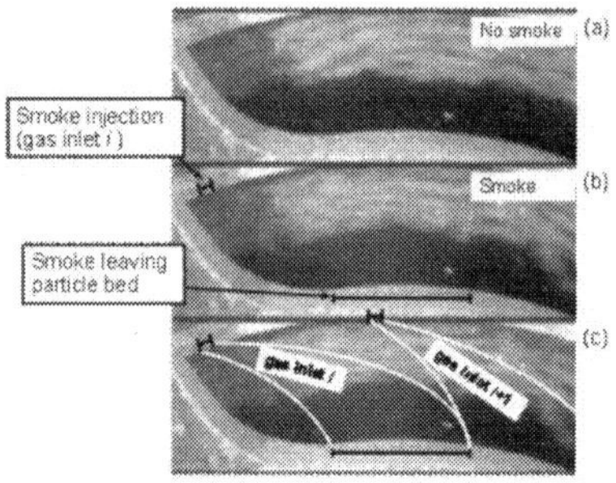

Fig. 4: Gas flow path and mixing visualization by means of smoke. (a) Before smoke injection; (b) During smoke injection; (c) Flow paths of gases injected via successive gas inlets.

REFERENCES

de Broqueville, Axel : Catalytic polymerization process in a vertical rotating fluidized bed : Belgian Patent 2004/0186, Internat. Classif. : B01J C08F B01F; publication number : 1015976A3.

de Broqueville, A., De Wilde, J., Central rotating chimney apparatus for obtaining a rotating fluidized bed in a cylindrical fixed or rotating fluidization chamber and processes using this apparatus, April 23, 2007: PCT/EP2007/053941 (F-995-WO), European and United States of America patent application.

De Wilde, J., de Broqueville, A., Rotating Fluidized Beds in a Static Geometry: Experimental Proof of Concept, AIChE J., 53(4), pp. 793-810, 2007.

De Wilde, J., de Broqueville, A., Experimental Investigation of a Rotating Fluidized Bed in a Static Geometry, Powder Technol., 183:(3), pp. 426-435, 2008.

De Wilde, J., de Broqueville, A., Experimental Study of Fluidization of 1G-Geldart D-type Particles in a Rotating Fluidized Bed with a Rotating Chimney, AIChE J., 54:(8), pp. 2029-2044, 2008b.

See more of this Session: Fundamentals of Fluidization - I
See more of this Group/Topical: Particle Technology Forum

Particle Clusters and Fluidized Beds

Ray A. Cocco, Particulate Solid Research, Inc., Chicago, IL

Frank D. Shaffer, National Energy Technology Laboratory, U.S. Department of Energy, Pittsburgh, PA

S.B. Reddy Karri, Particulate Solid Research, Inc., Chicago, IL

Roy C. Hays, Particulate Solid Research, Inc., Chicago, IL

Ted M. Knowlton, Particulate Solid Research, Inc., Chicago, IL

Accurately predicting the entrainment rate is important in designing a commercial fluidized bed. Yet, many correlations fall short in providing an accurate prediction of the entrainment rate. The issue is that many of these correlations assume that smaller particles (d_p < 44 μm) have a higher elutriation rate than larger particles. However, this is often not the case. Smaller particles can have lower effective entrainment rates than larger particles.

PSRI has collected entrainment data that suggest small particles may be clustering. Three cases are presented regarding the entrainment of various materials from fluidized beds in which clusters of particles affected the measured entrainment rates. First, the entrainment rate of FCC catalyst fines was measured at different fluidized bed heights. Higher entrainment fluxes were observed with lower bed heights (i.e., higher disengaging heights). The second case involved the replication of a phenomenon observed in an industrial scale process where the entrainment from the fluidized bed was highly variable and led to flooding of the recovery system. It was found in a batch entrainment test that with an initial high concentration of fines the entrainment flux was very low. As the fines were gradually elutriated away, the entrainment flux increased dramatically. Following the dramatic increase, the entrainment flux then exhibited the classical batch exponential decay as the fines were elutriated from the fluidized bed. Third, recent high speed video of a fluidized bed freeboard region was able to observe and track large clusters of particles in the range of 200 μm to 1000 μm. Yet, the bed material had only a mean particle size of 25 μm.

Thus, evidence suggests that fine particles in many materials may be clumping or clustering. This increases their effective particle diameter which reduces the entrainment rate. The clumps appear to be formed in the fluidized bed and can be ejected into the freeboard intact. High speed video observations through a borescope inserted into a fluidized bed at PSRI has confirmed the presence of clusters in fluidized beds. Such a phenomenon has many implications regarding how entrainment may be influenced by fines level, bed height, baffles, jet velocity at the distributor, etc.

See more of this Session: Fundamentals of Fluidization - I
See more of this Group/Topical: Particle Technology Forum

Electrostatic Charging in Fluidized Beds

Keith Forward, Dept. of Chemical Engineering, Case Western Reserve University, Cleveland, OH

Daniel J. Lacks, Department of Chemical Engineering, Case Western Reserve University, Cleveland, OH

R. Mohan Sankaran, Chemical Engineering, Case Western Reserve University, Cleveland, OH

It is well known that triboelectric charging can occur in fluidized particle systems, such that the particles become highly charged - however, the scientific understanding of this charging is very poorly understood. The charging can occur in general from both particle -particle and particle-wall interactions. To make progress in furthering the scientific understanding of this phenomenon, it is necessary to separate the particle -particle and particle-wall effects. In our work, we use a combination of experiments, theory and simulation to address the triboelectric charging that occurs solely from particle-particle interactions. An experimental methodology is developed to isolate the effects of particle-particle interactions, and the results show that smaller particles tend to charge negatively while larger particles tend to charge positively. We use theory and simulation to show that such charging can result from a nonequilibrium dynamics of electrons, whereby collision-induced electron transfer generates electron accumulation on a particle-size-dependent subset of the system.

K. M. Forward, D. J. Lacks and R. M. Sankaran, "Particle size dependent charge segregation in triboelectrically charged granular materials", Phys. Rev. Lett. 102, 028001 (2009).

D. J. Lacks, N. Duff and S. K. Kumar, "Nonequilibrium accumulation of surface species and triboelectric charging in single component particulate systems", Phys. Rev. Lett. 100, 188305 (2008).

See more of this Session: Fundamentals of Fluidization - I
See more of this Group/Topical: Particle Technology Forum

The Effect of Surface Modification On the Triboelectric Characteristics of Pharmaceutical Powders

Laila Jai Jallo, New Jersey Center for Engineered Particulates, New Jersey Institute of Technology, Newark, NJ

Utsav V Patel, Chemical Engineering, New Jersey Institute of Technology, Newark, NJ

Ajit Mujumdar, Chemical Engineering, New Jersey Institute of Technology, Newark, NJ

Rajesh N Dave, Chemical Engineering, New Jersey Institute of Technology, Newark, NJ

Powder handling processes involve contact of particles with each other and the surface of the transporting vessel, which normally lead to electrostatic charging of the particles. In many cases the electrostatic charging will affect the behavior of the powder during the handling process. This study is to understand the triboelectric characteristics of surface modified as compared to original untreated pharmaceutical active ingredient (API) powders. The triboelectric charging effect was quantified at different relative humidities (RH) and different surface coating levels; using stainless steel and polycarbonate as the contact surfaces. The results show that for untreated powders, the particles charge less at lower RH and charge more at high RH; while for the surface modified powders, the opposite is true. BET surface area measurement was used to determine the specific surface area of the particles, which is then used to calculate the charge density. Inverse Gas Chromatography (IGC) is used to determine the electron acceptor-donor characteristics and correlated with ones obtained from the triboelectric experiment for the unmodified and modified particles. Meanwhile, flow characterization results of these powders show a significant improvement in flow behavior that accompanies decrease of triboelectric charging for the surface modified powders.

See more of this Session: Fundamentals of Fluidization - I
See more of this Group/Topical: Particle Technology Forum

Effect of Electrostatic Charges On Hydrodynamics of Fluidized Beds

Ah-Hyung Alissa Park, Earth and Environmental Engineering and Chemical Engineering, Columbia University, New York, NY

Eliot Wycoff, Earth and Environmental Engineering and Chemical Engineering, Columbia University, New York, NY

Research in energy and environment areas, particularly alternative liquid fuel research, deals with the flow of dielectric liquids and/or particulate flows. During the operation of multiphase systems such as fluidized beds or high velocity transport systems, electrostatic charges are generated via frictional charging when the materials involved are dielectric in nature. The accumulation of electrostatic charges within the system can be operationally hazardous. Thus, in-depth knowledge in electrostatic phenomena occurring in multiphase flows is important in studying such technologies. Unfortunately, the dynamic electrostatic phenomena are highly complex, and as a result, the currently available measurement techniques cannot provide a full range of information of the charging effects in a flow system. This study focuses on the investigation of the effect of electrostatic charges on the hydrodynamics of fluidized beds containing dielectric liquids. A measurement technique providing a real time charge density, and a mathematical model of the charge generation and dissipation processes in a dynamic multiphase flow are also developed. A novel electrostatic measurement technique will greatly advance fundamental understanding of particle-particle interaction and interparticle forces, and in turn, improve the industrial operation of various multiphase reactors. Furthermore, the incorporation of this charge density function into the existing CFD code will also immensely advance the field of computational fluid dynamics.

See more of this Session: Fundamentals of Fluidization - I
See more of this Group/Topical: Particle Technology Forum

The Role of Fines in Controlling the Flow of Bulk Solids

Karl Jacob, The Dow Chemical Company, Midland, MI

Nestor A. Vasquez, R&D, The Dow Chemical Company, Midland, MI

It is widely assumed that fines control the flow of bulk solids but this has not been quantified in prior work. This study examines the effect of fines on the flow of bulk solids as function of fines loading in a bulk solid system. Using a Schulze ring shear tester, various systems of different types of bulk solids were studied including organic powders like sugar and inorganic systems such as sodium bicarbonate and sodium chloride across a wide variety of particle sizes. Key parameters such as unconfined yield stress, angle of internal friction and effective angle of internal friction as function of consolidation stress were determined for each system. Of particular interest are situations where the fines completely fill the interstices of the continuous coarse fraction. This work has application for bin design as well as design of materials for "tailored" flowability.

See more of this Session: Solids Handling and Processing
See more of this Group/Topical: Particle Technology Forum

A Study of Powder Flow Patterns (Dispersion vs. Convection) in a Continuous Annular Couette Cell Using Tracers

Gabriel Tardos, Department of Chemical Engineering, CUNY, New York, NY

Mehrdad Kheiripour Langroudi, Department of Chemical Engineering, The Graduate Center, The City University of New York, New York, NY

Paul Mort, Procter & Gamble, Cincinnati, OH

We studied the flow fields in a continuous Couette device obtained by superimposing a slow axial flow on radial shearing of particles thereby allowing the granular matter to move from the quasi-static to the intermediate regime of flow upon a slight decrease in solid volume fraction. This increase in porosity provides enough space for random collisions among particles so that the shear stress becomes dependent on the shear rate. In order to "see" the flow patterns we used colored particles and inserted them into the continuous flow. The insertion is made in such a way as to produce a narrow "pulse" of colored particles in an otherwise white background. The outflow of the mixture was collected at the bottom of the Couette cell at given time intervals of approximately one minute. Samples of approximately 80 grams were collected at a time and analyzed with a colorimetry technique.

Colorimetry entails the discharge of the sample mixture at very slow rate in front of a color-sensitive camera that assigns "colored" or "white" to each particle and gives the results in percentages. Residence time distributions of colored particles inside the bed show a clear peak and we were able to find the dispersion vs. convection pattern by transforming the RTD data. Another interesting trend is seen in the particle size of the tracer, when shown relative to the particle size of the white "base". We found that the coarser tracers are travelling axially faster compared to the finer tracer particles on the fine end of the distribution. The picture that emerges is that in the region where the bed is highly sheared, the axial velocity is somewhat higher (the solid fraction is lower) and the larger particles use this space to flow axially somewhat faster as compared to fine particles. Shearing extends into the bed to approximately 10 particle sizes depending on shape and size while beyond that, the bed is packed and moves axially very slowly. Diffusion is only important in the shearing zone and especially near the rotating wall where the shear is highest.

See more of this Session: Solids Handling and Processing
See more of this Group/Topical: Particle Technology Forum

Predicting Discharge Dynamics of Wet Cohesive Particles From a Rectangular Hopper Using Discrete Element Method (DEM)

Anshu Anand, Chemical Engineering, University of Florida, Gainesville, FL

Jennifer S. Curtis, University of Florida, Gainesville, FL

Carl R. Wassgren, Purdue University, West Lafayette, IN

Bruno C. Hancock, Pfizer Inc., Groton, CT

William R. Ketterhagen, Pfizer Global Research and Development, Pfizer Inc., Groton, CT

Accurate prediction of the discharge rate from hoppers is important in many industrial processes involving the handling of granular materials. The present work investigates the parameters affecting the discharge rate of a wet cohesive system from a quasi-3D, rectangular hopper using the Discrete Element Method (DEM). The cohesion between the particles is described by a pendular liquid bridge force model and the strength of the cohesive bond is characterized by a Bond number. The Beverloo correlation is applied to cohesive systems by modifying the Beverloo constant as a function of Bond number. The predictions obtained from this modified correlation fit the simulation data reasonably well. In addition, the effect of hopper angle in cohesive systems is shown to follow a trend similar to cohesionless systems, where the discharge rate is insensitive to changes in hopper angle except below a critical angle (with respect to the vertical) where the discharge rate increases rapidly. This critical angle of flow decreases with increasing cohesion.

See more of this Session: Solids Handling and Processing
See more of this Group/Topical: Particle Technology Forum

Consolidation of Granular Material Subject to Thermal Cycling

Adel F. Alenzi, Department of Chemical and Petroleum Engineering, University of Pittsburgh, Pittsburgh, PA

J. J. McCarthy, Department of Chemical and Petroleum Engineering, University of Pittsburgh, Pittsburgh, PA

Granular materials are ubiquitous, impacting a variety of industrial processing in the pharmaceutical, chemical, and construction industries, but also having implications in natural phenomena such as the formation of lunar regolith, dunes and beach sand. Recently, it has been found that cyclic variation of the temperature of a granular bed can cause the ostensibly static bead to consolidate (increase its packing fraction) over time due to thermo-mechanical coupling. In this work, we employ experimental techniques and numerical simulation -- using the thermal particle dynamics method (TPD) -- to study the origins and limits of this phenomenon, with an emphasis on particle/bed material properties, particle uniformity, bed geometry, and resultant "flow pattern", among other factors.

See more of this Session: Solids Handling and Processing
See more of this Group/Topical: Particle Technology Forum

Simulation and Experimental Studies of Jamming for Model Two-Dimensional Particles Under Flow

Alfredo Guariguata, Chemical Engineering, Colorado School of Mines, Golden, CO

David T. Wu, Chemical Engineering and Chemistry Departments, Colorado School Of Mines, Golden, CO

Carolyn A. Koh, Chemical Engineering, Colorado School of Mines, Golden, CO

Amadeu K. Sum, Chemical Engineering, Colorado School of Mines, Golden, CO

E. Dendy Sloan, Chemical Engineering, Colorado School of Mines, Golden, CO

Jamming and plugging of flowlines with gas hydrates is the most critical issue in the flow assurance of oil and gas production lines. Because solid hydrate particles are often suspended in a fluid, the pipeline jamming and flow constriction formed by hydrates depend not only on particle/wall properties, such as friction, binding forces and mechanical characteristics, but also on the concentration of particles upstream of the restriction, flow velocity, fluid viscosity, and forces between the particles. Therefore, to gain insight into the jamming phenomena, both experiments and computer simulations on two-dimensional model systems have been carried out to characterize the flow of particles in a channel, with the eventual goal of applying that knowledge to gas hydrates jamming. Using the simulation software PFC2d®, we studied the effect of restriction geometry and flow velocity on the jamming process of particles. Results from the simulations were compared to experimental measurements on polyethylene discs floating on water flowing in an open channel.

See more of this Session: Solids Handling and Processing
See more of this Group/Topical: Particle Technology Forum

Wall Friction Tests On Limestone Using An Annular Shear Cell

Clive E. Davies, School of Engineering and Advanced Technology, Massey University, Palmerston North, New Zealand

Miles C. E. Grafton, School of Engineering and Advanced Technology, Massey University, Palmerston North, New Zealand

Emile S. Webster, School of Engineering and Advanced Technology, Massey University, Palmerston North, New Zealand

Limestone is a commodity material in many economic sectors including the chemical and agricultural industries where large quantities are transported, stored and handled. Although nominally calcium carbonate, the properties of limestone, as quarried, are influenced by its geological origins which affect its mineralogy and overall chemical composition.

In a campaign of experiments to measure the flow parameters of three different limestones, using an annular shear cell, we have logged shear stress measurements over extended periods of time for a number of different normal stresses. Our data are indicative of stick-slip friction, and also suggest that the method of filling the trough of the annular shear cell has a strong influence on the form of the shear-stress/time record.

Results are presented for an algal Oligocene limestone containing small amounts of quartz and mica, an Oligocene sandy coquina, and an Ordovician marble containing small amounts of mica; the size distributions of each were similar.

See more of this Session: Solids Handling and Processing
See more of this Group/Topical: Particle Technology Forum

Optimization of Breakage in a Ball Mill

Matthew J. Metzger, Chemical and Biochemical Engineering, Rutgers University, Piscataway, NJ

Brendon Hausberger, School of Materials and Process Synthesis, Centre of Material and Process Synthesis (COMPS), Johannesburg, South Africa

Diane Hildebrandt, School of Materials and Process Synthesis, Centre of Material and Process Synthesis (COMPS), Johannesburg, South Africa

David Glasser, School of Materials and Process Synthesis, Centre of Material and Process Synthesis (COMPS), Johannesburg, South Africa

Benjamin J. Glasser, Chemical and Biochemical Engineering, Rutgers University, Piscataway, NJ

Handling a mixture consisting of two or more solid components dissimilar in size is no easy task. Particles of varying size may react differently to identical processing conditions resulting in a blend consisting of regions lean in one component and rich in another. A common approach to minimize this segregation is to maintain a consistent particle size through processing of the material, and thus remove the size differences that tend to cause the components to segregate in the first place. An operation that is particularly useful at accomplishing this size "standardization" is milling, which is regularly used in the food, mining, construction and pharmaceutical industries. However, despite its widespread use milling is a very crude operation, and adequate control and knowledge of the behavior inside the mill remains elusive and is the subject of much ongoing research.

In this work we present a systematic approach to optimize production of product falling within a specified size range, known as the Attainable Region (AR). Specific attention is paid to the total volume of grinding media inside the mill, the size of that media and the rotation rate of the mill and how each affects breakage within the mill. Suggestions are presented on how to reduce the overall time required to achieve the desired product as well as minimize the overall energy usage of the equipment. Finally, preliminary results from a DEM simulation including both charge motion and particle breakage are presented.

See more of this Session: Solids Handling and Processing
See more of this Group/Topical: Particle Technology Forum

MEASUREMENT AND PREDICTION OF PROCESS RELEVANT SEGREGATION FOR MULTI-COMPONENT BULK MATERIALS

Dr. Kerry Johanson, Material Flow Solutions Inc., Gainesville, FL, 32607

Introduction

One of the jobs a typical process engineer faces is how to create a product that performs a consumer dictated task while still preventing process hang-ups and maintaining product quality. Typically, much time is spent getting the consumer dictated task correct. Tasks like adding enhanced optical brightening to detergents, developing the drug that cures disease, or producing a food product that tastes good are high on the engineers priority list. Tasks focused on creating a material that is free flowing and will not segregate during typical handling are typically afterthoughts. However, these tasks maybe as important to the successful deployment of product as the core function of the product. Obviously, a detailed discussion of all these areas is well outside the scope of this extended abstract. Thus, this paper will focus on measurement and prediction of segregation effects with products.

Bulk segregation can occur due to a variety of mechanisms [1,2]. Finer particles sift down through a matrix of coarse particles [3], separating as they slide down a pile. Air currents generated by free fall carry fines to regions in the bin where gas velocities decrease enough to precipitate particles [4]. Differences in repose angles can cause materials to slide down the pile at different velocities, creating a separation of materials based on particle friction characteristics [5,6,7]. In order to mitigate segregation issues existing with bulk materials, it is necessary to understand the mechanisms causing segregation as well as the magnitude of the segregation. It is also valuable to know the pattern of segregation in process vessels. Attempting to cure a segregation problem caused by air entrainment of fines requires modifying the amount of air entrained in the free falling solid. However, modifying the free fall height will not affect segregation caused by sifting. In order to solve a segregation issue, the practicing engineer must understand the root causes of separation in processes and have a means of generating segregation measurements for the material mixtures of interest with minimal material sample. It is also important to understand how segregation may propagate through the unit operations in any given process. The question to be answered is: how bad will the segregation be at the end of the line?

Segregation Measurement

There are two uses for segregation data. Engineers may use segregation data to optimize product design creating a product with minimal segregation. Engineers may also wish to modify the processing to minimize the effect of segregation in their plant packaging process or handling facility. In either case, the segregation pattern, segregation mechanism, and magnitude of segregation are key parameters in process or product design recommendations.

In the past researchers have fed bulk materials onto a pile, sectioned a conical pile into annular sections, and performed chemical or size analysis on the material collected in each annular section [8,9]. This is a long and tedious method which results in only a very few measurements collected along the pile. Plant personnel often use thieves to sample piles in process equipment.

This method also yields only a few measurements along the pile and disturbs the surface, thereby changing the segregation pattern. Other researchers measure the segregation pattern by placing material in a funnel flow hopper and then discharging the hopper, collecting the output stream and measuring the quality of the mixture by chemical composition or particle size [10,11]. This method convolutes the segregation with the flow pattern from a particular hopper or bin. Unless the bin design in the tester is precisely the same as the bin design in the process, there is no guarantee that the composition during discharge from the tester will be similar to that found in the actual process. In addition, the cohesive properties of the bulk material can impede flow from the segregation tester hopper, even though hang-ups may not occur in the process with the same material.

One problem with current segregation testers is that they do not relate well to the process. Any two or three materials can segregate if there is a difference in properties and the mixture is subject to enough external stimuli. Adding enough fluidization gas will separate materials of different densities, even though these extreme conditions may not occur in the actual process. The real question to be answered is: will the mixture segregate when exposed to a feed behavior similar to that present in the process? Therefore, any measurement of segregation tendency should have three key elements. First, the feed should be controlled so as to allow the measured segregation to be scalable to process conditions. Second, the pattern of the segregation should be included as part of the measurement in order to predict the expected concentration leaving process equipment. Finally, the magnitude of the segregation should be quantified to provide guidance in determining if the mechanism is actually a potential problem in the process.

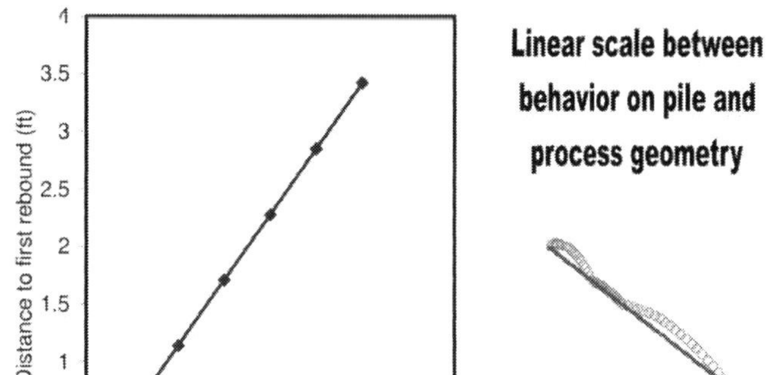

Figure 1. Typical rebound of free fall particles down pile surface

Material entering the process typically free-falls into the process vessel from some fall height at some process velocity. The dynamic effects caused by particle rebound or sliding down the pile scale linearly with the geometry of the process equipment (see figure 1). The same cannot be said of the case where gas is carried with the input stream. However, similar models can be created for this case and a scale law developed for this case. The small scale segregation tester can then be run at conditions that mimic the process, allowing the results of the segregation test to apply to real process conditions. Materials with cohesion will require other scaling laws and will not be addressed here.

The pattern of segregation requires that segregation be measured by analyzing spatial concentration data. This spatial analysis always has a sample size. It must be decided beforehand what volume of material represents a reasonable segregation basis. If that volume occupies only one or two particles, then any segregation measurements are doomed to failure since the segregation measurement volume cannot represent an average sample. If that sample volume is too large, then every segregation measurement will represent the global average concentration placed in the process vessel. The proper segregation volume should be chosen somewhere in between.

Obviously, these spatial segregation measurements require access to some spatial view of the material after it has segregated due to initial process filling. One way to gain access to the cross section of a pile is to fill a slice model with material and observe the segregation pattern through the side of the slice model using optical techniques (Figure 2).

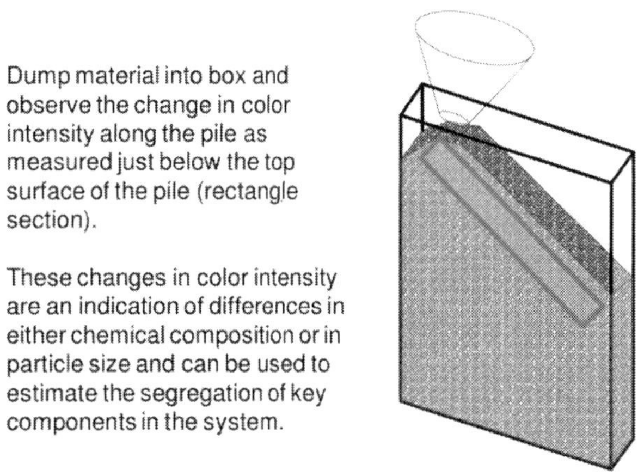

Figure 2. Schematic of segregation tester

Observation of segregation through the sides of a slice model will be biased to the optical pattern that exists at the wall of the segregation tester. Therefore, care must be taken when loading the tester to assure a representative sample is visible through the side of the tester. Filling the tester across the width of the slice model will distribute the material to the tester wall, creating a uniform material across the tester. Limiting the thickness of the slice model will also help with the distribution of representative material to the tester wall. However, one should be aware that extremely thin slice models are subject to banding due to wall effects that would not be present in wider slice models. For typical materials, this limits the slice model to a minimum of approximately 25 mm thick. This allows a visual observation of the spatial segregation pattern where the appropriate view port can be selected.

Reflectance spectroscopic methods can be employed to measure subtle differences in color. Since we are dealing with a discrete particle system, multiple measurements are required to determine the average concentration within a given viewport. Segregation measurement is also a scale issue. The size of the chosen viewport should be large enough to contain a representative number of particles, yet small enough that differences in local compositions are not lost in the averaging scheme (Figure 3).

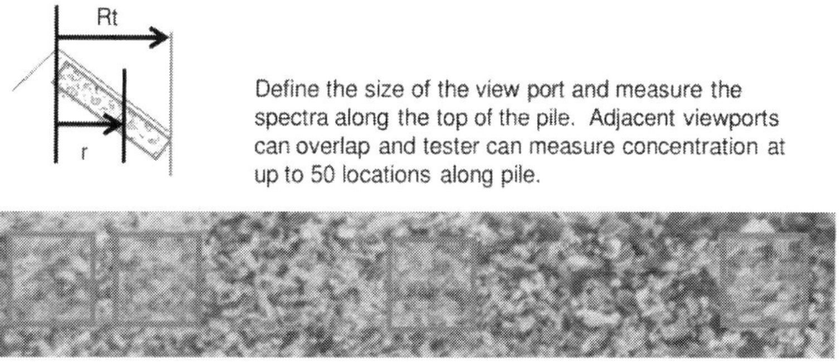

Figure 3. Measurement zone along pile top surface

If the spectra of the pure components are known, and the spectra of the mixture in different view port boxes along the pile are known then pure component spectra can be used to compute the concentration of the components along the length of the pile. One drawback of this technique is that good data requires many spectral measurements collected at multiple points along the pile. Manually collecting this data is tedious and time consuming. Therefore, an automatic instrument was constructed to control the feed, obtain the pure component spectra, and measure spatial concentration profiles for spectral measurements (see Figure 4).

Figure 4. Segregation tester used in experiment showing the variable speed vibratory feed system and the pile formed in the tester for analysis (note right door is open to expose the segregation hopper

Steak seasoning consisting of 47% salt, 28% minced garlic, 19% black pepper, 2% red pepper, and 4% dill seed was used in this experiment. The following procedure was used in the generation of steak seasoning segregation data.

1. Create a mixture of steak seasoning using one of three possible grades of salt.
2. Feed this mixture into the segregation tester slice model at a rate of 1.5 lit/min through a drop height of 152 mm inches.
3. Place salt, garlic, black pepper, red pepper, and dill seed in the individual component measurement trays and measure the reflective spectral response of each material for 1825 wavelengths between 420 nm to 850 nm.

4. Select a measurement zone that is parallel to the pile surface and positioned 20 mm below the top surface of the pile.
5. Select the viewport to be a square 15 mm by 15 mm and measure the spectra of 36 (a 6x6 matrix) distinct points within this viewport. Average these 36 values to create an average spectral reading of the material within the selected viewport.
6. Select 20 points along the pile to measure the average viewport spectra (total 720 complete spectral measurements per pile).
7. Using the spectra of the pure components and the average mixture spectra along the pile, compute the concentration of the pure components from numerical spectral de-mixing techniques. This will give the concentration of the different materials along the pile as a function of radial distance to the pile's edge.
8. Repeat steps 1 through 7 for each salt grade used.

Figure 5 shows the averaged spectra of the five components used in the mixture. Notice the different peaks at various wavelengths. The garlic and salt have similar spectra. The dill seed and black pepper are similar, and the red pepper has a unique spectra. These average spectra were created by filling the component holders with pure materials and averaging 10 individual spectral images of each pure component.

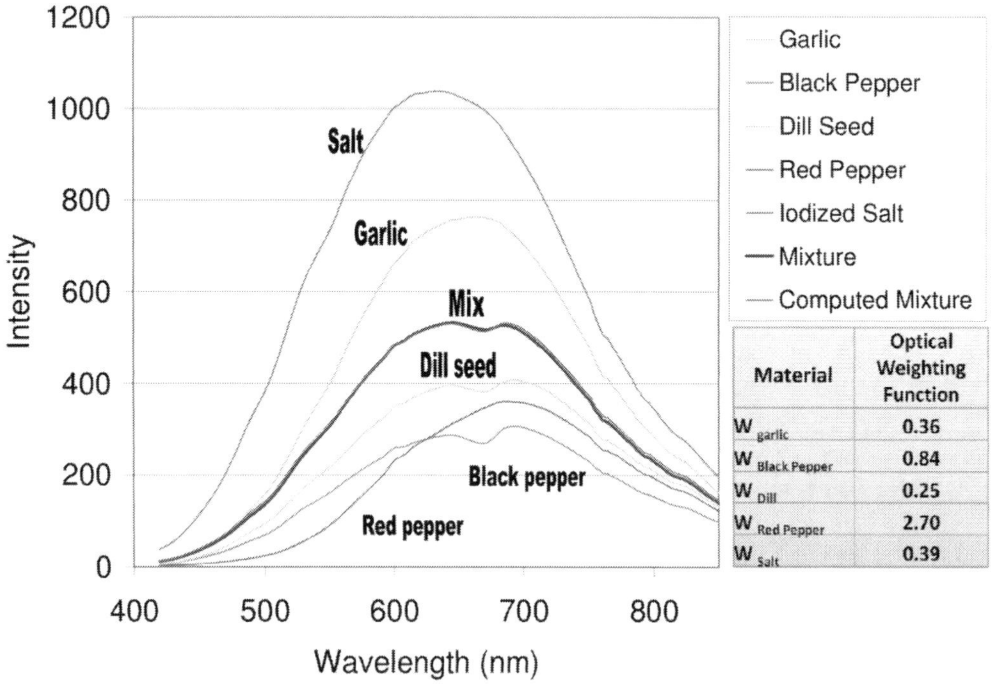

Figure 5. ***Spectra of steak seasoning components and sample mixture spectra resulting in a composition of 27.1% Garlic, 19.7% Black pepper, 4.0% Dill seed, 2.0% Red pepper, and 47.2% iodized Morton® salt***

These pure component spectra include the effect of surface texture, color and particle size. The mixture spectra also are affected by particle orientation. If the particles were spherical and all the same size, then there would be no optical difference between the particles in the mixture and pure particles placed in the component trays. In this case, the mixture spectral intensity $Fmix_j(\lambda)$ would be a simple linear combination of the spectral intensity of pure components ($F_i(\lambda)$) based on the local fraction ($x_{i,j}$) of each component. However, smaller particles fill the voids between coarse particles, creating a shadow effect for the coarse particles. These fine particles within the voids occupy a proportionally greater percent of area than the volume fraction would suggest. This indicates that a linear combination of pure spectra will not account for the mixture spectra. The mixture spectra will be more biased towards the fine materials. This effect can be modeled by adding a weighting factor (W_i) to the linear combination of pure spectra (equation 1).

$$Fmix_j(\lambda) = \sum_i W_i \cdot x_{i,j} \cdot F_i(\lambda) \qquad (1)$$

Figure 5 also contains a list of the computed optical weighting factors for the steak seasoning. The tester compares the computed intensity curve with the actual measured curve and adjusts the weighting factors and local fractions to minimize the error between the two curves. The data in Figure 5 shows the actual measured intensity at one point in the tester and the computed weighted spectral combination for steak seasoning using the overall spectral weighting factors. There is good agreement between the measured curve and the computed weighted linear combination.

The above analysis allows the calculation of local concentration of a complex mixture using spectral techniques. Consider the concentration profiles (Figure 6) for the various components of steak seasoning. The garlic accumulates at the bottom of the pile. The salt accumulates at the top of the pile. The black pepper also accumulates at the top of the pile. Both the red pepper and dill seed show little or no segregation.

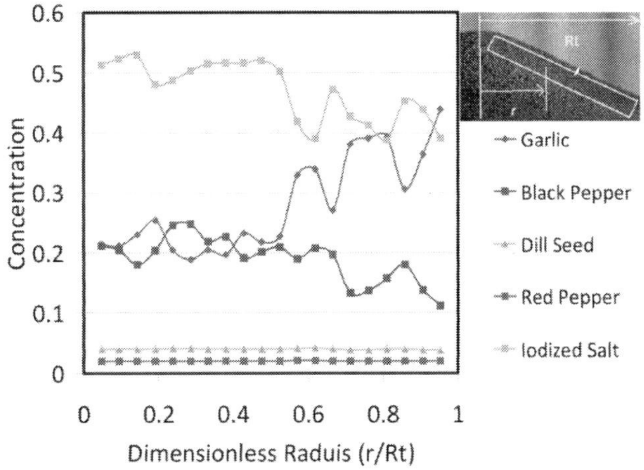

Figure 6. Segregation Test results for steak seasoning with iodized salt

We also measured 6 repose angles for each of these materials and selected the maximum repose angle to be representative of the bulk material (see Figure 7). If repose angle were the only mechanism of segregation with this material, then the materials should segregate in order of repose angle down the pile. The top of the pile should be rich in black pepper, followed by dill seed, then red pepper, garlic and finally iodized salt. However the reality is that the red pepper and dill seed do not segregate. This is likely due to the low percentage of these materials in the mix. There is not enough particulate material to effectively form a unique repose angle and the repose angle of these two ingredients are close to the median response angle. Thus, there is little driving force to cause these two materials to segregate. The black pepper has the largest repose angle and always shows a positive deviation from the cumulative concentration plots, indicating that black pepper accumulates near the top of the pile. This is validated by the repose angle measurements. However, the repose angle data suggests that the salt should accumulate at the bottom of the pile. In reality the segregation tests indicate that iodized salt accumulates at the top of the pile. This is due to the fact that the iodized salt has a finer particle size and will also sift down through the void of coarse particles depositing near the pile top. Thus, salt segregates by sifting. Pepper segregates by both sifting and angle of repose. Garlic segregates by angle of repose and sifting.

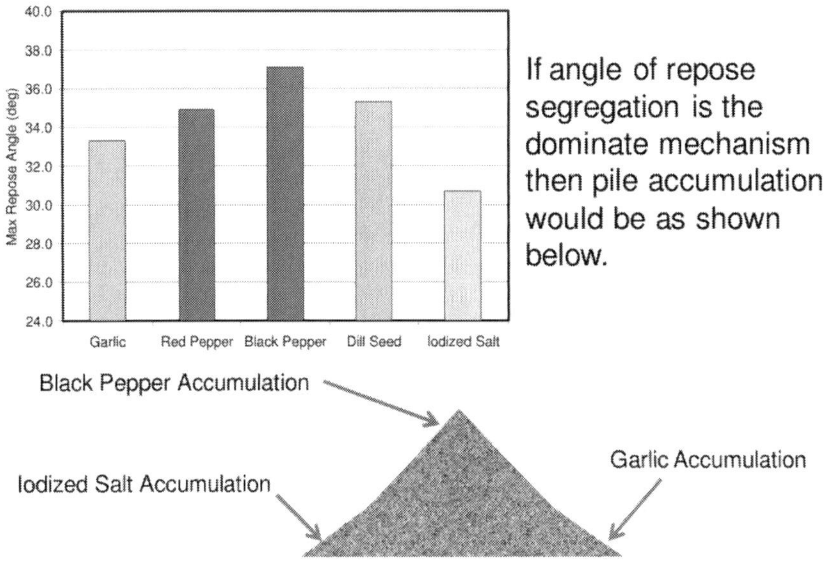

Figure 7. Angle of repose for steak seasoning components.

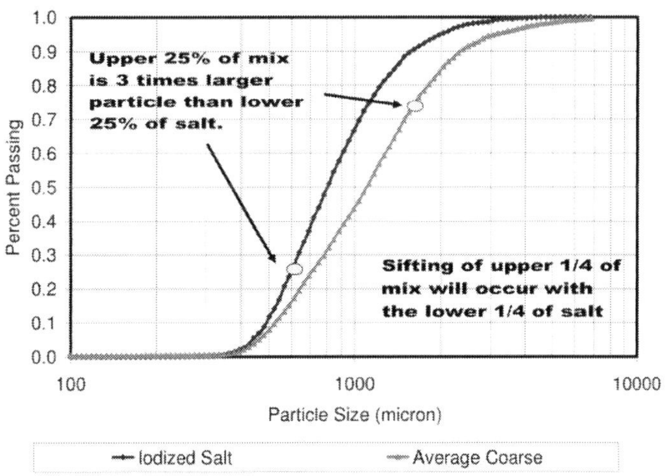

Figure 8. Particle size of salt and average particle size of steak seasoning mixture.

Nomenclature

$Fmix_j(\lambda)$	is the reflectance spectral intensity of the mixture in the tester as a function of wavelength.
$F_i(\lambda)$	is the reflective spectral intensity of the pure component as a function of wavelength.
r	is the radial position from the top of the pile to the point of interest.
Rt	is the radial dimension of the bottom of the pile.
W_i	is the optical weighting factors for each particle based on all data measure on the pile.
$x_{i,j}$	is the fraction of the ith pure component at the jth position on the pile.
λ	is the wavelength of light

Conclusions

The new spectral method of measuring segregation presented above is a valuable tool for enhancing product design. This test method provides an easy way to measure the potential segregation behavior of prospective or current mixtures. The measured segregation tendencies can then be used to optimize the mixture by selecting the right components for combination. This should facilitate the production of new materials using lab scale quantities.

An important side note is that segregation of multi-component mixtures is often due to multiple segregation mechanisms. Being able to measure the segregation profile of milti-component mixtures allows the identification of the active mechanisms for each component.

One of the keys to successful process control is the ability to directly measure the influence of key properties on critical bulk behavior. Wherever possible, direct measurements of critical process behavior should be used as the control variable. In other words, if product segregation is a key process behavior to control, then it is much more advantageous to measure segregation directly than measure wall friction angle and cohesion and infer segregation behavior. This test method allows direct measurement of product segregation.

References

1. Jerry Johanson, Solids segregation: causes and solutions, Powder and Bulk Engineering, August 1988
2. J. M. Ottino, and D. V. Khakhar, Fundamental research in heaping, mixing, and segregation of granular materials: challenges and perspectives , Powder Technology, Volume 121, Issues 2-3, 26 November 2001, Pages 117-122
3. Anjani K. Jha, and Virendra M. Puri, Percolation segregation of binary mixtures under periodic movement, Powder Tecnology, Available online 7 May 2009, doi:10.1016/j.powtec.2009.04.013
4. Are Dyrøy, Morten Karlsen, Gisle G. Enstad, Sunil de Silva, A system for the reduction of air current segregation in silos, Handbook of Powder Technology, Volume 10, 2001, Pages 623-630
5. Y. L. Ding, R. Forster, J. P. K. Seville, and D. J. Parker, Segregation of granular flow in the transverse plane of a rolling mode rotating drum, International Journal of Multiphase Flow, Volume 28, Issue 4, April 2002, Pages 635-663
6. D. V. Khakhar, Ashish V. Orpe, J. M. Ottino, Continuum model of mixing and size segregation in a rotating cylinder: concentration-flow coupling and streak formation, Powder Technology, Volume 116, Issues 2-3, 23 May 2001, Pages 232-245
7. Florence Cantelaube, Daniel Bideau, Stéphane Roux, Kinetics of segregation of granular media in a two-dimensional rotating drum, Powder Technology, Volume 93, Issue 1, September 1997, Pages 1-11
8. S. Massol-Chaudeur, H. Berthiaux, , and J. Dodds, The Development and Use of a Static Segregation Test to Evaluate the Robustness of Various Types of Powder Mixtures, Food and Bioproducts Processing Volume 81, Issue 2, June 2003, Pages 106-118
9. D. McGlinchey, Quantifying segregation in heaps: an experimental study, Powder Technology, Volume 145, Issue 2, 27 July 2004, Pages 106-112
10. William R. Ketterhagen, Jennifer S. Curtis, Carl R. Wassgren, and Bruno C. Hancock Modeling granular segregation in flow from quasi-three-dimensional, wedge-shaped hoppers, Powder Technology, Available online 3 July 2007. doi:10.1016/j.powtec.2007.06.023
11. Thomas Baxter, James Prescott, Process Development, Optimization, and Scale-up: Powder Handling and Segregation Concerns, Developing Solid Oral Dosage Forms, 2009, Pages 637-665

Shear and Flow Properties of Pharmaceutical Binary Mixtures

Alisa Vasilenko, Chemical and Biochemical Engineering, Rutgers University, Piscataway, NJ

Fernando J. Muzzio, Department of Chemical and Biochemical Engineering, Rutgers University, Piscataway, NJ

Benjamin J. Glasser, Chemical and Biochemical engineering, Rutgers University, Piscataway, NJ

Majority of pharmaceutical products today are solid dosage forms, such as tablets, capsules, and powders for inhalation, thus making powder technology critical for the manufacturing process. The complex behavior of pharmaceutical powders is not well understood and characterized; suitable processing conditions and formulations are determined by extensive trial-and-error procedures, and even the small changes in the said parameters can cause great variations in powder. The properties of the materials may change dramatically during processing introducing variability into the process and affecting final product quality and characteristics. It is therefore important to meaningfully characterize and quantify the powder flow.

In this study the flow behavior of twelve well-mixed two-component blends with varying concentrations of micronized acetaminophen, microcrystalline cellulose and milled lactose was characterized by a gravitational displacement rheometer (GDR) and a rotational shear cell. Principal stresses, cohesion and flow functions were obtained via shear cell techniques; flow and dilation index were determined with the GDR and compared with the mixture compressibility. At lower consolidations the mixtures displayed non-linear shear stress behavior; however, when normalized, the major principal stresses were found to be proportional to the mixture composition and particle density and size. Values of cohesion and flow factors obtained from the shear cell showed dependence on material compressibility and correlated discernibly with the flow and dilation indices obtained from the GDR.

See more of this Session: Characterization and Measurement in Powder Processing
See more of this Group/Topical: Particle Technology Forum

Measurement of Solid Fraction and Its Fluctuations in Dense Granular Flows Using a Capacitance Probe

Mehrdad Kheiripour Langroudi, Department of Chemical Engineering, The Graduate Center, The City University of New York, New York, NY

Gabriel I. Tardos, Department of Chemical Engineering, The City College of the City University of New York, New York, NY

We measured the solid fraction (porosity) and its fluctuations in dense granular flows using a capacitance probe employed earlier by Louge at al (1996). The probe, originally developed to measure distance (Capacitec, Inc., Boston, Ma), is used here to measure the dielectric constant of the granular bed as it changes due to variations in solid fraction. Two geometries that of a Couette device, where granular material is sheared between two concentric cylinders and a flat-bottom bin-hopper device, were used. The annular Couette cell was modified to allow changes in solid fraction inside the bed by super-imposing a slow axial flow on the radial shearing.

As a result, a small decrease in solid fraction partially breaks the force chains between particles and provides enough space for random collision to occur inside the bed. This results in a significant decrease in shear stresses on the rotating cylinder and shifts the regime of flow from quasi-static to intermediate. We were able to find the actual change in solid fraction required for this change to happen. During batch Couette experiments, where there is no axial flow, the solid fraction is at its maximum while by continuous removal of small amounts of material, the solid fraction decreases and the shear stress becomes shear rate dependent. We found that fluctuations of solid fraction only occur with axial-flow while in a no-flow (static) experiment, there are no fluctuations. We also observed that increasing the axial flow rate does not significantly change the solid fraction. The capacitance probe was originally mounted flush to the outer stationary wall but more recent experiments were carried out by pushing the probe inside the shearing zone. As the probe moves from the stationary wall towards the rotating wall, the fluctuations of solid fractions increase. Inside the shearing zone, the solid fraction is also shear rate dependent.

In the flat-bottom hopper (funnel flow), the solid fraction is mostly constant when the material is stagnant (not moving), but as soon as the material starts to flow, the solid fraction decreases and remains approximately constant for all flow rates. It fluctuates around an average value but when the flow is stopped the solid fraction does not change and the fluctuations also stop. It appears that the material remains in its "flowing" arrangement and when the flow is restarted, the solid fraction remains essentially unchanged and starts to fluctuate again. It seems that when the solid fraction on the wall reaches a certain value the material keeps its arrangement no matter if it is flowing or is static. This behavior is consistent with our previous stress measurements (Kheiripour Langroudi, et. al., 2008) that show that as soon as the material stops flowing (shearing) the stress "freezes" at its existing value and only changes if the flow is restarted. This appears to also confirm that stresses measured in a static powder are random and depend on the previous history and stressing of the powder as particles are arranged in a certain geometrical matrix.

Reference

1.M. Louge, M. Tuccio, E. Lander, and P. Connors, Rev. Sci. Instrum. 67, 1869-1877

2.M. Kheiripour Langroudi, S. Turek, A. Ouazzi, and G. Tardos; Submitted to Powder Technology, 2009

See more of this Session: Characterization and Measurement in Powder Processing
See more of this Group/Topical: Particle Technology Forum

Characterization of Pharmaceutical Powders Modified by Dry Coating

Lauren Beach, Otto H. York Department of Chemical, Biochemical and Pharmaceutical Engineering, New Jersey Institute of Technology, Newark, NJ

Matthew P. Mullarney, Material Science and Oral Products CoE, Pfizer Inc., Groton, CT

Bruno C. Hancock, Pfizer Inc., Groton, CT

Ajit Mujumdar, Otto H. York Department of Chemical, Biological and Pharmaceutical Engineering, New Jersey Institute of Technology, Newark, NJ

Laila Jai Jallo, New Jersey Center for Engineered Particulates, New Jersey Institute of Technology, Newark, NJ

Rajesh Davé, Chemical Engineering, New Jersey Institute of Technology, Newark, NJ

Engineered particles, which are surface modified via dry coating have been shown to have improved flow and handling properties, and hence are very useful for a number of industrial applications. Recent work in our group has indicated that dry coating makes it possible to impart improved flowability for blends with very high API contents (higher than 75 %) even for very cohesive micronized API powders. This is very useful for pharmaceutical processing including flow, handling and downstream processing. Although it is clear that these surface modified powders flow better as indicated by reduced angle of repose and faster or more uniform flow out of a funnel, there is an unresolved issue; the extent of powder attrition during surface modification and the influence of size reduction on flow properties that are specifically relevant to industry practice.

Acetaminophen and ibuprofen are considered as model active pharmaceutical ingredients (APIs), and are surface modified via dry coating of nano-additives in several dry coating devices. The study considers use of hydrophobic as well as hydrophilic nano-additives. The resulting powders are analyzed for their particle size distribution, coating coverage, and dispersibility; the latter is evaluated using the Sympatec Helos/Rodos system under varying dispersion conditions. In addition, a suite of techniques are employed for powder flow and other properties which are important for large scale pharmaceutical manufacturing. Extensive powder flow property characterization is done, ranging from novel methods such as vibrated packed density measurements to industry standard technique such as shear testing via the Schulze tester. In addition, other powder properties such as sticking tendency and electrostatic properties are evaluated, and several flow testers are also employed, including the Aeroflow, Freeman Rheometer, and the Hosokawa Powder Tester. The results of these tests are compared with those for untreated API powders in order to examine the influence of surface modification on property enhancements.

See more of this Session: Characterization and Measurement in Powder Processing
See more of this Group/Topical: Particle Technology Forum

Analysis of the Effects of Shearing Conditions On the Distribution of MgSt Using Laser-Induced-Breakdown-Spectroscopy (LIBS) and near Infrared Spectroscopy (NIR)

Aditya U. Vanarase, Chemical and Biochemical Engineering, Rutgers University, Piscataway, NJ

Patricia M. Portillo Brieva, Chemical and Biochemical Engineering, Rutgers University, Piscataway, NJ

Fernando J. Muzzio, Chemical and Biochemical Engineering, Rutgers University, Piscataway, NJ

The development of new analytical validation techniques in addition to sampling procedures that are well understood are some of the main topics discussed in the industry guidance (FDA, 2003). In this work we have used model formulations of Magnesium Stearate (MgSt), Lactose, and micro-crystalline cellulose exposed to different shear conditions. The operating parameters studied include total shear and strain rate. The effect of varying the MgSt percentage was also studied, considering the effect lubricants have on powder flow can greatly facilitate processing steps such as mixing and tableting (Faqih et al., 2007).

Samples from each formulation were retrieved and analyzed for MgSt content in order to determine the composition uniformity within the samples. The powder samples are initially scanned using NIR, and than compressed into tablets using a Carver Press with the intention to ensure a constant compression force on each tablet. This is crucial because the effect compression has on hardness is also studied since different hardness levels will affect the NIR and LIBS results for the acquisitions. NIR acquisition is also performed on tablets, with the intention to compare the variability in the prediction models between the powder samples and tablets using NIR. One of the disadvantages of NIR is that at low percentages of one of the components, NIR performance worsens, thus finding alternatives as well as other techniques that can validate the analysis results is greatly desired. Once the tablets have been analyzed using NIR, they are subsequently examined using LIBS.

LIBS has been used to determine the content of MgSt (Green et al., 2005), interestingly they found that increasing the number of shots and including additional sites did not improve there quantitative results. Considering there are several parameters to set when operating the process, in this work we will discuss the statistical significance of these effects. In addition to compare the statistical significance between the analysis methods (LIB, NIR) in terms of concentration and homogeneity predictions. An Analysis of Variance (ANOVA) will be used to examine the different sources (ie. MgSt %, Analysis Method, Shear Rate, Total Shear) and varying levels for each source.

In terms of predicting the concentration of MgSt within samples, the findings show that there is a statistically significant distinction between the analysis methods. Interestingly at no or low MgSt the effect of varying RPM on the predicted concentration is not significant for both methods. The effect of RPM becomes significant on the quantitative concentration predictions at high levels of MgSt. In terms of relative variance between the formulations at a 2 % MgSt level, the variability between the methods is not statistically significant. The effect of RPM also is considered to be statistically insignificant for a majority of the studies. Total shear (strain) has been shown to be an important parameter affecting flow index, MgSt micro-distribution, and tablet hardness (Mehrotra et al., 2007). Characterizing physical effects using fast analytical tools such as LIBS and NIR measurements was performed on the tablets made from blends exposed to different levels of strain. Homogeneity improved for powder and tablets when the strain was increased, but this effect diminished by increasing the MgSt concentration. NIR calibration models have been developed for powder and tablet samples. The effect of strain on the LIBS signal has been shown to be significant. Thus, applicability of LIBS and NIR for characterizing lubrication will be demonstrated.

References

FDA, Guidance for Industry, Powder Blends and Finished Dosage Units - Stratified In-Process Dosage Unit Sampling and Assessment, Pharmaceutical CGMP's, October 2003.

Faqih A. M., Mehrotra A., Hammond S. V., Muzzio F. J., 2007, Effect of moisture and magnesium stearate concentration on flow properties of cohesive granular materials, International Journal of Pharmaceutics, 336, 2, 338-345.

Green R., Mowery M., Good J., Higgins J., Arrivo S., McColough K., Mateos A., Reed R., 2005, Comparison of Near-Infrared and Laser-Induced Breakdown Spectroscopy for Determination of Magnesium Stearate in Pharmaceutical Powders and Solid Dosage Forms, Applied Spectroscopy, 59, 2, 2005, 340-347.

Mehrotra A., Llusa M., Faqih A., Levin M., Muzzio F. J., Influence of shear intensity and total shear on properties of blends and tablets of lactose and cellulose lubricated with magnesium stearate, International Journal of Pharmaceutics, 2007, 336, 284-291.

See more of this Session: Characterization and Measurement in Powder Processing
See more of this Group/Topical: Particle Technology Forum

CREC-GS-Optiprobe for Measuring Radial Distribution of Clusters in a Downer Reactor

Mohammad Ashraful Islam, Chemical & Biochemical Engineering, University of Western Ontario, London, ON, Canada

Hugo I. de Lasa, Chemical & Biochemical Engineering, University of Western Ontario, London, ON, Canada

Stefen Krol, University of Western Ontario, London, ON, Canada

It has been well recognized that particles in downer reactors tend to aggregate into clusters. Therefore, understanding and predicting of cluster velocity and size is one of the fundamental needs for the hydrodynamics of gas-solid flow in such a kind of fluidized bed, especially when the catalyst residence time is a very important issue. With the potentially increasing industrial use of these reactors, the need for accurate local measurement techniques has become apparent. Reflective-type (back scattering) fiber-optic sensors are one of the most effective methods for measuring local particle properties.

The major drawbacks of back scattering fiber-optic sensors are short focal distance, distortion of flow stream due to probe intrusion, low light intensity of the captured reflected light by the receiver fibers and infinite measurement volumes from which the probes collect reflections. By integrating a GRIN lens in a novel and precisely designed CREC-GS-Optiprobe most of these shortcomings were surmounted. More reliable and enhanced results were obtained using this technique.

The radial profiles of local cluster velocities as well as cluster sizes are reported in this paper. The experiment was done in a 2.5 cm downer unit. The operating conditions of the present study were set according to a combination of independent variables such as gas volumetric flow rate and solid mass flux. The gas velocities and solid circulation rates were varied from 0.54 (m/s) to 3.1 (m/s) and from 3 (kg/m^{2}s) to 108 (kg/m^{2}s).

See more of this Session: Characterization and Measurement in Powder Processing
See more of this Group/Topical: Particle Technology Forum

Two and Three Dimensional Microparticle Characterization Using Digital Holography

Taslima Khanam, School of Chemical and Biomedical Engineering, Nanyang Technological University, Singapore, Singapore

Emmanouil Darakis, Department of Computer Science, National University of Ireland, Maynooth, Ireland

Arvind Rajendran, School of Chemical and Biomedical Engineering, Nanyang Technological University, Singapore, Singapore

Vinay Kariwala, Division of Chemical and Biomolecular Engineering, Nanyang Technological University, Singapore, Singapore

Anand K. Asundi, School of Mechanical and Aerospace Engineering, Nanyang Technological University, Singapore, Singapore

For particulate processes, such as crystallization, polymerization and granulation, particle imaging has been proven to be an effective technique for on-line particle size and shape measurement [1,2]. Imaging based methods extract size and shape information from acquired images using image analysis algorithms. The successful application of image analysis algorithms for conventional 2D imaging systems [1,2], however, is often hindered by the problems arising due to limited depth of focus imposed by required magnification as well as undetermined perspective position and random orientation of the particles [3]. Thus, conventional imaging systems provide 2D information about the 3D particles, which can result in inaccurate inference of particle's 3D size and shape [3,4].

In this paper, we propose a novel method to accurately characterize the properties of microparticles. The proposed method uses digital holography as the imaging tool. In digital holography, a hologram of the object is stored digitally. Subsequent numerical processing yields reconstructions focused at particular depths of the scene. Thus, 3D information can be obtained using a single hologram by reconstructing it at different depths [5]. This distinct advantage allows digital holography to overcome the aforesaid limitations of the 2D imaging systems. This paper highlights two key aspects of the application of digital holography. Firstly, we demonstrate that digital holography is a potential tool for microparticle analysis. Secondly, we develop an algorithm to accurately profile and to measure true lengths of microfibers.

Firstly we apply to measurement of particle size distributions of a population of microparticles. An automated image processing algorithm is proposed to extract particle size and shape from the reconstructed images [4,6]. The performance of the proposed method is studied using both spherical and needle shaped objects, which represent the morphological extremes encountered in practice. Furthermore, experiments are carried out under different conditions, such as static, suspended and flow-through system. In the static condition, dry particles are placed on a glass slide, whereas in the latter cases particles are contained in a volume (cuvette or flow-through cell) in order to study the out of focusing effect as well as effect of flow on the measurements. In all cases, the technique automatically determines the best focused depths and yields good measurement of the size and shape of the particles. These measurements have been validated using independent methods such as optical microscopy or scanning electron microscopy (SEM).

Secondly, we propose a novel method to accurately measure size, orientation and position of needle shaped microparticles. In this method, a tailor-made image analysis algorithm is applied to the reconstructed images to extract length, position and orientation of the microfibers simultaneously [7]. The proposed method is applied to experimentally recorded holograms of a single fiber of known length and orientation for benchmarking purpose. Here, a single microfiber is placed on a mount, which can be fixed at a pre-determined off-axis tilt. The measured length and tilt of the microfiber are found to be in excellent agreement with their true values. It is worth noting that the image analysis algorithm is automated, which provides the same accuracy for microfibers with different lengths (150 and 1320 microns) and different tilts along the optical axis (0-60 °). Finally, the proposed method is applied to a suspension, where the length, position and orientation of the multiple fibers are determined without any *a-priori* information. These results successfully demonstrate the potential of digital in-line holography for characterization of needle shaped particles and opens up avenues for particle analysis, where simultaneous measurement of both particle size and shape are of immense interest.

References

1.Eggers, J., Kempkes, M. and Mazzotti, M., Measurement of size and shape distributions of particles through image analysis. Chemical Engineering Science, vol. 63, pp. 5513-5521, 2008.

2. Larsen, P. A. and Rawlings, J. B., The potential of current high-resolution imaging-based particle size distribution measurements for crystallization monitoring, AIChE Journal, vol. 55, pp. 896-905, 2009.

3. Wang, X. Z., Roberts, K. J. and Ma, C., Crystal growth measurement using 2D and 3D imaging and the perspectives for shape control, Chemical Engineering Science, vol. 63, pp. 1173-1184, 2008.

4. Khanam, T., Darakis, E., Rajendran, A., Kariwala, V., Asundi, A. K. and Naughton, T. J., On-line digital holographic measurement of size and shape of microparticles for crystallization processes, in Ninth International Symposium on Laser Metrology, 7155, SPIE, 2008.

5.Schnars, U. and Juptner, W., Digital Holography: Digital Hologram Recording, Numerical Reconstruction, and Related Techniques, Springer, Berlin, 2005.

6. Darakis, E., Khanam, T., Rajendran, A., Kariwala, V., Asundi, A. K. and Naughton, T. J., "Processing of digital holograms for size measurements of microparticlese," in Ninth International Symposium on Laser Metrology, 7155, SPIE, 2008.

7 Kempkes, M., Darakis, E., Khanam, T., Rajendran, A., Kariwala, V., Asundi, A. K., Mazzotti M., and Naughton, T. J., Three dimensional digital holographic profiling of micro-fibers, Optics Express, vol. 17, pp. 2938-2943, 2009.

See more of this Session: Characterization and Measurement in Powder Processing
See more of this Group/Topical: Particle Technology Forum

Statistical Modeling of Image-Based Crystal Observation

Christian Borchert, Physical and Chemical Process Engineering, Max Planck Institute for Dynamics of Complex Technical Systems, Magdeburg, Germany

Kai Sundmacher, Max Planck Institute for Dynamics of Complex Technical Systems, Magdeburg, Germany

Solution crystallization produces materials with specific crystal shape and size distributions. Both the crystal size and shape play an important role in deciding the processing rate of material in various downstream operations, for instance filtration, washing, and drying. Crystals of different shapes have different dissolution rates and henceforth bioavailability, the properties that are important in engineering crystals of active pharmaceutical ingredients. Thus, the significant impact of crystal size and shape demands for its tighter control. Therefore, the measurement of crystal shape distributions becomes important.

Using advanced mathematical techniques, such as multidimensional population balance equations, models for crystallization processes can take into account the complex behavior of crystal shape evolution as a result of crystal growth [1,2]. In order to experimentally validate model predictions, the inherent multidimensionality of the modeling framework must be taken into account when a measurement system is selected. Due to the rich information contained in crystal photographs, the observation of crystallization processes using imaging techniques has drawn the attention of researchers in chemical engineering in recent years. The images acquired give a good qualitative visual impression of the crystal morphologies produced. Clearly, the quantitative evaluation of these images with regards to crystal morphology demands for well developed mathematical tools. Besides image processing tools, the quantitative analysis requires the linking between the geometrical crystal model and the information which can be extracted from an image. This contribution seeks to provide such a link.

The three-dimensional convex crystal is characterized by an n-dimensional geometrical state vector, where n is the number of faces exposed on the crystal surface. The crystal is projected on a two-dimensional surface from which the boundary curve is extracted. Subsequently, the boundary curve is characterized by a set of m scalar measures, for instance several shape factors or Fourier Descriptors. The shape of the boundary curve depends highly on the crystal orientation. That is, there is no one-to-one mapping from the m-dimensional vector describing the boundary curve and the n-dimensional geometrical state vector of the crystal model. However, the usage of an orientation model, for instance assuming randomly or preferentially oriented crystals, allows the design of state estimators for the geometrical state vector.

[1] Borchert, C.; Nere, N.; Ramkrishna, D.; Voigt, A.; Sundmacher, K., "On the Prediction of Crystal Shape Distributions in a Steady State Continuous Crystallizer", Chem. Eng. Sci. 64 (2008), 686-696.

[2] Borchert, C.; Voigt, A.; Sundmacher, K.; Nere, N.; Ramkrishna, D., "Evolution of Crystal Shape Distributions and Morphology Classification", Proceedings of the 17th International Symposium on Industrial Crystallization, 14-17 September 2008, Maastricht, Netherlands.

See more of this Session: Characterization and Measurement in Powder Processing
See more of this Group/Topical: Particle Technology Forum

Particle Size Distribution Estimation through Image-Based Sensors: Sequential Simplex Optimization for Selection of Optimal Parameters

Ying Zhou, Process Science and Modeling, Institute of Chemical and Engineering Sciences, Singapore, Singapore

Lakshminarayanan Samavedham, Chemical & Biomolecular Engineering, National University of Singapore, Singapore, Singapore

Rajagopalan Srinivasan, Chemical and Biomolecular Engineering, National University of Singapore, Singapore, Singapore

Particle size distribution (PSD) and shape are critical to product quality in crystallization processes. A narrow particle size distribution with specific particle shape is indicative of good product quality; however, the absence of robust online sensors for size distribution prevents real-time monitoring and control. With the developments in sensor technology, instruments such as the Particle Vision and Measurement (PVM) from Lasentec / Mettler Toledo are widely used in manufacturing units to monitor PSD and shape variation. These have triggered research interest in image-analysis techniques that convert inline images to information about the process state - especially particle size distribution. In this work, we study the robustness of image-analysis algorithms (IAA) and use design of experiments based techniques for tuning the parameters associated with the IAA.

Several approaches to estimate the process state or product quality (e.g. size, shape, and growth rate of particles) in real-time and in situ (Sarkar et al., 2009, Zhou et al., 2006, Wilkinson et al., 2000, Wang et al., 2008, Patience, 2002, Li et al., 2006, Larsen et al., 2007) have been reported recently. Crystal size estimation from video images, especially for high aspect ratio systems (e.g. L-glutamicacid), has received much attention since laser-diffraction based methods that are an alternate way to estimate size do not perform adequately for such systems. Any image-analysis algorithm involves several steps such as image selection, image enhancement, edge detection and morphology operation. A large number of user-defined parameters are involved in each of the above steps. Our studies (Zhou et al., 2009) show that with proper settings of these IA parameters, the estimated PSD can be obtained within ~10% error (when compared to results obtained by the human-eye). However, when these parameters are specified unsuitably, very large errors (~50%) could occur. Given the large variability in the quality of the results, a systematic way to select the user-defined parameters for IAA is critical. In this paper, we propose to use a systematic design of experiments methodology for selecting optimal IA parameters.

In this work, uniform design (UD) (Liang et al., 2001) and sequential simplex optimization (SSO) (Kamoun et al., 2009) are applied to find optimal IA parameters. In UD, it is important for the designed parameters to uniformly fill the whole space of parameter range. Hence, the range and the number of levels of each parameter are predefined. The results from these designed experimental runs are used to build a mathematical model relating the input parameters and output results. The the optimal IA parameters are obtained by solving this mathematical model for the optima. In SSO, experimental runs are performed with specific number of random guesses (within the allowable limits) and parameter settings for subsequent experiments are calculated iteratively based on the results from prior runs.

Mono-sodium Glutamate (MSG) seeded cooling crystallization process is selected as the case study. We focus on 7 IA parameters that pertain to the key image analysis steps of edge detection and morphology operations. Results from the two approaches will be reported and compared based on efficiency and accuracy.

References

Kamoun, A., Jaziri, M. & Chaaouni, M., (2009), The use of the simplex method and its derivatives to the on-line optimization of the parameters of an injection moulding process. Chemometrics and Intelligent Laboratory Systems 96, 117-122.

Larsen, P.A., Rawlings, J.B. & Ferrier, N.J., (2007), Model-based object recognition to measure crystal size and shape distributions from in situ video images. Chemical Engineering Science 62, 1430-1441.

Li, R. F., Thomson, G. B., White, G., Wang, X. Z., Calderon De Anda, J. & Roberts, K. J., (2006), Integration of crystal morphology modeling and on-line shape measurement. AIChE Journal, Vol. 52, No. 6, 2297-2305.

Liang, Y., Fang, K. & Xu, Q., (2001), Uniform design and its applications in chemistry and chemical engineering. Chemometrics and Intelligent Laboratory Systems 58, 43-57.

Patience, D. B., (2002), Crystal engineering through particle size and shape monitoring, modeling and control. University of Wisconsin-Madison, PhD Thesis.

Sarkar, D., Doan, X. T., Zhou, Y. & Srinivasan, R., (2009) "In-situ particle size estimation for crystallization processes by multivariate image analysis", Chemical Engineering Science 64(1), 9-19.

Wang, X. Z., Roberts, K. J. & Ma, C., (2008), Crystal growth measurement using 2D and 3D imaging and the perspectives for shape control. Chemical Engineering Science 63 (2008) 1173-1184.

Wilkinson, M. J., Jennings, K. H. & Hardy, M., (2000), Non-invasive video imaging for interrogating pharmaceutical crystallization. Microscopy and Microanalysis, Vol (6) (Supplement), 996-997.

Zhou, Y., Doan, X. T. & Srinivasan, R., (2006), Real-time imaging and product quality characterization for control of particulate processes. Joint 16th ESCAPE and 9th PSE: Computer-Aided Chemical Engineering 21A, W. Marquardt and C. Pantelides (Eds), 775-780. Elsevier, Garmish-Partenkirchen.

Zhou, Y., Srinivasan, R. & Samavedham, L. (2009) "Critical evaluation of imaging based techniques for real-time crystal size measurements", Computers & Chemical Engineering, volume 33, issue 5, 1022-1035.

See more of this Session: Characterization and Measurement in Powder Processing
See more of this Group/Topical: Particle Technology Forum

Static Mixing of Granular MATERIALS

Watson L. Vargas, Department of chemical Engineering, Universidad de los Andes, Bogotá, D. C, Colombia

Andrés F. Nieto, Department of Chemical Engineering, Universidad de los Andes, Bogotá, D. C, Colombia

Diana M. Lievano, Department of Chemical Engineering, Universidad de los Andes, Bogotá, D. C, Colombia

While an understanding of the dynamics of segregation has started to emerge, controlling segregation continues to be a complicated issue. In this study we explore the use of time-modulation - via selective insert placement as suggested by (Shi, Vargas and McCarthy., PRL, 2007) - in order to suppress segregation in two geometries intended to be used as static granular mixers. We study bidisperse (size or density), cohesionless granular materials in three and quasi-two dimensional devices by means of simulations and experiments. Results are presented for two configurations of the inserts. Both experiments and simulations indicate that devices whose design is based on the fundamental physics of granular flow are much more effective in reducing segregation, while baffles or inserts whose design is based on empirical observations are less effective in reducing both density and size segregation. Qualitative and quantitative evidence is presented, in terms of the intensity of segregation.

See more of this Session: Mixing and Segregation of Particulates
See more of this Group/Topical: Particle Technology Forum

Impact of Binary and Continuous Particle Size Distributions On Clustering, Granular Shear Flows

R. Brent Rice, Department of Chemical and Biological Engineering, University of Colorado at Boulder, Boulder, CO

Christine M. Hrenya, Department of Chemical and Biological Engineering, University of Colorado at Boulder, Boulder, CO

The dynamic clustering phenomenon exhibited by rapid granular shear flows is investigated with respect to the effect of both binary and continuous size distributions. Analysis of these systems employs a length scale based on the radial distribution function along with an evaluation of clustered- and dilute-region concentrations and temperatures. In particular, two questions are considered: (1) What is the effect of the particle size distributions on the prevalence of clusters? (2) Does preferential segregation of large or small species occur within the clustered regions? Results indicate that all investigated particle size distributions (binary, Gaussian, lognormal) behave similarly with respect to these questions. The prevalence of clusters increases with an increased deviation from the monodisperse limit. Moreover, large particles tend preferentially toward the clustered regions, which exhibit lower temperatures than surrounding dilute regions. Such segregation toward low-temperature regions is consistent with the well-known tendency for the segregation of large particles toward steady-state low temperature regions, in spite of the transient nature of the clustered regions and the pertinent temperature gradients in the current work.

See more of this Session: Mixing and Segregation of Particulates
See more of this Group/Topical: Particle Technology Forum

Optimizing the Reduction of Segregation in Baffled-Tumbler Mixers

Tathagata Bhattacharya, Department of Chemical and Petroleum Engineering, University of Pittsburgh, Pittsburgh, PA

J. J. McCarthy, Department of Chemical and Petroleum Engineering, University of Pittsburgh, Pittsburgh, PA

Segregation is a major problem for many solids processing industries. A small difference in any mechanical properties like particle size or density can lead to flow-induced segregation. Recently, our group has proposed a method of thwarting segregation in surface-dominated flows whereby segregation is dramatically reduced via periodic inversion of the surface/flowing layers. In this study we use the discrete element method (DEM) to systematically investigate how this approach may be used to enhance the mixing in a tumbler mixer, probing such operating parameters as baffle shape, baffle placement and the tumbler filling level. Previous work has suggested that a simple design heuristic may be used to effectively discriminate between a host of tumbler/baffle geometries. The detailed flow profiles that result from our simulations allow us to exhaustively test the proposed link between the asymptotic mixing behavior of the system and the suggested design heuristic -- the probability distribution of the number of layer passes a particle takes per tumbler rotation.

See more of this Session: Mixing and Segregation of Particulates
See more of this Group/Topical: Particle Technology Forum

Radial Segregation of Granular Materials in Rotating Cylinders

Suman K. Hajra, Computational Science Division, U. S. Department of Energy, National Energy Technology Laboratory,, Pittsburgh, PA

Mehrdad Massoudi, Computational Science Division, U. S. Department of Energy, National Energy Technology Laboratory,, Pittsburgh, PA

D. V. Khakhar, Department of Chemical Engineering, Indian Institute of Technology, Bombay, Mumbai - 400076, India

Flowing granular mixtures segregate, and this can have a significant impact in many industrial processes ranging from pharmaceutical tablet manufacturing to cement production. Rotating cylinders are used as kilns, mixers, dryers and granulators. However, when mixtures of two different types of particles are rotated in a rotating cylinder then they tend to segregate/demix. Segregation is studied in a rotating cylinder for mixtures of different size particles. In most of the systems studied, the measured radial weight fraction profiles in the bed show significant double segregation (a core of small particles as well as a thin layer of small particles at the periphery). The profiles are found to be sensitively dependent on the surface roughness of the particles in the mixture, and double segregation reduces with particle roughness. Double segregation is observed for all cylinder fill fraction. Double segregation is found decreased with increasing the cylinder speed of rotation and no double segregation is observed at higher cylinder speed of rotation. A continuum based model is developed and a good agreement is obtained.

See more of this Session: Mixing and Segregation of Particulates
See more of this Group/Topical: Particle Technology Forum

Predicting Segregation When Materials Are Sensitive to More Than One Segregation Mechanism

Kerry D. Johanson, Material Flow Solutions Inc., Gainesville, FL

Bulk materials segregate due to a variety of mechanisms. Any difference in particle scale properties can cause segregation in materials as they are subjected to flow and transport. Fines may sift down through a matrix of coarser particles, depositing below the process charge point. Angle of repose differences may cause components within the mixture to separate as they flow down a pile. Fines may also be carried by air currents, depositing where these air currents decrease. Segregation can be represented by the convection dispersion equation, provided the applicable region of convection and dispersion can be identified for each specific segregation mechanism, and the relative magnitude of segregation fluxes can be estimated for combinations of components in the mixture. This paper presents a methodology to estimate segregation fluxes, and applies this methodology to binary and tertiary mixtures. The effect of single segregation mechanisms, as well as combined mechanisms, will be examined. In addition, the effect of cohesion on each mechanism will be examined. The use of cohesion to mitigate segregation is common. However, the amount of cohesion required to mitigate segregation depends on the cause of segregation. In some cases, the addition of cohesion can induce more segregation by shifting segregation mechanisms. In other cases, the addition of cohesion helps to mitigate all segregation mechanisms. This paper examines the influence of particle scale properties on individual segregation mechanisms and on combined mechanisms. It also looks at the relationship between key bulk scale properties such as unconfined yield strength, bulk density, and internal friction on segregation potential.

See more of this Session: Mixing and Segregation of Particulates
See more of this Group/Topical: Particle Technology Forum

Impact of Particle Aspect Ratio On Tumble Mixing Performance

Bruce D. Hook Sr., Epoxy Process R&D, Dow Chemical Co., Freeport, TX

Daniel T. Hook, Physics Dept., Austin College, Sherman, TX

Matthew Varvir, Physics, Austin College, Sherman, TX

Particle Shape plays a consideration in tumble mixing. For flat or disc shaped particles, understanding when the mechanism of particles sliding dominates over rolling has an impact on flow regime and the mixing time of the segregated mixture. Data and models will be presented.

See more of this Session: Mixing and Segregation of Particulates
See more of this Group/Topical: Particle Technology Forum

Verification of a Binary Mixing Process of Particulate Solids in a Fluidized Bed Predicted with the Theory of Markov Chains

Y. Khokhlova, Department of Powder Science and Technology (POSTEC), Tel-Tek, Porsgrunn, Norway

S. Ding, Department of Powder Science and Technology (POSTEC), Tel-Tek, Porsgrunn, Norway

V. Mizonov, Department of Applied Mathematics, Ivanovo State Power Engineering University (ISPEU), Ivanovo, Russia

G. G. Enstad, Department of Powder Science and Technology (POSTEC), Tel-Tek, Porsgrunn, Norway

Experimental investigation has been carried out to study the evolution of concentration of components in a binary mixture of particulate solids in a laboratory cylindrical fluidized bed. The fluidized bed is 250 mm in diameter and 800 mm in height. It was initially loaded with two kinds of particulate solids in layers. The particulate solids used were almost spherical, but varied in size and/or density. The fluidization bed was operated with different gas flow rates; meanwhile, samples were taken at some designated locations in four levels along the fluidization column with a predefined interval. The components of solids in samples were measured with image processing and/or sieving technique; and the evolution of concentration of particulate solids (i.e. the mixing process of material) was therefore investigated.

A mathematical model to describe the process was also proposed. In the model, the volume of fluidized bed is represented as a sequence of perfectly mixed cells along the bed height, the mixing process of solids was modeled based on the theory of Markov chains. Comparisons were made between the results predicted with the mathematical model and the experimental data obtained. The model is subjected to further improvement, at the same time the processing of binary mixing in a fluidization bed was addressed with regard to the density and size of particulate solids.

See more of this Session: Mixing and Segregation of Particulates
See more of this Group/Topical: Particle Technology Forum

On the Effect of Sub-Grid Drag Models*

Sofiane Benyahia
National Energy Technology Laboratory
Morgantown, WV

*This study was published recently in I&ECR journal: http://pubs.acs.org/doi/abs/10.1021/ie900658k

Goals, Objectives and Challenges

- **Conduct fast and accurate gas-solids flow simulations using coarse grids.**

- **Implement EMMS and filtered drag models in MFIX**
- **Conduct comparative study of these drag models with experimental data**

- **Challenges:**
  - Sub-grid models are still being developed (models not yet available for all flow conditions)
  - Lack extensive validation
  - More model development to capture near-wall effects

NETL

2K-2571

Gas-solids flow model

Governing equations

Conservation of mass for phase m (s: solids and g: gas):

$$\frac{\partial(\rho_m \varepsilon_m)}{\partial t} + \nabla \cdot (\rho_m \varepsilon_m \mathbf{v}_m) = 0$$

Conservation of linear momentum:

$$\left[\frac{\partial(\rho_m \varepsilon_m \mathbf{v}_m)}{\partial t} + \nabla \cdot (\rho_m \varepsilon_m \mathbf{v}_m \mathbf{v}_m)\right] = -\varepsilon_m \nabla P_g + \nabla \cdot (\boldsymbol{\tau}_m) - \beta_{mn}(\mathbf{v}_m - \mathbf{v}_n) + \varepsilon_m \rho_m \mathbf{g}$$

Translational granular energy conservation equation:

$$\frac{3}{2}\rho_s\left[\frac{\partial(\varepsilon_s \Theta_s)}{\partial t} + \nabla \cdot (\varepsilon_s \Theta_s \mathbf{v}_s)\right] = -\nabla \cdot \mathbf{q} + \boldsymbol{\tau}_c : \nabla \mathbf{v}_s - J_s + \Pi_\Theta$$

Interphase momentum exchange term ($\beta_{gs} = \beta_{sg} = \beta$):

Wen-Yu correlation: $\beta = \frac{3}{4} C_D \frac{\rho_g \varepsilon_g \varepsilon_s |\mathbf{v}_g - \mathbf{v}_s|}{d_p} \varepsilon_g^{-2.65}$

$$C_D = \begin{cases} 24/\mathrm{Re}\left(1 + 0.15\,\mathrm{Re}^{0.687}\right) & \mathrm{Re} < 1000 \\ 0.44 & \mathrm{Re} \geq 1000 \end{cases}, \quad \mathrm{Re} = \frac{\rho_g \varepsilon_g |\mathbf{v}_g - \mathbf{v}_s| d_p}{\mu_g}$$

EMMS drag model: $\beta = \begin{cases} 150 \frac{\varepsilon_s(1-\varepsilon_g)\mu_g}{\varepsilon_g d_p^2} + 1.75 \frac{\rho_g}{d_p} \varepsilon_s |\mathbf{v}_g - \mathbf{v}_s| & \varepsilon_g < 0.7 \\ \frac{3}{4} C_D \frac{\rho_g \varepsilon_g \varepsilon_s |\mathbf{v}_g - \mathbf{v}_s|}{d_p} f(\varepsilon_g) & \varepsilon_g \geq 0.7 \end{cases}$

Where: $f(\varepsilon_g) = \begin{cases} -15.63 + 16.63 \times \varepsilon_g & \varepsilon_g > 0.94 \\ 516.1 - 2243 \times \varepsilon_g + 3658 \times \varepsilon_g^2 - 2652 \times \varepsilon_g^3 + 721.6 \times \varepsilon_g^4 & 0.94 \leq \varepsilon_g > 0.84 \\ 17127 \times \varepsilon_g + 148026 \times \varepsilon_g^2 - 118519 \times \varepsilon_g^3 + 35599 \times \varepsilon_g^4 & 0.84 \leq \varepsilon_g > 0.79 \\ -39898 + 199506 \times \varepsilon_g - 372449 \times \varepsilon_g^2 + 307602 \times \varepsilon_g^3 - 94803 \times \varepsilon_g^4 & 0.79 \leq \varepsilon_g > 0.72 \\ -276.9 + 764.0 \times \varepsilon_g - 520.1 \times \varepsilon_g^2 & 0.72 \leq \varepsilon_g > 0.7 \end{cases}$

Filtered drag model: For dimensionless filter size: $\frac{g\Delta}{V_t^2} = 2.056$

$$\frac{\beta V_t}{\rho_s \varepsilon_s \varepsilon_g g} = \begin{cases} \frac{4.277 \times \varepsilon_g^4 - 2.363 \times \varepsilon_g^3 - 4.324 \times \varepsilon_g^2 - 2.474 \times \varepsilon_g + 4.904}{\varepsilon_g^2 - 5.411 \times \varepsilon_g + 4.427} & \varepsilon_s \leq 0.04083 \\ \frac{0.4820}{\varepsilon_g^{3.574}} & 0.04083 < \varepsilon_s \leq 0.2589 \\ \frac{148.0 \times \varepsilon_s^3 - 124.8 \times \varepsilon_s^2 + 37.50 \times \varepsilon_s - 3.642}{\varepsilon_g \varepsilon_s} & \varepsilon_s > 0.2589 \end{cases}$$

For dimensionless filter size: $\frac{g\Delta}{V_t^2} = 4.112$

$$\frac{\beta V_t}{\rho_s \varepsilon_s \varepsilon_g g} = \begin{cases} \frac{5.301 \times \varepsilon_g^3 - 5.473 \times \varepsilon_g^2 - 5.723 \times \varepsilon_g + 5.906}{\varepsilon_g^3 - 3.820 \times \varepsilon_g^2 + 1.145 \times \varepsilon_g + 1.686} & \varepsilon_s \leq 0.0533 \\ \frac{0.3349}{\varepsilon_g^{3.872}} & 0.0533 < \varepsilon_s \leq 0.2922 \\ \frac{-839.0 \times \varepsilon_s^4 + 1574 \times \varepsilon_s^3 - 1012 \times \varepsilon_s^2 + 277.3 \times \varepsilon_s - 27.51}{\varepsilon_g \varepsilon_s} & \varepsilon_s > 0.2922 \end{cases}$$

NETL

2K-2571

Gas-solids flow model

Constitutive relations for gas and solids phases

Gas-phase stress tensor: $\boldsymbol{\tau}_g = 2\mu_g \mathbf{S}_g$, where $\mathbf{S}_m = \frac{1}{2}\left(\nabla \mathbf{v}_m + (\nabla \mathbf{v}_m)^T\right) - \frac{1}{3}\nabla \cdot \mathbf{v}_m \mathbf{I}$

Solids-phase stress tensor: $\boldsymbol{\tau}_s = \left[-(P_s + P_f) + \eta\mu_b \nabla \cdot \mathbf{v_s}\right]\mathbf{I} + 2(\mu_s + \mu_f)\mathbf{S_s}$

Solids kinetic-collisional pressure: $P_s = \varepsilon_s \rho_s \Theta_s \left[1 + 4\eta\varepsilon_s g_0\right]$

The radial distribution function at contact: $g_0 = \frac{1 - 0.5\varepsilon_s}{(1-\varepsilon_s)^3}$

Solids viscosity model:

$$\mu_s = \left(\frac{2+\alpha}{3}\right)\left[\frac{\mu^*}{g_0\eta(2-\eta)}\left(1 + \frac{8}{5}\eta\varepsilon_s g_0\right)\left(1 + \frac{8}{5}\eta(3\eta-2)\varepsilon_s g_0\right) + \frac{3}{5}\eta\mu_b\right]$$

$$\mu^* = \mu\left[1 + \frac{2\beta\mu}{(\varepsilon_s\rho_s)^2 g_0 \Theta_s}\right]^{-1}, \quad \mu = \frac{5}{96}\rho_s d_p \sqrt{\pi\Theta_s}, \quad \mu_b = \frac{256}{5\pi}\mu\varepsilon_s^2 g_0$$

Granular energy flux and conductivity: $\mathbf{q} = -\kappa_s \nabla\Theta_s$

$$\kappa_s = \left(\frac{\kappa^*}{g_0}\right)\left[\left(1 + \frac{12}{5}\eta\varepsilon_s g_0\right)\left(1 + \frac{12}{5}\eta^2(4\eta-3)\varepsilon_s g_0\right) + \frac{64}{25\pi}(41-33\eta)\eta^2(\varepsilon_s g_0)^2\right]$$

$$\kappa^* = \kappa\left[1 + \frac{6\beta\kappa}{5(\varepsilon_s\rho_s)^2 g_0\Theta_s}\right]^{-1}, \quad \kappa = \frac{75\rho_s d_p\sqrt{\pi\Theta_s}}{48\eta(41-33\eta)}$$

Frictional solids pressure and viscosity:

$$\frac{P_f}{P_c} = \left(1 - \frac{\nabla\cdot\mathbf{v_s}}{n\sqrt{2}\sin(\delta)\sqrt{\mathbf{S_s}:\mathbf{S_s} + \Theta_s / d_p^2}}\right)^{n-1}$$

$$\mu_f = \frac{\sin(\delta)}{\sqrt{2}}\frac{P_f}{\sqrt{\mathbf{S_s}:\mathbf{S_s} + \Theta_s / d_p^2}}\left\{n - (n-1)\left(\frac{P_f}{P_c}\right)^{\frac{1}{n-1}}\right\}$$

$$\text{Where, } P_c = \begin{cases} 10^{25}\left(\varepsilon_s - \varepsilon_s^{\max}\right)^{10} & \varepsilon_s > \varepsilon_s^{\max} \\ Fr\dfrac{\left(\varepsilon_s - \varepsilon_s^{\min}\right)^r}{\left(\varepsilon_s^{\max} - \varepsilon_s\right)^s} & \varepsilon_s^{\max} \ge \varepsilon_s > \varepsilon_s^{\min} \\ 0 & \varepsilon_s \le \varepsilon_s^{\min} \end{cases}, \text{ and } n = \begin{cases} \dfrac{\sqrt{3}}{2}\sin(\delta) & \nabla\cdot\mathbf{v_s} \ge 0 \\ 1.03 & \nabla\cdot\mathbf{v_s} < 0 \end{cases}$$

Collisional dissipation of granular energy and viscous damping terms:

$$J_s = \frac{48}{\sqrt{\pi}}\eta(1-\eta)\frac{\rho_s\varepsilon_s^2 g_0}{d_p}\Theta_s^{3/2}, \quad \Pi_\Theta = -3\beta\Theta_s + 81\frac{\varepsilon_s\mu_g^2\left|\mathbf{v}_g - \mathbf{v}_s\right|^2}{g_0 d_p^3 \rho_s\sqrt{\pi\Theta_s}}$$

Gas-solids wall boundary conditions

Gas-phase: free slip.

Solids phase: Johnson-Jackson partial slip[27]

$$\frac{\mathbf{v}_{sl}}{\left|\mathbf{v}_{sl}\right|}\cdot\left(\boldsymbol{\tau}_k + \boldsymbol{\tau}_f\right)\cdot\mathbf{n} + \frac{\phi\pi\rho_s\varepsilon_s g_0\sqrt{\Theta_s}}{2\sqrt{3}\varepsilon_s^{\max}}\mathbf{v}_{sl} + \left(\mathbf{n}\cdot\boldsymbol{\tau}_f\cdot\mathbf{n}\right)\tan\delta_w = 0$$

$$\mathbf{n}\cdot\mathbf{q} = \frac{\phi\pi\left|\mathbf{v}_{sl}\right|^2\rho_s\varepsilon_s g_0\sqrt{\Theta_s}}{2\sqrt{3}\varepsilon_s^{\max}} - \frac{\sqrt{3}\pi\rho_s\varepsilon_s g_0\left(1-e_w^2\right)\sqrt{\Theta_s}}{4\varepsilon_s^{\max}}\Theta_s$$

NETL

2K-2571

Simulation description

Process temperature	300 K
Process pressure	101325 Pa
Air density	~ 1.2 kg/m^3
Air viscosity	1.8×10^{-5} Pa·s
Solids density	1712 kg/m^3
Particle diameter	76 μm
Particle-particle restitution coefficient	0.95
Particle-wall restitution coefficient	0.7
Specularity coefficient	0.0001
Void fraction at maximum packing	0.4

Cartesian 2D mesh:
14 x 1017

Cylindrical 3D mesh:
4 x 10 x 408

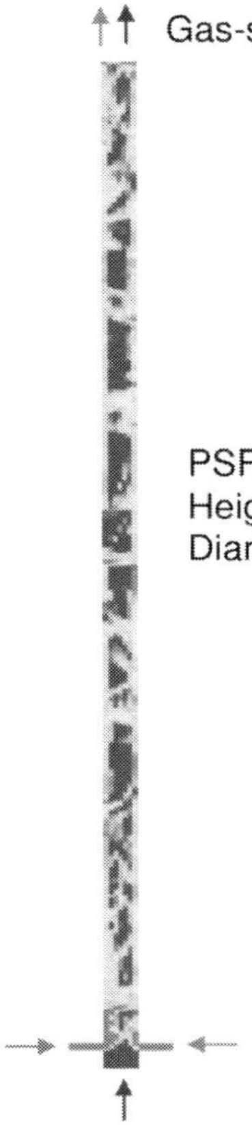

2K-2571

Instantaneous values of friction coefficient

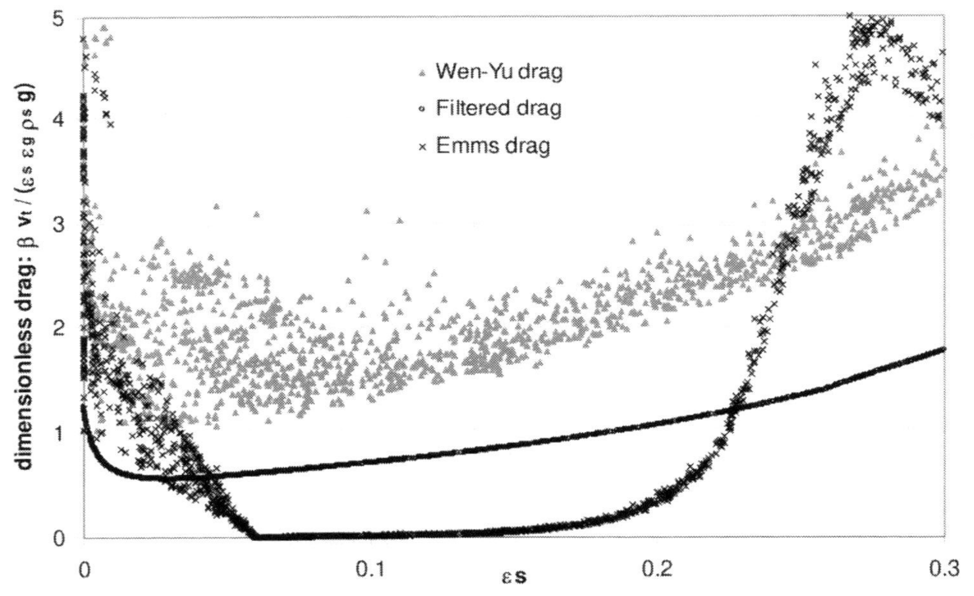

NETL

2K-2571

Instantaneous void fraction contour profiles

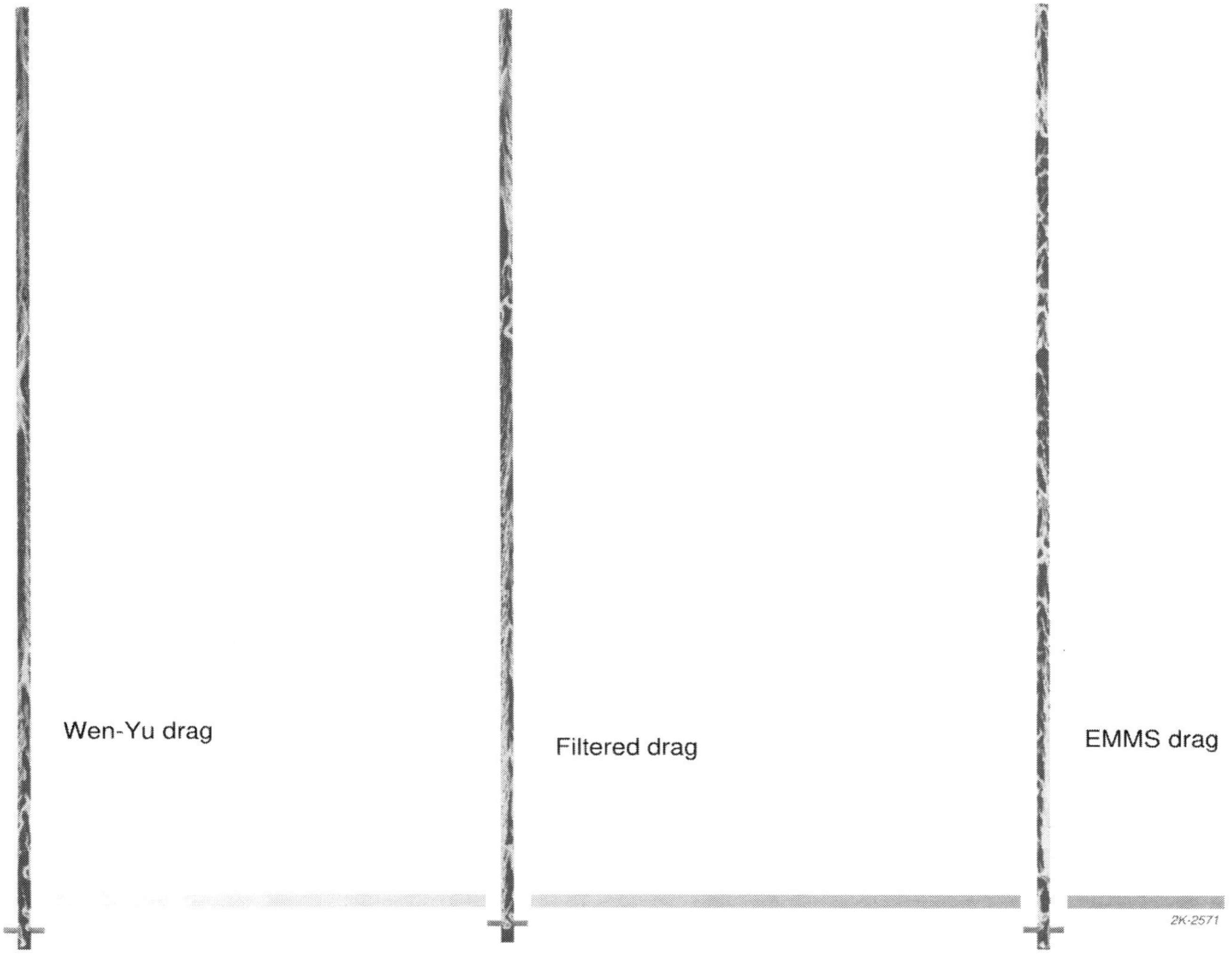

2K-2571

2D time-averaged profiles

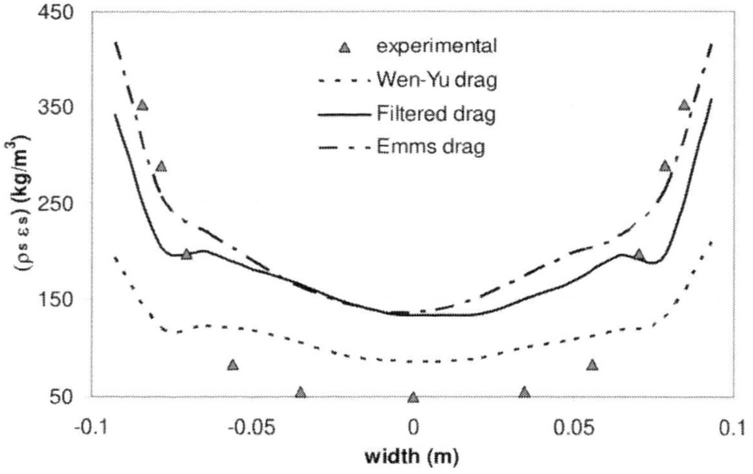

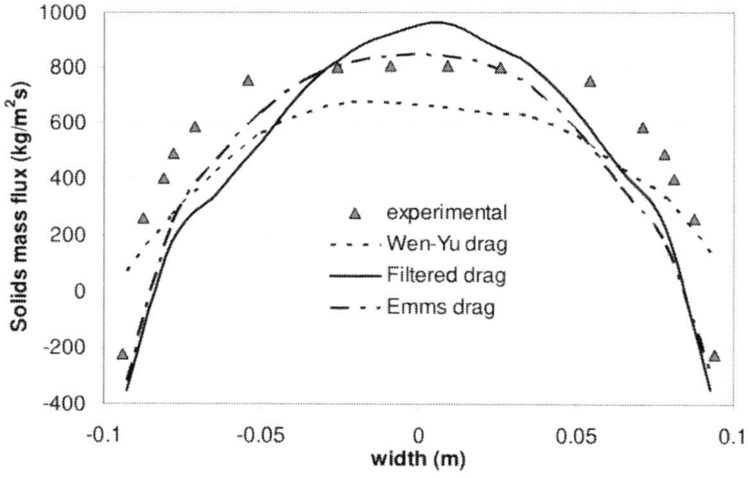

NETL

2K-2571

2D time-averaged profiles

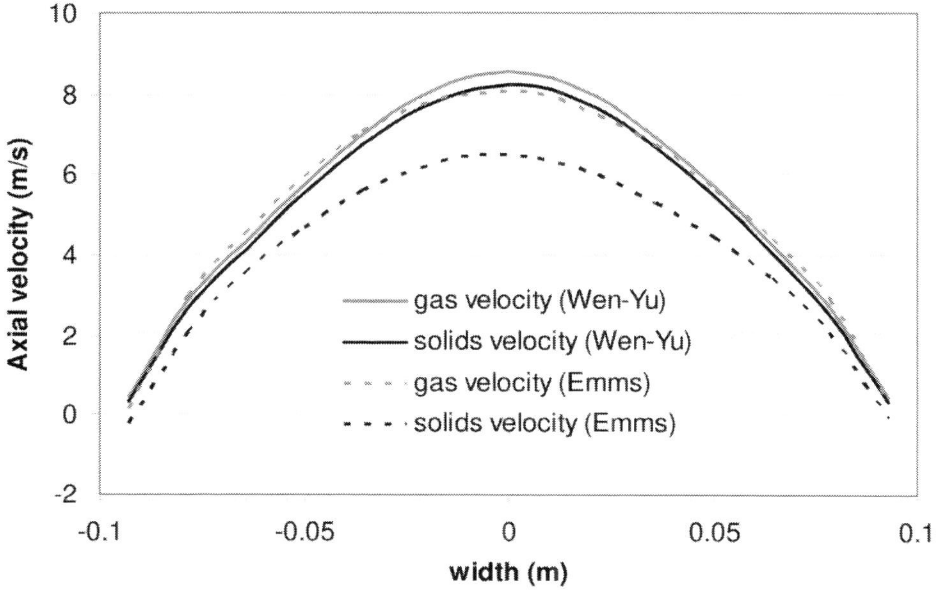

2K-2571

2D time-averaged profiles

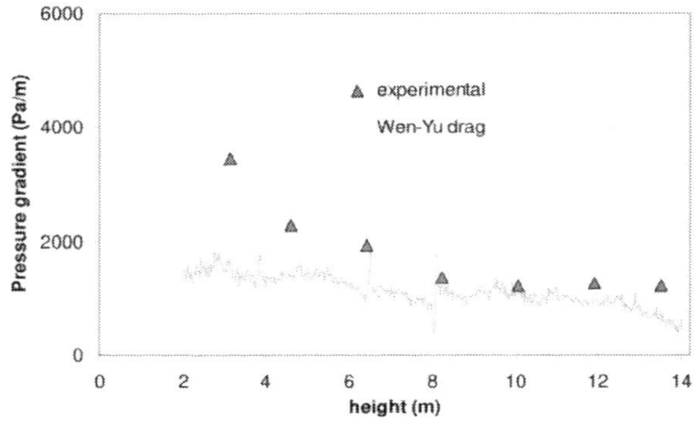

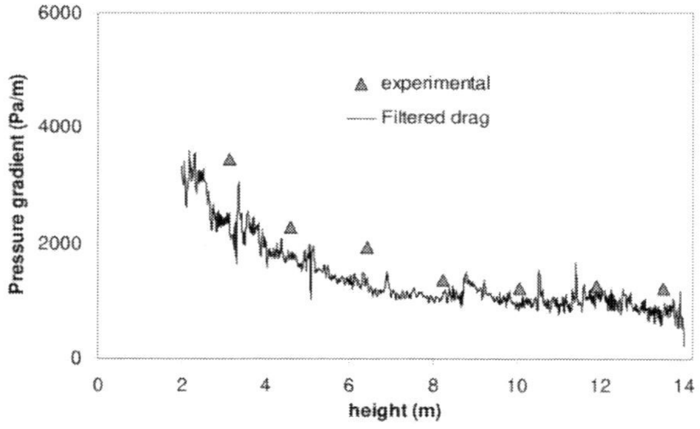

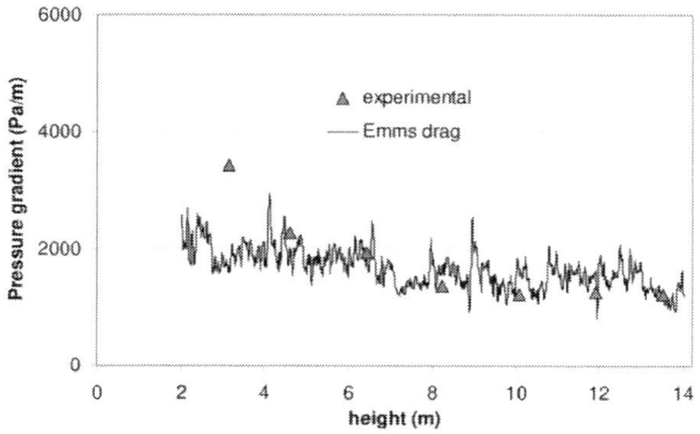

NETL

2K-2571

3D time-averaged profiles

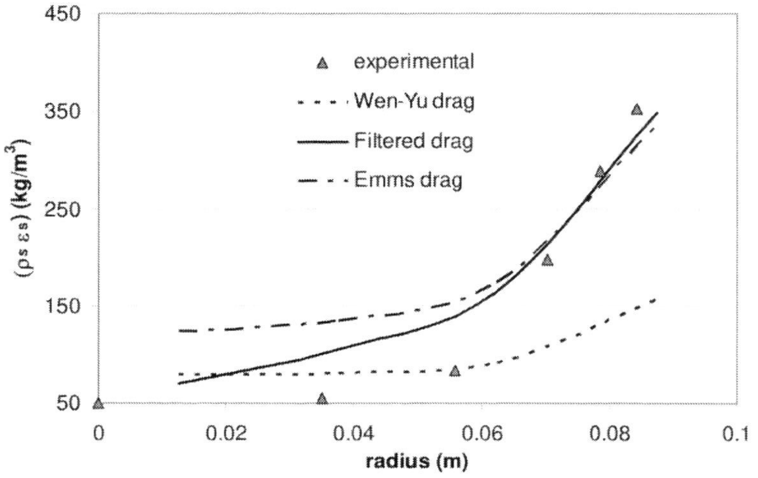

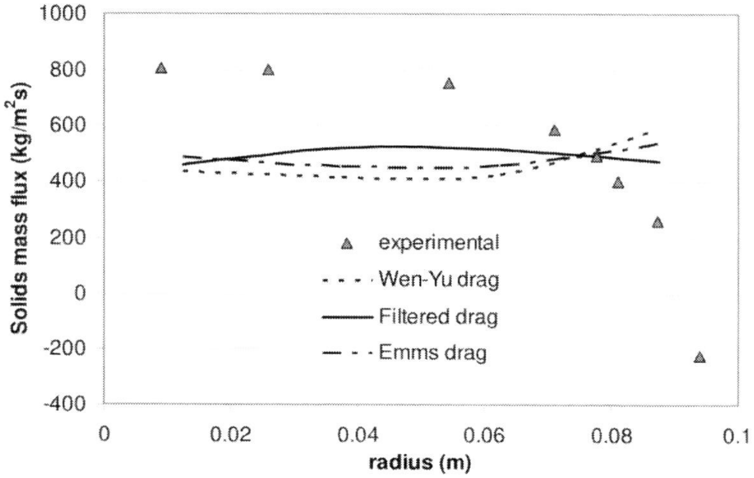

2K-2571

3D time-averaged profiles

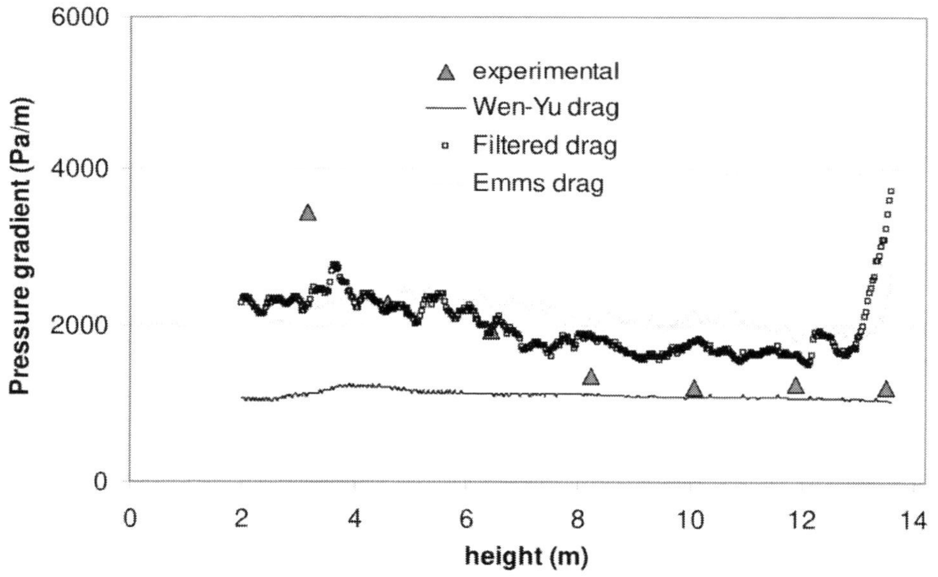

NETL

2K-2571

Summary

- Sub-grid drag models are demonstrated to be both needed and useful. Predictions show higher solids hold-up and pressure drop in the riser and yield better agreement with experimental data than the standard drag model.

NETL

2K-2571

The effective drag closure for gas-solid flows in vertical risers

Ali Ozel[1,2], Pascal Fede[1,2], Olivier Simonin[1,2]

[1] Université de Toulouse; INPT, UPS;
Institut de Mécanique des Fluides de Toulouse;
Allée du Professeur Camille Soula, 31400 Toulouse, France
[2] CNRS; IMFT; 31400 Toulouse, France
ozel@imft.fr

The Eulerian multi-fluid approach is a powerful numerical tool to investigate high-velocity gas-solid flows in vertical risers. The overall hydrodynamic behaviors of gas-solid flows are significantly modified by the existence of mesoscales, such as streamers and clusters in dilute flows. Clusters and streamers continuously occur in risers by a reason of local instabilities, such as damping of fluctuating motion of particles by the interstitial fluid, inelastic collisions, the non-linear drag between phases Agrawal et al. (2001). The typical length scale of mesoscales is in the order of $10-50$ particle diameters and the time scale varies between $0.001\ s$ and $0.1\ s$ De Wilde (2005). Although formations and breakage of mesoscales structures can be captured by multi-fluid formalism and transport equations developed in the frame of the kinetic theory of granular media on a small domain, large ranges of spatial and temporal variations of these structures restrict the resolution of calculations for large industrial units due to computational cost. To account for the effect of mesoscales on the macroscopic behavior in coarse grid simulations, the analogy can be constructed with large eddy simulation of single phase turbulent flow. These effects can be taken into account by subgrid scales through additional closure relations. The key feature in the present study is the proposition of the subgrid model for monodisperse gas-solid flow which takes into account the effects of mesoscales and validation of this model by experimental data.

In this paper, the modelling approach is based on the multi-fluid model formalism that involves mean separate transport equations of mass, momentum and energy for each phases. Interactions between phases are coupled through interphase transfers. Two transport equations, developed in the frame of kinetic theory of granular media supplemented by the interstitial fluid Balzer et al. (1996), were solved to model the effects of velocity fluctuations and inter-particle collisions on the dispersed phase hydrodynamic. The gas phase is predicted using Large Eddy Simulation. Concerning the transfers between the phases with non-reactive isothermal flow, the drag and Archimede forces are accounting for the momentum transfer. The point has to be stated that this study does not discuss the multi-fluid model, only proposes a different closure equation for an effective drag coefficient for the interfacial momentum transfer which is explicitly filtered by the grid size.

The effects of the mesoscale structures have been investigated by several authors. The pioneer work done by Agrawal et al. (2001) and it is remarked that coarse grid simulations overestimate the drag force. Wang (2006) proposed a structured dependent drag force by the energy-minimization multi-scale model. Addition to these works, Heynderick (2004) has taken into account the clusters effect by the concept of effective drag with considering the solid particles to be a part of a cluster and also to be present as individual particles. Igci (2007) proposes filter size dependent closures for the effective drag.

In the present modelling approach, we propose the following decomposition of the filtered interfacial momentum transfer:

$$\mathbf{I}_g = -\mathbf{I}_p = \frac{\overline{\alpha_p \rho_p}}{\tau_{gp}^F}\mathbf{V}_r = \frac{\overline{\alpha_p}\rho_p}{\tilde{\tau}_{gp}^F}\left(\tilde{\mathbf{U}}_p - \tilde{\mathbf{U}}_g + \mathbf{V}_d^*\right) \tag{1}$$

where $\overline{\alpha_p}$ is the filtered solid volume fraction (the "overline" defines the filtered quantity),

$$\tilde{U}_p = \frac{\overline{\alpha_p\, U_p}}{\overline{\alpha}_p} \quad \text{and} \quad \tilde{U}_g = \frac{\overline{\alpha_g\, U_g}}{\overline{\alpha}_g} \tag{2}$$

In (1), $\tilde{\tau}_{gp}^F$ is the particle relaxation time scale. The velocity $\mathbf{V}_\mathrm{d}^*$ is the subgrid contribution that we propose to model as:

$$\mathbf{V}_\mathrm{d}^* = \ln\left[\frac{\mathbf{\Delta}^*}{\mathbf{1} + \mathbf{\Delta}^*}\right] \times \sqrt{\frac{\mathbf{\Delta}^*}{\mathbf{1} + \mathbf{\Delta}^*}} \times \left(\tilde{\mathbf{U}}_\mathrm{p} - \tilde{\mathbf{U}}_\mathrm{g}\right) \tag{3}$$

where the non-dimensional grid size, Δ^*, is defined as follows:

$$\Delta^* = \overline{\alpha_p}\, \frac{\Delta}{\left(\frac{1}{2}\tilde{\tau}_{gp}^F \left|\tilde{\mathbf{U}}_p - \tilde{\mathbf{U}}_g\right|\right)^3} \tag{4}$$

where Δ is the control volume of cell. The subgrid model construction has been done by "a priori" tests. High-resolution numerical predictions have been computed with fine computational grid ensuring results without dependency on the grid resolution. These high-resolution data are filtered by several filter widths based on explicit grid sizes. Then, the subgrid drift velocity has been constructed by difference between filtered and high-resolution data.

The results obtained by the subgrid model are validated with experimental data of Andreux (2008) and Andreux (2001). The riser is a scaled cold circulating fluidized bed of $10\ m$ high having a square cross section of $11\ cm\ x\ 11\ cm$. Typical FCC particles (A-type) with density $1400\ kg/m^3$ and mean diameter $70\ \mu m$ were conducted. The density of air is set to $1.2\ kg/m^3$ and dynamic viscosity is equal to $1.8 \cdot 10^{-5} Pa \cdot s$. Averaged volume fractions of solid and pressure gradients obtained with and without subgrid model are compared in Fig.1. The unphysical accumulation at the bottom of the riser vanishes and the pressure gradient obtained by fine mesh is in a good agreement with the one obtained by subgrid model.

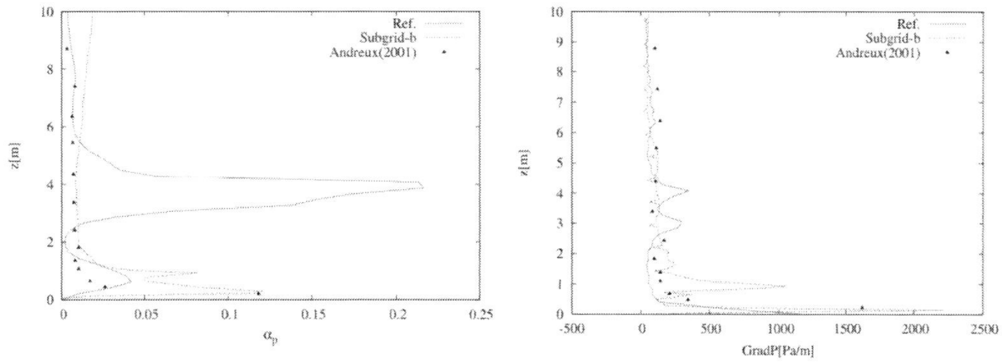

Figure 1: (a) Averaged volume fraction solid distribution and (b) pressure gradient along streamwise direction.

References

K. Agrawal, P. N. Loezos, M. Syamlal, and S. Sundaresan, "The role of mesoscales structures in rapid gas-solid flows", J. Fluid Mech. 445, 151 (2001).

J. De Wilde, "Reformulating and quantifying the generalized added mass in filtered gas-solid flow models", Phys. Fluids 17, 113304 (2005).

G. Balzer, O. Simonin, A. Boelle, J. Laviéville, "A unifying modelling approach for the numerical prediction of dilute and dense gas-solid two phase flows", Circulating Fluidized Bed Technology V, M. Kwauk, J. Li (editors), Science Press, Beijing(1996), 432, (1996).

O. Simonin, "Continuum Modelling of Dispersed Two-Phase Flows", in Combustion and Turbulence in Two-Phase Flows, Lecture Series 1996-02, von Karman Institute for Fluid Dynamics, Rhode Saint Genése (Belgium) (1996).

W. Wang, J. Li, "Simulation of gas-solid two phase flow a multi-scale CFD approach - Extension of the EMMS model to the sub-grid level", Chem. Eng. Sci. 62, 208 (2007).

G. J. Heynderickx, A. K. Das, J. De Wilde, and G. B. Marin, "Effect of clustering on gas-solid drag in dilute two-phase flow", Ind. Eng. Chem. Res. 43, 4635 (2004).

Y. Igci, A. T. Andrews IV, S. Sundaresan, S. Pannala, T. O'Brien, "Filtered two-fluid models for fluidized gas particle suspensions", AIChE J. 54, 1431 (2008).

R. Andreux, G. Petit, M. Hemati, O. Simonin, "Hydrodynamic and solid residence time distribution in a circulating fluidized bed: Experimental and 3D computational study", Chem. Eng. Proc. 47, 463 (2008).

R. Andreux, "Experimental and numerical study of a gas-solid seperator of a cold FCC moke-up", Ph.D. Thesis, Institut National Polytechnique de Toulouse, France, (2001).

A. Gobin, H. Neau, O. Simonin, J.R. Llinas, V. Reiling, J.L. Selo, "Fluid Dynamic Numerical Simulation of a Gas Phase Polymerisation Reactor", Int. J. Num. Met. Fluids 43, 1199 (2003).

E. Deutsch, O. Simonin, "LES applied to the motion of particles in stationary homogenous fluid", In Turbulence modification in multiphase flow, ASME FED, 110, 35 (1991).

Three-Dimensional Computational Fluid Dynamic Study of a Gas-Solid Riser Based on the Kinetic Theory of Granular Flow

Maximilian J. Hodapp[1], **Maria G. E. Silva**[1], **Fábio Marini**[1], **Milton Mori**[1*]
[1]University of Campinas, Campinas-SP, Brazil

Abstract

The description of a gas-solid flow through the Eulerian-Eulerian approach, or two-fluid model, requires the association of certain properties to the solid phase, which can not be measured physically, as solid pressure or viscosity. Therefore models had to be developed to account for these terms, relating them to physical properties. In this context the Kinetic Theory of Granular Flow (KTGF) seeks to represent a dispersed solid-phase, even on low porosity systems. In this study, a three-dimensional two-phase flow model based on the KTGF was used to predict the behavior of a gas-solid fluidized bed. Particular attention was given to the description of the near-wall particular phase flow. The model of Johnson and Jackson (1987), which contains a specularity coefficient, was used. The study of the influence of this parameter in the solids velocity profiles was the aim of this work. The turbulence was represented by the k-epsilon model and multiphase mixture was used. For the validation of the numerical model, the results obtained were compared to the experimental work of Samuelsberg and Hjertager (1996), who used a cold flow laboratory circulating fluidized bed and simulated a two-dimensional model. The geometry and numerical mesh were generated using Ansys Icem CFD software and the set of partial differential equations was discretized and solved using Ansys Fluent 6.3. The comparison of the results with the experimental data showed that the model agrees with the data and predicts a flow behavior similar to that found experimentally. It predicts the core annulus flow, which is known from the literature.

Keywords: Fluidized bed, CFD, Multiphase flow, Kinetic theory of granular flow

Introduction

Computational Fluid Dynamic (CFD) simulations of multiphase flow remain a challenging task, even with the large amount of research dedicated to this topic. The great, and constant, increase of computational power made possible the study of more complex systems. Several studies that were carried out in bi-dimensional grids, in order to be executable in a fair amount of time, can now be simulated in 3D. This is especially important for non-symmetrical geometries. One class of equipments that is frequently the interest of CFD investigations are fluidized beds. They are widely used in many operations in chemical, metallurgical, energy generation and especially petrochemical industries. Major applications are fluid catalytic cracking (FCC) risers and CFB combustor systems. Although fluidized beds are successfully and widely used in commercial industrial operations, much remains to be done due to the complexity of the gas-solid flow. Mathematical models have been applied to

---

* Corresponding author. Phone + 55-19-3521-3963. E-mail:mori@feq.unicamp.br

predict the behavior of a fluidized bed and several models have been proposed. Gidaspow *et al.* (2004) carried out an extensive review of the models developed for the fluidized bed reactors and Peirano and Leckner (1998) reviewed turbulent gas-solid flow modeling in circulating fluidized beds. However, several model features in predicting core-annulus flows, as boundary conditions and model parameters, are not well understood, as investigated by Benyahia and O'Brien (2007).

Gas-particle two-fluid models, which treated the particle phase as a continuous fluid based on the Eulerian approach, were widely employed in modeling gas-solid flow in the past several decades. A fundamental problem encountered on the modeling of the hydrodynamics of a gas-solid fluidized bed by the two-fluid method is how to include the stress of the particle phase in the particulate momentum equation. In the more recent two-fluid models, for the particle phase in dense gas-solid flow, the kinetic theory of granular flow has received attention and its use in Eulerian simulations has increased (Jenkins and Savage, 1983; Ding and Gidaspow, 1990; Gidaspow, 1994; Hrenya and Sinclair, 1997; Samuelsberg and Hjertager, 1996). The kinetic theory of granular flow is based on the kinetic theory of gases, first developed by Chapman and Cowling (1970). In this theory, the inter particle interactions are taken into account by adding the contribution of collisions between particles (Jenkins and Savage, 1983), which are the main mechanism of transport due to particulate phase properties.

Another important aspect of the correct description of such flows, is the treatment given to the near wall region, especially for the particulate phase. The particle-wall collisions, modeled only through the restitution coefficient, are not always accurate enough to describe the observed behavior of the solid phase. Therefore Johnson and Jackson (1987) presented a model based on the studies of Hui et al. (1984) for the boundary condition, which uses a specularity coefficient to account for the particle-wall interactions. This coefficient ranges from zero to unity, approaching the free-slip and no-slip conditions respectively.

The present study aims to simulate a three-dimensional two-phase flow model, based on the kinetic theory of granular flow (KTGF) and the near wall specularity coefficient model, to predict the behavior of a gas-solid fluidized bed, using computational fluid dynamics technique.

Model Description

Based on Eulerian-Eulerian description of the phases, a multiphase computational fluid dynamics model for turbulent gas-solid flow is presented. The Eulerian approach considers the two phases, gas and solid, as a continuum. The conservation equations for the solid phase are based on the kinetic theory for granular flow (Gidaspow, 1994).

Mathematical model

The governing equations for transient three-dimensional gas-solid flow are presented as follows:

The continuity equation for phase i (= g,s):

$$\frac{\partial}{\partial t}(\varepsilon_i \rho_i) + \nabla \cdot (\varepsilon_i \rho_i \vec{\upsilon}_i) = 0 \qquad (1)$$

where the subscript *i* can denote gas or solid phase, ε is the volume fraction, ρ is the density, t is the time, $\vec{\upsilon}$ is the velocity.

The gas phase momentum equation:

$$\frac{\partial}{\partial t}(\varepsilon_g\rho_g\vec{\upsilon}_g)+\nabla\cdot(\varepsilon_g\rho_g\vec{\upsilon}_g\vec{\upsilon}_g)=-\varepsilon_g\nabla p_g+\nabla\vec{\tau}_g+\varepsilon_g\rho_g\vec{g}+\beta(\vec{\upsilon}_s-\vec{\upsilon}_g) \quad (2)$$

where $\vec{g}$ is the gravity vector, p is the pressure and β_{gs}^{m} is the two-phase drag coefficient.

The stress tensor of the gas phase can be written as

$$\vec{\tau}_g=\varepsilon_g\mu_g\left(\left(\nabla\vec{\upsilon}_g+(\nabla\vec{\upsilon}_g)^T\right)-\frac{2}{3}(\nabla\cdot\vec{\upsilon}_g)\vec{I}\right) \quad (3)$$

Here, μ is the shear viscosity.

The gas phase turbulence is modeled with the realizable k-epsilon model.

The solid phase momentum equation:

$$\frac{\partial}{\partial t}(\varepsilon_s\rho_s\vec{\upsilon}_s)+\nabla\cdot(\varepsilon_s\rho_s\vec{\upsilon}_s\vec{\upsilon}_s)=\nabla\vec{T}_s+\varepsilon_s\rho_s\vec{g}+\beta(\vec{\upsilon}_g-\vec{\upsilon}_s) \quad (4)$$

where the solid stress tensor, $\vec{T}_s$, is the sum of a collisional part and a kinetic part and may be formulated as

$$\vec{T}_s=(-p_s+\lambda_s\nabla\cdot\vec{\upsilon}_s)\vec{I}+\mu_s\left(\left(\nabla\vec{\upsilon}_s+(\nabla\vec{\upsilon}_s)^T\right)-\frac{2}{3}(\nabla\cdot\vec{\upsilon}_s)\vec{I}\right) \quad (5)$$

where λ_s is the solid phase bulk viscosity.

The total solid phase pressure, p_s, consists of two terms; one of them, p_s^{k+t}, is determined from Lun *et al.* (1984) and includes both kinetic and collisional pressures, as follows:

$$p_s^{k+t}=\rho_s\varepsilon_s[1+2(1+e)\varepsilon_s g_o]\theta \quad (6)$$

where e is the coefficient of restitution and g_o is the radial distribution function.

The other term, p_s^f, is caused by friction and collisions in the solid phase due to highly dense flow, as proposed by Johnson and Jackson (1987), as follows:

$$P_s^f=Fr\frac{(\varepsilon_s-\varepsilon_{s,min})^n}{(\varepsilon_{s,max}-\varepsilon_s)^p} \quad (7)$$

where the values of the constants in Equation (7) are Fr = 0.05, n = 2, p = 3 and $\varepsilon_{s,min}$ = 0.55, as suggested by Van Wachen (1998).

The total solid phase pressure is then given by

$$P_s=P_s^{k+t}+P_s^f \quad (8)$$

The radial distribution function, adopted by Lun and Savage (1986), is

$$g_o=\left(1-\frac{\varepsilon_s}{\varepsilon_{s,max}}\right)^{-1/3} \quad (9)$$

where $\varepsilon_{s,max}$, the maximum particle packing chosen is 0.63.

Equation (9) is appropriate for dilute flow.

In Equation (6), θ is the granular temperature related to the kinetic turbulent energy of the particle motion. The transport equation derived from the kinetic theory is simplified to an algebraic formulation.

Solid phase bulk viscosity, λ_s, due to particle collisions (Lun *et al.*, 1984) and solid phase shear viscosity, μ_s, for dense and dilute flow (Gidaspow, 1994) are given by the following equations:

$$\lambda_s = \frac{4}{3}\varepsilon_s^2\rho_s d_s g_o(1+e)\sqrt{\frac{\theta}{\pi}} \tag{10}$$

$$\mu_s = \frac{4}{5}\varepsilon_s\rho_s d_s g_o(1+e)\left(\frac{\theta}{\pi}\right)^{1/2} + \frac{\varepsilon_s\rho_s d_s\sqrt{\theta\pi}}{6(3-e)}\left(1+\frac{2}{5}(1+e)(3e-1)\varepsilon_s g_o\right) \tag{11}$$

The gas-solid drag coefficient when $\varepsilon_g \leq 0.8$ is based on the Ergun equation and is given by

$$\beta = 150\frac{\varepsilon_s^2\mu_g}{\varepsilon_g d_p^2} + 1.75\frac{|\vec{\upsilon}_s - \vec{\upsilon}_g|\varepsilon_s\rho_g}{d_p} \tag{12}$$

And when $\varepsilon_g > 0.8$, based on the Wen and Yu equation:

$$\beta = \frac{3}{4}C_D\frac{|\vec{\upsilon}_s - \vec{\upsilon}_g|\varepsilon_s\rho_g}{d_p}\varepsilon_g^{-2.65} \tag{13}$$

where $f(\varepsilon_g) = \varepsilon_g^{-2.65}$, as used by Gidaspow and Ettehadieh (1983).

The drag coefficient, C_D, is related to the Reynolds number as follows:

$$C_D = \varepsilon_g^{-2.65}\max\left[\frac{24}{Re_p\,\varepsilon_g}\left(1+0.15(Re_p\,\varepsilon_g)^{0.687}\right)\right] \text{ for } Re_p < 1000 \tag{14}$$

$$C_D = 0.44 \text{ for } Re_p \geq 1000 \tag{15}$$

$$Re_p = \frac{|\vec{\upsilon}_s - \vec{\upsilon}_g|\rho_g d_p}{\mu_g} \tag{16}$$

The specularity coefficient, ϕ', is found as given by Johnson and Jackson (1987) in the bulk stress vector, Equation (20):

$$\frac{\vec{\upsilon}_{sl}(\sigma_c+\sigma_f)n}{|\vec{\upsilon}_{sl}|} + \frac{\phi'\sqrt{3}\pi\rho_p\varepsilon_p\theta^{1/2}|\vec{\upsilon}_{sl}|}{6\varepsilon_{p,max}\left(1-\left(\frac{\varepsilon_p}{\varepsilon_{p,max}}\right)^{1/3}\right)} + N_f\tan\delta = 0 \tag{17}$$

where $\vec{\upsilon}_{sl}$ is the slip velocity between the wall and the particle, σ_c and σ_f are respectively the collisional-translational and frictional contributions to the stress and the wall, δ is the angle of friction between the surface and the particle.

Physical Description

The experimental results of Samuelsberg and Hjertager (1996) were compared to the numerical results obtained in this work. The simulations were carried out for the riser section of a circulating fluidized bed similar to that used by Samuelsberg and Hjertager (1996) and Mathiesen *et al.* (2000), shown in Figure 1 (a). The riser section, as shown in Figure 1 (b), is modeled and simulated in a three-dimensional numerical mesh.

The riser section is 1m high with an inner diameter of d = 0.032m. A secondary air supplier, positioned 0.05m above the gas inlet, feeds the solid back into the riser. The secondary gas inlet and the gas outlet have a diameter of 0.008m (Figure 1 (b)).The initial bed height is 0.05m and the initial solid volume fraction is 0.61. The system properties are defined in Table 1.

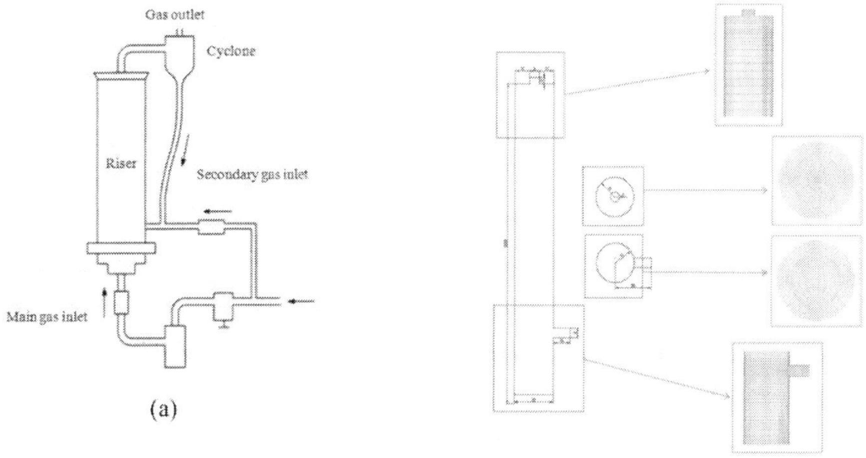

(a)

(b)

Figure 1 – (a) Geometry of Mathiesen *et al.* (2000). (b) Calculation domain and grid nodes.

Table 1 – System properties

Gas density	1.185kg/m^3
Particle density	1600kg/m^3
Particle size	60μm
Gas viscosity	1.831x10^{-5}kg/m s
Maximum solids volume fraction	0.63
Particle-particle restitution coefficient	0.95

Initial and Boundary Conditions.

At the primary gas inlet, the superficial gas velocity, V_{sup}, was 1.42m/s. At the secondary gas inlet, the gas velocity was 0.05m/s. In order to permit all particles at the outlet back into the riser at the secondary gas inlet, a subroutine was implemented in the Fluent code. At the outlet, atmospheric pressure was prescribed. At the walls, no-slip conditions were used for the gas phase and different specularity coefficients were set for solid phase.

Results and Discussion

The finite volume method was used to resolve the set of partial differential equations through the commercial software Ansys Fluent 6.3 and the Icem CFD code to generate the numerical grid, shown in Figure 1 (b). From previews studies of mesh-results independence, Marini (2008), a grid of approximately 300,000 control volumes was considered satisfactory for the precision of the results and the computation time consumed. The time step used in all simulations was 10^{-5}s. The simulation results were computed through time-averaged distribution of flow variables. All the simulations were run for 15s of real time.

Figure 2 shows a comparison of experimental results from Samuelsberg and Herjtager (1996) and the present numerical results to the solid radial velocity profiles for heights of 0.16, 0.32 and 0.48m in the riser section and superficial gas velocity, υ_{sup}, of 1.42m/s in the primary gas inlet.

As can be seen in Figure 2, at all three heights the coefficient of 0.4 agreed well with the velocity profile at the center region of the flow. Although the tendency of the flow was depicted by all the specularity coefficients studied, the simulations over-predicted the velocity of the particulate phase. Some deviation was also seen at the lower height, 0.16m, probably due to the perturbation caused by the presence of the secondary inlet, just below this measure line.

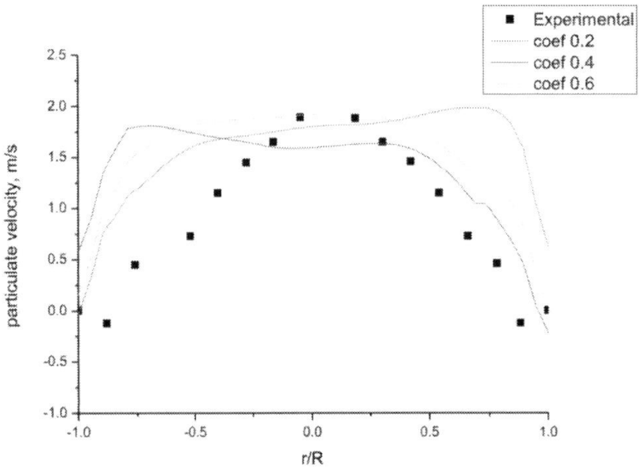

a)

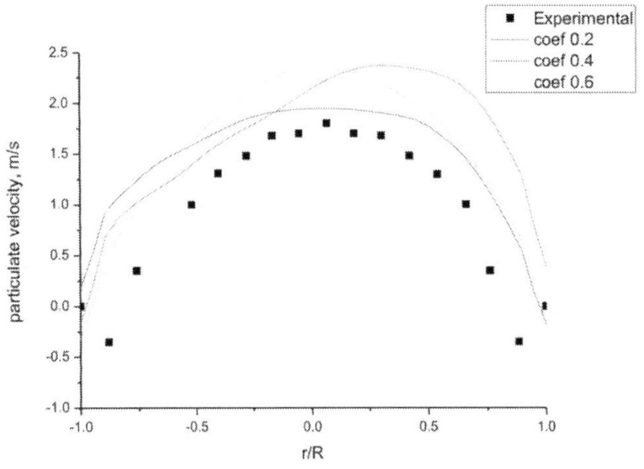

b)

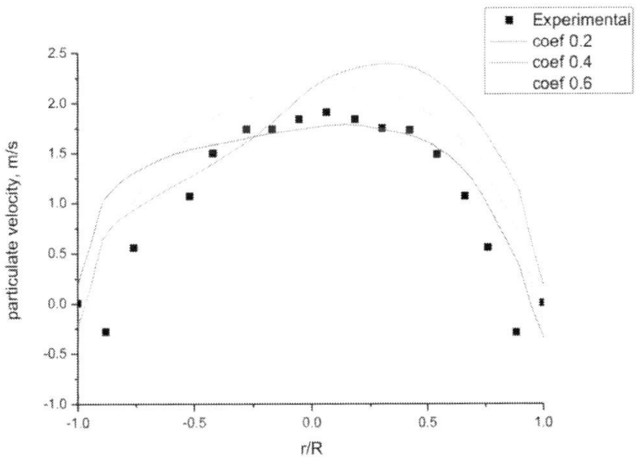

c)

Figure 2 – A comparison of experimental data due to Samuelsberg and Herjtager (1996) and computed solid velocity profiles at different heights in the riser section and superficial gas velocity, υ_{sup}, of 1.42m/s: a) 0,16m; b) 0,32m; c) 0,48m;.

Conclusions

A three-dimensional two-phase flow model based on the kinetic theory of granular flow (KTGF) was used to predict the behavior of a gas-solid fluidized bed. A wall boundary condition of partial slip for the solid phase was also studied. The numerical results were compared against the experimental results of Samuelsberg and Hjertager (1996). The results for the specularity coefficient of 0.4 showed the best agreement with experimental data and were found to be satisfactory especially in the riser central region

References

1. Benyahia, S., Syamlal, M., O' Brien, T. J. (2007) Study of the ability of multiphase continuum models to predict core-annulus flow. *Particle Technology and Fluidization.* v. 53, n. 10, p. 2549-2568.
2. Chapman, S., Cowling, T.G. (1970). *The Mathematical Theory of Non-Uniform Gases*, Cambridge University Press, London.
3. Ding, J., Gidaspow, D. (1990). A bubbling fluidization model using kinetic theory of granular flow. *AICHE J.*, 36, 523-538.
4. Gidaspow, D. (1994). *Multiphase Flow and Fluidization*, Academic Press, London.
5. Gidaspow, D., Ettehadieh, B. (1983). Fluidization in two-dimensional beds with a jet. 2. hydrodynamic modeling. *Industrial and Engineering Chemistry Fundamentals*, 22, 193-201.
6. Gidaspow, D., Jung, J., Singh, R.K. (2004). Hydrodynamics of fluidization using kinetic theory: an emergin paradigm 2002 flour-daniel lecture. *Powder Technology*, 148, 123-141.
7. Hrenya, C.M., Sinclair, J.L. (1997). Effects of particle-phase turbulence in gas-solids flows.

AIChE Journal, 43, 853-869.
8. Hui, K., Haff, P.K., Ungar,J.E., Jackson, R. (1984). Boundary condition for high-shear grain flows. *J. Fluid Mech.* 145, 223-233.
9. Jenkins, J.T., Savage, S.B. (1983). A Theory for the rapid flow of identical, smooth, nearly elastic, spherical particles. *AIChE Journal*, 42, 1536-1546.
10. Johnson, P.C., Jackson, R. (1987). Frictional-collisional constitutive relations for granular materials, with application to plane shearing. *Journal of Fluid Mechanics*, 176, 67-93.
11. Lun, C.K.K., Savage, S.B. (1986). The effects of an impact velocity dependent coefficient of restitution on stresses developed by sheared granular materials. *Acta Mechanica*, 63, 15-44.
12. Lun, C.K.K., Savage, S.B., Jeffrey, D.J., Chepurniy, N. (1984). Kinetic theories for granular flow: inelastic particles in couette flow and slightly inelastic particles in a general flowfield. *Journal of Fluid Mechanics*, 140, 223-256.
13. Marini, F. (2008), Simulation of a fluidized bed applying the CFD technique based in the kinetic theory of granular flow, M.Sc Thesis, University of Campinas, Campinas, 2008.
14. Mathiesen, V., Solberg, T., Hjertager, B.H. (2000). An experimental and computational study of multiphase flow behavior in a circulating fluidized bed. *International Journal of Multiphase Flow*, 26, 387-419.
15. Peirano, E., Leckner, B. (1998). Fundamentals of turbulent gas-solid flows applied to circulating fluidized bed combustion. *Progress Energy Combustion Science*, 24, 259-296.
16. Samuelsberg, A., Hjertager, B.H. (1996) Computational modeling of gas/particle flow in a riser. *AIChE Journal*, 42, 1536-1546.
17. VanWachen, B.G.M. (1998). Derivation, Implementation and Validation of Computer Simulation Models for Gas-Solid Fluidized Beds, PhD Thesis - Delft University.
18. Wen, C.-Y., Yu, Y.H. (1966). Mechanics of Fluidization. *Chem. Eng. Prog. Symp. Series*, 62,100-111.

Riser-Flow Measurements of Polydisperse Geldart Group B Particles

Jia Wei Chew, Chemical and Biological Engineering, University of Colorado at Boulder, Boulder, CO

Christine M. Hrenya, Chemical and Biological Engineering, University of Colorado at Boulder, Boulder, CO

Ray A. Cocco, Particulate Solid Research, Inc., Chicago, IL

John G. Findlay, Particulate Solid Research, Inc., Chicago, IL

Ted M. Knowlton, Particulate Solid Research, Inc., Chicago, IL

Flows involving solid particles are used extensively in industry including areas such as pollution control, pharmaceuticals, energy production, and materials synthesis. However, a predictive understanding of polydispersity in riser flows, particularly for continuous size distributions, has not been established. This work is part of an overall aim to develop validated, continuum models for polydisperse, gas-solid systems using a combination of theory, simulations, and experiments. The focus of this effort is to generate a comprehensive experimental dataset that will provide a testing ground for first-principle models. Experiments involving Geldart Group B particles with a continuous size distribution have been carried out over a range of parameters: superficial gas velocity, overall solids flux, and various distribution widths. Measurements include pressure drop, solid concentration, solid flux, solid velocity and segregation. The results here expand on the previous findings of Karri and Knowlton (1998) at lower velocity, which showed (i) radial size-segregation was observed only in downflow annulus and not upflow annulus, and (ii) the lack of axial size-segregation. Surprisingly, at the higher velocities in this work, both axial and radial size-segregation was observed for both upflow and downflow annulus.

Reference:

Karri, S.B. and T.M. Knowlton, "Flow direction and size segregation of annulus solids in a riser," Fluidization IX, Durango, CO (1998).

See more of this Session: Circulating Fluidized Beds
See more of this Group/Topical: Particle Technology Forum

Evaluation of CFB Riser and Downer Reactor Performance by Catalytic Ozone Decomposition

Dongbing Li, Chemical and Biochemical Engineering, The University of Western Ontario, London, ON, Canada

Jesse Zhu, Chemical and Biochemical Engineering, The University of Western Ontario, London, ON, Canada

Madhumita B. Ray, Chemical and Biochemical Engineering, The University of Western Ontario, London, ON, Canada

Ajay K. Ray, Chemical and Biochemical Engineering, The University of Western Ontario, London, ON, Canada

Radial and axial ozone concentration profiles in a CFB riser (3" i.d., 10.2 m in height) and two downers (2" i.d. and 3" i.d., 5.0 m in height) were measured. Experiments were conducted under different solids circulation rates and gas superficial velocities, using fresh FCC particles impregnated with iron oxide as the solid phase and catalyst for ozone decomposition. Results show that the acceleration region gives significant contribution to the overall conversion. Also, conversions from the experiments were compared with numerical predictions.

See more of this Session: Circulating Fluidized Beds
See more of this Group/Topical: Particle Technology Forum

Comparison study of riser geometry effect on the solids distribution in CFBs

Jing Xu and J.-X Zhu*

Particle Technology Research Centre
Department of Chemical & Biochemical Engineering
The University of Western Ontario, London, Ontario, Canada N6A 5B9

Abstract

The axial and radial solids holdup profiles were studied by measuring with an optical fibre probe in a 7.6 m rectangular circulating fluidized bed with a narrow 19×114 mm cross-section riser under a wide range operating conditions. Comparing the rectangular riser with the other cylindrical columns, it is found that the general shapes of the axial and radial profiles of solids holdup in rectangular riser are quite similar to that in cylindrical risers, but more uniform. It is concluded that the geometry of the riser is a factor affects the solids distribution due to the perimeter per unit cross-sectional area and the distance from the wall to the centre. In some extent, the two-dimensional and three-dimensional risers are more comparable under fast fluidization conditions. Generally, the solids distribution along the axial and the radial direction in rectangular riser are close to that in cylindrical risers, and the main differences have been known by the study of this paper.

Keywords: Fluidization, Hydrodynamics, Rectangular CFB, Riser geometry, Scale-up, Solids holdup

Introduction

Circulating fluidized bed reactors have been used for a wide range of industrial applications over the past 50 years. A significant number of scientific studies have been done focused on the behavior of gas-solid two-phase flow in risers, such as axial and radial voidage distributions, gas and solids mixing, as well as flow regime transition, while very few were relevant to geometry effect of the riser cross-section.

Different from the common cylindrical column, the riser employed in this study was a narrow rectangular cross-sectional column. To some extent, the rectangular riser can be considered as two-dimensional column and has been proved to be especially useful for studies of bubble properties, as an educational tool and as a means of qualitatively viewing the fluidization characteristics of powder (Arena et al., 1989b; Grace and Baeyens, 1986; Pallares and Johnsson, 2006; Park et al., 2002).

However, some differences exist between the rectangular two-dimensional and the conventional three-dimensional fluidized beds, which arise from the column geometry effect, relative with bubble velocity, bubble coalescence and other matters in bubbling or turbulent beds (Almendros-Ibanez et al., 2006; Fan, 1990; Gera and Gautam, 1995). Very few studies, relevant with solids holdup and particle velocity, have been done in rectangular fluidized bed under fast fluidization conditions in which the gas velocities are higher than 3 m/s. This paper presents the investigation of solids concentration profile in a rectangular CFB system measured by optical fiber probes and the relationship to the solids distribution in three-dimensional columns, as well as the influence of column geometry to the flow pattern.

1. Experimental apparatus

A rectangular CFB system was designed and setup, as shown in Fig. 1, to conduct the experiments. The riser is a rectangular column with 19×114 mm (0.75×4.5 inch) cross-section, 7.6 m height and can be considered as a 2-inch-equivalent diameter cylindrical column. See Fig. 2. Besides the riser, a 203 mm (8 inch) i.d. downcomer storage and 38×203 mm (1.5×8 inch) downcomer, two cyclones, a bag filter, flapper valves for solids circulation rate measurement and a perforated distributor at the bottom of the riser compose the CFB system. FCC particles with a Sauter mean diameter of 67 μm and a particle density of 1884 kg/m^3 were used. The total solids inventory was approximately 70 kg, which was equivalent to 5 m solids height in the downcomer. The primary air at ambient temperature and pressure was supplied to the windbox located in the riser bottom with a small stream of steam merged in to control the humidity value of the gas fed into the riser to be between 70 to 80%, which is considered as the optimal value required to avoid misleading effects due to electrostatic forces (Park et al., 2002). The gas fed into the riser was monitored by a rotameter and a thermo-hygrometer. The solids circulation rate was measured by a device which is located at the top of the downcomer, dividing the column into two halves with a central vertical plate. Two half butterfly valves are fixed at the top and the bottom of the two-half sections. By appropriately flipping over the two valves from one side to the other, solids circulated through the system can be accumulated on one side for a given time period and then the solids circulation rate can be obtained.

The local solids concentrations were measured with a multi-fibre optical probe, PV6, which was developed by Institute of Process Engineering, Chinese Academy of Sciences. The 3.8 mm diameter probe tip consists of two bundles of optical fibres, which arrange precisely in vertical line. Each of bundles consists of approximately 8000 emitting and receiving quartz fibres with a diameter of 15 μm for each fibre. The active area, where the fibres are located, is approximately 1 mm × 1 mm. With precise calibration and cross-correlation applied, solids concentration and particle velocity can be obtained. Details of the calibration process and velocity calculation were referred in literatures (Zhang *et al.*, 1998; Liu *et al.*, 2003a).

The solids concentrations were measured at 9 lateral positions located at y/Y = -0.98, -0.75, -0.50, -0.25, 0, 0.25, 0.50, 0.75, 0.98, in which y/Y is the dimensionless distance from the riser axis, and at 6 axial levels at H = 1.27, 2.29, 3.30, 4.32, 5.33, 6.35 m, where H is the height from the riser distributor. In these experiments, the superficial gas velocity and solids circulation rate were in the range of 3.5 to 8 m/s and 50 to 200 $kg/m^2 \cdot s$ respectively.

2. Results and discussion

The motion of particles in the current riser corresponded to that observed in cylindrical risers, which is what one would expect if the rectangular column was a diametral slice of a cylindrical riser. Solids were seen to be dilute and moving faster in the centre but were much denser and slower close to the side walls. Almost all particles moved upward while some flowed downward at the side walls under higher flux conditions. It was observed that there was no evidence of solids buildup or fall down on both face walls.

2.1 Comparison between different risers

2.1.1 Geometry and dimension of different risers

"Two-dimensional" columns are considered as an effective means to view the flow pattern in a rectangular cross-section where the width is considerably greater than the depth. There are somewhat quantitative differences between two- and three-dimensional fluidized beds (Grace and Baeyens, 1986). It is of great importance to know these differences in particle flow for the future investigation of particle properties in this experiment apparatus. Any comparison between the results obtained from the rectangular riser and from the other researchers' data must be based on the same reference basis.

Table 1 shows the dimensions for the riser of the current study and some other columns referred in literatures. Compared to the riser used in this study, risers #3 to #5 are cylindrical columns with different diameters investigated by Yan *et al.* (2004) and Qi *et al.* (2008) respectively. Riser #2 is an imaginative column, has an 2-inch-equivalent diameter calculated according to the equation for transfer of area,

$D_e = \sqrt{4A_r / \pi}$. The cross-sectional area of the current rectangular riser is equivalent to the 2-inch-diameter cylindrical column but has a much larger perimeter per unit cross-section (See Fig. 2). For the four cylindrical columns, as the diameter increases, the specific surface area decreases, meaning that less surface area is in contact with the particles and more particles may be present in the same wall surface.

Table 1 Dimensions of risers

No.	*Riser name*	*Riser geometry*	*D_r (inch)*	*Area equivalent diameter $D_e = \sqrt{4A_r / \pi}$ (inch)*	*Perimeter per unit cross-sectional area P_A (1/inch)*
1	Current riser	Rectangular	0.75×4.5	2	3.1
2	Equivalent riser	Cylindrical	2	2	2
3	Yan's (2004)	Cylindrical	3	3	1.3
4	Qi's (2005)	Cylindrical	4	4	1.0
5	Yan's (2004)	Cylindrical	8	8	0.5

2.1.2 Axial and radial profile of solids holdups

Two typical conditions were selected to compare the profiles of solids holdup in the rectangular riser and in the cylindrical risers studied by the other researchers. Fig. and

Fig. 4 show the radial profiles of solids holdup in the developed region of the five risers listed in Table 1. Under both operating conditions, the distributions of solids holdup were both parabolic, low in the centre of columns and increase when close to the wall. For cylindrical risers #3, #4 and #5, the local average solids concentration increased with the increase of riser diameter. At the same solids circulation rate and superficial gas velocity, the solids holdups in the centre regime ($y/Y < 0.5$) were comparable, but quite different in the wall area. In Fig. 3, the solids concentration close to the wall of the 8-inch riser was significantly higher than that of the smaller risers while the 4-inch riser has slightly higher concentration over the 3-inch one. This verified the conclusion stated by Yan (2004) that the radial profiles of solids concentration of a large riser increase more sharply towards the wall than in a small riser. According to this rule, the profile of solids holdup for riser #2 which has a 2 inch diameter has been anticipated as the dash line shown in Figures. The radial solids distribution of the current rectangular riser with an equivalent diameter of 2 inch was much flatter compared with that of the 2 inch cylindrical riser, especially under higher solids flux conditions, as Fig. 4 shows. The solids holdup of the rectangular riser was higher than that of cylindrical risers in the centre region but comparably lower closer to the wall.

For this section of comparison, two effects exist. One is the scale-up effect and the other is the influence of the column shape. The scale-up effect, according to the bed diameter, makes the solids concentration decrease as the bed diameter decreasing, with the rate of decrease of concentration declining from the wall to the centre. Consequently, the solids holdup in the rectangular riser is supposed to be sharply decreased from the wall to the centre and lower than that in the larger risers in each radial position. Whereas, the dash lines in Fig. 3 and Fig. 4 predicted by the rule of the scale-up effect are incapable of anticipating the solids holdup profile in the rectangular riser with 2-inch-equivalent diameter. Therefore, the effect of riser structure impacts the solids holdup in a different way from the scale-up effect does, flattening the radial distribution of solids concentration by increasing the solids holdup in the column centre while reducing it close to the wall.

Fig. 5 shows the comparison of typical axial profiles of solids holdup and corresponding radial profiles for #1, #3 and #5 risers with 2-inch-equivalent, 3-inch and 8-inch diameter respectively. The operating conditions for 3-inch and 8-inch diameter risers were U_g = 5.5 m/s and G_s = 100 kg/m^{2}s, for 2-inch-equivalent diameter riser was U_g = 5.0 m/s and G_s = 100 kg/m^{2}s. For axial solids holdup profiles, the profile was slightly higher in the bottom sections than that in the upper sections. Smaller risers had a lower cross sectional average of solids holdup than larger risers had on each axial level. In conformity with the axial profile of solids holdup, the radial distribution of solids concentrations were decreasing along the longitudinal level from the riser bottom to the top and more uniform in the upper sections. Comparing among the three risers, it can be found that the solids holdups were increasing at each longitudinal level with the riser size increasing and the slope of the solids concentration distribution were sharper. The Fig. 5 also shows that the 2-inch-equivalent diameter riser had much more uniform radial profiles of solids holdup than the three-dimensional risers had, which were higher than that of the 3-inch riser in the centre region, even higher than that in the 8-inch riser, while much lower than that in the cylindrical risers in the close-to-wall region.

The cause for this unusual distribution of solids holdup in current rectangular riser has been summarized into two effects, scale-up effect and the column shape effect. To the in-depth understanding , the perimeter per unit cross-sectional area is an important factor contributing significantly to the two effects mentioned above and mainly affect the solids distribution on the surface wall. In larger columns, there will be less wall surface available for the particles buildup and more particles move slowly, downwards, or stagnantly to accumulate on the wall surface to increase wall coverage. Consequently, the cross-sectional average of solids holdup is higher with the riser size increasing even under the same solids circulation rate which is a net flux escaping from the riser. This can also be agreed with the axial and radial profile of solids holdup. From Table 1, it can be seen that the P_A for 2-inch-equivalent diameter, 3-inch and 8-inch riser are 3.1, 1.3 and 0.5 /inch respectively. As such, the area of the wall surface is much greater in current rectangular riser than in 8-inch cylindrical riser. Thus, at the specified solids circulation rate and gas velocity, the axial solids holdup increases with the riser size increase, so as the radial solids holdup does. As a result, the perimeter per unit cross-sectional area is the main reason causing the reducing of the solids accumulation on the wall and lowering the solids holdup closer to the wall.

From the other aspect, the distance between the wall and the riser centre also plays an important role in determining the distribution of solids holdup, especially in the centre region of the riser. The gas distribution is restricted by the distance between the wall and the centre since the resistance to the gas is mainly encountered at the wall due to the wall friction. Consequently, the gas velocity on the face walls is slower than in the centre. Yan (2004) also stated that the greater the distance between the wall and the centre under the same superficial gas velocity and solids circulation rate, the bigger the difference of gas velocity in these two regions becomes. Therefore, as a rectangular column, the two face walls are so close to each other that there is no significant difference between the gas velocity on the face walls and in the centre. As moving with the gas, the motion of particles on the face walls is as fast as in the centre. As a result, the solids tend not to cling to the face wall but to distribute uniformly along the depth direction. For the side walls, though they are further to the centre of the column, the area is so small that the amount of particles which are contacting with the side walls is small as well. Therefore, the rectangular geometry leads to increase the solids concentration in the centre region.

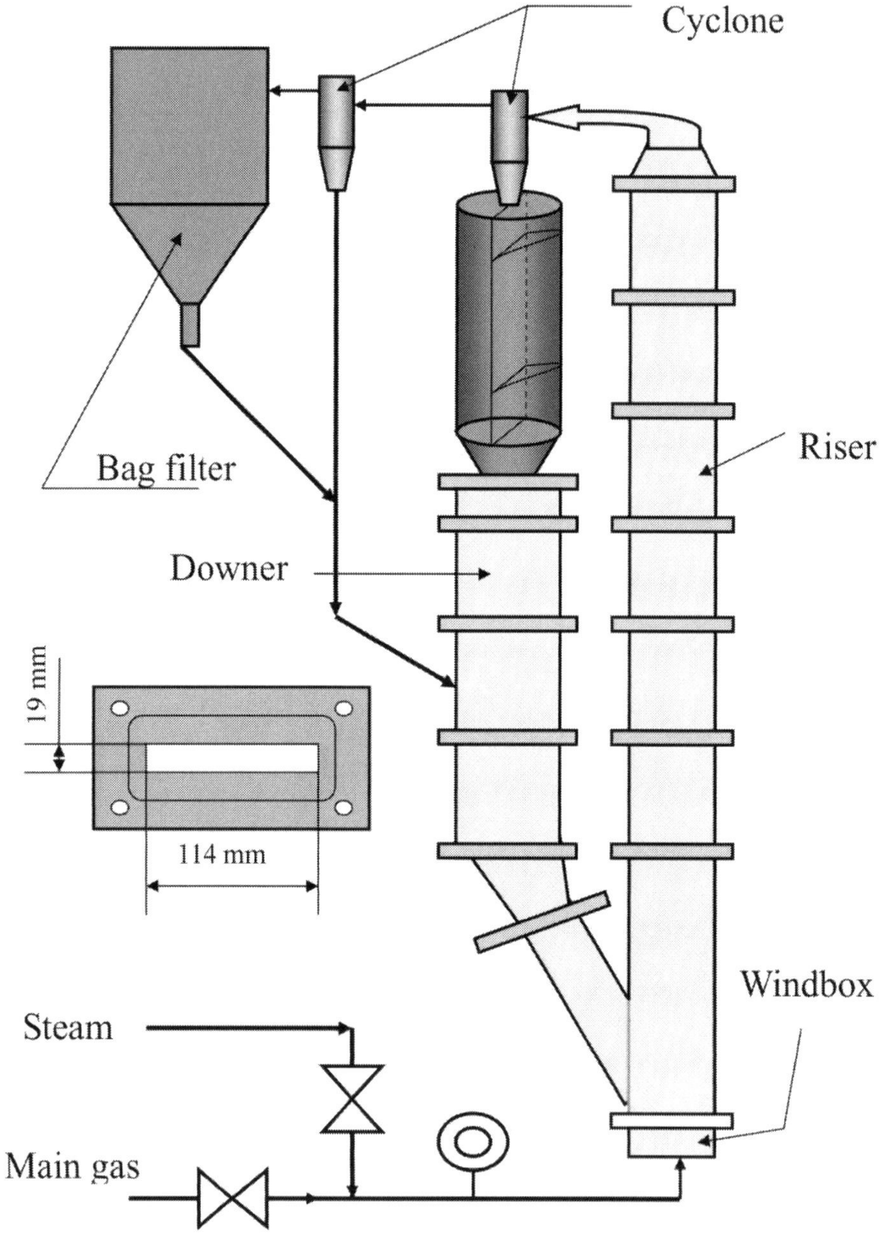

Fig. 1 Narrow rectangular Circulating Fluidized Bed unit

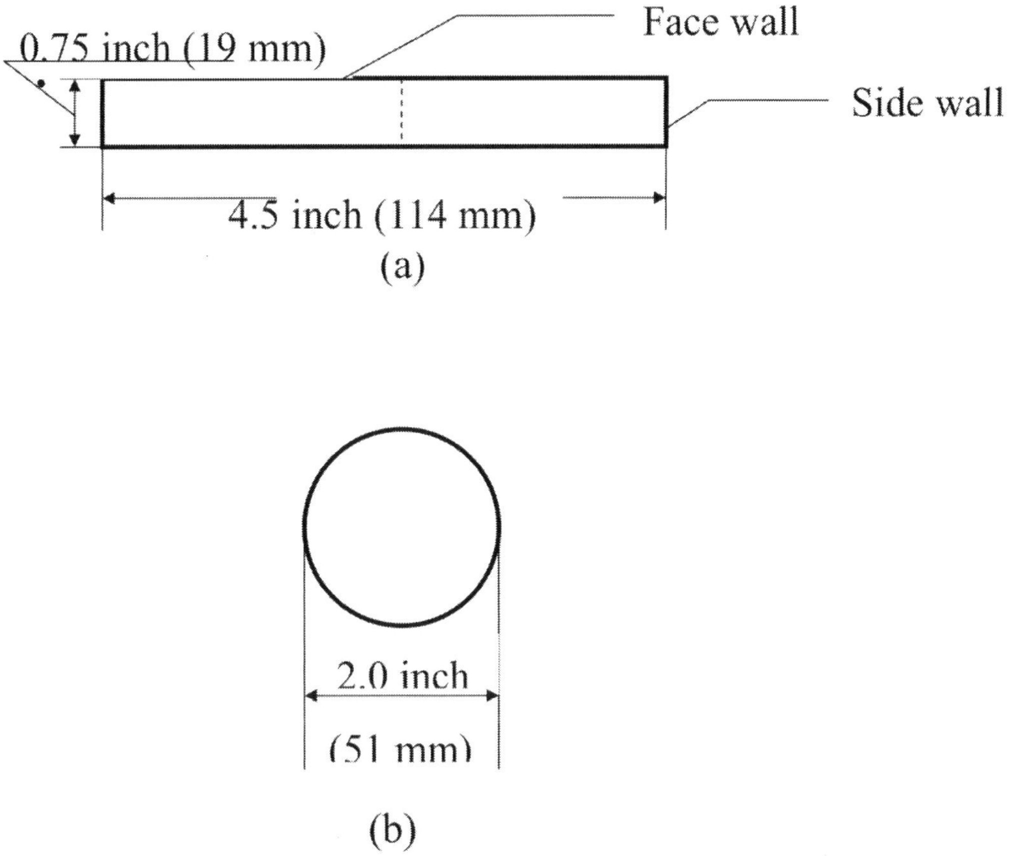

Fig. 2 Schematic of the cross section of the rectangular riser and the equivalent cylindrical riser

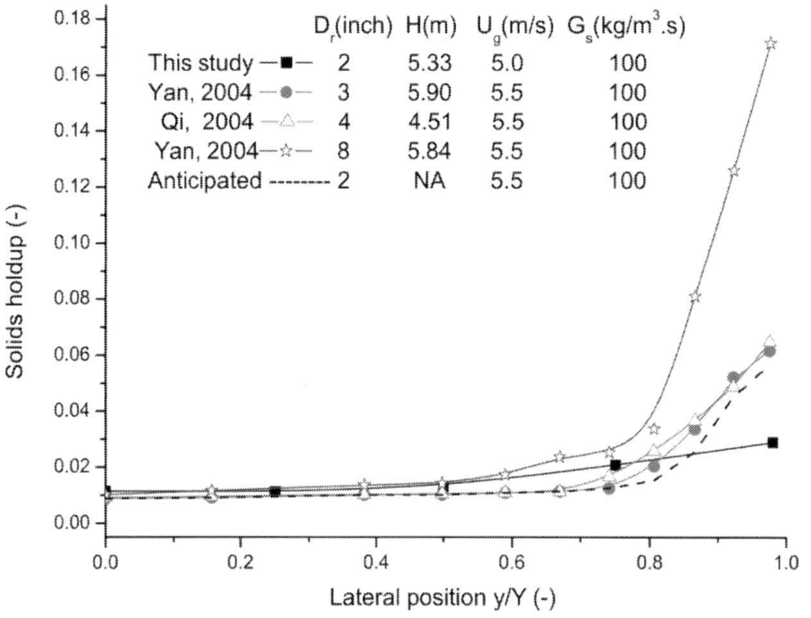

Fig. 3 Comparison of lateral/radial solids holdup profiles of different bed column structure

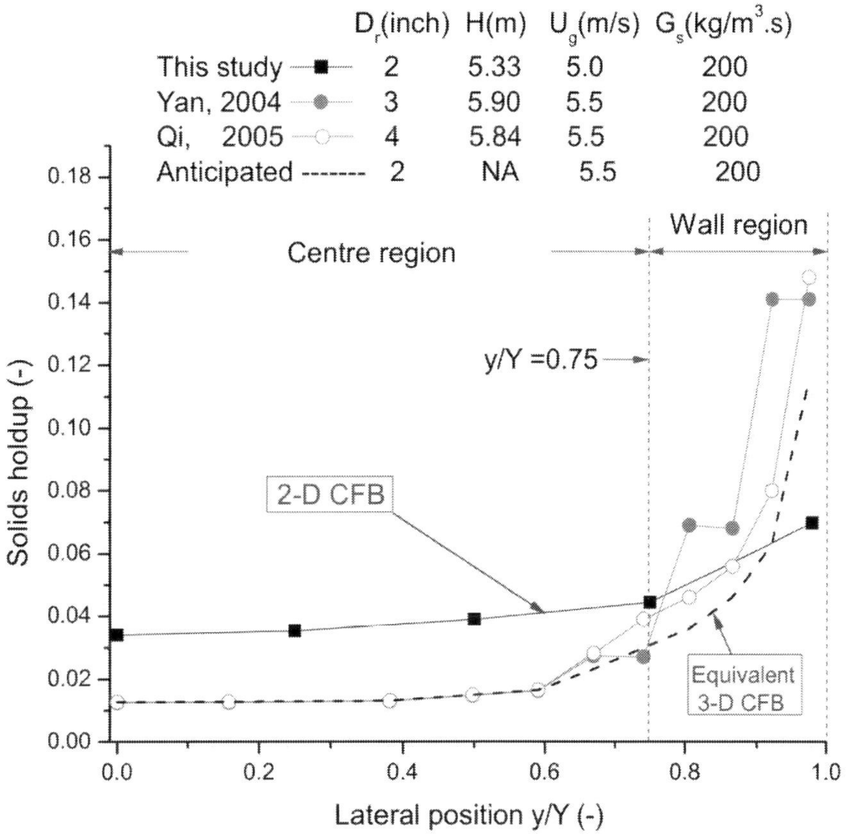

Fig. 4 Comparison of lateral/radial solids holdup profiles of different bed column structure

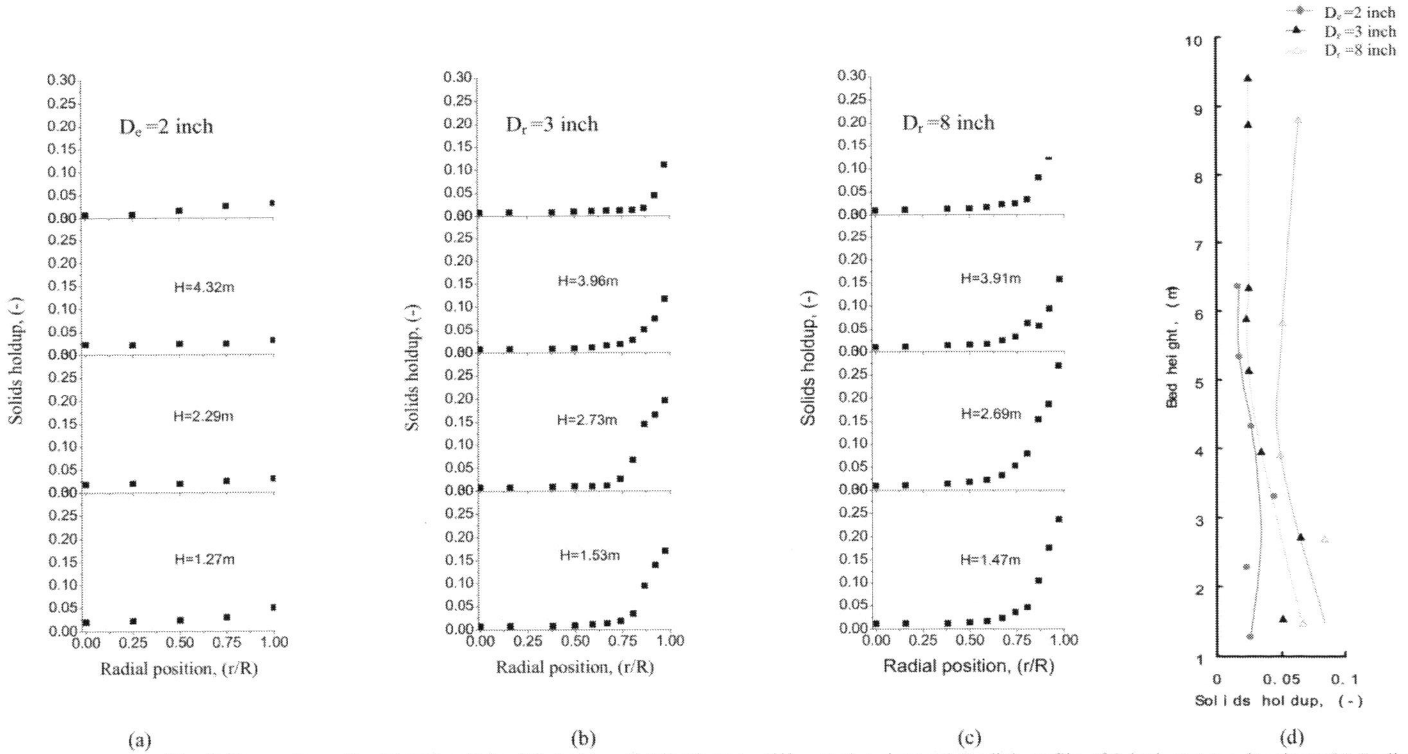

Fig. 5 Comparison of axial and radial solids holdup distributions in different size riser: (a) Radial profile of 2-inch rectangular riser; (b) Radial profile of 3-inch cylindrical riser; (c) Radial profile of 8-inch cylindrical riser; (d) Axial solids holdup profiles of the three risers

Nomenclature

A_r	cross-section area of riser	[sq. inch]
D_e	area equivalent riser diameter	[inch]
D_r	riser diameter	[inch]
G_s	solids circulation rate	[kg/m^2s]
H	height from the riser bottom	[m]
P_A	perimeter per unit cross-sectional area	[1/inch]
r	radial distance from riser axis	[m]
R	radius of riser	[m]
U_g	superficial gas velocity	[m/s]
V_p	particle velocity	[m/s]
y	lateral distance from riser axis	[m]
Y	lateral distance between riser axis and wall	[m]
ε_s	local solids holdup	[-]
ρ_s	particle density	[kg/m^3]

References

[1] Adanez, J., Gayan, P., Garcia-Labiano, F. and Diego, L. F., 1994. Axial voidage profiles in fast fluidized beds. Powder Technology 81, 259-268.

[2] Almendros-Ibanez, J. A., Sobrino, C., de Vega, M. and Santana, D., 2006. A new model for ejected particle velocity from erupting bubbles in 2-D fluidized beds. Chemical Engineering Science 61, 5981-5990.

[3] Arena, U., Cammarota, A. and Marzocchella, A., 1989a. Solids flow structures in a two-dimensional riser of a circulating fluidized bed. Journal of Chemical Engineering of Japan 22, 236-241.

[4] Bai, D.-R., Jin, Y., Yu, Z.-Q. and Zhu, J.-X., 1992a. The axial distribution of the cross-sectionally averaged voidage in fast fluidized beds. Powder Technology 71, 51-58.

[5] Bai, D. R., Shibuyav, E., Masuda, Y., Nakagawa, N. and Kato, K., 1996. Flow structure in a fast fluidized bed. Chemical Engineering Science 51(6), 957-966.

[6] Bi, H. T., 2007. A critical review of the complex pressure fluctuation phenomenon in gas-solids fluidized beds. Chemical Engineering Science 62, 3473-3493.

[7] Bi, H. T., Zhu, J. and Yu, Z.-Q., 1993. Forms of particle aggregations in CFBs. In: Proceeding of the Sixth Chinese Conference on Fluidization. Wuhan, China. pp. 162-167.

[8] Fan, Z., 1990. Analysis of pressure fluctuations in a 2-D fluidized bed. Powder Technology 62, 139-145.

[9] Gera, D. and Gautam, M., 1995. Effect of bubble coalescence on throughflow velocity in a 2-D fluidized bed. Powder Technology 83, 49-53.

[10] Grace, J. R., 1996. Influences of riser geometry on CFB particle and fluid dynamics. In: Kwauk, M. and Li, J. Circulating Fluidized Bed Technology V. Beijing. pp.

[11] Grace, J. R. and Baeyens, J., 1986. Instrumentation and experimental techniques. In: Geldart, D., Gas fluidization technology, Wiley & Sons Ltd., pp. 415-463.

[12] Issangya, A. S., Bai, D., Bi, H. T., Lim, K. S., Zhu, J. and Grace, J. R., 1999. Suspension densities in a high-density circulating fluidized bed riser. Chemical Engineering Science 54, 5451-5460.

[13] Liu, J., Grace, J. R. and Bi, X., 2003a. Novel Multifunctional Optical-Fibre Probe I. AIChE Journal 49, 1405-1420.

[14] Liu, J., Grace, J. R. and Bi, X., 2003b. Novel Multifunctional Optical-Fibre Probe II. AIChE Journal 49, 1421-1432.

[15] Mostoufi, N. and Chaouki, J., 2001. Local solid mixing in gas–solid fluidized beds. Powder Technol 114, 23-31.

[16] Pallares, D. and Johnsson, F., 2006. A novel technique for particle tracking in cold 2-dimensional fluidized beds—simulating fuel dispersion. Chemical Engineering Science 61, 2710-2720.

[17] Park, A.-H., Bi, H. and Grace, J. R., 2002. Reduction of electrostatic charges in gas - solid fluidized beds. Chemical Engineering Science 57, 153-162.

[18] Parssinen, J. H. and Zhu, J. X., 2001b. Axial and radial solids distribution in a long and high-flux CFB riser. AIChE Journal 47(10), 2197-2205.

[19] Parssinen, J. H. and Zhu, J. X., 2001c. Particle velocity and flow development in a long and high-flux circulating fluidized bed riser. Chemical Engineering Science 56, 5295-5303.

[20] Qi, X.-B., Huang, W.-X. and Zhu, J., 2008. Comparative study of flow structure in circulating fluidized bed risers with FCC and sand particles. Chmical Engineering Technology 31, 542-553.

[21] Schlichthaerle, P. and Werther, J., 1999. Axial pressure profiles and solids concentration distributions in the CFB bottom zone. Chemical Engineering Science 54, 5485-5493.

[22] Yan, A., 2004. Hydrodynamics and scale-up of circulating fluidized beds. In: Chemical Engineering, The University of Western Ontario, London, Ontario, Canada, pp. 6-14.

[23] Yan, A., Ball, J. and Zhu, J. J., 2005. Scale-up effect of riser reactors (3) axial and radial solids flux distribution and flow development. Chemical Engineering Science 109, 97-106.

[24] Yang, Y.-L., Jin, Y., Yu, Z.-Q. and Wang, Z.-W., 1992. Investigation on slip velocity distributions in the riser of dilute circulating fluidized bed. Powder Technology 73, 67-73.

[25] Yang, Y.-L., Y. Jin, Z.-Q. Yu and Z.-W. Wang, 1991. Particle Flow Pattern in a Dilute Concurrent Upflow and Downflow Circulating Fluidized Bed. In: Hasatani, M. K. a. M., Fluidization'91:Science and Technology, Science Press, Beijing, pp. 66-75.

[26] Yang, Y. L., Jin, Y., Yu, Z. Q., Zhu, J. X. and Bi, H. T., 1993. Local slip behaviors in the circulating fluidized bed. AIChE Symposium Series 296, 81-90.

[27] Zhang, H., Johnston, P. M., Zhu, J.-X., de Lasa, H. I. and Bergougnou, M. A., 1998. A novel calibration procedure for a fibre optic solids concentration probe. Powder Technology 100, 260-272.

[28] Zhou, H., Flamant, G. and Gauthier, G., 2007. Modelling of the turbulent gas–particle flow structure in a two-dimensional circulating fluidized bed riser. Chemical Engineering Science 62, 269-280.

[29] Zhu, J., Salah, M. and Zhou, Y., 1997. Radial and axial voidage distributions in circulating fluidized bed with ring-type internals. Journal of Chemical Engineering of Japan 30, 928-937.

Separation of Oil and Other Organics From Water Using Inverse Fluidization of Hydrophobic Aerogels

Ding Wang, Chemical Engineering, Arizona State University, Tempe, AZ

Trent Silbaugh, Chemical Engineering, Arizona State University, Tempe, AZ

Robert Pfeffer, Chemical Engineering, Arizona State University, Tempe, AZ

Jerry YS Lin, Chemical Engineering, Arizona State University, Tempe, AZ

Surface-treated hydrophobic aerogel (Cabot Nanogel®) granules of size range of about 1 mm are fluidized by a downward flow of oil-contaminated water in an inverse fluidization mode. Nanogel particles are nanostructured, extremely light and porous, have a very large surface area per unit mass, and are sufficiently robust to be fluidized. Their hydrophobic surface gives them a strong affinity for oil and other organic molecules, with the exclusion of water. Furthermore, as they adsorb/absorb oil they become heavier, lose their buoyancy and eventually leave the column. Hence the fluidized bed can be operated continuously by adding fresh Nanogel at the same rate as the saturated Nanogel is removed from the bed.

The hydrodynamic characteristics of the inverse fluidized bed of Nanogel and the granule density were obtained by measuring the pressure drop and bed expansion as a function of the superficial water velocity. Either pure vegetable oil, or an oil-in-water emulsion using 1 and 4 % Tween 80 as a surfactant to keep the emulsion stable, were mixed with de-ionized water in a static mixer before entering the inverse fluidized bed. Microscopic examination of the inlet mixture showed that the oil droplets were less than 20 microns in size for the emulsified oil, but larger than 20 microns for the pure oil However, the oil removal efficiency (breakthrough curve) and capacity of the Nanogel granules in the inverse fluidized bed were found to depend on the amount of stabilizer added, indicating that the Tween 80 was affecting the aerogel surface and allowing some water to enter as well as oil.

See more of this Session: Applications of Fluidization
See more of this Group/Topical: Particle Technology Forum

FUNDAMENTAL STUDY ON WASHING CHARACTERISTICS OF A NOVEL MULTIPHASE FLOW SYSTEM

Yusumi Nagahashi, Kochi National College of Technology, Nankoku, Kochi, JAPAN
John R. Grace, Norman Epstein, University of British Columbia, Vancouver, CANADA
Yuataka Asako, Tokyo Metropolitan University, Hachioji, Tokyo, JAPAN
Akira Yokogawa, Kochi University of Technology, Kami, Kochi, JAPAN

Introduction

When small quantities of water (typically 10% by vol.) are added to a gas-fluidized bed of relatively large non-wettable particles, such as 6 mm Polystyrene spheres, the added water can cause a dramatic improvement in the extent of fluidization, contrary to most conventional gas-solid fluidized beds where liquid causes immobilization. This novel multiphase flow system, somewhat similar to wetted gas-fluidized beds with smaller bed pressure drop than liquid-solid beds, was first reported by us in 2003 [1] and then extended work has been reported [2], including a spouted bed system [3].

The resulting unique fluidized bed has advantages for soft-abrasion, due to the motion of light rounded particles, and also for reducing the water needed for washing because only a small quantity of liquid is required to enhance bed mobilization. These characteristics are desirable for industrial washing of some farm products or fragile industrial devices in need of soft abrasive washing. A possible application is the nursing bathing system [4] illustrated in **Fig. 1-a;** another is washing farm products [5] as shown in **Fig. 1-b**.

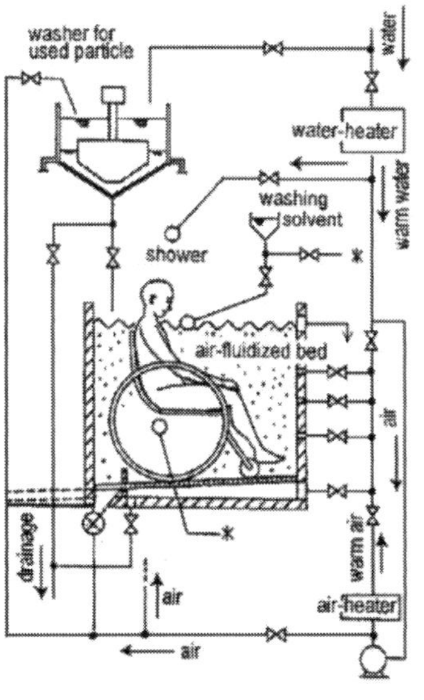

Fig.1-a Nursing bathing system

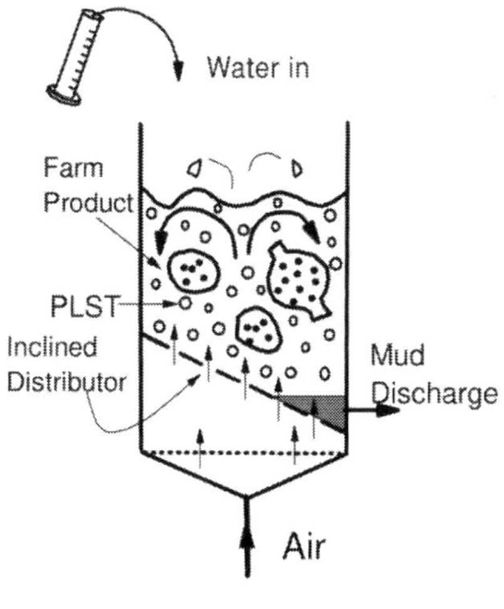

Fig.1-b Farm products washing devices

In this paper, the washing characteristics of this novel multiphase flow system are investigated in a fundamental study for the application of fluidization in farm products washing. The effect of small particle addition is also discussed. Although the small solid particles reduce the extent of fluidization enhancement[1], the effect should be investigated because the small particles simulate those from mud waste, and also because small solids could improve the washing performance. We have already confirmed[5] that fluidization enhancement by addition of water is not significantly weakened by the existence of small high-density particles, at least for solid volume fractions less than or equal to 10% (**Fig. 2**).

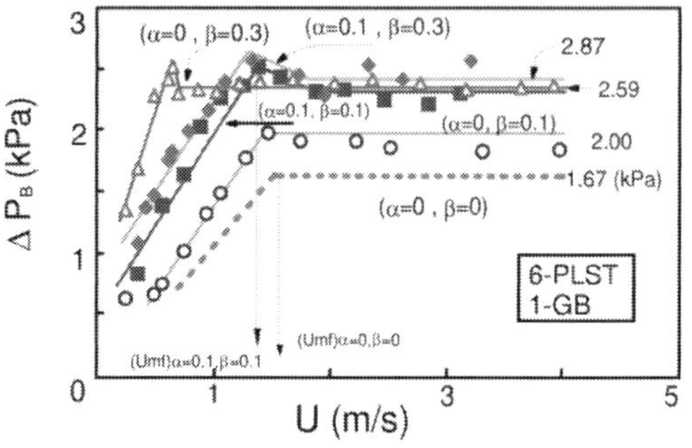

Fig.2 ΔP_B-U_{mf} diagram[5] for wetted binary solids fluidized beds; U_{mf} for α=0.1 and β=0.1 is smaller than for the original dry bed; α is water volume fraction, and β is the small particle volume fraction.

The mechanism of washing in this unique fluidized bed is assumed to be a combination of calm abrasive washing by liquid movement and hard attrition washing due to particle collisions. This understanding is reached from results, presented below, of four kinds of experiments: particle collision force on a cylinder surface, attrition of a plaster cylinder, whitening of a test cylinder covered with soiled cloth, and observations of liquid distribution on the cylinder surface.

Particle Collision Force on a Cylinder in the Novel Fluidized Bed

Experimental apparatus and measurement

The experimental apparatus and measurement system are shown in **Fig. 3**. An acrylic half-column of 180 mm diameter and 1500 mm height is used. Polystyrene particles (PLST: d_p=6 mm, ρ_s=1010 kg/m^3) provided the major washing medium, with two sizes of glass beads to simulate mud or a second washing medium (GB: d_p=0.6, 2.1 mm, ρ_s=2450 kg/m^3).

Experiments were carried out after water addition, for ratios of bed height to effective bed diameter, z_0/D_e, of 1.1 and 1.5.

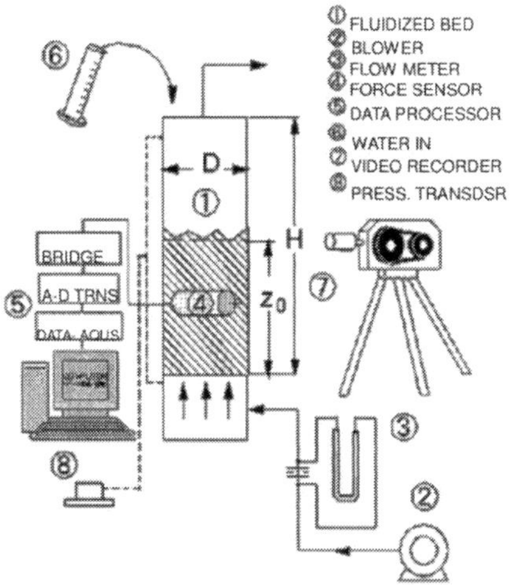

Fig.3 Experimental set-up

Force detector:

Details of the force sensor are shown in **Fig. 4**. The dynamic forces are detected by a strain gage attached behind the surface of a cantilever plate (stainless steel: 5 × 6.5 × 0.1 mm thick) mounted on an acrylic tube of 30 mm diameter. The dynamic forces are measured

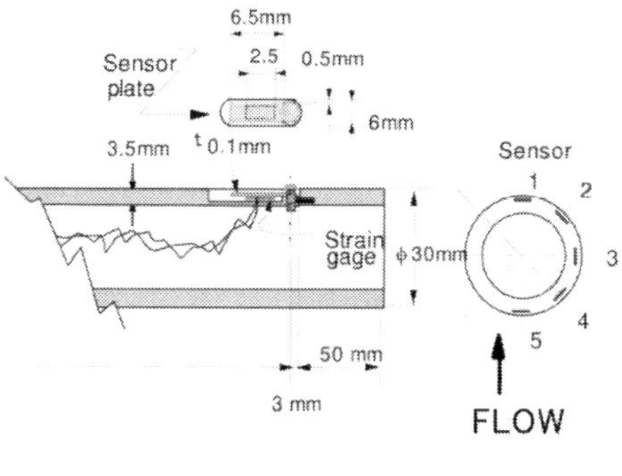

Fig.4 Sensor tube for particle collision force

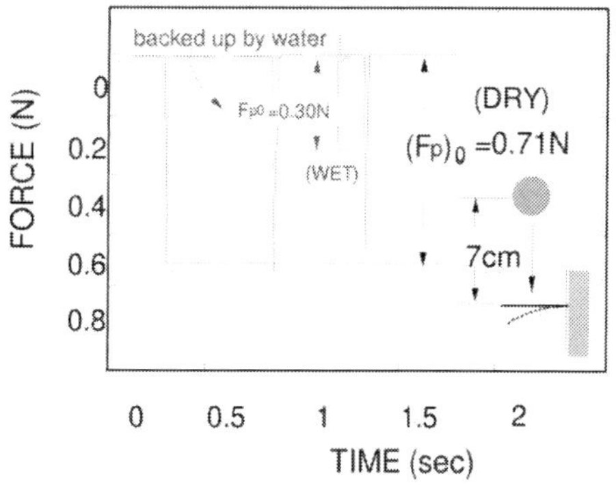

Fig.5 Impulse force for a single free-falling particle

along the circumference at 5 points (Nos. 1 - 5) on the half-tube surface. The force detectors were calibrated both statistically and dynamically. The linearity and fast response were confirmed to be adequate for the experiments. For this force sensor, the detected force is the statistically averaged impulse force relative to that which comes from a single particle of diameter 6 mm striking the center of the plate of 5 mm width and 6.5 mm length .

The impulse force $(F_p)_0$ for a single PLST particle free-falling from 70 mm above the plate is shown in **Fig. 5**. The average value of three such runs was used as the normalized base value for the results which follow.

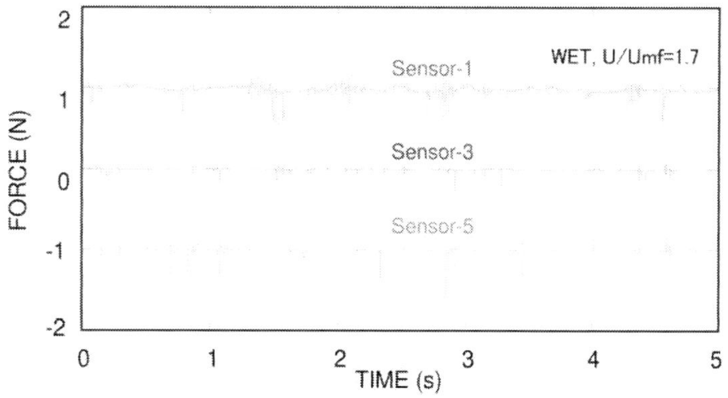

Fig.6 Typical time trace for each sensor output

Results for Particle Collision Force

Fig.6 shows typical time traces of the impulse signal for the fluidized bed promoted by addition of 10 vol. % water. The signal of No.5 sensor indicates the impulse force at the bottom of the test tube, whereas No.3 sensor indicates the force at the side of the tube, and No.1 is that at the top of the tube. The peak data for the impulse force existing at the upper 50% region was chosen for the force data, except for a maximum peak signal measured with a 5 kHz sampling time, assuming that the washing performance, such as mud removal, depends on high-intensity collision forces. Force data were averaged with 50% -upper-peak signals during 5 s of measurement and were normalized by $(F_p)_0$. From the analysis of VTR observations, it can be identified that the force at the top of tube is the impact force caused by falling aggregates of wetted particles, whereas the force at the side of tube is a thrust force by descending particles associated with wake shedding, and the force at the bottom of tube is due to impact by passing wake particles.

Fig. 7 shows the relationship between the normalized mean force $(F_p)/(F_p)_0$ at the bottom of the tube and dimensionless air velocity, U/U_{mf}, for three test runs, whereas **Fig. 8** shows the force distribution around the tube surface. In order to confirm the effect of water addition and/or addition of small large-density particles, results of four types of test are shown in **Fig. 7**. From this figure, it is found that: (1) Forces are proportional to the air velocity in the dry bed, but in the wetted bed they level off beyond $U/U_{mf}=2$. (2) Forces are reduced by adding small GB particles or water because the free motion of the PLST particles (major washing media) is damped by them. (3) The magnitude of the intensity is almost of the same order as the impulse for particles free falling from a height of 70 mm, except for the force at the bottom of the dry bed, which is 3 times higher. (4) The circumferential profile of attrition has a slender shape, with the force at the bottom is always larger than in any other direction, i.e. $(Fp)_{bottom} > (Fp)_{side} \approx (Fp)_{top}$. For a dry bed, the results are similar to previous work[(6),(7)], indicating that the impulse force due to particle impacts is approximately proportional to the square of the velocity.

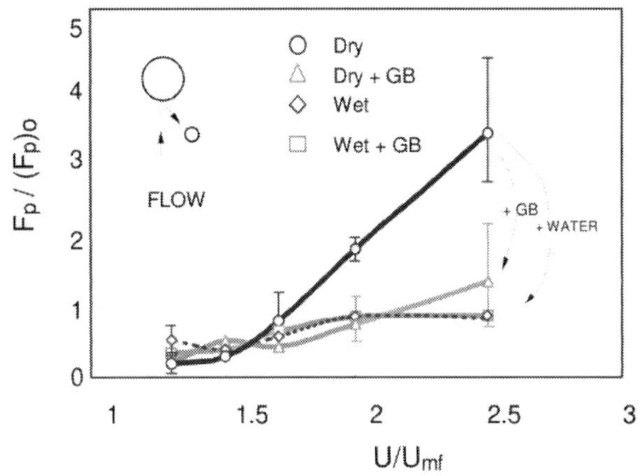

Fig.7 Normalized mean force vs. dimensionless gas velocity

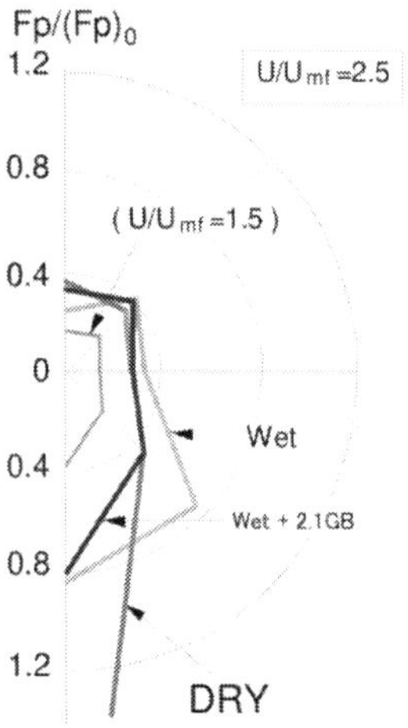

Fig.8 Particle collision force around a cylinder

Attrition of Plaster Coating on a Cylinder in the Novel Fluidized Bed

Experimental apparatus and measurement technique

Since the mechanism of washing in this unique fluidized bed is a combination of calm abrasive washing and hard attrition washing due to particle collisions, the relationship between particle collision forces and attrition of the internals is discussed here.

To assess the attrition washing, e.g. for mud removal, a plaster-coated cylinder was installed horizontally as a model washing object in the bed, shown in Fig. 3, replacing the force sensor. A craftwork plaster (Tokyo Sun Home Co. Ltd., YS-1) was painted as a layer of 1.5 mm thickness on an aluminum tube of 30 mm diameter mounted on the acrylic base cylinder, and then dried for 6 h at 60°C. Attrition was measured by an eddy type coating thickness gauge Surfix FN (resolution < 0.03 mm, PHYNIX GmbH & Co.) at each of 8 blocks circumferentially on the cylinder surface. The attrition extent, μ, is defined as

$$\mu = (t_0 - t)/t_0 \qquad (1)$$

where t_0 = original thickness, t = thickness after operation.

Fig. 9 shows the original surface of the plaster cylinder (**Fig.9-a**) and the surface after 15 minutes of operation (**Fig.9-b**). It can be observed that there was enough attrition to provide successful measurements.

To show the validity of this measuring method, reproducibility of attrition of the plaster cylinder is shown in **Fig. 10**. The result was obtained under the condition of low bed height (L/D_e =1.1) and volumetric water content α=0.1, U/U_{mf}=2.5. From the figure, the experimental accuracy (degree of reproducibility) is approximately less than ±20 %. This is not sufficiently accurate to discuss quantitatively, but is enough to discuss qualitatively.

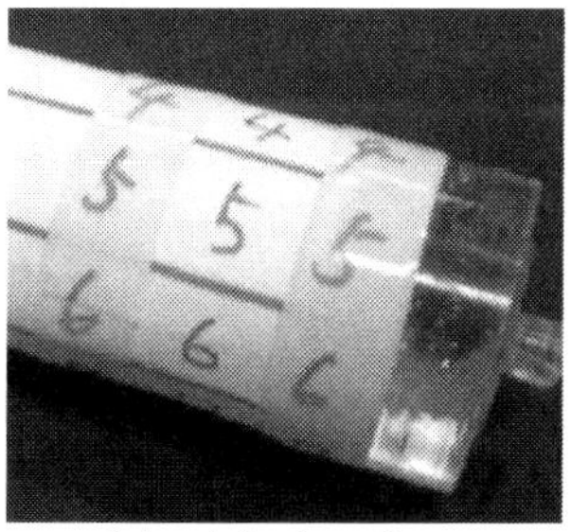

(a) original surface

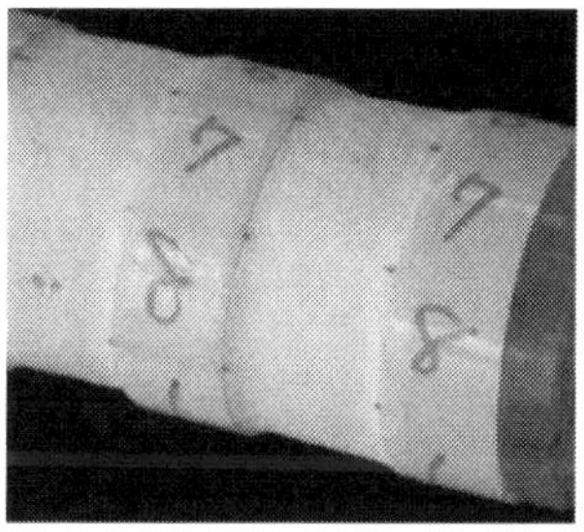

(b) after 15 minutes operation

Fig.9 Attrition of plaster cylinder in wetted bed, α=0.1, U/U_{mf}=2.0

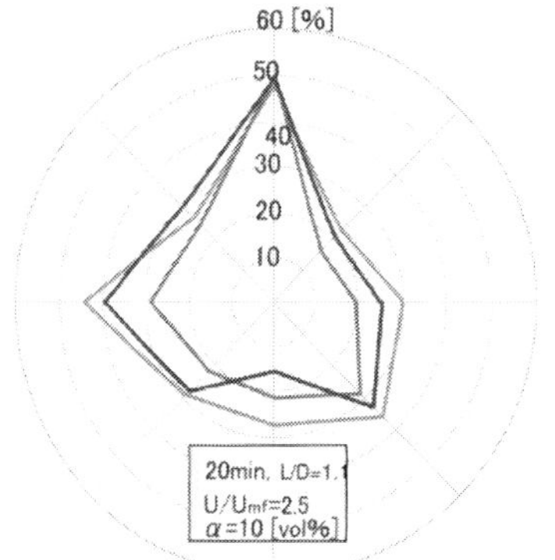

Fig.10 Reproducibility for attrition of plaster cylinder

Attrition of plaster cylinder in the novel fluidized bed

Attrition rate profiles around plaster cylinder are shown in **Figs. 11-a - 11-d**. Upper figures indicate results for the standard depth bed (L/D_e=1.1) and lower figures for the deeper bed (L/D_e=1.5) resulting in more violent slugging. From **Fig.11-a**, the attrition tendency of profile transition can be seen; attrition linearly increased with increasing operation time for every location on the surface. **Fig. 11-b** shows the change of attrition rate profile as a function of air velocity ratio U/U_{mf}; it changes from uniform attrition in all directions to a longitudinal slender profile, meaning that the attrition at the tube bottom becomes much larger than at the sides. The effect of bed depth can be seen from **Fig. 11-d**; the bed movement becomes more vigorous in a wet deeper bed, and therefore attrition rates becomes larger than for the standard bed. If $U/U_{mf}>2.5$ and the operation time exceeded 15 min, most of the mounted plaster was stripped off the cylinder surface.

The attrition tendency in bubbling water, ie., for water without solid particles, is shown in **Fig. 11-c.** (In this case, $U/U_{mf}=1.0$ means air velocity U=1.45 m/s). From this figure, it can be seen that directionality of the attrition disappears and the rate of attrition is small. If solid particles are not present in the bed, the attrition rate is only one-quarter of that in a corresponding wetted fluidized bed (**Fig. 11-d**). If the bed is dry, the attrition rate is only 10-15 % at most at the bottom of the cylinder, although this is not presented here.

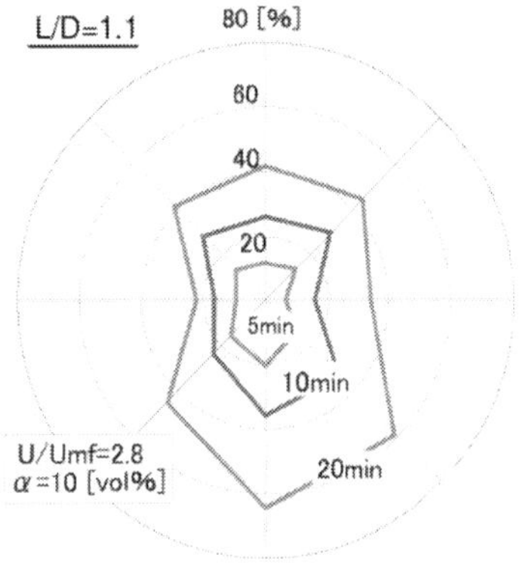

Fig.11-a Attrition profile (transition during time of operation)

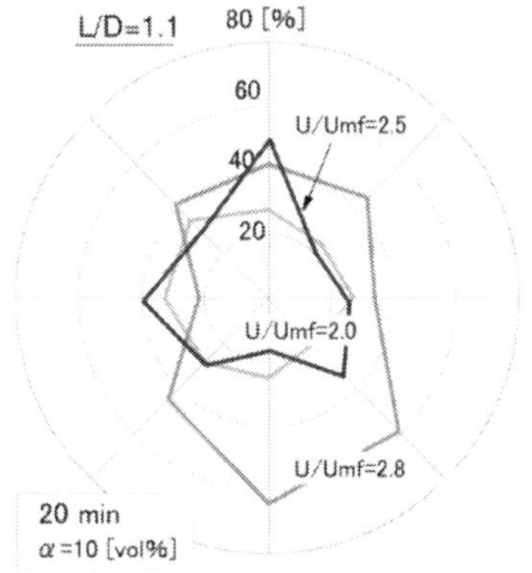

Fig.11-b Attrition profile (influence of gas velocity)

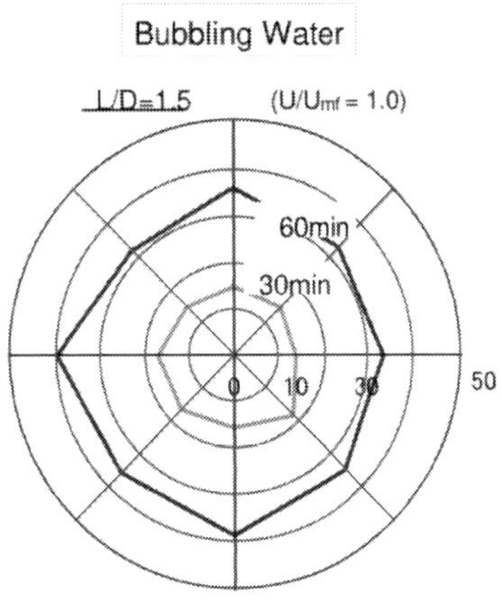

Fig.11-c Attrition profile (in bubbling water)

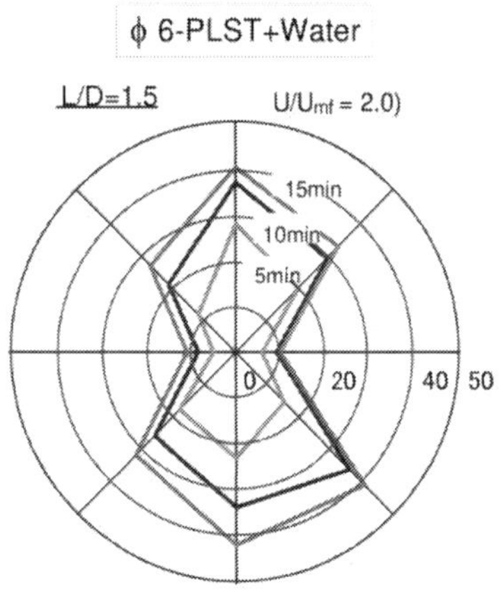

Fig.11-d Attrition profile (transition over time of operation for L/D=1.5)

The above results obtained in the wetted novel bed are somewhat similar to those from previous conventional attrition studies[8] in dry beds, but there are differences: The attrition at the side of the cylinder is smaller, and that at top of the cylinder is more prominent, with attrition at the top being negligible in conventional dry beds. This could be due to slight solubility of the plaster and also to differences in solid and liquid behavior around the top of the cylinder.

Whitening of the Artificial Soiled Cloth in the Novel Fluidized Bed

A washing test of an artificial waste cloth was carried out to prepare the materials for another washing factor which relates to calm abrasive washing by liquid movement, such as oil removal.

Fig. 12 shows artificial soiled cloths (Washing Science Association, Japan), (**Fig. 12-a**) commonly used to evaluate washing performance[9] (fats or oil removal etc.) of commercial detergents or washing machines, and the washed cloth after 15 minutes operation (**Fig. 12-b**). The lightness of each cloth was measured from a photograph of the washed cloth via image analysis and measurement software Win Roof (MITANI Co. Ltd., Ver.5.6). Washing performance is evaluated as the whitening extent, η, according to

$$\eta = (b - b_0)/(b_s - b_0) \qquad (2)$$

where b is the lightness of washed cloth, b_s that of standard white cloth and b_0 is that of the original soiled cloth.

Fig. 12-a Tested soiled cloth and white cloth

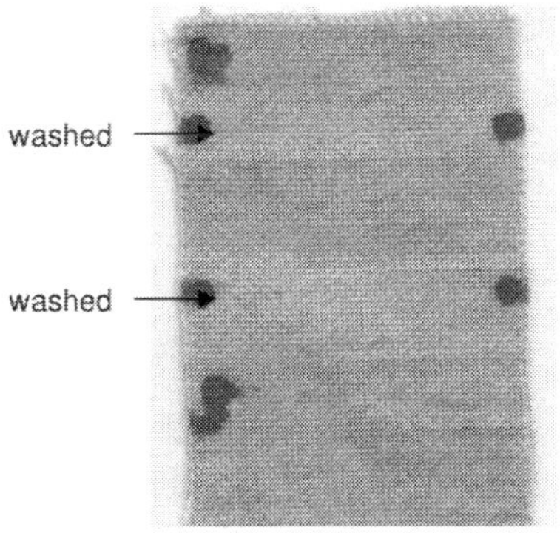

Fig. 12-b Washed cloth after 15 minutes operation

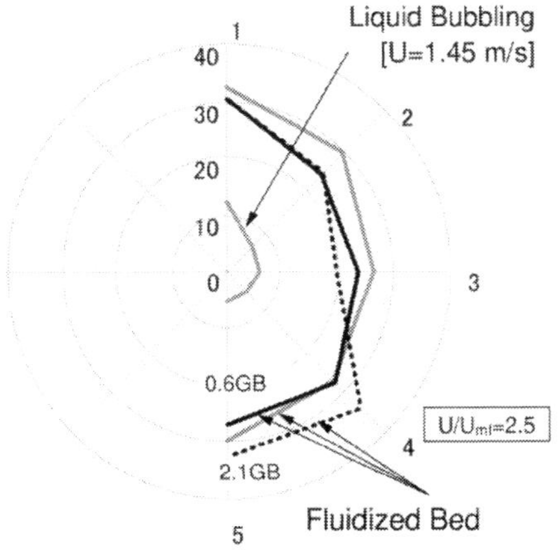

Fig. 13 Distribution of whitening extent, η, around the cylinder surface after 15 min of operation.

The whitening test results are shown in **Fig. 13**. For this test, the soiled cloth was attached to the ϕ30 mm acrylic cylinder and the lightness was measured on 5 blocks in the half side of the cylinder. Experimental conditions were L/D_e=1.5, U/U_{mf}=2.5, α (water content)=0.1, operation time=30 min. The figure indicates little directionality (compared with the attrition and impulse force results), and some influence of the addition of small particles. Detailed discussion is provided in the next section.

Relationship between Impulse Force, Attrition and Whitening

To discuss the effect of particle and water motion on washing and to identify the washing mechanism in this unique fluidized bed, the findings are summarized in **Fig. 14**. (1) Firstly, the particle collision force and attrition profiles are similar with bottom position > top position > side position of the cylinder for the plotted values. This shows that hard attrition washing is strongly related to particle collision. (2) Although addition of small particles (ϕ2.1 mm GB) led to a reduction in impulse force, the attrition was not changed. On the other hand, in the case of smaller particles (ϕ0.6 mm GB) addition, the attrition was clearly enhanced. (Due to limitations of the force detector, force data could not be obtained). From this, other factors, such as improved water movement or water residence fraction, appear to be provided by the presence of the smaller particles.

Next, consider the relationship between attrition and whitening. (3)The profile of whitening extent differs from that of attrition extent, with less directionality, ie., almost uniform whitening over the whole surface of the cylinder. Moreover, the effect of the addition of small particles cannot be seen clearly. Therefore in washing for soiling by fats or industrial oils, another factor which differs from attrition washing is predominant. It appears clearly at the top of the cylinder, where the washing performance is best, while this is the weakest location for particle collisions. These results imply that water movement is important rather than particle collisions, for the calm abrasive washing needed for oil removal. To confirm this finding, observations of the wetting around the tube are reported in the next section.

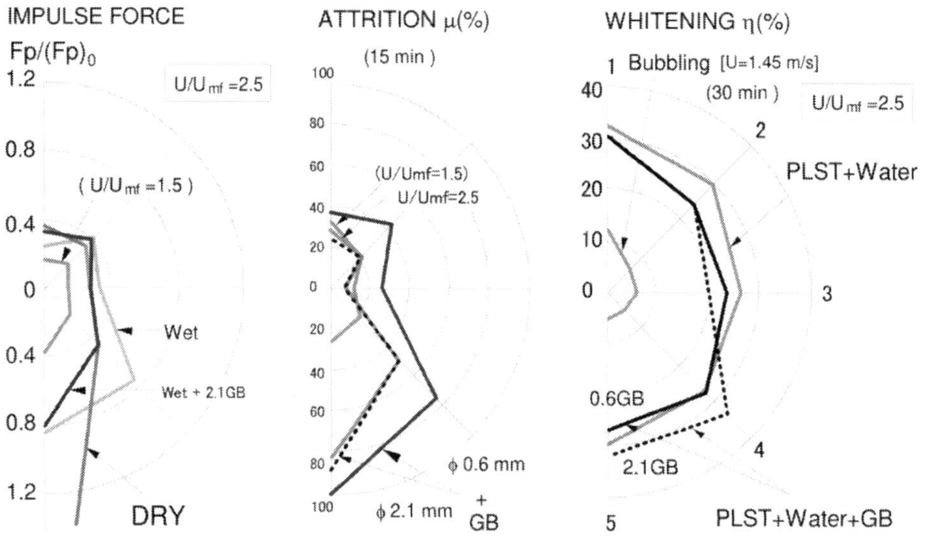

Fig. 14 Relationship between impulse force, attrition extent and whitening extent

Observation of Wetting around Object Circumference

The wetted area fraction around the tube surface was next measured at five circumferential measuring points by a borescope from inside the acrylic tube. The digital VTR data, recorded over a period of about 5 s, were analyzed by WinRoof visual analysis software (MITANI Co. Ltd.). Typical photographs are shown in **Fig. 15**. The colored portion is the area wetted by colored water.

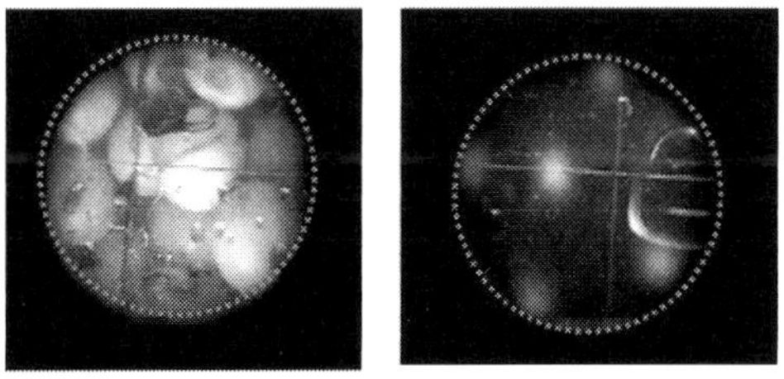

BOTTOM U/U_{mf}=0.7, α=0.05, TOP U/U_{mf}=2.0, α=0.2

Fig.15 Observation of wetted state from the tube inside; colored portions are wetted.

From the figure, it is found that the wetted area fraction increases with increasing water content α and air velocity ratio, U/U_{mf}, as particle movement becomes vigorous. The fraction was calculated simply as the colored area divided by the projected circle area (with no consideration of the tube curvature). This fraction is almost independent of air velocity when α >0.1. The fractional values reach a maximum at the top (80-95%), next at angle 45° from the top position (60-75%), then at 90° (nearly equal to that at 135°) (50%), and falling to a minimum value (20-45%) at the bottom. For α ≥ 0.1, sufficient quantity of water is delivered to the tube surface, except at the bottom of the tube, even though the water added into the bed is only 20 vol % at most.

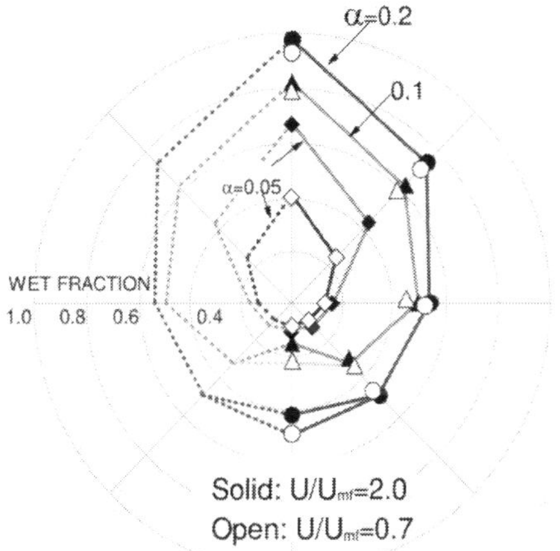

Fig.16 Distribution of wetted fraction around the cylinder surface.

This could be why washing performance is favorable for a small water supply, and the washing performance, such as for oil removal, is best even though the weak point for particle collision is at the top of the tube.

In earlier work the explanation about the washing mechanism was insufficient by using results of force and attrition and the results of attrition and whitening, individually, but now it can be explained successfully. One key factor is the small quantities of added water: the time and local volume fraction of water on the object surface determines the washing performance. This makes this system unusual. There is sufficient water at the top of the tube for gentle washing while this is the weakest area for particle attack. On the contrary, the tube bottom has relatively poor water supply, but provides the best area for attrition.

From these findings, two main factors for the washing are considered to exist in this novel multiphase flow: One is hard attrition washing due to particle collision, which promotes mud removal, while the other is calm abrasive washing by liquid movement, helpful for fats and industrial waste oil removal. These two factors interact and compensate for each other, enabling high overall washing performance. In the previous sections, the overall washing and whitening results could not be explained in terms of either the force or attrition results alone, but these results can now be explained.

Conclusion

Fundamental characteristics of washing were investigated on objects immersed in a novel multiphase flow (water-added large-light-particle gas-fluidized bed). The attrition, which is related to hard attrition washing, is directly affected by particle collisions. The attrition extent is largest at the bottom and small at the sides and top of the immersed object. It is enhanced by the addition of small (<~1 mm diameter) and large-density particles. The weaker areas for particle attack are, however, well supplied with water, so they are favorable areas for washing removal of fats and oils.

From the experimental findings, the washing mechanism involving two main factors is clarified for this unique multiphase flow system.

Acknowledgement

Financial support from both the Natural Sciences and Engineering Research Council of Canada and the Japan Society for the Promotion of Science (# C-18560728) are gratefully acknowledged. The authors also thank Y. Nishii, S. Takaoka and A. Ikeda for their assistance in experiments and in analyzing data.

References

1. Y.Nagahashi, D.H.Lee, J.R.Grace, N.Epstein, A.Yokogawa and Y.Asako (2003), "Enhancement of Large-Particle Gas-Fluidization by Adding Liquid", AIChE Journal, **49 -3**, pp.675-681.
2. Y.Nagahashi, J.R.Grace, Y.Asako, N.Epstein, D.H.Lee and A.Yokogawa (2003), "Effect of Liquid Addition on Heat Transfer in Gas-Fluidized Beds of Large Light Particles", Journal of Process Mechanical Engineering, **217**, pp.279-286.
3. Y. Nagahashi, N. Epstein, J.R. Grace, Y. Asako and A. Yokogawa (2006), "Spouting Enhancement by Addition of Small Quantities of Liquid to Gas-Spouted Beds", Can.J.Chem.Eng.,**84-5**, pp.527-531.
4. A.Yokogawa (1999), "The Method and Facility for Human Washing", Japan Patent, No.JP-11-37627.
5. Y. Nagahashi, J.R. Grace, N. Epstein, Y. Asako and A. Yokogawa (2007)," Flow Characteristics of Fully Wetted Binary Solid Mixtures in Gas Fluidized Beds with Inclined Gas Distributors", Fluidization 12, ed. X, Bi, F. Berruti and T. Pugsley, UEC, NY, pp. 337-344.
6. J.Zhu, J.R.Grace and C.J.Lim (1989),"Erosion-Causing Particle Impacts on Tubes in Fluidized Beds" , Fluidization 6, ed. J.R.Grace, L.W.Shemilt and M.A.Bergougnou, Eng. Found., NY, pp.605-611.
7. R.R.Rathbone, M.Ghadiri and R.Clift (1989), "Measurement of Particle Velocities and Associated Stress on Immersed Surfaces in Fluidized Beds", ibid, pp.629-635.
8. J. Zhu (1988), "Tube Erosion in Fluidized Beds", PhD Dissertation, Univ. British Columbia, Canada.
9. N. Obata (2004), "Application of Electrolyzed Water to Washing System", Bulletin of Toyama University of International Studies, **4**, pp.121-125.

CFD Modeling of Electrostatic Effects On Catalyst and Fine Particles in Fluidized-Bed Reactors

Ram G. Rokkam, Department of Chemical & Biological Engineering, Iowa State University, Ames, IA

Rodney O. Fox, Iowa State University, Ames, IA

Michael E. Muhle, Michael E, Houston, TX

Electrostatic charges are known to influence the hydrodynamics of gas solid fluidized bed reactors. Excessive charges on particles cause agglomeration and wall sheeting (Hendrickson, 2006). Charges on particles are formed due to particle-particle and particle-wall collisions. When particles collide and separate there is charge formed on the particles and this phenomena is known as Triboelectrification. The development and verification of the electrostatic model with a simple test case is presented in AICHE fall meeting, 2008 (Rokkam et al, 2008). In this work the application of the electrostatic model on pilot plant reactor will be presented and two cases are considered here:

(1) Continuous Catalyst Injection: The objective is to study the electrostatic effects on catalyst particles. The multi-fluid model solves for one gas and four solid phases and the catalyst is injected after the bed is fully fluidized. Three cases are run with different charge on catalyst (positive, zero, negative). The charge magnitude is approximated to realistic conditions.

(2) Effect of Electrostatics on Entrainment: The objective is to study the electrostatic effects on fine polymer particles. Multi-fluid model solves for one gas phase and four solid phases. The bed is initialized with a uniform distribution of fine particles and other solid phases.

An Eulerian-Eulerian based multi-fluid model is used to solve the hydrodynamics of gas-solid fluidized-bed reactor. The multi-fluid model is solved in commercial CFD code Fluent 6.3 and uses the kinetic theory granular flow to calculate the solids pressure and viscosities. The volume fraction and diameter of the solid phases are obtained from Chemical reaction engineering model and quadrature method of moments (Fan et al, 2007). Time dependent two-dimensional simulations are done on a pilot-plant scale fluidized bed reactor.

REFERENCES

Fan R, Fox R O, Muhle M E. Role of intrinsic Kinetic and Catalyst Particle Size Distribution in CFD Simulations of Polymerization Reactors. 12th International Conference in Fluidization in Vancouver, Canada. 2007

Hendrickson G. Electrostatics and gas-phase fluidized bed polymerization reactor wall sheeting. Chemical Engineering Science. 2006; 61(4), 1041-1064.

Rokkam R G, Fox R O, Muhle M E. Cfd Modeling of Polymerization Fluidization Reactor. Presented at Annual AICHE Meeting,Philadelphia. 2008

See more of this Session: Applications of Fluidization
See more of this Group/Topical: Particle Technology Forum

Numerical Evaluation of the Process Intensification Potential of Rotating Fluidized Beds in a Static Geometry: Application to Fluid Catalytic Cracking

Philippe Eliaers, Imap, UCL, Louvain-la-Neuve, Belgium

Waldo Rosales Trujillo, Imap, UCL, Louvain-la-Neuve, Belgium

Juray De Wilde, Materials and Process Engineering, UCL, Louvain-la-Neuve, Belgium

1. Introduction

The operating conditions in conventional, that is, gravitational fluidized bed reactors are limited by the use of earth gravity. Internal mass and heat transfer limitations can be encountered, the van der Waals forces becoming too important when trying to fluidize smaller size particles. External mass and heat transfer limitations can be encountered and are related to limitations on the gas-solid slip velocities. With heterogeneous (catalytic) reactions, as well external mass or heat transfer limitations as internal mass transfer limitations may impose limitations on reaction rates and on the activity of the catalysts that can be used.

Fluidization in a centrifugal field may allow overcoming the above mentioned limitations and may, as such, allow fluidized bed process intensification and in particular the use of more active catalysts [1,2]. Current technology for fluidization in a centrifugal field is based on a rotating fluidization chamber [1,2] and its industrial application faces challenges with respect to sealing, mechanical vibrations and continuous operation. Recently, a novel route for fluidization in a centrifugal field was presented, the rotating fluidized bed in a static geometry [3-5], facilitating industrial process application and offering additional specific advantages.

2. Concept

In rotating fluidized beds in a static geometry, the centrifugal field is generated by injecting the fluidization gas tangentially in the fluidization chamber via multiple gas inlet slots in its outer cylindrical wall [3,4,5] (Figure 1). A combined tangential-radial fluidization of the particle bed can be obtained by forcing the fluidization gas to leave the fluidization chamber via a central chimney. The fluidization gas flow rate influencing both the centrifugal force and the counteracting radial gas-solid drag force in a similar way renders rotating fluidized beds in a static geometry extremely flexible with respect to the fluidization gas flow rate and the gas-solid contact time. In particular, dense operation at high fluidization gas velocities is possible, allowing fast and highly endothermic or exothermic reactions to be carried out in a relatively small volume rotating fluidized bed in a static geometry.

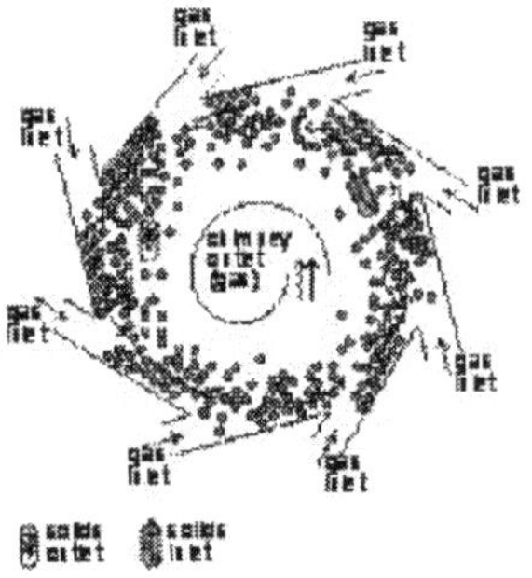

Figure 1. Rotating fluidized bed in a static geometry and rotating chimney concepts.

Rotating fluidized beds in a static geometry furthermore allow angular multi-zone operation, the fluidization gases injected via successive gas inlet slots hardly being mixed in the rotating particle bed. Such multi-zone operation opens perspectives for carrying out catalytic reactions and catalyst regeneration within the same reactor vessel.

3. Numerical evaluation of the fluid catalytic cracking process intensification potential

The potential process intensification of the fluid catalytic cracking (FCC) process using a rotating fluidized bed in a static geometry is numerically evaluated. A comparison with the performance of a riser type reactor as described in Froment and Bischoff [6] is made.

3.1. Hydrodynamic model

For the riser reactor simulations, a 1D plug flow model with slip between the gas and solids is used [6].

For the rotating fluidized bed in a static geometry, experimental observations and computational fluid dynamics (CFD) type simulations show an essentially cross-flow type flow pattern for the gas and solids - radial for the gas and angular for the solids [7]. Furthermore, plug flow can be assumed [7]. The relations between the gas and solids phase velocities and between the gas phase velocity and the solids volume fraction (or bed density) is taken from experimental observations [7]. The hydrodynamic model for the rotating fluidized bed in a static geometry is schematically shown in Figure 2.

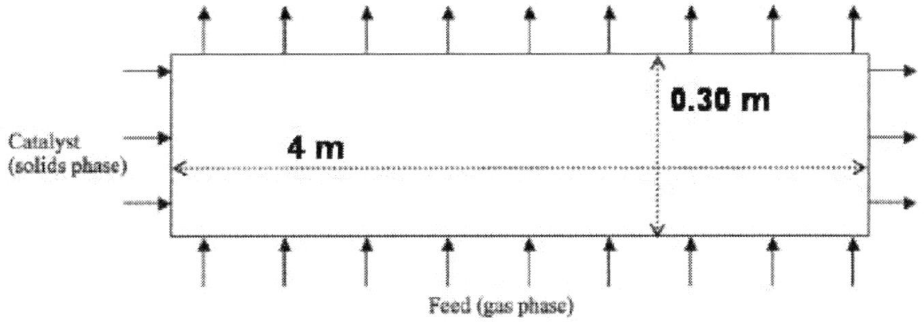

Figure 2. Geometry and hydrodynamic model for a rotating fluidized bed in a static geometry.

3.2. Kinetic model

The 10-lump model proposed by Jacobs et al. (see [6]) is adopted. A distinction is made between heavy paraffins, heavy naphtenic molecules, heavy aromatic substituent groups, heavy carbon atoms among aromatic rings, light paraffins, light naphtenic molecules, light aromatic substituent groups, light carbon atoms among aromatic rings, gasoline, and a lump of coke and light gases (C1-C4).

3.3 Simulation cases

First, identical catalyst is assumed to evaluate the intrinsic process intensification by rotating fluidized beds in a static geometry (Figure 3). Particular attention is paid to the cat-to-oil-ratios that can be achieved. Next, the possibility of using more active catalyst is investigated. Multi-zone aspects are also discussed.

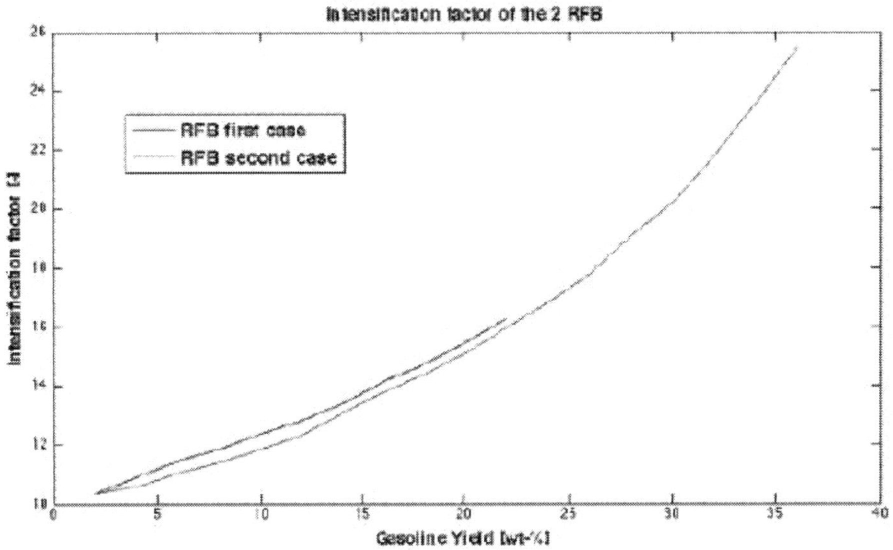

Figure 3. Fluid catalytic cracking process intensification by a rotating fluidized bed in a static geometry. Comparison with conventional riser technology using identical catalyst. Intensification factor defined as the ratio of the reactor volumes required to convert a given flow rate to a given gasoline yield.

References

1. Qian, G.-H., Bagyi, I., Burdick, I.W., Pfeffer, R., Shaw, H., Stevens, J.G. *AIChE J.* 47(5), p. 1022-1034 (2001).

2. Watano, S., Nakamura, H., Hamada, K., Wakamatsu, Y., Tanabe, Y., Dave, R.N., Pfeffer, R. *Powder Technol.*, 141, p. 172-176, (2004).

3. de Broqueville, Axel : Belgian Patent 2004/0186, Internat. Classif. : B01J C08F B01F; publication number : 1015976A3.

4. De Wilde, J., de Broqueville, A. *AIChE J.*, 53(4), p. 793-810, (2007).

5. De Wilde, J., de Broqueville, A. *Powder Technol.*, 183:(3), p. 426-435, (2008).

6. Froment, G.F., Bischoff, K.B., *Chemical Reactor Analysis and Design*, Wiley, (1990).

7. Staudt, N., De Wilde, J., UCL report (2008).

See more of this Session: Applications of Fluidization
See more of this Group/Topical: Particle Technology Forum

Liquid and Solids Entrainments in Slurry Bubble Columns

Fei Wang, William G. Lowrie Department of Chemical and Biomolecular Engineering, The Ohio State University, Columbus, OH

Zhao Yu, William G. Lowrie Department of Chemical and Biomolecular Engineering, The Ohio State University, Columbus, OH

Bing Du, William G. Lowrie Department of Chemical and Biomolecular Engineering, The Ohio State University, Columbus, OH

Raymond Lau, William G. Lowrie Department of Chemical and Biomolecular Engineering, The Ohio State University, Columbus, OH

L. -S. Fan, William G. Lowrie Department of Chemical and Biomolecular Engineering, The Ohio State University, Columbus, OH

The liquid and solids entrainment using hydrocarbon fluids that mimic the Fischer-Tropsch reactor behavior and industrial catalysts are experimentally studied in a 4-in ID high pressure slurry bubble column and a 6-in ID ambient slurry bubble column. Paratherm NF heat-transfer fluid is used as the liquid phase, with ConocoPhillips aluminum oxide catalyst particles as the solids phase and nitrogen and air as the gas phase. The effects of superficial gas velocity, pressure, temperature, solids loading and freeboard height on the liquid and solids entrainment rates in the slurry bubble columns are examined. The operating superficial gas velocities, pressures, temperatures, solids loading, and freeboard heights vary by up to 0.55 m/s, 30 bar, 90 ^{0}C, 25%, and 0.95 m, respectively. The study indicates that the liquid entrainment rate increases with superficial gas velocity, pressure, temperature, and solids loading. The solids entrainment rate increases with pressure and superficial gas velocity but decreases with temperature. At high superficial gas velocities, solids entrainment rate increases with solids loading. Liquid and solids entrainment rates remain almost unchanged when the freeboard height is bigger than a critical height. Correlations for the critical heights are developed for the liquid and solids respectively.

See more of this Session: Applications of Fluidization
See more of this Group/Topical: Particle Technology Forum

Heat Transfer Simulations in Multi Size Particle Riser Flow

YeRhee Cha, Chemical and Biological Engineering, Illinois Institute of Technology, Chicago, IL

Hamid Arastoopour, Chemical and Biological Engineering, Illinois Institute of Technology, Chicago, IL

In practical applications of gas-multi size particle flow, heat transfer is mostly important. In this work, heat transfer between particles with different sizes in a riser flow was evaluated. For the numerical validation of the particle hydrodynamics, Particulate Solid Research, Inc. (PSRI) challenge problem 2 experimental data were used. In this step, simulations of single size and three different size particle flow in a two dimensional riser were performed using MFIX. The simulations are based on Eulerian view for all phases in the flow and employed different kinetic theory of granular flow (KTGF) models for each of the simulations. The single size simulations are based on the kinetic theory of granular flow (KTGF) model developed by Lun *et al.*[1]; the multi size simulations are based on the Iddir&Arastoopour kinetic theory[2] for multi size particles. The numerical results are compared with the available experimental data for pressure drop, solid axial velocity, solid concentration, and solid mass flux presented in PSRI challenge problem 2. Heat transfer in the riser is compared to published experimental data showing the effect of the presence of multi-size particles on heat transfer. Combined Eulerian-Lagrangian approach is applied for the simulations using MFIX in that Lagrangian approach provides detailed information of each solid particle. It is assumed that particle-particle collisions follow a hard sphere model, and that heat transfers from particle to particle is due to conduction through the gas layer near the contact point. The amount of heat transferred between different size particles is given by the simulation results and the effect of particle size on heat transfer in the riser flow is examined.

[1] Lun CKK, Savage SB, Jeffrey DJ, Chepurniy N. "Kinetic theories of granular flows: inelastic particles in Couette flow and slightly inelastic particles in a general flow field", J Fluid Mech., 140, pages 223-256, (1984)

[2] Iddir, Hadjira, "Modeling of the multiphase mixture of particles using the kinetic theory approach", Ph.D. Thesis, Illinois Institute of Technology, (2004)

See more of this Session: Applications of Fluidization
See more of this Group/Topical: Particle Technology Forum

Cultivation of Microalgae Using Draft-Tube Bioreactor

Xue Chen, School of Chemical and Biomedical Engineering, Nanyang Technological University, Singapore, Singapore

Wei-Ning, William Chen, School of Chemical and Biomedical Engineering, Nanyang Technological University, Singapore, Singapore

Raymond Lau, School of Chemical and Biomedical Engineering, Nanyang Technological University, Singapore, Singapore

Microalgae have been gaining great interests as a feedstock for biomass and an alternative method for carbon dioxide removal. In this study, a draft-tube bioreactor is used to cultivate the microalgae, chlorella.sp. A lipid probe using Nile Red is employed to determine the intracellular lipid content of the microalgae. Effects of carbon dioxide concentration, light intensity, and system pressure on the growth and lipid content of chlorella.sp are studied.

See more of this Session: Applications of Fluidization
See more of this Group/Topical: Particle Technology Forum

Enhancement of Particle Collection Efficiency in Electrohydrodynamic Atomization Processes for Pharmaceutical Particle Fabrication

Abstract ID: 162231

Alireza Rezvanpour, Chi-Hwa Wang

Department of Chemical & Biomolecular Engineering, National University of Singapore
4 Engineering Drive 4, Singapore 117576

INTRODUCTION

EHDA is a well-known phenomenon for production of fine droplets of uniform [1-2]. EHDA in the cone-jet mode has been studied for many years [3]. The main concept here is applying an electric field on pendant and flowing droplets (through a capillary) to atomize them. The droplet profile as a function of applied voltage can be studied as the main data extracted from EHDA experiments.

Particle collection efficiency and residual solvent in collected particles are two important issues in particle fabrication using EHDA process which are not addressed in the literature. These two factors are exceedingly important for designing EHDA system especially from the point of view of operating costs. Since the materials used for pharmaceutical particle fabrication are expensive, higher particle collection efficiency is extremely desirable. Moreover, allowable concentration of residual solvent in collected particles after EHDA process and before freeze-drying is desirable because it eliminates freeze-drying as an additional energy-consuming process. The present work aims to employ EHDA technology to fabricate particles in a new generation of shuttle chamber. Here, we investigated the production process with focus on the particle collection efficiency and residual solvent in the collected particles. The experiments are designed based on the Taguchi (combined statistical and experimental) optimization method. Subsequently, the results are summarized in different tables and figures. In the present study, we aim to develop a deeper understanding of the EHDA process for polymeric particle fabrication and also to achieve better collection efficiency and solvent evaporation by adjusting some of the operational and geometrical parameters.

EXPERIMENTAL METHODS

In the experimental apparatus used in this study (shown in Fig. 1), a shuttle glass chamber was used to enclose a nozzle and a ring electrode, which were both applied with high electrical potential relative to the ground plate, together with a grounded collecting plate. The nozzle and the ring were connected to two high-voltage DC power supplies independently forming the electric field for EHDA. A high positive electrical potential relative to ground was applied to the nozzle while a lower positive electrical potential was applied to the ring.

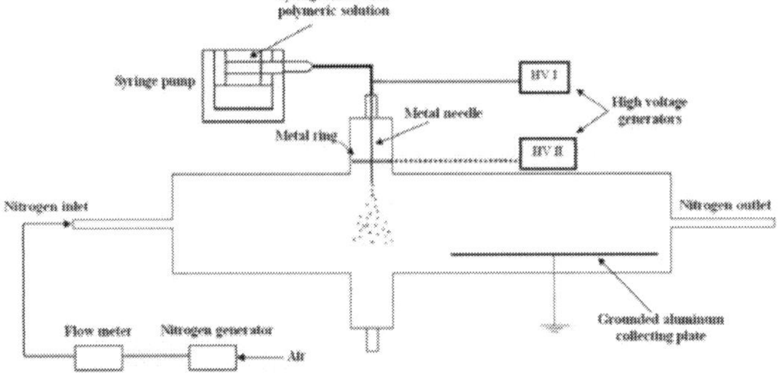

Figure 1: Apparatus used in the experiments.

In this study four different voltages are applied to both nozzle and ring. These applied voltage difference values are listed in Table 1. The ground plate is also the collecting location of the particles formed. Solution was pumped out using a programmable syringe pump. Nitrogen as inert gas was connected to the left end of the chamber and the other end was also open as nitrogen outlet such that the collection rate of the particles could be adjusted through a pneumatic conveying method. Details of the nitrogen and solution flow rates are listed in Table 1. In contrast, relevant experimental conditions and setup characteristics are listed in Table 2.

Table 1: Controllable factors and their levels used in the experiments.

	Factors	Level 1	Level 2	Level 3	Level 4
1	Nozzle Voltage (kV)*	9.0	8.5	8.0	7.5
2	Ring Voltage (kV)*	7.0	6.5	6.0	5.5
3	Soln. Flow Rate (ml/hr)	2.5	2.0	1.5	1.0
4	Nitrogen Flow Rate (l/min)	35	30	25	20

* Values shown are relative to the ground.

Table 2: Experimental conditions, setup characteristics and physical properties of the used materials.

Inner diameter of the nozzle	0.41 mm (27 Gauge)
Diameter of the ring	33 mm
Thickness of the ring	3 mm
Inner diameter of the chamber	100 mm
Length of the chamber	800 mm
Diameter of the chamber outlet and inlet	15 mm
Collecting plate material	Aluminum
Collecting plate dimensions	90 mm×100 mm
DCM density	1326.6 kg/m^3
DCM electrical conductivity	0.0025 μS/cm
DMF density	944 kg/m^3
Vacuum permittivity	8.85×10^{-12} C^2.s^2/kg.m^3

The electrospray solution would contain PLGA at 8% w/v of DCM and Paclitaxel at 20% w/w of PLGA. An Ultrasonicator was used to dissolve the components of the mixture homogeneously for about 15 minutes. The physical properties are listed in Table 2. Gas chromatography/mass spectrometry (GC/MS) was used to determine the residual content of DCM in the fabricated particles. Standard solutions were prepared with the level of DCM concentrations in DMF ranging from 0.5 to 10×10^{-6} ml DCM per ml DMF and stored in the refrigerator before analysis to prevent evaporation of DCM in the mixture [4].

A statistical experimental optimization method (Taguchi) is used to determine the optimal collection conditions and to find the main parameters affecting the particle collection efficiency [5]. In this study four different variables were chosen as controllable factors and particle collection efficiency as the response factor. The choice of the factors was based on a literature survey [6-9]. The choice of the levels was based on prior testing of EHDA process to achieve a stable cone-jet mode at the tip of the nozzle. All factors and their levels were combined to sixteen trials. Collection efficiency (CE) can be calculated using the following equation.

$$\text{CE, \%} = \frac{\text{Mass of the collected particles}}{\text{Total mass of the initial PLGA and Paclitaxel}} \times 100\% \qquad (1)$$

RESULTS AND DISCUSSION

Optimization of experimental trials

A parameter which is defined for comparing the trials and the influence of a factor in the Taguchi method is the ratio of the desired factor to the noise created during the experiment, S/N. This ratio was used for comparing the trials and the influence of a factor in the Taguchi method [10]. Since a higher collection efficiency is desirable in this experiment, "the bigger the better" criterion was chosen for the S/N ratio. In general, a smaller noise is desirable so a larger S/N ratio yields results that fulfill the objectives [11].

The quantitative analysis of the influences and the relative importance of the controllable factors which may be provided via ANOVA (Analysis of Variance) confirm that solution flow rate appears to have the biggest contribution to the results. Based on ANOVA results the important factors in order are: (i) solution flow rate, (ii) nitrogen flow rate and (iii) ring voltage and (iv) nozzle voltage.

ANOVA may be used to estimate the process performance at optimum conditions. The results of estimated performance are listed in Table 3. The collection efficiency at these conditions is given by this analysis as 82.04%. Since the optimum conditions are not belonging to any of the trials, a confirmation experiment was performed to verify the predicted result. If the average results of the experiments are within the confidence limit then the prediction is acceptable. The value of the collection efficiency at optimum conditions was measured as 79.4% and it may be seen that it is within the range of the confidence limit.

Table 3: Estimated performance at optimum conditions

	Factors	Level description	Level	Contribution
1	Nozzle Voltage (kV)	8.5	2	0.209
2	Ring Voltage (kV)	6.5	2	0.428
3	Soln. Flow Rate (ml/hr)	1	4	0.710
4	Nitrogen Flow Rate (l/min)	35	1	0.511

Concentration of residual DCM in particles

Using the integration equation resulted from GC/MS experiment, the concentration of DCM could be found for all the trials. The obtained results verify that the residual DCM content of the particles fabricated using the EHDA method was below the safety standards margin (600 ppm) at the end of the process [4]. This is achieved even without using freeze-drying process which is used to evaporate the residual solvent in the fabricated particles.

Effect of solution flow rate

Fig. 2 shows the variation of the collection efficiency versus solution flow rate at optimum conditions listed in Table 3. This figure indicates that collection efficiency drastically decreases with increasing solution flow rate from 1 to 1.8 ml/hr. However, the collection efficiency increases as the solution flow rate increases from 1.8 to 2.5 ml/hr. At high solution flow rates, the charges on the liquid at the nozzle tip could not be concentrated as there is more liquid per unit time and there is not enough electrical force to overcome the mechanical forces of liquid motion.

In contrast, at low solution flow rates the electric charges on the drop at the nozzle tip are sufficient to overcome the surface tension to achieve jet spraying. Therefore, the spraying is more stable and more particles with smaller size are produced on the collecting plate [8]. As the optimum solution flow rate in this study is 1ml/hr (as shown in Table 3), it shows that this solution flow rate is low enough to produce stable cone jet. As solution flow rate increases, high surface tension on the Taylor cone ceases the formation of cone jet mode and converts it to a dripping mode [12]. Moreover, high collection efficiency at low solution flow rates can also be related to the small size of the fabricated particles. Drag force and gravitational force have direct relation with size of the particle and, thus, smaller size of the particles results in lower drag, gravitational and buoyancy forces. As such, the importance of these forces decreases with decreasing size of the particle and electrical field force, therefore, predominant in these conditions. Electrical field force dominance is favorable for achieving high particle collection efficiency. On the other hand, when the solution flow rate is increased further from 1.8 to 2.5 ml/hr, the formation of a simple jet can actually occur at higher solution flow rates [13]. This phenomenon occurs because at high solution flow rates, the kinetic energy of the liquid at the outlet of the drop at nozzle tip is greater than the surface tension of the jet and simple jet occurs [14]. As the solution flow rate increases, the simple jet mode becomes more stable [13]. The simple jet allows the formation of more particles and the production yield is increased and more particles would be produced on the collecting plate.

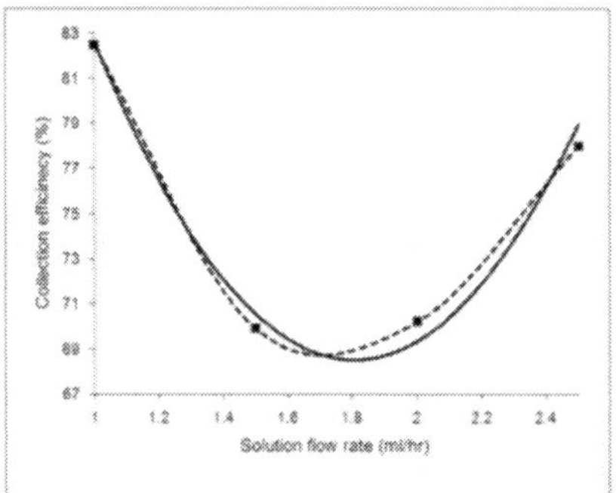

Figure 2: Variation of the collection efficiency versus solution flow rate at optimum conditions. (__ Predicted by Taguchi, ---- Experimental results)

Effect of nitrogen flow rate

Fig. 3 (a) shows the variation of the collection efficiency with nitrogen flow rate at optimum conditions listed in Table 3. Furthermore, Fig.3 (b) and (c) also show the same trend for the other conditions. These figures indicate that collection efficiency slowly decreases with increasing nitrogen flow rate from 20 to 26 lit/min. However, the collection efficiency increases as the nitrogen flow rate increases from 26 to 35 lit/min.

The variation range of the collection efficiency versus nitrogen flow rate is smaller than the corresponding solution flow rate presented in Table 3. Increasing in nitrogen flow rate can indeed increase the corresponding drag force, which is almost parallel to the electrical field force in this system. However, increasing nitrogen flow rate below 26 lit/min does not have observable effect on drag force. In contrast, it results in enhanced turbulence and eddy of nitrogen flow inside the chamber. This turbulence can cause the particles to circulate inside the chamber and increase the chance of particle deposition on the chamber's walls before reaching the collecting plate. Conversely, increasing nitrogen flow rate from 26 lit/min onwards has significant effect on enhancing the drag force. In fact, the effect of pneumatic conveying of the particles overcomes the effect of turbulent flow inside the chamber. Furthermore, the cylindrical shape of the glass chamber with an open and wide outlet effectively prevents the reverse flow and circulation of the nitrogen inside the glass chamber. Additionally, the increase in nitrogen flow would also evaporate the solvent on the particles faster and the particles would be less sticky for pre-mature deposition on the chamber's walls.

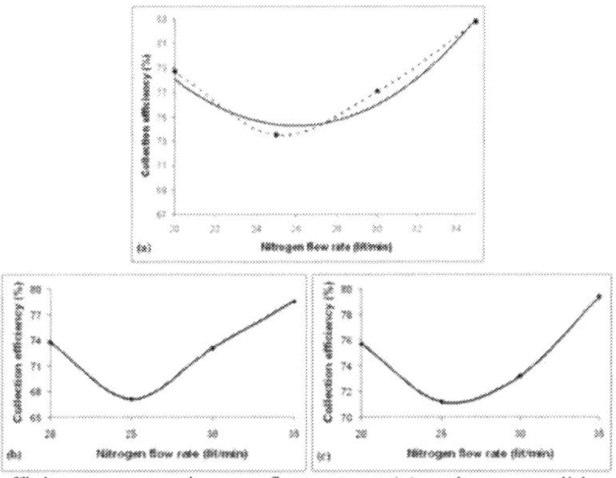

Figure 3: Variation of the collection efficiency versus nitrogen flow rate at (a) optimum conditions shown in Table 6 (__ Predicted by Taguchi, ---- Experimental results), (b) Nozzle voltage 8.5 kV, ring voltage 5.5 kV and solution flow rate 1.5 ml/hr (c) Nozzle voltage 8.5 kV, ring voltage 6.5 kV and solution flow rate 2.5 ml/hr.

Effect of nozzle and ring voltages

Figs. 4 (a) and (b) show the variation of the collection efficiency versus nozzle and ring voltages respectively at the optimum conditions listed in Table 3. This figure illustrates that collection efficiency increases with increasing nozzle voltage and decreases with increasing ring voltage in a non-monotonic fashion. The collection efficiency slowly changes with variations in nozzle and ring voltages. Weak effect of these factors on collection efficiency was previously declared by Taguchi results (Table 3). The collection efficiency is not strongly influenced by these settings because the combined nozzle and ring system allows the production of a single cone-jet over a wide range of voltage differences.

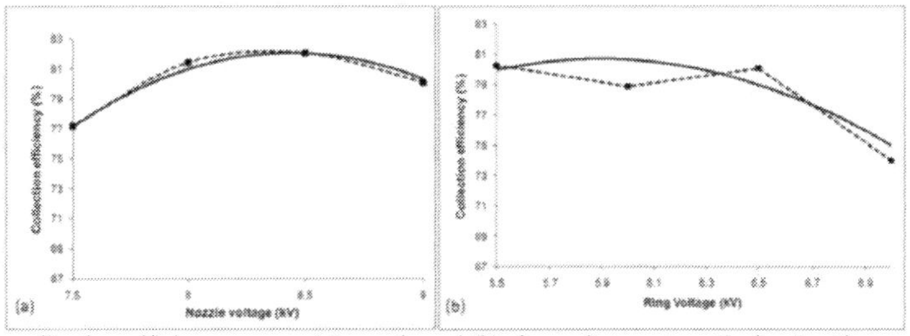

Figure 4: Variation of the collection efficiency versus (a) nozzle and (b) ring voltages respectively at optimum conditions. (__ Predicted by Taguchi, ---- Experimental results).

It is noteworthy that ring voltage has more obvious effect on collection efficiency which is also declared by Taguchi results shown in Table 3. This is because the polarized ring electrode stabilizes the EHDA spray and it could be better controlled by reducing the radial electric field component on the jet [15]. The stability of the jet spraying to produce particles produced on the collecting plate would be more dependent on the ring voltage than nozzle voltage for the range of electrical potential difference tested in this study.

The most important feature in determining the electric fields in the EHDA process would be the voltage difference between the ring and nozzle [16]. As the voltage difference between the ring and nozzle decreases, the spray mode could change from a multiple cone spray to a single spray cone, then finally to a dripping model. The dripping mode appears when the voltage difference is small and this can cause a significant reduction in collection efficiency as spraying of particles would be disrupted. In fact, the polymer and drug solution would be lost in drippings instead of being atomized.

In another words, difference between nozzle and ring voltages has more significant effect on both particle size and collection efficiency. In fact, higher collection efficiency occurs at higher voltage difference between nozzle and ring. In addition to the formation of the stable cone-jet at high voltage difference between nozzle and ring, this phenomenon may be related to the size of the fabricated particles. Smaller size of the particles causes electrical field force to be predominant. Therefore, increasing voltage difference between nozzle and ring can produce smaller particles and higher particle collection efficiency.

It can be observed in fig. 5 that with increasing voltage difference between nozzle and ring, the sizes of particles decreases. In this figure, the trial with maximum voltage difference between nozzle and ring (figure 5-a) has the highest collection efficiency.

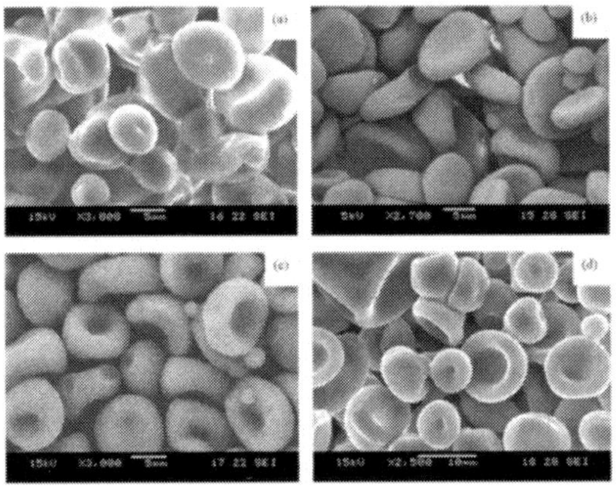

Figure 5: Effect of voltage difference between nozzle and ring on the particle size and particle collection efficiency. (a) trial 7: voltage difference: 2.5 kV; average CE: 78.6%; particle size: 4-5 μm (b) trial 1; voltage difference: 2 kV; average CE: 75.8%; particle size: 5-6 μm (c) trial 15; voltage difference: 1.5 kV; average CE: 62.2%; particle size: 6-8 μm (d) trial 9; voltage difference: 1 kV; average CE: 63.5%; particle size: 8-10 μm. Detailed operating parameters for the respective trial numbers are shown in Table 7.

Residual DCM content in pharmaceutical particles

The low concentration of residual DCM on the particles could be achieved without freeze-drying because of the following reasons. Firstly, the distance from the nozzle tip to the edge of collecting plate is longer than the corresponding distance in the previous prototype of the chamber. Because of this longer distance, the droplets from the cone-jet at the nozzle tip can have more time to travel through the electric field to the collecting plate. Therefore the residual organic solvents in the droplets have more time to evaporate. Secondly, the collecting plate is designed in such a way that the particles stay on the collecting plate throughout the EHDA process. The particles deposit on the collecting plate in the cross-flow of flow of nitrogen in the encapsulation chamber. Consequently, the organic solvent on the particles would evaporate easily due to the prolonged exposure to the nitrogen cross flow.

It is predicted that the residual amount of organic solvent in the particles would depend on the nitrogen flow rate in the encapsulation process and would be independent of other factors. Different values of DCM concentration for the same nitrogen flow rates in the trials can probably be related to the different durations of time in the encapsulation process for the different trials. Fig. 6 shows that the concentration of DCM in particles decreases with increasing nitrogen flow rate from 20 to 25 l/min, but it increases with increasing nitrogen flow rate from 25 to 35 l/min and approaches to a constant value for higher nitrogen flow rates.

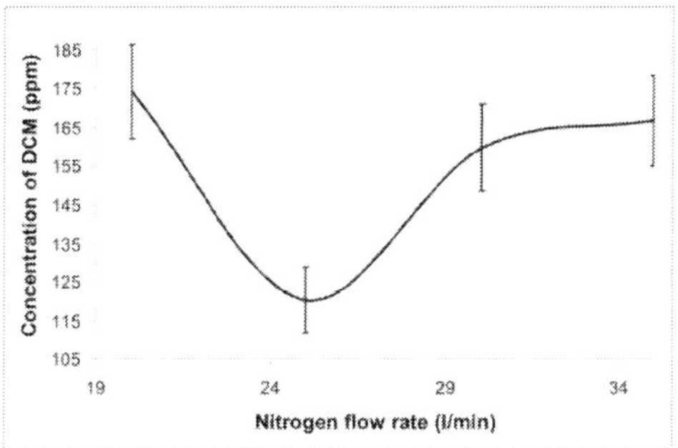

Figure 6: Variation of residual concentration of DCM in collected particles versus nitrogen flow rate.

Increase in nitrogen flow rate from 20 to 25 ml/min improves the convective mass transfer around the particles and expedites the evaporation of DCM from the particles without affecting the flight time of the particles in the encapsulation chamber. In contrast, the convective mass transfer around the particles is also improved with increasing nitrogen flow rates from 25 to 35 l/min but the flight time of the particles is drastically shortened and this results in the increase in the DCM concentration in particles.

CONCLUSIONS

A new generation of shuttle glass chamber was used in this study to fabricate biodegradable polymeric microparticles for drug delivery applications. This new shuttle glass chamber, in contrast to the previous prototypes, had a cylindrical shape with larger diameter of nitrogen inlet and outlet in which particles were collected inside the chamber. The effect of solution flow rate, nitrogen flow rate, nozzle and ring voltage on the particle collection efficiency and residual amount of organic solvent in collected particles were investigated. The experimental study was designed and analysed by the Taguchi statistical method. The results shown in table 5 verified that most effective factors had been considered. The Taguchi analysis showed that the most important factor affecting particle collection efficiency was the solution flow rate. Moreover, based on the quantitative analysis, ANOVA, nitrogen flow rate and ring voltage appeared to contribute the most to the results. An F-ratio less than 1 showed that the effect of the nozzle voltage could be regarded as insignificant and the important factors were listed in the following order: solution flow rate, nitrogen flow rate and ring voltage. ANOVA was also used to estimate the collection efficiency at the optimum condition of 1 ml/hr solution, 35 lit/min nitrogen and 6 kV ring voltage. The calculated collection efficiency, 82.04% was, within the range of the confidence limit of the measured value, 79.4%.

In addition to the quantified factors mentioned before, duration of the encapsulation process was considered as a minor adjustable factor which can also affect the electrical field in the chamber and therefore can influence the particle collection efficiency. Another minor factor, the low electrical conductivity of DCM, can also affect the cone-jet mode and thus affecting the particle collection efficiency. Residual solvent in collected particles was another issue perused in this study. Gas chromatography/mass spectrometry (GC/MS) was used to determine the residual content of DCM in the collected particles. For all the trials, the residual DCM content of the particles fabricated using the EHDA method was well within the safety standards (< 600 ppm) at the end of process without applying freeze-drying which is usually used after EHDA process. Long distance from the spray zone to the collecting plate is the main reason for achieving the low concentration of residual DCM in the collected particles. Furthermore, the design of the system for allowing particles to reach the collecting plate for a long exposure time throughout the encapsulation chamber is another reason for achieving low quantity of residual DCM in the collected particles. The findings of the present study clearly show that high collection efficiency and low concentration of residual solvent in collected particles can be achieved with good adjustment on the operating parameter of the EHDA encapsulation chamber.

ACKNOWLEDGEMENT

This work is supported by the National University of Singapore under the grant number R279-000-222-112.

REFERENCES

[1] Almekinders, J. C., Jones, C., 1999. Multiple jet electrohydrodynamic spraying and applications. Journal of Aerosol Science 30, 969-971.

[2] Tatemoto, Y., Ishikawa, R., Takeuchi, M., Takeshita, T., Noda, K., Okazaki, T., 2007. An electrospray methos using a multi-capillary nozzle emitter. Chemical Engineering and Technology 30, 1274–1279.

[3] Hartman, R. P. A., Brunner, D. J., Camelot, D. M. A., Marijnissen, J. C. M., Scarlett, B., 1999. Electrohydrodynamic atomization in the cone-jet mode; physical moeling of the liquid cone and jet. Journal of Aerosol Science 30, 823-849.

[4] Hemin, N., Wang, C. H., 2007. Fabrication and characterization of PLGA/HAp composite scaffolds for delivery of BMP-2 plasmid DNA. Journal of Controlled Release 120, 111–121.

[5] Park, S. K., Kim, K. D., Kim, H. T., 2002. Preparation of silica nanoparticles: determination of the optimal synthesis conditions for small and uniform particles. A: Physicochemical and Engineering Aspects 197, 7–17.

[6] Ciach, T., 2006. Microencapsulation of drugs by electro-hydro-dynamic atomization. International Journal of Pharmaceutics 324, 51–55.

[7] Hong, Y., Li, Y., Yin, Y., Li, D., Zou, G., 2008. Electrohydrodynamic atomization of quasi-monodisperse drug-loaded spherical/wrinkled microparticles. Journal of Aerosol Science 39, 525 – 536.

[8] Xie, J., Marijnissen, J. C. M., Wang, C. H., 2006a. Microparticles developed by electrohydrodynamic atomization for the local delivery of anticancer drug to treat C6 glioma in vitro. Biomaterials 27, 3321–3332.

[9] Xie, J., Lim, L. K., Phua, Y., Hua, J., Wang, C. H., 2006b. Electrohydrodynamic atomization for biodegradable polymeric particle production. Journal of Colloid and Interface Science 302, 103–112.

[10] Taguchi, G.,1987. System of Experimental Design. Engineering Methods to optimize quality and minimize costs, first ed. White Plains, New York.

[11] Grace J. M., Marijnissen, J. C. M., 1994. A review of liquid atomization by electrical means. Journal of Aerosol Science 25, 1005-1019.

[12] Hartman, R. P. A., Brunner, D. J., Camelot, D. M. A., Marijnissen J. C. M., Scarlett, B., 2000. Jet break-up in Electrohydrodynamic Atomization in the cone-jet mode. Journal of Aerosol Science 31, 65-95.

[13] Cloupeau, M., Prunet-Foch, B., 1994. Electrohydrodynamic spraying functioning modes: a critical review. Journal of Aerosol Science 25, 1021-1036.

[14] Lindblad N. R., Schneider, J. M., 1965. Production of uniform-sized liquid droplets, Journal of science and instruments 42, 635-638.

[15] Borra, J. P., Hartmann, R., Marijnissen, J., Scarlett, B., 1996. Destabilisation of sprays in the cone-jet mode by electrical discharges on the jet. Journal of Aerosol Science 27, Supp. 1, 203-204.

[16] Ding, L., Lee, T., Wang, C. H., 2005. Fabrication of monodispersed Taxol-loaded particles using electrohydrodynamic atomization. Journal of Controlled Release 102, 395–413.

Small-Angle Neutron Scattering On Hydrophobically Stabilized Metal Nanoparticles in Gas-Expanded Liquids

Gregory Von White II, Chemical and Biomolecular Engineering, Clemson University, Clemson, SC

Christopher L. Kitchens, Chemical and Biomolecular Engineering, Clemson University, Clemson, SC

We have examined the structural properties of alkane thiol stabilizing ligands using Small Angle Neutron Scattering (SANS) as a function of CO2 composition in Gas-Expanded Liquids (GXLs). SANS was used to determine the solvation effects of nanoparticle ligands as well as the effective ligand length as a function of the solvent properties. Recent advances in metallic nanoparticle synthesis and processing have opened the door for new opportunities with novel applications. For many of these nanoparticle applications, monodispersity is critical because of their size-dependent functionality. CO2 has been shown to work as an anti-solvent for organic solvent dispersed metal nanoparticles providing a means to size-fractionate them into monodisperse populations. Here, the stabilizing ligands play a major role in the synthesis and application of metallic nanoparticles and allow for dispersion in various solvents. Manipulation of the nanoparticle surface chemistry through synthesis or ligand exchange affords significant opportunity for solution based "green" processing and impacts their potential applications; the solvent - ligand interactions are deterministic of the nanoparticle behavior in solution. Previous work demonstrates the use of an interaction energy model to predict the dispersibility of nanoparticles in gas-expanded liquids (GXLs) as a function of the solvent - ligand interactions; however, questions remain regarding the stabilizing ligand structure.

See more of this Session: Characterization of Engineered Particles and Nano-Structured Particles
See more of this Group/Topical: Particle Technology Forum

Vibro-Spring Particle Sizer

Haroun Mahgerefteh, Chemical Engineering, University College London, London, United Kingdom

Aisha Salihu, Chemical Engineering, University College London, London, United Kingdom

Powder technology is a very important field in the world today. In fact over, 75% of all materials processed in industry are in particulate form, and knowledge of particle size is important to the manufacture, handling and application of many of these powders. (Merkus, 2009)

Several hundred instruments have been developed for the measurement of particle size distribution. Widely used techniques include microscopy, sieving and laser diffraction. However, the majority are either expensive, not robust, require complicated operating procedures or capable of handling only small amount of samples as a result raising sample representative problems.

In 1996, Mahgerefteh et al. developed a novel Spring Particle Sizer which overcomes most of the problems associated with conventional techniques. In the most basic form, the instrument comprises a horizontally held closed coil helical spring partly filled with powder; the size of the discharged particles is directly related to the spring extension.

The results of experimental investigations carried out on the Spring Particle Sizer are described. This task also includes a search for an optimal spring design in terms of uniform spring opening. The development of methodologies for characterising the spring performance in terms of uniform openings and reproducibility using image analysis is also considered. Test results are presented based on springs supplied by three different manufactures each fabricated to produce uniform extension. These investigations revealed that out of the three manufacturers, one (Airedale Ltd), produced the best performance when compared to sieves. However, these springs cannot be fabricated in a reproducible manner hence compromising the commercial potential.

A number of methodologies are presented in pursuit of obtaining sample mass within the sping in-situ to produce particle size distribution data. Data relating the pressure drop versus air flow rate are presented to deduce the mass of particles within the spring from the minimum fluidisation velocity. A viable correlation between the minimum fluidization velocity and the mass was not obtained. Thus, a more direct measurement approach is investigated, with the aid of a strain gauge.

Finally, the designs and development of a hand-held model of spring particle sizer is described next. Calibrations and performance evaluations are carried out using powder samples with different characteristics. The results show a close correspondence with sieve data, and thus the establishing the device's potential for quality control assessment in the powder industry.

References

1. Mahgerefteh H and Shaeri A Vibro-Spring Particle Size Analyser, PhD Thesis, University College London, 1996.

2. Merkus H.G. Particle Size Measurements: Fundamentals, Practice, Quality Series, Particle Technology Series, Vol 17, XII, p.536, Springer, 2009

See more of this Session: Characterization of Engineered Particles and Nano-Structured Particles
See more of this Group/Topical: Particle Technology Forum

Effect of Substrate Roughness On Nano-Silica Coated Micron-Scale Particle Adhesion

Dave Balachandran, Chemical Engineering, Purdue Univesity, West Lafayette, IN

Laila Jai Jallo, New Jersey Center for Engineered Particulates, New Jersey Institute of Technology, Newark, NJ

Rajesh Davé, Chemical Engineering, New Jersey Institute of Technology, Newark, NJ

Stephen P. Beaudoin, School of Chemical Engineering, Purdue University, West Lafayette, IN

Industrial and pharmaceutical particle manufacturing can include processes like particle flow, mixing, storage, fluidization, and compaction which are greatly affected by the cohesive forces of particles within powders and adhesive forces between powder particles and process surfaces. Dry coatings and fillers are sometimes added to a powder formulation to modify interaction forces by altering particle surface morphology and the chemical properties of the material in the area of contact. However, if the process surface is sufficiently rough, the effect of dry powder particle coatings on adhesion can be reduced. In this study, the adhesion of uncoated and nano-silica coated alumina particles to stainless steel substrates with different roughness regimes was measured using and atomic force microscope (AFM). Surface roughness and particle geometry were measured using an AFM and scanning electron microscope, respectively. The van der Waals force of adhesion was characterized and simulated using a predictive model.

See more of this Session: Characterization of Engineered Particles and Nano-Structured Particles
See more of this Group/Topical: Particle Technology Forum

Dispersion of Surface Modified Powders for Inhalation Application

Ajit Mujumdar, Otto H. York Department of Chemical, Biological and Pharmaceutical Engineering, New Jersey Institute of Technology, Newark, NJ

Xi Han, Otto H. York Department of Chemical, Biological and Pharmaceutical Engineering, New Jersey Institute of Technology, Newark, NJ

Daniel To, New Jersey Center for Engineered Particulates, New Jersey Institute of Technology, Newark, NJ

Rajesh Dave, Otto H. York Department of Chemical, Biological and Pharmaceutical Engineering, New Jersey Institute of Technology, Newark, NJ

Stefan Steigerwald, Systems of Particle Technology, SYMPATEC, Lawrenceville, NJ

George Foster, Systems for Particle Technology, SYMPATEC, Lawrenceville, NJ

Active pharmaceutical ingredients used in the inhaler application have aerodynamic diameters typically in the size range of 2 ~ 5μm. Such powders are extremely cohesive due to strong Vander Waal's forces and very difficult to flow, thus create severe problems for the effective dispersion for the treatment of respiratory diseases. In addition, the drug dosage uniformity is also a critical issue because of the formation of agglomerates. Excipients such as monohydrate lactose are commonly employed, which act as drug carriers in these applications. Understanding of interaction of drug particles with excipients is very important for the effective performance and depends on various parameters such as blending procedures, application of high/low shear, particle size distribution, properties of materials, etc. In this work, various tools to improve flow of powders through surface modification methods, which may be dry coating of original poorly flowing powders with nano silica (physically altering the surface structure), surface treatment (chemically altering the surface structure) or continuous grinding and coating techniques are presented. Specifically, application of a low shear and high shear blending using a hybridization system (Nara machineries Ltd.) is discussed to investigate the particle size reduction of the carrier lactose particles. The effect of various rotational speeds and the duration of processing time are investigated. In order to increase the dispersion and dissolution of drugs while maintaining the flowability of drugs, nano-size guest particles are coated on the surface of drugs (host particles) through a novel approach of simultaneous size reduction and surface modification by using Fluid Energy mill (FEM). The novelty of the method lies in the simultaneous coating and size reduction and use of hydrophobic as well as hydrophilic materials for surface modification. Various Sympatec dispersion instruments, including the Rodos and the Inhaler are utilized for careful characterization of the particle size as well as dispersibility of the surface modified particles. It is found that powders in size range 1-10 microns are produced with high API content (>95 %), having good dispersion and flowability, suitable for the inhaler application.

See more of this Session: Characterization of Engineered Particles and Nano-Structured Particles
See more of this Group/Topical: Particle Technology Forum

Microscale Surface Energy Measurements of SAM-Coated Particle Surfaces Using a Modified Flotation-Based Approach

Chrysanty Tedjo, Chemical and Biomolecular Engineering, Vanderbilt University, Nashville, TN

Paul E. Laibinis, Chemical and Biomolecular Engineering, Vanderbilt University, Nashville, TN

Flotation methods are widely used in metallurgical engineering for measuring the surface energy (or hydrophobicity) of fine solids such as coal dusts and various types of colloidal particles. In a typical flotation experiment, particulate materials are suspended in solutions of different surface tensions while air is bubbled into the mixtures. The fraction of particles that float in each solution is determined, often by physically separating these particles from the solution and determining their contribution by mass. The surface energy of the particles is obtained by classical approaches that determine a critical surface tension from the above experiments. Typically, these methods require large amounts of the particulate sample in order to achieve reliable results. Their suitability in assessing the properties of custom synthesized particles that are often available in small quantities has been limited. To address this gap, we have developed a flotation approach for measuring particulate surface energies using a spectroscopic method that then allows measurements on small amounts of particles. In our experiments, we used porous silica particles (5 μm of diameter) that have been functionalized with mixed self-assembled monolayers (SAMs) of Cl3Si(CH2)11(EG)3OCH3 (EG3OMe) and Cl3Si(CH2)7CH3 (C8). These mixed SAMs allowed the surface energy of the particles to be varied. The SAMs were similarly formed on flat silica surfaces, where contact angle measurements could be used for characterization, thereby allowing a direct comparison of the surface energetics for these mixed SAMs on the flat and particle surfaces. The result showed that this flotation method can be used to interrogate the surface energies of small quantities of functionalized particles.

See more of this Session: Characterization of Engineered Particles and Nano-Structured Particles
See more of this Group/Topical: Particle Technology Forum

Collective Behavior of Self-Propelled Microcapsules

Amitabh Bhattacharya, Chemical Engineering, University of Pittsburgh, Pittsburgh, PA

O. Berk Usta, Chemical Engineering, University of Pittsburgh, Pittsburgh, PA

Victor. Y. Yashin, Chemical Engineering, University of Pittsburgh, Pittsburgh, PA

Anna C. Balazs, Chemical Engineering, University of Pittsburgh, Pittsburgh, PA

We design a system of synthetic microcapsules that can mimic the behavior of cells. The microcapsules can 'signal' to each other via nanoparticles, and they can undergo motion via a self-induced adhesion gradient on the substrate where they are localized. Hydrodynamic interaction amongst the capsules plays a crucial role in initiating motion. For a train of N capsules moving with velocity U in a straight line, we show that U is proportional to $N^{-1/2}$. We illustrate how triggering of the initial motion can be controlled using temperature-sensitive capsules. We also present studies on the collective behavior of these self-propelled capsules when they are allowed to move unconstrained on a surface.

See more of this Session: Characterization of Engineered Particles and Nano-Structured Particles
See more of this Group/Topical: Particle Technology Forum

Mechanical Properties and Microstructure of the Nanoparticle Silica/Polymer Composite Coatings

Karan Jindal, Chemical Engineering and Materials Science, University of Minnesota, Minneapolis, MN

Lorraine F. Francis, Chemical Engineering and Materials Science, University of Minnesota, Minneapolis, MN

Alon V. McCormick, Chemical Engineering and Materials Science, University of Minnesota, Minneapolis, MN

Nanocomposite coatings containing ceramic nanoparticles and polymers are used in variety of applications such as paints, ink-jet print media, abrasion resistance coatings and optical coatings. Cracking from drying-induced stresses is frequently a problem in these coatings, which are prepared from dispersions of ceramic nanoparticles. The polymer is added to prevent cracking. The polymer provides strength to the porous coating, but changes the microstructure by filling in pore space. The polymer also affects the stress and the mechanical properties. Therefore, understanding the effect of polymer content on stress, microstructure and mechanical properties is necessary to create functional nanocomposite coatings. The goal of this research is to characterize microstructure, stress, and elastic modulus of nanocomposite coatings composed of silica nanoparticles and polyvinyl alcohol (PVA) polymer over the entire range.

Nanocomposite coatings were fabricated from suspension of nanosized silicon oxide particles (20 nm primary particle diameters within branched aggregates of 150 nm diameters) and polyvinyl alcohol (PVA) polymer (molecular weight range of 85,000-140,000 & degree of hydrolysis of 87.8%). The suspensions were deposited as coatings onto silicon substrates and dried at room temperature and nitrogen gas flow of ~ 3L/min. The final dried coating stress as measured by a cantilever deflection device indicated that the stress goes through maximum as the amount of polymer in the coating increases. The highest final stress was found to be ~ 110MPa at a PVA content of 60 wt%. The nanoparticle coating microstructure was characterized by imaging the dried samples using scanning electron microscopy (SEM). Porosity measurement using the nitrogen gas adsorption (micrometrics ASAP 2000) reveals that all the pores in the coatings are filled at 50wt% PVA content. The mechanical properties of silica nanocomposite coatings of varying composition were studied by the nanoindentation. The modulus of the composite coating increased as the polymer was added, reached maximum and decreased with further addition of polymer to the coating. The research demonstrates that the mechanical properties and the microstructure can be tuned be varying the relative amount of polymer in the silica nanoparticles coating. For example porous but strong and crack free nanocomposite coatings can be prepared with 30 wt% polymer. Furthermore, nonporous coatings with high modulus can be prepared with 50 wt% polymer content.

See more of this Session: Fluid-Particle Interactions and Processing
See more of this Group/Topical: Particle Technology Forum

Using Shear Flow to Assist the Coating of Microparticles with Nanoparticles

Kapil V. Deshpande, Department of Chemical and Biochemical Engineering, Rutgers University, Piscataway, NJ

Nina C. Shapley, Department of Chemical and Biochemical Engineering, Rutgers University, Piscataway, NJ

Previous studies involving dry powders and static aqueous solutions have demonstrated dramatic changes in aggregation and flowability properties when microparticles are coated with nanoparticles, particularly in the presence of significant charge interactions between the two particle types. In this work, we examine the influence of shear flow on the extent and morphology of coating during mixing of microparticles and nanoparticles of opposite charge in an aqueous solution. Shear flow is generally presumed to counteract aggregation and therefore may allow higher particle concentrations to be combined than in a static solution.

The experimental system involves cationic chitosan gel nanoparticles and anionic alginate microparticles mixed in a narrow-gap, concentric cylinder Couette cell. Optical and electron microscopy are used to characterize the resulting population of multiscale particles. The impact of the mixture composition and average shear rate will be discussed, as well as steady shear compared to oscillatory shear flow. Furthermore, using known shear-induced particle migration rates for the geometry, we can also explore tailoring the flow to create the particle segregation conditions that are best suited for the assembly of multiscale particles.

See more of this Session: Fluid-Particle Interactions and Processing
See more of this Group/Topical: Particle Technology Forum

Effect of Particle and Fluid Properties On the Pickup Velocity of Fine Particles

Kimberly H. Henthorn, Missouri University of Science and Technology, Rolla, MO

Systems involving fluid-particle flows are a key component of many industrial processes, but they are not well-understood. One important parameter to consider when designing a conveying system is pickup velocity, the minimum fluid velocity required for particle entrainment. Many theoretical and experimental analyses have been performed to better understand pickup velocity, but there is little consistency with regard to system conditions, fluid properties, and particle characteristics, which makes comparisons between these studies very difficult. Although the proper design of many conveying systems requires the utilization of expressions that are applicable across a broad range of operating parameters, most expressions are system specific, which means that they are not extendable to other conditions. Also, there is currently an absence of a universal expression to predict particle entrainment in both gases and liquids.

In this work, the pickup velocity of glass spheres, crushed glass, and stainless steel spheres in water have been measured for particles less than 450 microns. The effects of particle size, particle shape, and particle density will be discussed and compared to the pickup velocity trends previously determined for similar gas-phase systems. In addition, the experimental data have been used to assess an existing force balance model previously developed for gas-phase systems.

See more of this Session: Fluid-Particle Interactions and Processing
See more of this Group/Topical: Particle Technology Forum

SUCCESSES AND EMERGING ISSUES IN SIMULATING THE PROCESSING BEHAVIOR OF LIQUID-PARTICLE NUCLEAR WASTE SLURRIES AT THE SAVANNAH RIVER SITE – 205e

David C. Koopman, Savannah River National Laboratory, Aiken, SC
Daniel P. Lambert, Savannah River National Laboratory, Aiken, SC
Michael E. Stone, Savannah River National Laboratory, Aiken, SC

ABSTRACT

Slurries of inorganic solids, containing both stable and radioactive elements, were produced during the cold war as by-products of the production of plutonium and enriched uranium and stored in large tanks at the Savannah River Site. Some of this high level waste is being processed into a stable glass waste form today. Waste processing involves various large scale operations such as tank mixing, inter-tank transfers, washing, gravity settling and decanting, chemical adjustment, and vitrification. The rheological properties of waste slurries are of particular interest. Methods for modeling flow curve data and predicting the properties of slurry blends are particularly important during certain operational phases. Several methods have been evaluated to predict the rheological properties of sludge slurry blends from the data on the individual slurries. These have been relatively successful.

INTRODUCTION

Legacy radioactive high level waste (HLW), generated at the U. S. Department of Energy's Savannah River Site (SRS) from production of enriched uranium and plutonium during the Cold War, is currently being processed into a stable borosilicate glass waste form for long term storage. The majority of the waste is stored as a mixture of hydroxide and hydrous oxide insoluble solids in large cylindrical storage tanks at SRS. Over 90% of the waste solids are non-radioactive chemical byproducts derived from fuel rod targets and purification chemistry. The 3,500-4,900 cubic meter (~1,000,000 gallon) carbon steel waste storage tanks also contain 5-7M sodium solutions rich in hydroxide, nitrate, and nitrite anions that are contaminated with soluble radioactive isotopes of cesium and strontium. Insoluble solids have settled to the bottom of the tanks and have been aging for 25-50 years. Waste solids must be mobilized and transferred between tanks as the first step in waste treatment. Subsequent steps separate dissolved radionuclides from insoluble radionuclides for separate waste treatment and downstream processing.

A partial summary of processing issues includes:

1) Rheological properties of suspended waste solids degrade with time under shear (slurries becomes more viscous).

2) Slurry pump flow fields in the large, cooling coil filled, tanks have numerous stagnant zones where radioactive waste mounds form creating closure issues.

3) Settling times during tank washing operations cannot be accurately predicted. Settling times also do not always mimic those of small scale radioactive settling test samples.

4) Radiolytic hydrogen generation and bubble accumulation during gravity settling constrains washing volumes.

5) Chemical adjustment (pretreatment) of the waste solids upstream of the waste vitrification melter results in various issues such as excessive foaming, catalytic production of hydrogen from formic acid via noble metal fission product catalysis, occasional problems with tackiness or excessive air entrainment, loss of mixing and heat transfer during concentration, etc.

HLW slurry must first be moved from the large waste tanks prior to any processing to stabilize the waste for geologically long term storage. The insoluble sludge solids have settled and packed into a clay-like consistency over time. Local episodes of over-heating have also hardened some of the insoluble solids and formed moderately strong bridges between individual particles. Centrifugal slurry mixing pumps are lowered into the settled sludge layer in order to break down and mobilize insoluble solids for transfer. Four pumps are typically deployed to resuspend the solids, Figure 1.

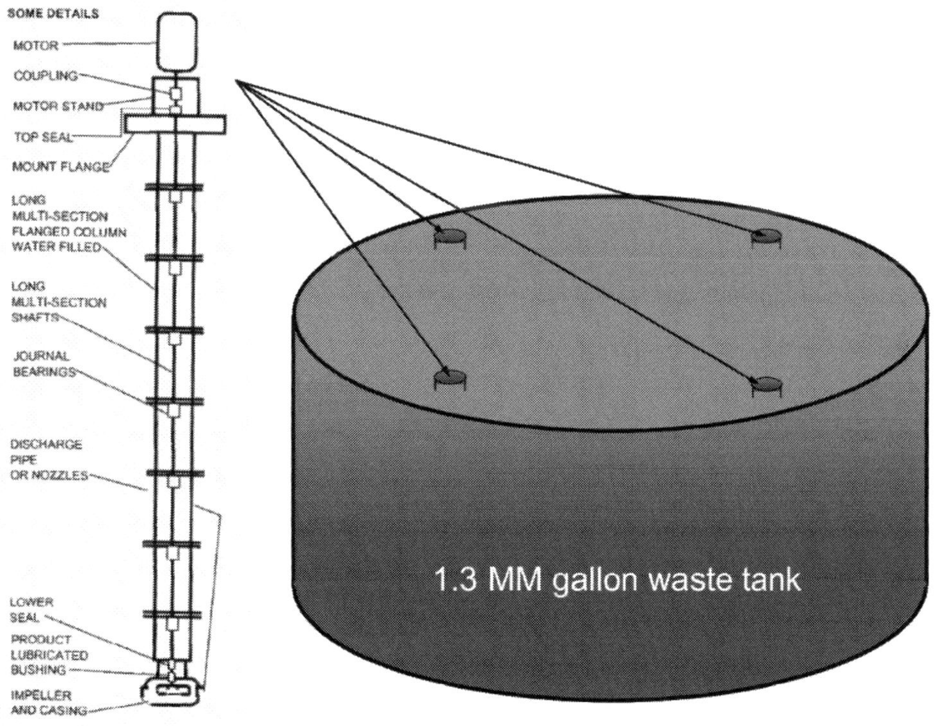

Figure 1. Placement of slurry mixing pumps in large waste storage tanks at SRS.

The Savannah River National Laboratory (SRNL) has a program that produces non-radioactive "simulants" or analogs of the waste slurries in the SRS tank farm. Simulant slurries are used to evaluate the potential feasibility of various proposed operations. Although simulant preparation chemistry generally follows that of the original radioactive wastes, the physical properties (rheology, particle size, cohesiveness, foaming tendency, etc.) often do not adequately match those of the actual wastes. The radioactive and simulant slurries generally behave like thixotropic fluids while being mixed or pumped. Typical particles sizes for the precipitated solids are less than 40 micrometers.

RESEARCH PROGRAM

HLW slurries in the SRS tank farm often behave like pseudo-homogeneous fluids. The insoluble solids concentration is kept high enough to keep particles in suspension long enough to perform typical processes such as inter-tank transfers. Significant progress has been made recently in producing rheologically similar simulants. Testing has shown that batch precipitated simulants tend to be more viscous than those produced using continuous stirred tank crystallizers. High shear mixers (such as the Silverson L4RT-A) have been used to increase the apparent viscosity of simulants that were less viscous than their corresponding radioactive target slurries. This can come at the expense of creating a mismatch in particle size distribution.

The latest advance has been the addition of an ultrasonicator (Vibracell VCX 750) coupled to a flow cell. The ultrasonicator has produced order of magnitude increases in the Bingham plastic yield stress of simulant using a short residence time, whereas the high shear mixers have typically produced factor of 2-3 increases in yield stress in relatively longer times. The continuous stirred tank crystallizer coupled with the ultrasonicator permits low yield stress slurries to be prepared initially. The rheological properties can then be adjusted using the ultrasonicator to match the properties of the corresponding radioactive waste samples. The remaining issue with this approach is to investigate the relative particle size distributions obtained and to determine if any significant processing properties have been lost if the particle size distributions are significantly different. A typical particle size distribution is given in Figure 2. Data were obtained on a MicroTrac instrument.

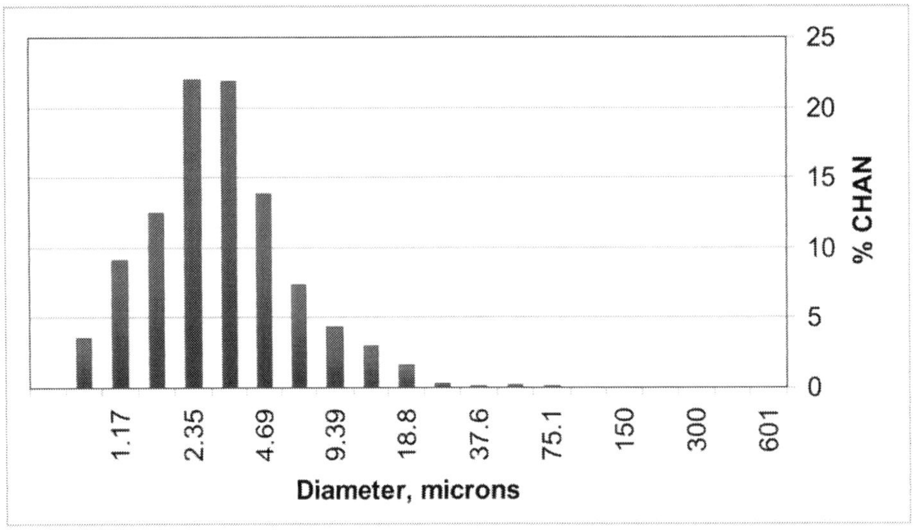

Figure 2. Typical particle size distribution of the insoluble solids in SRS waste slurries.

Often SRNL will receive individual samples of two or more slurries in the tank farm that are ultimately going to be blended together. Due to the difficulty and cost of pulling representative samples from the HLW tanks, there may be insufficient sample mass to characterize the individual slurries and then prepare blend samples for characterization. When this occurs, the properties of the blend often need to be calculated from the individual slurry properties. This is generally not an issue for chemical composition, but presents more of a challenge for certain physical properties such as rheology.

A fairly successful method for predicting the Bingham plastic yield stress and consistency of waste slurry blends has been developed. Rheometric shear stress-shear rate data were obtained for two slurries. Data are obtained using Haake rotational rheometers operated in the Searle mode (a vertical concentric cylinder geometry with a rotating bob in centered inside a fixed cup). SRNL has one Haake rotational rheometer that can be operated remotely inside one of the Shielded Cells where HLW samples can be handled.

The apparent viscosity of a non-Newtonian fluid is a function of shear rate. Apparent viscosity was selected as the mixing parameter rather than the fitted yield stress and consistency. Apparent viscosity, μ_i = shear stress/shear rate of slurry at a selected shear rate. SRS waste sludge feeds for the Defense Waste Processing Facility were expected to be in the range of 2.5-10 Pa in Bingham plastic yield stress and 4-12 cP in Bingham plastic consistency. In apparent viscosity versus shear rate these ranges are equivalent to the following, Figure 3.

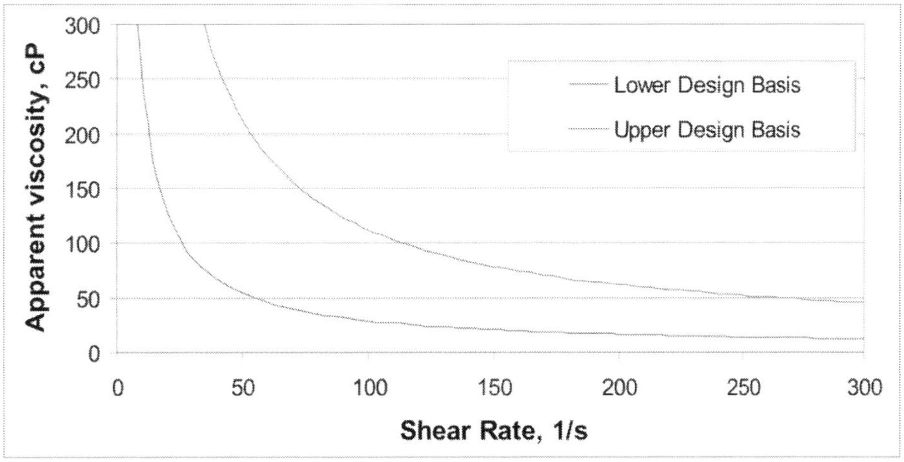

Figure 3. Rheological design bases for sludge slurries.

The empirical Kendall-Monroe rule was used locally on the individual apparent viscosities to produce the apparent mixture viscosity: $\mu_m^{1/3} = x_1\mu_1^{1/3} + x_2\mu_2^{1/3}$. The two x_i were taken as the mass fractions of the two slurries being blended. An example fit of the measured flow curve of a blend of two radioactive waste slurries to that predicted from adaptation of the Kendall-Monroe equation is shown below, Figure 4.

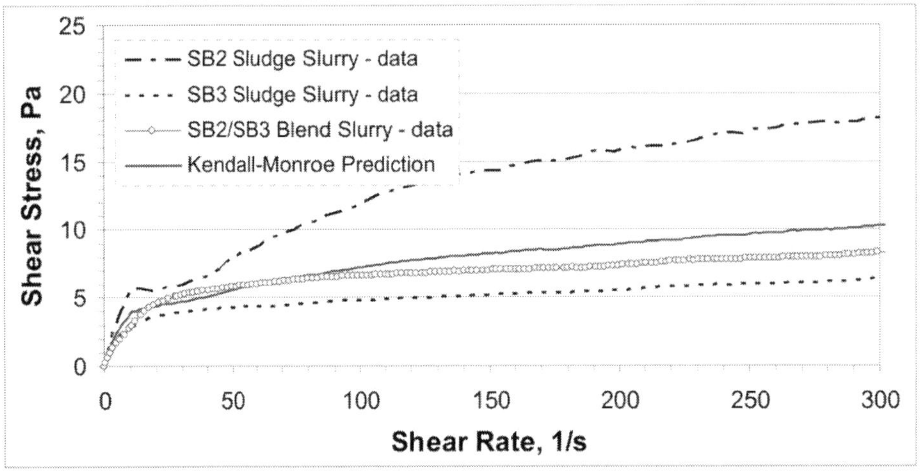

Figure 4. Application of the Kendall-Monroe viscosity blending algorithm to HLW.

The rheological data are obtained as shear rate-shear stress data points (γ, τ). Each curve in Figure 4 contains approximately 300 data points. At each shear rate data point, $\mu_i = \tau_i/\gamma_i$ is found for the two starting slurries and for the mixture. The prediction curve in Figure 3 came from approximately 300 applications of the Kendall-Monroe model, one for each shear rate in the data. The predicted curve was particularly good for shear rates from 0-100/s. The predicted curve from 100-300/s is biased high relative to the measured values, but is not off by much more than the normal reproducibility of flow curve data obtained on radioactive samples using the remotely manipulated equipment in the SRNL Shielded Cells.

The Olney-Carlson model for predicting mixture viscosities was also applied to the apparent viscosities of the waste slurries at each shear rate. The Olney-Carlson model is given by: $\mu_m = \mu_1^{x1} + \mu_2^{x2}$. The results of applying the Olney-Carlson model at each shear rate in a method analogous to that used in the Kendall-Monroe prediction are given in Figure 5.

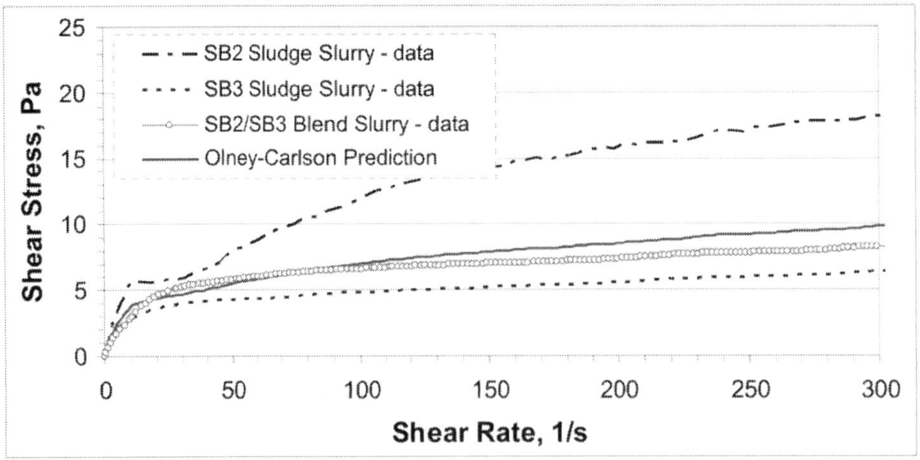

Figure 5. Application of the Olney-Carlson viscosity blending algorithm to HLW.

The Olney-Carlson algorithm was comparable to the Kendall-Monroe equation at low shear rates and slightly better at higher shear rates. Comparable agreements have been obtained on several subsequent slurry blend systems.

An alternative perspective on the model performance in predicting the measured values can be obtained by plotting the apparent viscosity versus the shear rate. Figure 6 shows the values obtained on the blended waste sample compared to the two empirical viscosity mixing models.

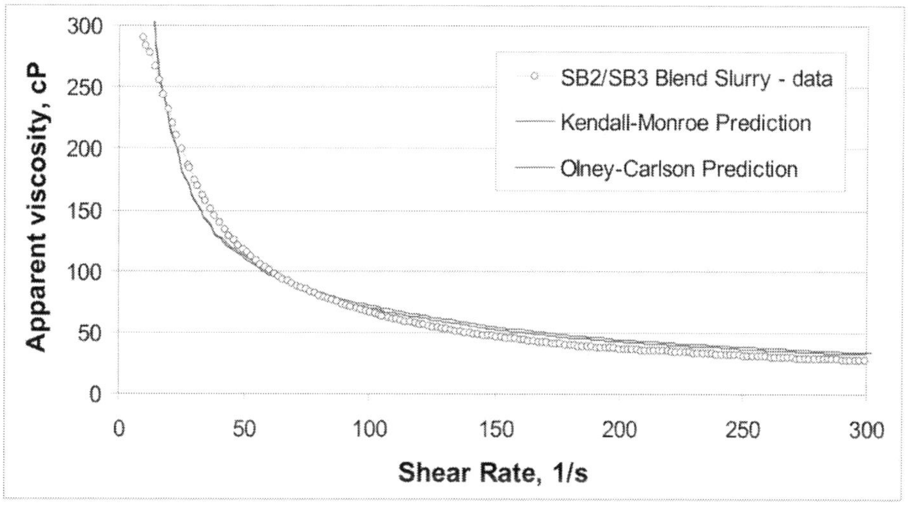

Figure 6. Comparison of predicted to measured apparent viscosity.

The results in Figure 6 indicate that the predicted flow curve data will be adequate for most engineering applications such as predicting pressure drop in pipe flow.

One on-going R&D program is investigating rheology modifiers. Pre-screening and preliminary experiments are all being conducted with non-radioactive simulants of the HLW slurries. Both ionic and non-ionic surfactants are being investigated. The challenges in this task include the variations in the chemical composition of individual waste tanks, the changes in pH during washing and chemical pre-treatment of the waste prior to vitrification, and the impact of adding significantly larger glass frit particles to the waste slurry just prior to the melter. Another issue is the use of high temperature sodium hydroxide treatments to leach much of the aluminum content of the insoluble solids into the aqueous phase for eventual separation from the bulk of the insoluble species.

One final tank farm issue is the preparation of former HLW tanks for ultimate closure. The slurry pumps in Figure 1 do not mobilize all of the insoluble solids in a given tank. Large cooling coil banks inside the tanks create obstructions to flow and local low velocity regions where solids continually settle out. The current program in this area is based on the use of oxalic acid solutions to break down the residual solids through dissolution. Once a tank has been sufficiently cleaned, the tank and internal coil are both to be filled with grout to immobilize any remaining material.

The bulk of the insoluble HLW is processed through the Defense Waste Processing Facility. Chemical pre-treatment is performed there to convert the waste to an acidic condition that permits separation of mercury which is present at concentrations that are not stable in glass. A combination of waste slurry with glass frit is ultimately fed into a cylindrical, joule-heated melter to produce a stable borosilicate glass waste form. The water content of the feed impacts melter throughput, but it is often constrained by processing issues such as loss of mixing and heat transfer during the final concentration by boiling. The impact of glass frit on rheological properties is a separate issue from the pseudo-homogeneous behavior of the waste slurry. The mechanism of rheological properties modification is more like the effect obtained when adding small glass beads to water. These issues were presented in a separate talk at the 2009 AIChE Annual meeting.

REFERENCES

Kendall and Monroe, *Journal of the American Chemical Society*, **43**, 115 (1921).

Olney and Carlson, *Chemical Engineering Progress*, **43**, 473 (1947).

Koopman, D., B. Pickenheim, D. Lambert, and M. Stone, "Successes and emerging issues in simulating the mixing behavior of liquid-particle nuclear waste slurries at the Savannah River Site", Proceedings of the 2009 Annual AIChE Meeting, New York, NY: AIChE. 2009.

How Intrusive Are Our Intrusive Probes?

Ray A. Cocco, Particulate Solid Research, Inc., Chicago, IL

John G. Findlay, Particulate Solid Research, Inc., Chicago, IL

Ted M. Knowlton, Particulate Solid Research, Inc., Chicago, IL

Many of the probes used to understand hydrodynamics in risers are intrusive. In other words, these probes are interacting with the very environment being measured. Pitot tubes, extraction probes, optical fiber probes, γ-ray densitometer probes, capacitance probes, etc. are all intrusively inserted into the riser flow stream to obtain radial profiles. Yet, the assumption they are not intrusive is almost always made. PSRI has found that this could be a poor assumption. Fiber optic probe measurements were found to consistently over predict the solids concentration compared to local ΔP/L values obtained from differential pressure measurements in the fully developed flow region of a 12-inch (0.3-meter) diameter by 60-foot (18-meter) tall riser containing FCC catalyst powder. The magnitude of this difference was sensitive to both superficial gas velocity and solids circulation rates. Barracuda™ CFD simulations confirmed this, and suggested that particle momentum is lost at the bottom of the probe and particles are spilling over (deflecting) to the sensor face. Based on these observations, new probe designs are proposed.

See more of this Session: Fundamentals of Fluidization - II
See more of this Group/Topical: Particle Technology Forum

Effect of Column Height and Diameter On Minimum Fluidization Velocity

Akhil A. Rao, University of Florida, Gainesville, FL

Jennifer Sinclair Curtis, University of Florida, Gainesville, FL

Bruno C. Hancock, Pfizer Inc., Groton, CT

Carl R. Wassgren, Purdue University, West Lafayette, IN

Reducing the size of the fluidized bed is gaining interest because it requires small quantities of material, better gas distribution and good operability. Experiments show that minimum fluidization velocity of particles increases as the diameter of the column is reduced or if the height of the fixed bed is increased. These effects are related to the effect of the wall. A new semi-correlated model is proposed which incorporates Janssen's wall effect in Ergun's pressure drop equation to the calculation of the minimum fluidization velocity. The predictions of the model are compared to some existing correlations and experimental data.

See more of this Session: Fundamentals of Fluidization - II
See more of this Group/Topical: Particle Technology Forum

Effects of System Pressure and Imposed Solids Mass Flux on Gas Bypassing in Deep Fluidized Beds of Group A Materials

Allan Issangya, Reddy S. B. Karri, Ted M. Knowlton and Ray Cocco
Particulate Solid Research, Inc., Chicago, IL

Bubbling and turbulent beds are widely used in commercial processes because of their intimate gas-solids contacting. Fluidized beds have high mass transfer rates and rapid solids mixing that leads to a nearly homogeneous temperature in the bed. However, recent studies conducted in Plexiglas™ units have revealed that significant bypassing of the fluidizing gas can occur in tall, fluidized beds containing Geldart Group A particles. Typically, the gas was observed to preferentially flow in one or more rapidly-moving streams of bubbles near the wall. The remainder of the bed was a mass of defluidized solids. If gas bypassing occurs in industrial beds, such as FCC regenerators, FCC strippers or in CFB combustors, it can result in afterburning in the freeboard, poor gas/solids contacting, poorly fluidized entrances to standpipes and the discharge regions of cyclone diplegs.

The gas bypassing phenomenon has been rarely reported in the literature, because laboratory beds are generally not tall enough and commercial units have limited instrumentation. Experimental and CFD work at PSRI have found that gas bypassing is due to increased gas compression at the bottom of tall fluidized beds. Reduction in the solids phase permeability at high gas compression ratios results in a biasing of the gas flow - even with proper grid and sparger designs. Previous studies have shown that gas bypassing can be eliminated or significantly reduced by increasing the fines content, increasing the gas velocity, lowering the bed height or installing properly-spaced horizontal baffles in the entire fluid bed.

Commercial fluidized beds often operate with solids flowing through the bed. PSRI has found that imposing a solids flux on the bed can promote gas bypassing. Solids fluxes over the range of 0 to 80 kg/s-m^2 were imposed on a 0.9-meter diameter fluidized bed by continuously recycling the solids around the test unit. A Computational Fluid Dynamic (CFD) analysis of the imposed flux testing using the Barracuda™ CFD code also showed that gas bypassing was more likely to occur at higher solids fluxes.

Increasing system pressure from atmospheric to 280 kPa was also found to lower or eliminate gas bypassing in a 0.6-meter diameter test unit. In both types of tests, gas bypassing was determined using traversing bubble probes and by high-frequency monitoring of the differential pressure fluctuations in the fluidized bed. This paper discusses the effects of pressure and imposed solids flux on gas bypassing in deep fluidized beds of Group A materials.

Validation of 3-D ECVT Velocimetry in Multi-Phase Flow Systems

Fei Wang, William G. Lowrie Department of Chemical and Biomolecular Engineering, The Ohio State University, Columbus, OH

Qussai Marashdeh, William G. Lowrie Department of Chemical and Biomolecular Engineering, The Ohio State University, Columbus, OH

W. Warsito, School of Physics, University of Indonesia, Indonesia

L.-S. Fan, William G. Lowrie Department of Chemical and Biomolecular Engineering, The Ohio State University, Columbus, OH

Real-time 3-D imaging of multi-phase flow systems and velocimetry tracking play an important role in understanding hydrodynamics phenomena, reaction modeling, and CFD verification in such systems. Electrical capacitance volume tomography (ECVT) is a three-dimensional real-time non-intrusive technique for fast volume imaging of multi-phase flow systems that provides three-dimensional real-time phase concentrations (solids, liquid, and gas). In this study, an ECVT system is applied to provide 3-D images of a multi-phase flow system. The real-time ECVT volume images contain quantitative information of solids concentration distribution of the flow. A velocity extraction method based on the 3-D ECVT quantitative images is applied to track local average volume velocities of solids in the multi-phase flow system. The ECVT velocimetry calculations are verified using fiber optic probe system, which is used to measure the local solids concentration and solids velocity.

See more of this Session: Fundamentals of Fluidization - II
See more of this Group/Topical: Particle Technology Forum

Characterization of Gas-Solid Fluidization for Fine Powders - A New Characteristic Fluidization Gas Velocity

Qing Huang, Department of Chemical and Biochemical Engineering, University of Western Ontario, London, ON, Canada

Jingxu Zhu, Department of Chemical and Biochemical Engineering, The University of Western Ontario, London, ON, Canada

H. Zhang, The University of Western Ontario, London, ON, Canada

Fluidization of over 100 Geldart group C and A/C powder samples with various flow properties were studied by measuring pressure drops and bed expansion ratios with respect to gas velocities. It was found that bed expansion ratio associates closely with the fluidization of fine powders. As a result, a new characteristic velocity, breakthrough fluidization gas velocity (Ubf), was proposed to demarcate the transition between poor and smooth fluidization. Ubf was defined as the gas velocity at which the bed expansion ratio has the highest rate of increase. Therefore Ubf can be identified through locating the inflection point on the plot of bed expansion ratio with respect to superficial gas velocity. Ubf is different from other characteristic gas velocities because it is developed from bed expansion while others are based on pressure drop measurement and/or force balance. Ubf is able to characterize the fluidizability of Geldart group C and A/C powders well, but not necessarily for groups A, B and D powders. The validity has been confirmed by comparing it with angle of repose. Based on Ubf, it is promising to develop a practical and applicable powder characterization technique, which is desirable for applications where the fluidization of fine powders is required.

See more of this Session: Fundamentals of Fluidization - II
See more of this Group/Topical: Particle Technology Forum

Effect of Carrier Morphology On Drug Loading and in Vitro Aerosolization and Deposition Properties in Dry Powder Inhalation

Meer Saiful Hassan, School of Chemical and Biomedical Engineering, Nanyang Technological University, Singapore, Singapore

Raymond Lau, School of Chemical and Biomedical Engineering, Nanyang Technological University, Singapore, Singapore

For dry powder inhalation, large carrier particles are normally mixed with smaller sized active ingredients (API) to improve their inhalation properties. The large carrier particles would experience lower interaction force and can be dispersed from the inhaler easily. As the particle mixture is inhaled, drug particles would detach from the carriers in the initial lung airways and travel further to the lower airways. Therefore, the physical and geometrical properties of the carrier particles are important for efficient drug delivery.

Lactose (LA) particles are commonly used as carriers in dry power formulations. They are of rocklike shape and have limited drug loading capacity. In this study, biocompatible and biodegradable hydroxyapatite (HA) particles with pollen-like surface are proposed as drug carriers. The pollen-like surface provides high drug loading capabilities. The depositions of drugs are studied in vitro and the performance of HA carriers are compared against LA carriers. Budesonide (Bd) is chosen as the model drug and it is blended with different carriers with carrier to drug weight ratio of 2:1, 10:1 and 45:1. Formulations with HA carriers show good blending homogeneity and stability. The characterization of the formulations shows that HA carriers are capable of higher drug loading than LA carriers. A large amount of the Bd particles are found unattached in the formulations with LA carriers. In the in vitro deposition study, experiments are conducted at 30 and 60 liter/min via Rotahaler® using an Anderson Cascade impactor (ACI). Formulations of HA and LA carriers with different drug loadings are tested. At a flow rate of 30 liter/min, the unattached Bd particles from LA carriers are found to coagulate with each other and stick to the surface of the container and inhaler. The emitted dose (ED) and fine particle fraction (FPF) of Bd using HA carriers are higher than that using LA carriers with all blending ratios. As the flow rate is increased to 60 liter/min, even though the ED and FPF of Bd using HA carriers are still higher than that using LA carriers, the difference between the two carriers are less significant. The findings of this study confirm that the delivery efficiency of inhaled therapeutics can be improved by introducing a unique pollen-like morphology of carrier particles which would reduce the drug loss at higher drug loading and improve the aerosolization and deposition properties significantly.

See more of this Session: Synthesis, Characterization and Modelling of Nanoparticle Systems with Pharmaceutical Applications
See more of this Group/Topical: Particle Technology Forum

Atomistic Simulation Studies On the Stability of Fenofibrate Nanocrystals in An Aqueous System by the Addition of Surfactants, Polymers, and Surfactant-Polymer Mixtures

M. Silvina Tomassone, Rutgers University, Piscataway, NJ

Wusheng Zhu, Chemical and Biochemical Engineering, Rutgers University, Piscataway, NJ

Frank Romanski, Chemical and Biochemical Engineering, Rutgers University, Piscataway, NJ

In the pharmaceutical industry high-throughput screening and combinatorial chemistry are routinely used to select good drug candidates. A large number of poorly water soluble drugs are typically rejected due to poor bioavailability. Nanosizing the drug increases its bioavailability but typically nanocrystals agglomerate and lose their stability. It is therefore of crucial importance to develop methods that help maintain a stable nanocrystal suspension system without agglomeration. Over the past few decades, a number of different methods have been developed for this purpose. A widely used technique is to add surfactants or water-soluble polymers to the drug suspension system.

It is generally believed that surfactant-coated or polymer-coated drug nanocrystals are less likely to agglomerate due to the strong quenching effects of the additives. However, experiments found that not all kinds of surfactants or polymers work well to stabilize drug suspension systems, and the effectiveness of stabilizers is typically analyzed by trial and error. In this work atomistic simulations are employed to study the stability of a Fenofibrate nanocrystal suspension system by adding surfactant, polymers, and their mixtures. Fenofibrate is a poorly water soluble drug used to reduce cholesterol levels in patients at risk of cardiovascular disease. Various surfaces (100), (010) and (001) of the drug nanocrystal are cleaved for studies. The attachment energies between a specific drug crystal surface and the pure solvent as well as between the drug and additives are studied. Our systems consist of the Fenofibrate molecule, the surfactants Tween 80 and Sodium Dodecyl Sulfate (SDS), as well as the polymers Pullulan and Hydroxylpropyl Methyl Cellulose(HPMC) in an aqueous system. We perform a systematic study of the two selected drugs in presence of Tween 80, SDS, Pullulan or HPMC, and for different polymer-surfactant mixtures. Our results show that addition of Tween 80 or SDS, or the polymer (Pullulan or HPMC) substantially enhance the stability of Fenofibrate nanocrystal in an aqueous system. Each additive has different quenching capabilities on different crystal faces. HPMC is overall the most efficient sole stabilizer, while Tween 80 shows more efficiency to block the growth of Fenofibrate on the surface (010). For the mixtures of the polymer (Pullulan or HPMC) and the surfactant (Tween 80 or SDS), we observe a synergistic effect between polymers and surfactants. Only in a few cases such as side-by-side configuration of the mixtures of Pullulan-SDS, Pullulan-Tween 80, and HPMC-Tween 80 on Fenofibrate crystal surface (010), the stabilization capability is not improved. Our analysis also shows that both geometry and chemistry factors between the polymer, the surfactant, and the drug crystal surface play important roles on the stability enhancement of the suspension system. Our simulation results are consistent with recent experimental findings at NJIT.

See more of this Session: Synthesis, Characterization and Modelling of Nanoparticle Systems with Pharmaceutical Applications
See more of this Group/Topical: Particle Technology Forum

Precipitation of Poorly Water Soluble Drugs Using Ultrasound and Stabilizers: Controlling Process Parameters

Sameer Dalvi, New Jersey Institute of Technology, Newark, NJ

Rajesh Davé, Chemical Engineering, New Jersey Institute of Technology, Newark, NJ

Precipitation of ultra-fine particles of poorly water soluble BCS class II drugs is attempted to improve bioavailability. A technique of liquid antisolvent (LAS) precipitation is chosen as it offers flexibility to control the particle size and distribution by manipulating physicochemical properties of solution and antisolvent phases using the additives. The objective of this work is to develop better understanding through a systematic and detailed study of a potentially scalable, LAS precipitation process for preparation of stable aqueous suspensions of ultra-fine particles of poorly water soluble active pharmaceutical ingredients (APIs). A combination of jets, ultrasound, and stabilizers has been used to achieve Damkohler number (Da) less than 1, which is a ratio of mixing time and precipitation time, for precipitation and stabilization of ultra-fine particles of six different APIs. Use of ultrasound and high stream velocities enhances micromixing and decreases mixing time, whereas addition of stabilizers increases precipitation time by delaying nucleation and inhibiting the particle growth. Particles with mean size ranging from 500 nm- 10 µm have been precipitated. A rationale understanding of precipitation is developed based on governing material and process parameters. The favorable process conditions have been identified to precipitate and stabilize submicron particles.

See more of this Session: Synthesis, Characterization and Modelling of Nanoparticle Systems with Pharmaceutical Applications
See more of this Group/Topical: Particle Technology Forum

PLGA-Zwitterionic Block Copolymers: Synthesis, Characterization, and Application for Drug Delivery

Zhiqiang Cao, Department of Chemical Engineering, University of Washington, Seattle, WA

Shaoyi Jiang, Chemical Engineering, University of Washington, Seattle, WA

Stability and functionability are two key issues for nanoparticle (NP) -based pharmaceutical applications. At present, polyethylene glycol (PEG) is widely used as coating materials for this purpose. For example, poly(D,L-lactic-co-glycolic acid) (PLGA) -PEG copolymers are developed to encapsulate hydrophobic drug into NPs with targeting ligands immobilized on the NP surface. Recently, it is found that zwitterionic poly(carboxybetaine) (pCB) has good stealthy properties like PEG. Furthermore, pCB has an abundance of functional groups (each monomer has one COOH group in its side chain) for ligand functionalization while each PEG polymer has only one functional group (e.g., COOH) at its end. These unique features render pCB a promising NP coating platform to solve existing stability and multi-functional problems in pharmaceutical industry.

In this work, PLGA is chosen as base materials to package drug for controlled release purposes. PLGA-b-PCB is synthesized and is used to form NPs (with docetaxel, a model drug) via nanoprecipitation method. A novel synthetic route and modified nanoprecipitation protocols are used to fabricate copolymers and NPs, respectively, due to the huge difference in polarity between hydrophobic PLGA and zwitterionic PCB. The resulted NPs are well controlled in sizes (e.g. 150 nm) and super stable in biological medium. Furthermore, The NPs can efficiently retain the sizes after high-speed centrifugation and even after lyophilization, only with brief re-suspension (no sonication required). This superiority can hardly be achieved by any existing NP systems (e.g. PLGA-PEG NPs) without additives such as 10% sucrose. Drug loading and releasing profiles are also studied to tailor PLGA-PCB NPs for pharmaceutical applications. How zwitterionic nature contributes to the super stability and influences drug delivery behaviors are also discussed.

See more of this Session: Synthesis, Characterization and Modelling of Nanoparticle Systems with Pharmaceutical Applications
See more of this Group/Topical: Particle Technology Forum

Polymer Stabilized Drug Nanoparticles Via Flash Nanoprecipitation

Zhengxi Zhu, Chemical Engineering and Materials Science, University of Minnesota, Minneapolis, MN

Haitao Qian, Chemistry, University of Minnesota, Minneapolis, MN

Thomas R. Hoye, Chemistry, University of Minnesota, Minneapolis, MN

Christopher W. Macosko, Dept. of Chemical Engineering and Materials Science, University of Minnesota, Minneapolis, MN

Nanoparticles of highly insoluble drugs can be created by flash nanoprecipitation, FNP. Turbulent mixing combined with a polymer stabilizer can produce particles in the 100 nm size range, ideal for passive targeting of tumors and advantageous to increase solubility. Biodegradable block copolymers (e.g., PLGA-b-PEG) or polyelectrolytes (e.g., chitosan) were used to stabilize beta-carotene particles. Particle size and distribution were measured by dynamic light scattering and particles were imaged via scanning electron microscopy (SEM) and cryogenic transmission electron microscopy (cryo-TEM). Drug loading capacity of at least 83 wt% (drug loading ratio carotene/stabilizer = 5/1) was achieved. X-ray diffraction (XRD) patterns showed that beta-carotene was trapped in the amorphous state.

See more of this Session: Synthesis, Characterization and Modelling of Nanoparticle Systems with Pharmaceutical Applications
See more of this Group/Topical: Particle Technology Forum

Controlled Liquid Antisolvent Precipitation of Poorly Water Soluble Drugs Using a Scalable Rapid Mixing Device

Christian Beck, Department of Chemical, Biological and Pharmaceutical engineering, New Jersey Institute of Technology, Newark, NJ

Sameer Dalvi, New Jersey Institute of Technology, Newark, NJ

Rajesh Davé, Chemical Engineering, New Jersey Institute of Technology, Newark, NJ

Nano/micro-particulate formulation is a promising way to improve bioavailability of poorly water soluble BCS class II drugs. Among the various techniques available for production of ultra-fine particles, liquid antisolvent (LAS) precipitation offers flexibility to control the particle size and distribution by manipulating physicochemical properties of solution and antisolvent phases using the additives. Particle formation by LAS involves two steps, namely the mixing of solution-antisolvent streams to generate supersaturation and the precipitation (which involves nucleation and growth by coagulation and condensation) of the particles. Thus, there are two main time scales associated with it, mixing time (τ_{mix}) and precipitation time or induction time ($\tau_{precipitation}$). Dimensionless Damkohler number (Da) is the ratio of τ_{mix} and $\tau_{precipitation}$. Value of Da >1 implies the mixing controlled process whereas Da < 1 implies that the precipitation step controls the process. Under a condition of Da < 1, mixing time is significantly less; mixing is uniform, and particle size is not affected by the mixing conditions. This produces rapid and uniform supersaturation and results in the precipitation of ultrafine particles. Thus, it is necessary to reduce τ_{mix} and increase $\tau_{precipitation}$ in order to keep Da < 1. In this work, we demonstrate the use of an ultrasonically driven T-shaped mixing device to precipitate ultra-fine particles. The greater control over particle size and its distribution has been achieved through enhanced micromixing and stabilization using various polymers and surfactants. Analysis of micro-mixing and stabilization has also been carried out to identify the extent of influence of above two steps and develop a methodology to control particle size and size distribution. The methodology developed has been illustrated for precipitation of ultra-fine particles of multiple APIs in the size range of 100 nm - 10 µm as a function of material and process parameters.

See more of this Session: Synthesis, Characterization and Modelling of Nanoparticle Systems with Pharmaceutical Applications
See more of this Group/Topical: Particle Technology Forum

Nanocarriers for Pressurized Metered-Dose Inhalers: Formulation, Aerosol Performance, and in-Vitro Transport Studies

Balaji Bharatwaj, Chemical Engineering and Materials Science, Wayne State University, Detroit, MI

Sandro R.P. da Rocha, Wayne State University, Detroit, MI

Oral inhalation is a promising alternative to traditional routes of administration for the regional (lung) and systemic delivery of therapeutics. Even greater opportunities may lie ahead if recent advances in nanomedicine can be translated to such non-invasive delivery routes. Despite recent advances in the formulation of both small molecules and therapeutic biomolecules for pulmonary delivery, our knowledge of the interaction of these inhaled formulations with the lung is very limited. The formulation of nanotherapeutics in the propellants used in pressurized metered-dose inhalers (pMDIs) is also very challenging, given the size range required for optimum deep lung deposition - particles smaller than 0.5 μm may be efficiently exhaled, while those larger than 4 μm will be deposited preferentially in the mouth and throat.

In this work we discuss the formulation of a general platform for the delivery of nanocarriers (NCs) to the lungs using pMDIs. The physical stability (in the propellant) and the aerosol performance of the NC-based formulations will be discussed, as well as the fundamentals leading to the ability to generate stable dispersions in pMDIs. The interaction of the NCs with a model lung epithelial membrane in vitro, and the transport of solutes across the epithelial barrier will be also discussed.

Key Words: metered-dose inhalers, pMDIs, oral inhalation, nanocarriers, in vitro cellular transport, nanocarrier cell interaction.

See more of this Session: Synthesis, Characterization and Modelling of Nanoparticle Systems with Pharmaceutical Applications
See more of this Group/Topical: Particle Technology Forum

UNDERSTANDING PARTICLE BREAKAGE USING POPULATION BALANCE MODELING AND GRAIN TEXTURE ANALYSIS

Dr. Kerry Johanson, Material Flow Solutions Inc., Gainesville, FL, 32607

Introduction

There is a movement in industry to create large size agglomerates of a high degree of porosity and large surface areas where nano-particles on the surface perform various functions. Large particles are preferred due to the lack of cohesive flow problems. However, large particles have low surface area and, hence, limited chemical functionality. Herein lies the dilemma. One solution to this particle design process quandary is to use an agglomeration process that creates a large porous particle from a series of small nano-particles. Unfortunately, these engineered materials are generally both highly porous and friable, resulting in significant degradation during use. Degradation produces fines that limit the life of catalyst in fluid bed operations, causes fines production in pharmaceutical products, and causes consumer acceptance issues with many consumer products from food to bath products. Whatever the reason, degradation of particles is a problem in today's production facilities.

It is critical to match the degradation measured by a particular method to the type of degradation present in the unique plant production facility. For example, if impact during processing dominates the process flow behavior, then degradation due to impact behavior is the critical property to measure. Using a degradation tester that causes size reduction due to stress and strain behavior may give erroneous results when applied to processes which are impact dominated. To mitigate degradation of a given product, one must understand the relationship between structure and the causes of degradation. Population balance models coupled with particle structure analysis can help explain degradation problems.

Although these degradation principles are common to many industries, we have chosen to examine bran cereal as an example product. Two types of bran particles are considered in this paper. We performed degradation analysis on two bran agglomerates: rounded particles measuring $\pm$ 4500 microns with an aspect ratio of 1:1 (referred to as pearls) and 2) extruded bran product with a diameter of $\pm$2200 microns and a length $\pm$14,000 microns (bran fibers).

Determining degradation mechanism for bran pearls

Particles, especially agglomerates, consist of a set of smaller sized particles held together by inter-particle or crystalline forces. The strength of the grain boundaries is one governing factor that limits particle size degradation. To understand this from a theoretical point of view, one requires a model describing the breaking of particles from a structural approach. However, measurement of particle breakage mechanisms is equally important. Therefore, it would be helpful to understand the fundamental mechanisms that create smaller particles from a larger one. Population balance modeling provides a means of identifying critical particle breakage mechanisms through the computation of breakage selection coefficients (Epstein [2], Broadbent and Calcott [2]). When this is coupled with a structural examination of the agglomerate particle, a powerful tool evolves which allows enhanced particle design. The first step in using this population balance model is to divide the particle size distribution function into regions called particle size bins. These bins are chosen so that an adjacent bin contains particles that are roughly about half the size of the next larger bin. The modeling method monitors the rate of transfer between different particle size bins.

Since the population balance model is at the heart of understanding degradation rates, it is useful to define some parameters and functions that describe breakage (Sedlatschek & Bass [3]).

- Cumulative density: $R(x, t)$
- Breakage distribution function: $B(x, y)$
- Specific breakage rate function: $S(x)$

$B(x, y)$ describes the fraction of mass transported from y to x upon breakage of y. The breakage distribution function (B) is an indication of what particle size bin the particle fragments end up in during breakage – often described as a matrix of numbers indicating the fractional distribution of transfer to bin sizes smaller than the original. The B(x,y) function plays the role of a weighting function or stoichiometric constant in a reaction. The S(x) function represents the probability that a given particle impact or contact will result in breakage. It represents an overall rate of one particle size breaking into any adjacent particles size bins. S functions are the rate constants for degradation from one particle size to all others (Kapur [4]). Equations governing the rate of transfer between particle bin sizes can be written for cumulative distributions (see equation 1).

$$\frac{dR_i(t)}{dt} = -S_i R_i(t) + \sum_{j=1}^{i-1} \left(S_{j+1} B_{ij+1} - S_j B_{ij} \right) R_j(t), \quad R_i(0) = R_{i0} \tag{1}$$

In general, we must solve the equations above for the cumulative distribution (R) and find the rate constants from the solution of these equations. For the purposes of this work we assume the rate constants are constant and not a function of time. This simplifies the analysis somewhat. However, other assumptions are possible. Consider the work of Bilgili et al [5].

B functions are found simultaneously, from the solution of these equations. Since mass cannot be destroyed, the sum of the B functions for any particle size bin must equal one. This solution of the above equation set is not a trivial task, but it provides much more information concerning the degradation phenomena experienced by the bulk material. If the B values indicate a large fraction for the next adjacent bin size, then fracture of primary particles is occurring. If the B values indicate significant fraction of material transferring more than one particle size bin, then abrasion is likely occurring. Abrasion is creation of fine particles directly from larger particles.

The probability S-function is a value that identifies the likelihood that a collision will result in a breakage of particles. This probability is a function of the size of the particles which typically shows an increasing functionality between the particle size and the probability that a given size particle will break. The solution of the cumulative equations given in equation 1 can be calculated numerically (Reid [6]) but can also be approximated by the following equations if the rate constants are assumed time invariant (Kapur [4]).

$$R_i(t) = R_i(0) \exp\left(G_i t + H_i t^2/2 \right)$$

$$G_i = -S_i + \sum_{j=1}^{i-1} \lambda_{ij} \frac{R_j(0)}{R_i(0)}, \quad H_i = \sum_{j=1}^{i-1} \lambda_{ij} \left(G_j - G_i \right) \frac{R_j(0)}{R_i(0)}, \quad \lambda_{ij} = S_{j+1} B_{ij+1} - S_j B_{ij} \tag{2}$$

The particle size distribution of bran pearls when subjected to repeated impacts at velocities of 5.42 m/sec was measured at various times to generate the time sequence data required for population balance modeling (Figure 1). The equations listed above were solved with this time sequence PSD data to determine the B(x,y) distribution matrix and the S(x) breakage rate function. The distribution function and the breakage rate function were used to compute the predicated cumulative particle size bin distributions for bran pearls. Figure 2 show good agreement between the measured cumulative particle size bins distribution and those predicted by the population balance as a function of time.

The S-values represent the overall rate of particle size degradation as seen in Figure 3.. The coarse particles around 4300 microns break at a rate that is about eleven times the rate of 2000 micron particles. This indicates a significant weakness in large particles or bran pearls.

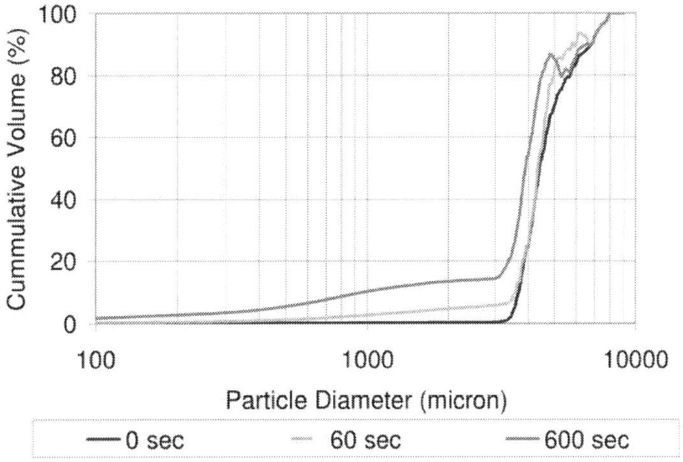

Figure 1. ***Time sequence particle size distribution data for repeated impacts of bran pearls at impact velocities of 5.42 m/sec.***

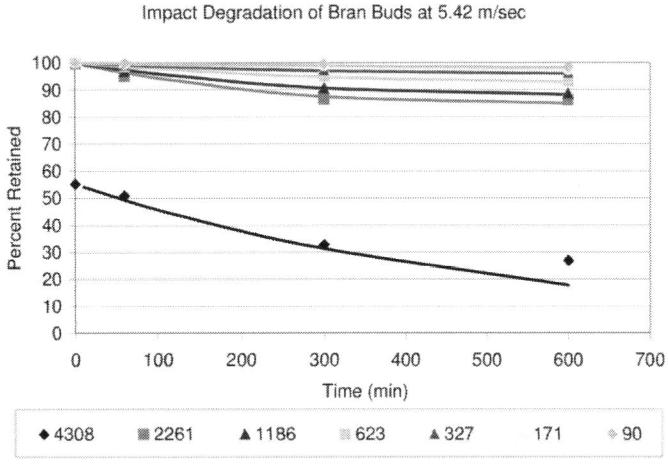

Figure 2. ***Cumulative particle size distributions as a function of time for bran pearls subjected to repeated impacts at 5.42 m/sec impact velocities.***

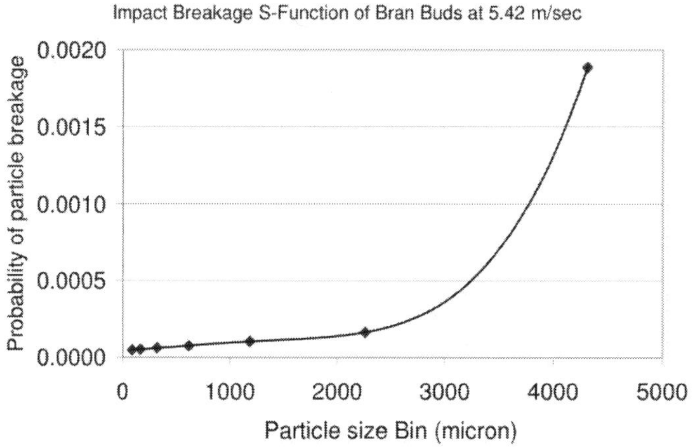

Figure 3. ***S-Values for bran pearls subjected to repeated impacts at 5.42 m/sec impact velocities.***

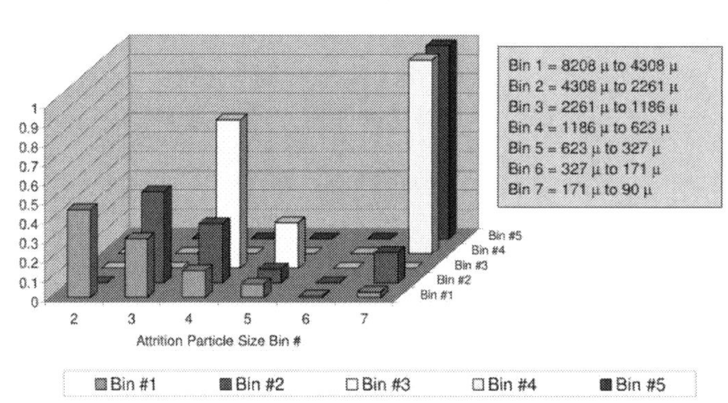

Figure 4. ***B-Values for bran pearls subjected to repeated impacts at 5.42 m/sec impact velocities.***

The S-value data indicate overall breakage rate. However, B-value data indicate the type of breakage occurring in the system (Figure 4). This data suggest that 45% of material initially in bin #1 (average particle size 6258 micron) end up in bin #2 (3284 micron). Thus, 45% of the breakage of coarse particles is due to fracture (particles split in half.) The remainder of the coarse particles is distributed from bin #3 through bin #7, indicating that some limited abrasion is occurring with this material. This fracture pattern, with some secondary abrasion, also occurs for bins #2 (3284 micron) and #3 (1650 micron). However, particles < 904 microns show an abrupt change in breakage mechanism. The primary cause of breakage with mid-size particles is abrasion (very small particles are broken off edges and corners.) The population balance model gives a quantitative measure of the type of breakage that can occur with bran pearls.

This can be coupled with a structural study of the bran pearls to determine the particle structures that cause size degradation. It is important to point out that the bran pearls are cylindrical with a height-to-diameter ratio of about 1.0 to 1.1. These agglomerate particles are about 4000 microns to 4500 microns with a series of doublets and triplets. The first particle size bin goes from about 4300 to 8300 microns and includes these doublets and triplets. Optical analysis based on the aspect ratio of the original particles indicates that 8.4% of the particles are triplets and 25.3% of the particles are doublets. Optical analysis of the coarse fraction was performed after subjecting this material to repeated impacts at 5.42 m/sec. This analysis indicated 3.5% triplets and 21.1% doublets, giving an overall reduction in doublets and triplets of about 9.1%. Part of the fracture observed in bin size #1 is due to breakage of doublets and triplets. The results of optical analysis combined with population balance model provide a clearer picture of the breakage phenomena.

The selectivity plot indicates that all the material in bin #2 (3284 micron) ends up as particles smaller than or equal to bin #4 (904 microns). Almost all particles in bin #3 (1650 micron) end up as particles around 904 microns. The population balance model indicates that particles initially 904 microns and 475 microns crumble into particles that are about 130 microns. This suggests that bran pearls break into individual particles of about 4300 microns which, in turn, fracture into particles around 904 microns. Abrasion then causes these particles to crumble into particles around 130 microns.

A close look at the structure of the particle indicates that bran pearl is comprised of large flakes that have a D_{50} particle size of about 640 microns (Figure 5). The population balance model suggests that these bran pearls (4500 microns) fracture into smaller bran particles that are around 904 microns average particle size. Once this occurs, smaller particles are knocked off the corners of larger particle during repeated impact to yield fine bran powder (130 microns).

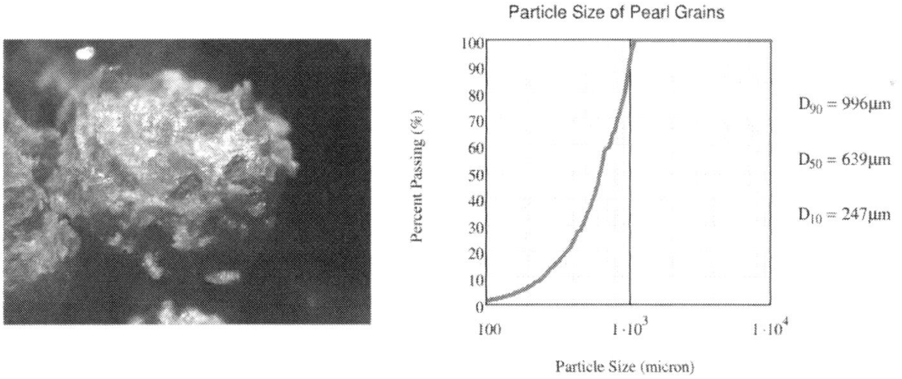

Figure 5. Bran pearls showing primary particles about 640 micron.

Determining degradation mechanism for bran fibers

Bran cereal is also formulated in another shape consisting of long extruded particles with a length of 14,000 micron and a diameter of 2100 microns. Particle size degradation was measured after subjecting the material to repeated impacts at 5.42 m/sec. The data in this case showed substantial change in particle size within the first 10 seconds followed by a smaller shift in particle size (Figure 9). This trend is also evident in the cumulative particle size distributions plots as a function of impact time (Figure 6). This figure indicates that the amount of particles less than 4300 micron drops 20% within the first 10 seconds of impact.

A population balance study of this material suggests that a breakage rate constant independent of time is not sufficient to predict the observed fast size degradation behavior. Thus, the population balance model approach was used to analyze data for the first 60 seconds as well as for data between 60 seconds and 600 seconds. Two different sets of breakage rate constants were obtained for this data. Figure 7 shows good agreement between this dual rate population model and cumulative particle size data. Notice the very step change in concentration for after just a few seconds of repeated impact.

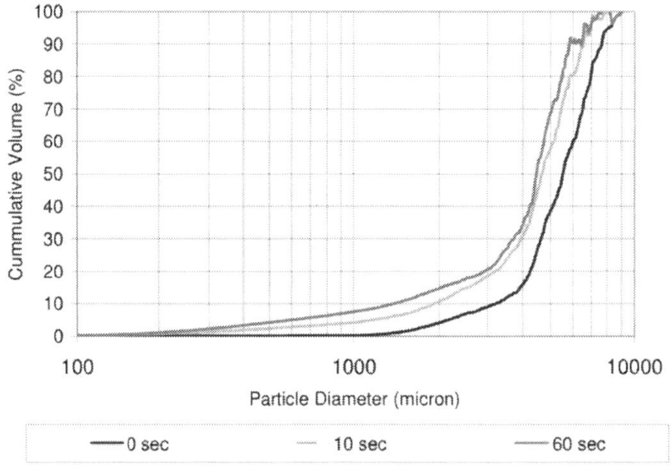

Figure 6. ***Time sequence particle size distribution data for repeated impacts of bran fibers at impact velocities of 5.42 m/sec.***

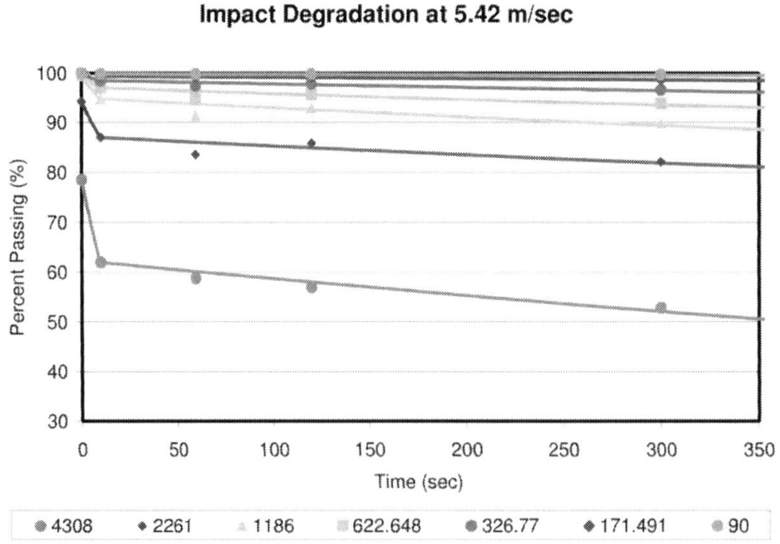

Figure 7. ***Cumulative particle size distributions as a function of time for bran fibers subjected to repeated impacts at 5.42 m/sec impact velocities.***

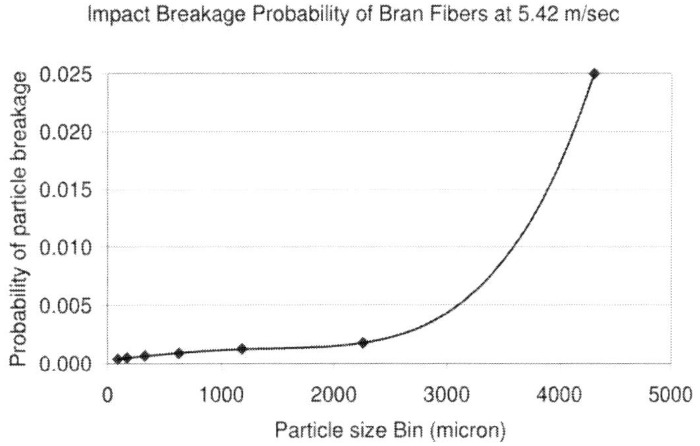

For times less than 60 seconds

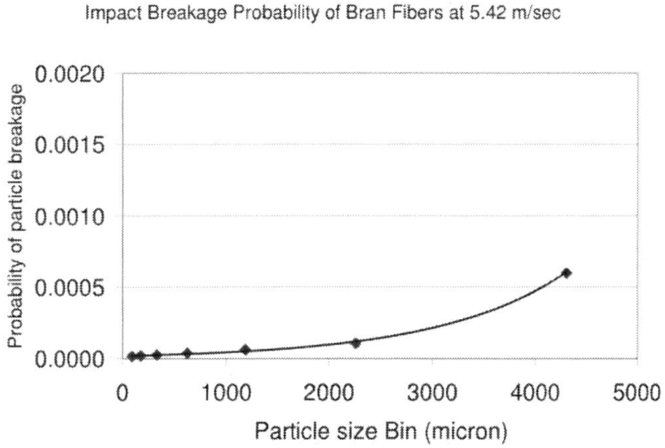

For times greater than 60 seconds

Figure 8. ***S-Values for bran fibers subjected to repeated impacts at 5.42 m/sec impact velocities.***

The breakage rate of large particles occurs very quickly and then slows to a rate about 43 times smaller than the initial breakage rate. This trend persists for smaller size particles where initial breakage rate is about 20 times the breakage rate observed after 10 seconds of repeated impacts. Population balance modeling provides a way of quantifying this effect. The breakage selectivity function B-values for the bran fibers (Figure 9) show strong tendency to form particles in the next adjacent bin size during the initial impact phase, indicating fracture of large particles. Thus, bran fibers tend to break initially by fracture. There is a shift towards abrasion near the end of degradation testing, indicating that, after initial fracture, particles require repeated impacts to knock off edges. A close look at bran fiber particles with a large aspect ratio suggests that these particles likely break into two or three pieces due to bending during an impact event to produce particles with an aspect ration closer to one. These more rounded particles then tend to behave similarly to the bran pearls and break because of both fracture and abrasion.

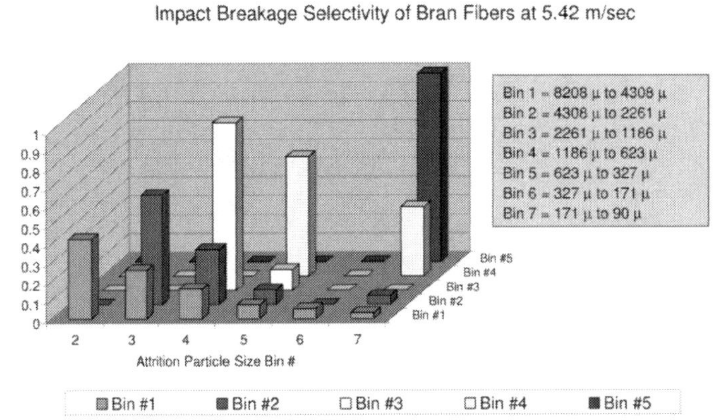

For times less than 60 seconds

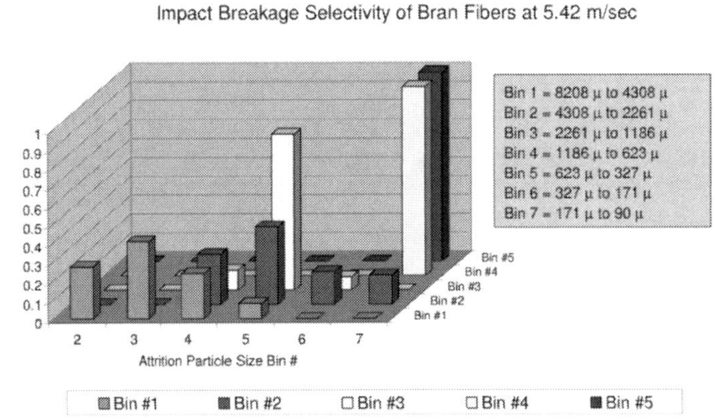

For times greater than 60 seconds

Figure 9. ***B-Values for bran fibers subjected to repeated impacts at 5.42 m/sec impact velocities.***

Conclusion

If a simple estimate of particle size breakage is required for a typical process, then performing a degradation test with a tester that generates the same type of behavior as in the process can give information necessary to compute an estimate of degradation in the prescribed process. However, simple industrial size degradation tests are often not sufficient to provide information about the cause of particle size degradation of a given material. A more detailed analysis such as the population balance model study is required to help understand the mechanics of particle breakage. When this tool is combined with a structural study of the particle, then the means of preventing breakage becomes evident and provides guidance to create more robust particles. In the case of bran pearls, the main breakage mechanism for large particles was fracture to eventually form particles about 900 microns in diameter. These larger particles then break down by attrition to form fine bran powder (130 microns). If the process that produces these particles could be changed to create a stronger bond between the individual 900 micron bran particles, then these agglomerates would be more robust. Conversely, the bran fibers which consists of large aspect ratio material initially fractures rapidly into particles that are half or one third the size of the original particles. These smaller particles then undergo both fracture and abrasion to generate fines. The maximum rate step is the production of mid-size particles due to the initial fracture of these long particles. In this case, the particle should be designed with some flexibility to withstand bending moments that cause initial fracture of large bran fiber particles. Population balance modeling then provides a powerful means of guiding product design.

Nomenclature

B is the breakage selectivity constants from the population balance model.
b is the differential breakage selectivity constants from the population balance model.
G is an internal parameter used to compute the predicted cumulative particle size.
H is an internal parameter used to compute the predicted cumulative particle size.
m is the differential mass particle size density
R is the cumulative particles size.
S is the breakage probability function for the particle size bins.
s is the differential breakage probability function for the particle size bins.
λ is an internal parameter used to compute the predicted cumulative particle size.

References

1. Broadbent and Calcott, J. *Inst. Fuel, Vol 30*, 1957, pp 13-15
2. Epstein, B., "Logarithmico – Normal Distribution in Breakage of Solids", *Ind. Eng. Chem. Vol 40*, pp 2281-1191
3. Sedlatschek K., Bass L., Contribution to the theory of milling processes, *Powder Metal, Bull 6*, 1953, pp 148-153
4. Kapur, P.C. and Agrawal P.K., Approximate solutions to the discretized batch grinding equation, *Chemical Engineering Science, Volume 25, Issue 6*, 1970, pp 1111-1113
5. Bilgili E. and Scarlett B., Nonlinear effects in parti culate processes *Nonlinear Analysis, Volume 63, Issues 5-7, 30* November 2005 to 15 December 2005, pp e1131-e1141
6. Reid, K.J., A solution to the batch grinding equation, Chem Eng Sci 20, 1965, pp 953-963

Formulation of discrete models for non-linear breakage: flow induced alteration of breakage rates

Jayanta Chakraborty and Doraiswami Ramkrishna
School of Chemical Engineering, Purdue University, West Lafayette, IN 47906

Introduction:

One of the main difficulties in formulating quantitative models for milling is the complex nature of the interaction between the particles inside a mill. Most models have circumvented this difficulty by viewing the breakage of a solid particle as independent of its environment (Austin, 1971) whereas in reality it depends on the nature of the dynamic granular environment (Fuerstenau and Abouzeid, 1991) present in the mill. In a few recent studies (Bilgili et al., 2006) such interactions have been considered by letting the breakage rate of a particle be a functional of the particle size distribution (PSD). However, such an approach would require more guidelines for accommodating the phenomenological description of powder flow, segregation and mixing and so on. In this work, we formulate a discrete model for non-linear breakage which is much simpler than the non-linear models previously proposed and yet contains the essential features of the process.

Discrete formulations such as a Markov chain model (Cox, 1965) have the potential to be a computationally simplified tool for modeling size reduction of particles. For example, the effects of transport can be accommodated in a simple way through transitions from one spatial zone to another. (Chakraborty and Ramkrishna, 2009). In this communication we shall restrict ourselves to discrete models for non-linear breakage. In particular, we will be concerned with a Markov chain model,which is very useful in describing a complicated mill with relatively simple mathematics. In this model (Berthiaux 2004) the PSD in the mill is described using the *discrete particle state vector* where each state corresponds to a discrete size. The mill is described by a *transition matrix* which represents the probability of transition between each pair of states. The transition matrix along with an initial state vector describes the dynamics of the system completely. Breakage is viewed as a transition from particles of larger size to the particles of smaller size and transport of powder among various zones can also be easily incorporated in this model. If the condition in the mill (transition matrix) remains invariant with time, the model leads to a *homogeneous Markov model* and its parameters can be extracted rather easily from the observed data even in the presence of transport (Chakraborty and Ramkrishna, 2009). However, for non-linear milling the transition matrix changes continuously with time leading to *non-homogeneous Markov models*. We shall next discuss such Markov models for non-linear milling.

Model Development:

Breakage Phenomenon

Breakage of brittle material depends on propagation of cracks (Rhodes, 1998). In a milling environment, energy is supplied to individual particles and the strain produced by absorption of this energy help in the propagation of cracks and finally the particle breaks. In a typical mill a particle is supplied with energy either by compression or impact or both. Generally a mill is composed of moving and stationary parts and the energy supplied to moving parts is transmitted to individual particles through other particles present in the mill.

The granular media present in the mill strongly influences the amount of energy transmitted to individual particles and therefore affects their breakage rates. As mills can vary widely in terms of the granular media present, the nature and degree of this influence also vary widely. For example, the granular media can be much diluted and the interstitial fluid may fill most of the space. Such condition is encountered in a fluid energy mill. Such mills do not contain any moving part and collision between particles is the only mechanism to break the particles. In this extreme, non-linear interaction (through aggregation) is the only breakage mechanism. On the other hand, if the granular media is very densely packed where no movement is possible, the media acts like a rigid body and hardly any interaction occurs among particles except for the neighboring ones. Such a situation is encountered in equipments like a jaw crusher.

Apart from these two cases, many milling processes involve moderate powder flow. The flow characteristics of the powder are very important because more fluid the powder is the less the energy available for breakage. If the time scale for flow is smaller than the time scale for absorption of sufficient energy in a mill, the material will flow rather than break.

It is clear from the above discussion that a single non-linear model cannot represent the widely varying situations present in various mills. Different models should be developed for different situations keeping the essential features in mind. For example, the non-linear model applicable for a fluid energy mill will be very different from the corresponding model for a ball mill. In the next section we shall discuss a few basic concepts of discrete formulation and then develop the model for the two extreme cases, namely, the breakage through aggregation which happen in a gas like granular environment and breakage in a dense granular media.

Breakage through aggregation:

Breakage through aggregation is a well known phenomenon where collision between particles are responsible for breakage. The general population balance model can be formulated as:

$$\frac{\partial n(x,t)}{\partial t} = \int_x^\infty \eta(x \mid x') b(n,x') n(x',t) dx' - b(n,x) n(x,t) \tag{1.1}$$

where $b(n,x)$ is the breakage rate of a particle of size x which also depends on the temporal particle size distribution n and $\eta(x \mid x')$ is the daughter distribution function which is considered to be independent of the particle size distribution. In the simplest case, the collision may be viewed as binary collision and for this case $b(n,x)$ can be written as:

$$b(n,x) = \int_0^\infty Q(x,x') n(x',t) dx' \tag{1.2}$$

where $Q(x,x')n(x',t)$ is the contribution of particles in the range x' to $x'+dx'$ in the breakage of particle of size x. Hence, eq 1.1 can be written as (Kostoglou and Karabelas, 2000):

$$\frac{\partial n(x,t)}{\partial t} = \int_0^\infty \int_x^\infty \eta(x \mid x') Q(x',x'') n(x',t) n(x'',t) dx' dx'' - n(x,t) \int_0^\infty Q(x,x') n(x',t) dx' \tag{1.3}$$

It may be noted that this equation is similar to the PBE for pure aggregation with a different physical interpretation of the terms Q and η. The discrete form of this equation was derived by Kumar and Ramkrishna (1996) which is given by:

$$\frac{dn_i}{dt} = \sum_{j,k} \eta_{ijk} Q_{jk} n_j n_k - n_i \sum_{k=1}^{M} Q_{ik} n_k \tag{1.4}$$

where n_i represents the fraction of particles in the ith discrete domain. Introducing

$$D_{ijk} \equiv \eta_{ijk} .Q_{jk} \tag{1.5}$$

this equation can be written as:

$$\frac{dn_i}{dt} = \sum_{j} \sum_{k} D_{ijk} n_j n_k - n_i \sum_{k=1}^{M} Q_{ik} n_k \tag{1.6}$$

which can be written in vector-matrix notation as:

$$\frac{d\mathbf{n}}{dt} = \mathbf{D} : \mathbf{nn} - \mathbf{n}^D.\mathbf{Q}.\mathbf{n} \tag{1.7}$$

where $\mathbf{n}^D$ represent the diagonal matrix of $\mathbf{n}$. Now, in order to obtain the time discrete version of this equation we need to multiply the above equation with an appropriate time scale (say τ) which leads to:

$$\mathbf{n}^{i+1} - \mathbf{n}^i = \tau(\mathbf{D} : \mathbf{n}^i\mathbf{n}^i - \mathbf{n}^{iD}.\mathbf{Q}.\mathbf{n}^i) \tag{1.8}$$

where $\mathbf{n}^{i+1}$ and $\mathbf{n}^i$ are the particle size vector in two successive times. Rearranging the above equation:

$$\mathbf{n}^{i+1} = \tau(\mathbf{D} : \mathbf{n}^i\mathbf{n}^i - \mathbf{n}^{iD}.\mathbf{Q}.\mathbf{n}^i) + \mathbf{n}^i \tag{1.9}$$

It can be readily seen that it is not possible to write a non-homogeneous Markov model for this case due to the first term in the right hand side which represents the non-linear interaction. Hence, this model should be used in this form itself to estimate the parameters.

Breakage in a granular environment:

For many grinding mills the particle density is relatively high and collisions are rather weak to provide sufficient energy for breakage. However, the media play a vital role in transmitting the power from the rotating part of the mill to the individual particles. The flowability of the powder determines the fraction of energy that a particle may absorb in order to break. The more easily the particulate assembly flows, the more difficult it is to provide the energy for breakage.

Unlike the previous case, direct interaction among particles is not present in this case and hence the time inhomogeneous Markov chain model may be employed. As grinding proceeds in a batch mode, the nature of the media changes due to the change in the PSD and hence progressive change in the flow characteristics of the powder occurs. Hence, the Markov chain model for the system can be written as:

$$\mathbf{n}^{(i+1)} = \mathbf{P}(\mathbf{n}^{(i)}).\mathbf{n}^{(i)} \tag{1.10}$$

where the transition matrix $\mathbf{P}$ changes with change in the particulate environment.

The effect of PSD of granular flow of irregular particles can be very complex. However, we will forsake a detailed description in favor of a key factor for flow. It is known that the mobility of a granular bed is mostly influenced by the amount of fines present in the bed (Phillips et al 2006). In many cases, such materials are added to improve the flow of powders. In the case of grinding such flow conditioners are produced due to grinding itself. Hence in our model we shall consider that the transition probability matrix is influenced only by the amount of fines which we shall define as the particle in the smallest size class.

In light of the above discussion, the Markov chain model can be written as:

$$\mathbf{n}^{(i+1)} = \mathbf{P}(f_i).\mathbf{n}^{(i)} = \mathbf{P}_i.\mathbf{n}^{(i)} \tag{1.11}$$

where f_i is the fraction of particles in the smallest size class and $\mathbf{P}_i$ is the time dependent transition probability matrix.

As discussed earlier, the nature of the granular media can influence only the availability of energy for grinding. The breakage of a particle can happen only at individual particle level and hence the daughter size distribution is unlikely to be influenced by such factors. This feature is also observed in experiments (Fuerstanau and Abouzeid, 1991). Hence in the current model we shall consider that only the selection matrix (breakage frequency) part of the transition matrix is affected by the change in the granular media.

Now, because the environment is the same for particles of different size, we shall assume that the breakages of particles of various size classes are retarded by the same factor. Hence, we define the retardation factor for breakage as:

$$r_i = af_i^{\,b} \tag{1.12}$$

where a and b are constants and $r_i > 1$. Now, the transition matrix can be written as:

$$\mathbf{P}_i = \mathbf{B}\mathbf{S}^{(i)} + (\mathbf{I} - \mathbf{S}^{(i)}) = \mathbf{I} + (\mathbf{B} - \mathbf{I})\mathbf{S}^{(i)} = \mathbf{I} + \mathbf{B}'\mathbf{S}^{(i)} \tag{1.13}$$

where the matrix $\mathbf{B}$ and $\mathbf{S}^{(i)}$ are given by:

$$\begin{aligned} S_{jj}^{(i)} &= 1 - P_{jj}^{(i)} \\ S_{ij}^{(i)} &= 0 \, (i \neq j) \\ B_{ij} &= P_{ij}^{(0)} / (1 - P_{jj}^{(0)}) \end{aligned} \tag{1.14}$$

Because the elements of any column of the matrix $\mathbf{B}$ add up to 1.0, $\mathbf{P}_i$ remains a stochastic matrix irrespective of the variation in $\mathbf{S}_i$. However, the maximum value of the factor r_i is restricted by the limit $S_{ij}^{(i)}.r_i \leq 1.0$.

Now, using eq 1.12 we obtain:

$$\mathbf{S}^{(i)} = r^{(i-1)}\mathbf{S}^{(i-1)} = \left(\prod_{k=0}^{i-1} r^k\right)\mathbf{S}^{(0)} = \pi_i\mathbf{S}^{(0)} \tag{1.15}$$

Now, the time inhomogeneous Markov chain can be written as:

$$\begin{aligned} \mathbf{n}^{(1)} &= \mathbf{P}_0\mathbf{n}^{(0)} \\ \mathbf{n}^{(2)} &= \mathbf{P}_1\mathbf{n}^{(1)} = \mathbf{P}_1\mathbf{P}_0\mathbf{n}^{(0)} \\ \mathbf{n}^{(3)} &= \mathbf{P}_2\mathbf{n}^{(2)} = \mathbf{P}_2\mathbf{P}_1\mathbf{P}_0\mathbf{n}^{(0)} \\ &\vdots \\ \mathbf{n}^{(n)} &= \mathbf{P}_{n-1}\mathbf{n}^{(n-1)} = \mathbf{P}_{n-1}\mathbf{P}_{n-2}\cdots\mathbf{P}_0\mathbf{n}^{(0)} \end{aligned} \tag{1.16}$$

Substituting $\mathbf{P}_i$ from eq 1.13:

$$\mathbf{n}^{(n)} = (\mathbf{I} + \mathbf{B}'\mathbf{S}^{(n-1)})(\mathbf{I} + \mathbf{B}'\mathbf{S}^{(n-2)})\cdots(\mathbf{I} + \mathbf{B}'\mathbf{S}^{(0)})\mathbf{n}^{(0)} \tag{1.17}$$

Using eq 1.15,

$$\mathbf{n}^{(n)} = (\mathbf{I} + \pi_{n-1}\mathbf{B}'\mathbf{S}^{(0)})(\mathbf{I} + \pi_{n-2}\mathbf{B}'\mathbf{S}^{(0)})\cdots(\mathbf{I} + \mathbf{B}'\mathbf{S}^{(0)})\mathbf{n}^{(0)} \tag{1.18}$$

Now, let us examine whether this simple model can capture non-linear behavior observed in practical situations. To demonstrate this we consider a mill the linear breakage behavior of which is shown in Fig. 1. Now, we use $a = 4.0$ and $b = 1.0$ in eq 1.12 and the retarded breakage is shown in Fig. 2.

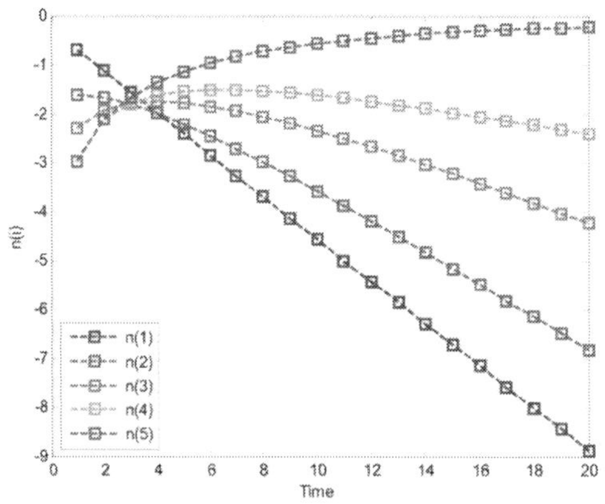

Figure 1: Evolution of the particle state vector: linear breakage behavior.

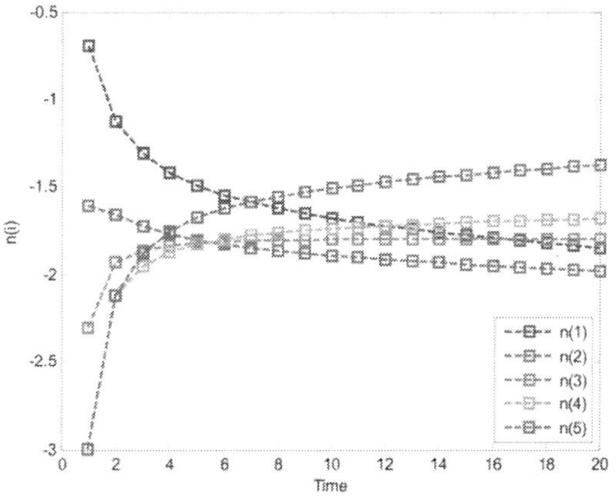

Figure 2: Evolution of particle state vector: non-linear breakage (retardation)

It can be noted that the function described in eq 1.12 can also be used to predict acceleration of breakage rate with accumulation of fines which can also be observed in few cases. Figure 3 shows such behavior.

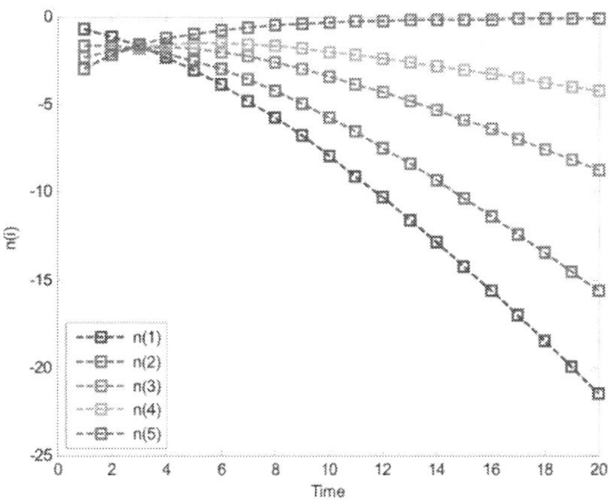

Figure 3: Evolution of particle state vector: non-linear breakage (acceleration)

Conclusion and outlook:

Discrete models for non-linear breakage have been developed for two cases, namely the breakage through aggregation and breakage in a granular environment. It has been shown that the breakage through aggregation is inherently non-linear and it cannot be modeled using an approximate time inhomogeneous Markov chain. However, the breakage of particles in a granular media can be modeled with time inhomogeneous Markov chains where the presence of fine powder is considered to be responsible for the retardation of breakage rates. Both the models consider that only the breakage rate is affected by the non-linear interaction while the daughter size distribution remains unaffected. The models can be used with the experimental data in order to estimate various parameters, the appropriate methodology for which will be suggested.

References:

1. Austin, L. G. and Bagga, P., (1981), "An analysis of fine dry grinding in ball mills," *Powder Technology*, **28,** pp. 83-90.
2. Fuerstenau, D. W. and Abouzeid, A. –Z. M., (1991), "Effect of fine particles on the kinetics and energetic of grinding coarse particles," *Int. J. Min. Proc.*, **31,** pp. 151-162.
3. Bilgili, E., Yepes, J. and Scarlett, B., (2006), "Formulation of a non-linear framework for population balance modeling of batch grinding: Beyond first order kinetics," *Chem. Eng. Sci.*, **61,** pp. 33-44.
4. Cox, D. R. and Miller H. D., (1965), *The Theory of Stochastic Processes.*, John Wiley and Sons.
5. Chakraborty, J. and Ramkrishna, D. (2009), "Identification of the Markov matrices of the milling models," *Ind. Eng. Chem. Res.* (In press, **DOI:** 10.1021/ie900456j)
6. Berthiaux, H. and Mizonov, V., (2004), "Applications of Markov chains in particulate process engineering: a review," *Can. J. Chem. Eng.*, **85,** pp.1143-1168.
7. Rhodes, M., (1998), *Introduction to particle technology*, John Wiley and Sons.
8. Kostoglou, M. and Karabelas, A. J., (2000), "A study of the nonlinear breakage equation: analytical and asymptotic solutions," *J. Phys. A: Math. Gen.*, **33,** pp. 1221-1232.
9. Kumar, S. and Ramkrishna, D., (1996), "On the solution of population balance equations by discretization-I. A fixed pivot technique." *Chem. Eng. Sci.*, **51,** pp. 1311-1332.
10. Phillips, J. C., Hogg, A. J., Kerswell, R. R. and Thomas, N. H., (2006), "Enhanced mobility of granular mixtures of fine and coarse particles," *Earth and Planetary Science Letters*, **246,** pp. 466-480.

THE EFFECT OF THE SHEAR RATE ON THE SURFACE PROPERTIES OF EGG-PHOSPHATIDYLCHOLINE LIPOSOMES

Julia Exaltação Trevisan, Lucimara Gaziola de la Torre, Maria Helena Andrade Santana.
Laboratory for Development of Biotechnological Processes, School of Chemical Engeineering, University of Campinas-UNICAMP, Campinas, Brazil

Introduction

Liposome or phospholipid vesicles are phospholipids aggregates structured in bilayers with an aqueous nucleus surrounded by one or multiple concentric lamella. They are unilamelar or multilamellar particles with a diameter that vary from the order nanometers to micrometers. These colloidal structures have been attracting the attention for its applicability as a drug delivery system and as a membrane model. Nowadays, a variety of lipid-based drug formulations is being commercialized or under clinical studies (LASIC, 1993; BARNADAS-RODRÍGUEZ & SABÉS, 2001).

A variety of liposome preparation techniques has been developed since liposome was first prepared by the Bangham Method, in 1965. Such methods are sonication, detergent dialysis, ethanol injection, extrusion, high-pressure homogenization and others. Because of the promising results obtained by the use of lipid vesicles in food and pharmaceutical industries, since the 1990's the laboratory scale production began to be insufficient to meet the demand for this type of material, being necessary the development of large scale techniques (GREGORIADIS, 1993; LASIC, 1993). High-shear treatments have been used as high-energy, easy, clean and scalable processing, leading to the homogeneous and nanometric sizes of liposomes required for specific applications (MAYHEW, et al., 1987; TALSMA, ÖZER, VAN BLOOIS, & CROMMELIN, 1989). Modern equipment such as homogenizers and multichannel microfluidization provide diferent ranges of shear and are commonly used in large-scale liposome production. In these processes, the comprehention of the influence that the shear has on the surface properties of the particle is a key element for the prediction of the best operational conditions of the process.

Considering the great importance of the shear range on the surface properties of lipid vesicles and the lack of sistematic studies in this subject, the objective of this work is to investigate the effects of the shear rate and flow rate on the size, polidispersity and zeta potencial of liposomes composed of egg-phosphatidylcholine (Egg-PC). The effects were produced from two levels of shear, using Ultra-Turrax equipment (2,560 to 21,430 1/s), and a mechanical disperser with cowles impeller (1,000 to 2,530 1/s), chosen for their distinct and complementary range of shear. In order to reduce the total number of experiments and statistically optimize the processes variables, we have used factorial design. This tool has proven to be a useful instrument of analysis in the study of the response surface (BARROS NETO, SCARMINIO, & BRUNS, 2003).

Experimental

The scheme of the experimental set up used for the liposome production is shown in Figure 1. Liposomes were obtained by pouring a dispersion of Egg-PC Degussa with 60% purity diluted in ethanol (400 mM) (3) into aqueous medium at room temperature (2). The lipid dispersion was added into the aqueous phase through a peristaltic pump (4) with pre-determined flow rate in order to obtain a liposome dispersion of 64 mM. The stirring rate was adjusted in rpm for both Ultra-Turrax IKA T25 (Ika Works) and mechanical disperser RW20 (Ika Labortchnik) with cowles impeller (1). For the Ultra-Turrax disperser the shear rate was calculated by approximation with rotational viscometer, using Equation (1).

$$\dot{\gamma} = 2\omega \frac{\alpha^2}{\alpha^2 - 2} \quad \text{(Eq 1)}$$

where ω is the angular velocity (rad/s) and α is determined by Equation (2).

$$\alpha = \frac{R_C}{R_B} \quad \text{(Eq 2)}$$

R_C and R_B are the outer radius and the inner radius of the Turrax, respectively (STEFFE, 1992). The shear provided by the mechanical disperser with cowles impeller was calculated according to the following equation.

$$\dot{\gamma} = k\omega \quad \text{(Eq 3)}$$

where k is a constant dependent on impeller design. For cowles impeller we considered k equal to 11 (NAGATA, 1975).

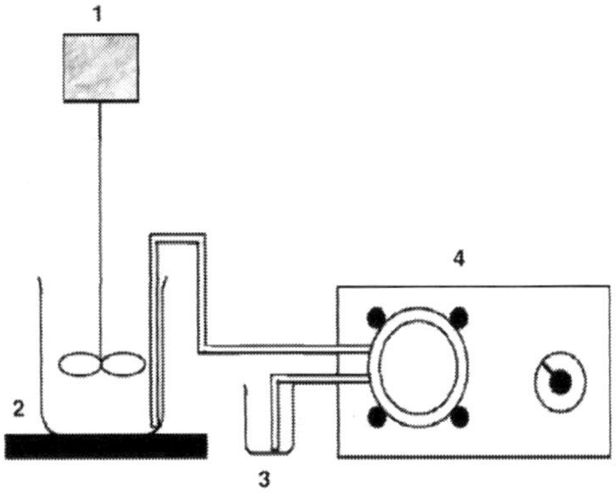

Figure 1 Scheme of experimental set up (1) High shear blender, (2) aqueous solution, (3) lipid solution and (4) peristaltic pump.

The liposomes were characterized by hydrodynamic diameter, distribution and zeta potential using a Zetasizer Nano ZS (Malvern Instruments Ltd.). The mean diameter and polidispersity were determined by dynamic light scattering. The experiments were planned according to the levels shown on the Table 1.

Results

The results obtained for each experiment, using both dispersion systems, are displayed on Table 1.

Table 1 Experimental planning and the obtained results for each type of disperser

IKA ULTRA-TURRAX T25 Digital

Factors	-	0	+
1: Shear rate (s^{-1})	2,860	12,140	21,430
2: Inlet lipid solution flow rate (mL/s)	0.09	0.54	0.96

Experiment / Factor	1	2	Diameter (nm)	Polydispersity	Zeta Potential (mV)
1	-1	-1	386.10	0.37	-53.73
2	1	-1	254.73	0.31	-56.33
3	-1	1	550.20	0.58	-53.73
4	1	1	246.50	0.35	-48.57
5	0	0	374.83	0.47	-55.90
6	0	0	359.00	0.36	-54.60
7	0	0	358.90	0.45	-54.40

Mechanical dispersor with cowles impeller

Factors	-	0	+
1: Shear rate (s^{-1})	1,000	1,750	2,530
2: Inlet lipid solution flow rate (mL/s)	0.09	0.54	0.96

Experiment / Factor	1	2	Diameter (nm)	Polydispersity	Zeta Potential (mV)
1	-1	-1	626.17	0.65	-55.00
2	1	-1	414.87	0.51	-71.27
3	-1	1	616.47	0.68	-57.50
4	1	1	446.77	0.51	-52.60
5	0	0	582.77	0.65	-57.27
6	0	0	642.67	0.58	-55.80
7	0	0	567.40	0.62	-56.00

Figure 2 shows the effects of the shear rate and flow rate on the mean diameter, polydispersity and zeta potential of the liposomes prepared using the Ultra-Turrax disperser.

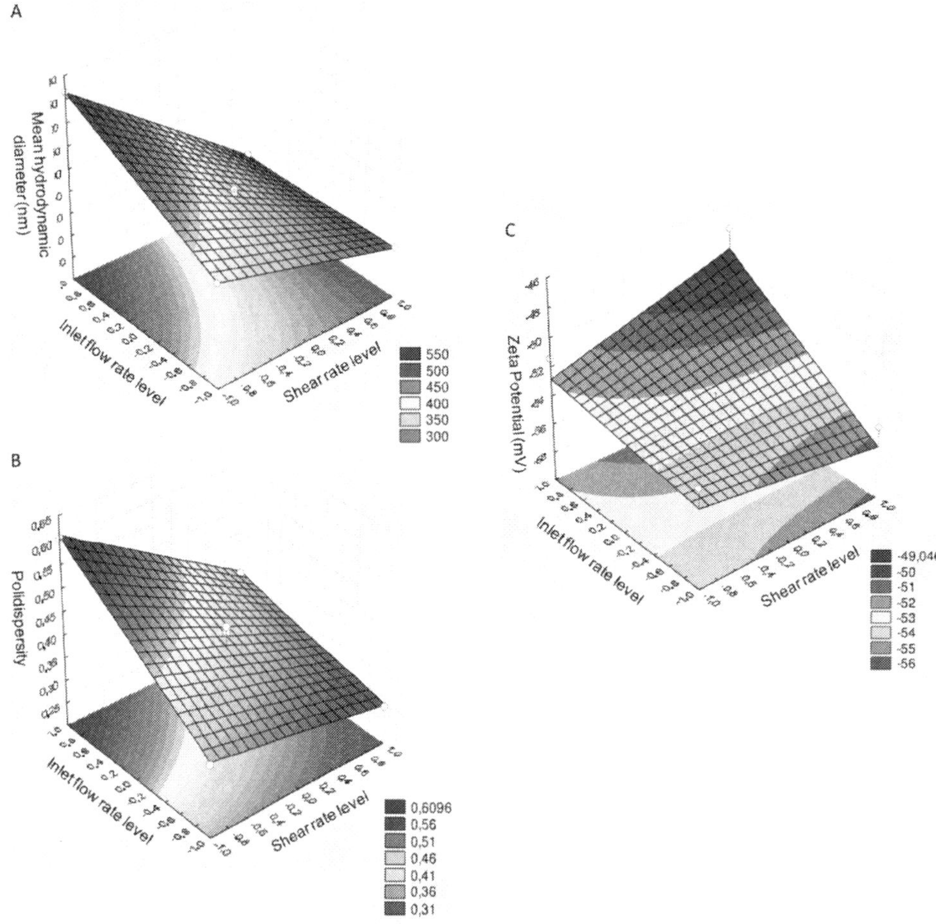

Figure 2 Response surfaces for dispersion in Ultra-Turrax. (A) Mean hydrodynamic diameter, (B) polydispersity and (C) zeta potential. Total phospholipid concentration of 64 mM.

As can be seen in Figure 2 and in Table 2, the liposomes obtained with dispersion using Ultra-Turrax had mean hydrodynamic diameters varying from 250 nm to 550 nm, polydispersity from 0.30 to 0.58 and zeta potential from -57 mV to -49 mV. In lower shear rates, the influence of the inlet flow rate of lipid dispersion on the mean hydrodynamic diameter and on the polydispersity is more significant, whereas in lower inlet flow rates, the shear rate influence is less significant. The zeta potential had no major variations in the studied range. This means that, though all parameters studied had variations in higher flow rates, the mean hydrodynamic diameter and the polydispersity of the liposomes produced with Turrax were more affected in lower shear rates, whereas the zeta potential had more important variations in higher shear rates.

The influence of the independent variables, inlet flow rate and shear rate, on the studied responses can be more clearly analyzed through a Pareto´s chart (Figure 3). Here, the significant influence of both factors and their interactions on the hydrodynamic diameter and polydispersity can be confirmed, as they provide P-values greater than 0.05 in a confidence level of 95%. For the diameter, the influence of the shear rate is higher than the flow rate and this can be verified with the more emphasized slope of the surface studied in

Figure 2. The Pareto´s chart also explains the little variation of the zeta potential observed in Figure 2, seeing that the inlet flow rate and shear rate are not significant for this parameter (P-values lower that 0.05 in a confidence level of 95%).

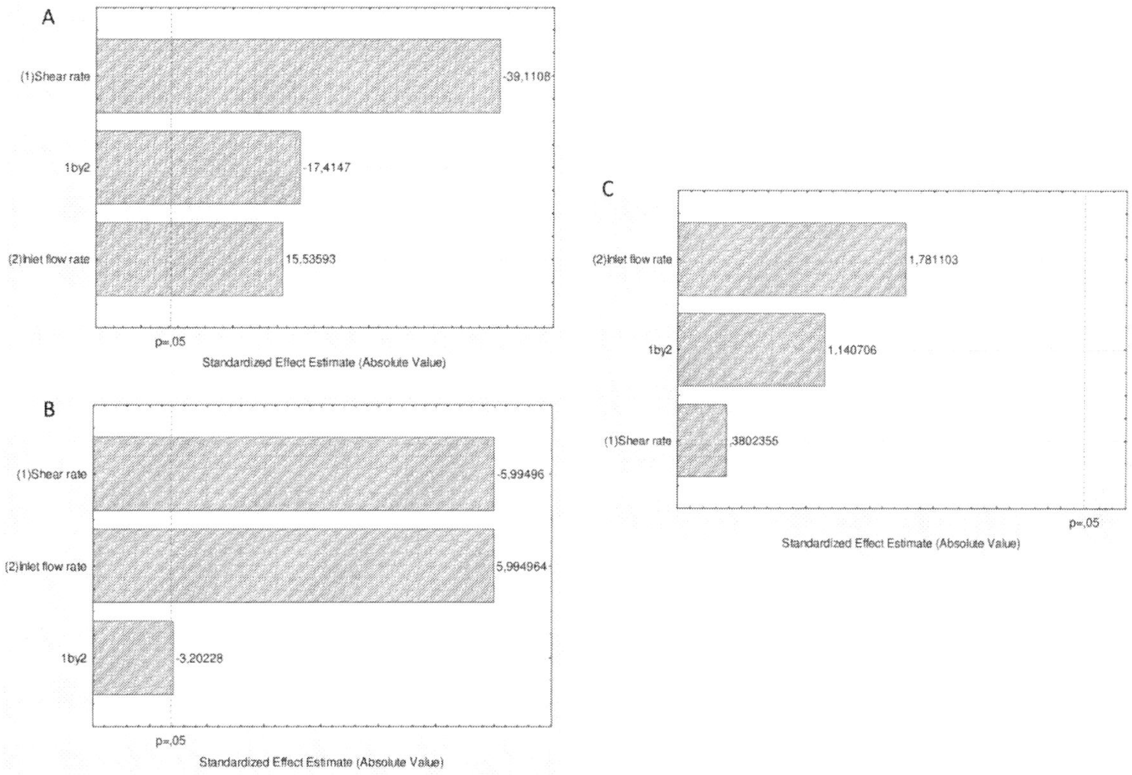

Figure 3 Pareto´s chart for dispersion in Ultra-Turrax. (A) Mean hydrodynamic diameter, (B) polydispersity and (C) zeta potential.

The behavior of the samples when the lipid dispersion was processed through the mechanical disperser is presented in Figure 4.

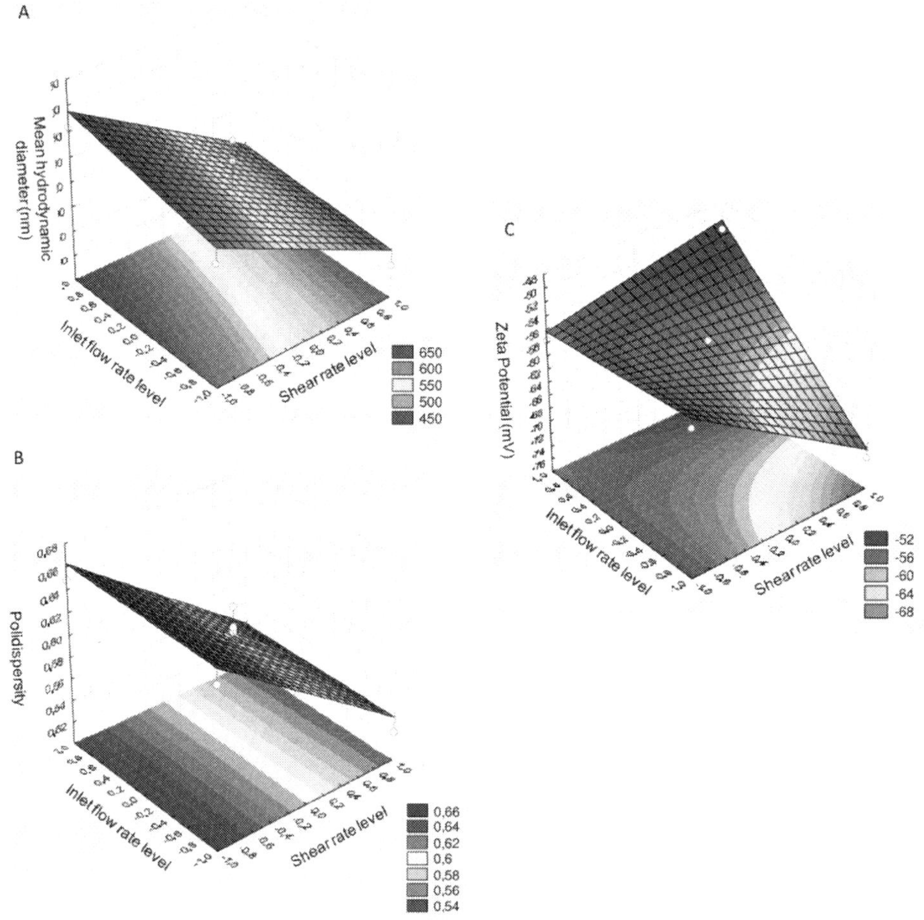

Figure 4 Response surfaces for mechanical dispersions with cowls impeller. (A) Mean hydrodynamic diameter, (B) polidispersity and (C) zeta potential. Total phospholipid concentration of 64 mM.

Using the mechanical disperser with cowles impeller (Figure 4), liposomes were formed with a larger mean hydrodynamic diameter (430 nm to 700 nm), polydispersity (0.50 to 0.68) and absolute zeta potential (-70 mV to -53 mV). The mean hydrodynamic diameter and the polydispersity had similar tendencies, with no apparent variation with the inlet flow rate of lipid dispersion and slight increase values with lower shear rate applied. For the zeta potential, there is an important decrease with higher shear rate and lower inlet flow rate and an increase with high flow rate and with low shear rate.

The Pareto´s chart for the dispersion with mechanical disperser is shown in Figure 5. It confirms that, for the diameter, although the importance of the shear rate if greater than the flow rate, both are statistically insignificant in a confidence level of 95%. For the polydispersity, the shear applied is the only significant factor, with P-value over 0.05, as has been verified in Figure 4. The zeta potential is the only parameter with both factors being significant, which confirms the observations of the response surface.

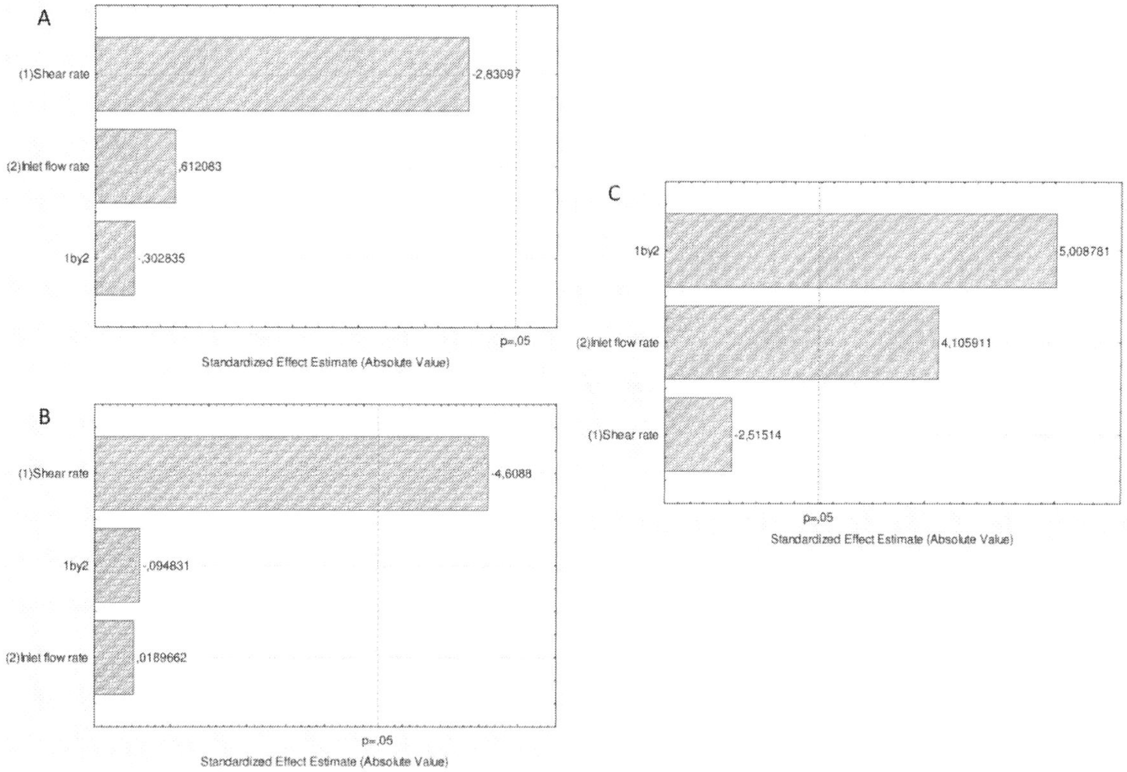

Figure 5 Pareto´s chart for dispersion in mechanical disperser with cowls impeller. (A) Mean hydrodynamic diameter, (B) polidispersity and (C) zeta potential.

For the studied situations, in both dispersion conditions, a larger mean hydrodynamic diameter and polydispersity were achieved with low shear rate and high inlet flow rate, and a larger zeta potential was obtained with high shear rate and high flow rate. Also, for both situations, a smaller diameter and polydispersity were a result of the combination of high shear rate and low inlet flow rate, and a smaller zeta potential was obtained with low shear rate and low inlet flow rate.

In both cases, we were able to modulate the mean hydrodynamic diameter as a function of shear rate and inlet flow rate in a useful range for applications as drug delivery systems. Also, it is able to guarantee a large range of colloidal stability due to the values obtained of zeta potential.

The effects of thermal dissipation during dispersion may alter the liposome packing. This result has direct influence in the particle's zeta potential. Additional research will be taken place with the purpose of studying this effect. The parameters were analyzed after one day and again after three days of the preparation, showing no statistical difference between the two measurements. This is an indicative that the process is able to form liposomes with certain stability. However, in order to obtain more information, it is necessary to valuate these data after a longer period of time.

Considering the experimental behavior of vesicle breakage observed in this work, it is possible to link the mean hydrodynamic diameters with the shear applied. The operational limits of both disperser systems used in this study are complementary, seeing that the maximum capacity of the mechanical disperser is equivalent to the minimum capacity of the Ultra-Turrax. This characteristic gives a full extension of the diameters obtained in the range of shear varying from 1,000 s^{-1} to 21,430 s^{-1}. Figure 6 shows how the shear rate applied affects the mean hydrodynamic diameter of the liposomes. A clear tendency can be observed, where the diameter decreases as the shear increases, despite the approximations used for the calculated shear in both systems. This behavior is expected and has already been established in previews studies (LIU, 2000; LINDEN, et al., 1992 and DIAT, et al., 1993). However, there is a tendency that has not already been stated: a difference in the slope at the point near 500 nm. This behavior can be explained by the fact that above 500 nm the particles are closer to the microscale range and, therefore, have a different behavior than the particles that are closer to the nanoscale range. The effects of shear rates are more effective for the microscale range.

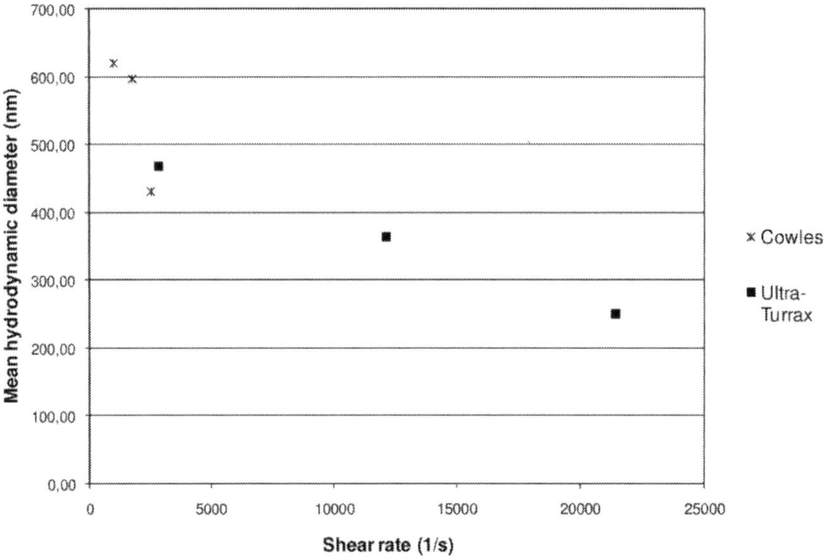

Figure 6 Mean hydrodynamic diameter of liposomes as a function of the shear rate applied for dispersions in Ultra-Turrax and cowles impeller

The relationship between shear rate, liposome radius, superficial tention and lamellar deformation has been well stated in previews works (LINDEN, et al., 1992; DIAT, et al., 1993 and RICHARDSON, et al., 2007), where the theoretical treatment used for emulsion droplets is extended to liposome formation. The balance of the superficial tension and shear rate can give the critical radius where liposomes are more stable. This balance can be represented by the capillary number shown in Equation (4) (RICHARDSON, et al. 2007):

$$Ca = \frac{Shear\ forces}{Surface\ tension\ forces} = \frac{\mu R \dot{\gamma}}{\delta} \qquad \text{(Eq 4)}$$

where μ is the liquid viscosity, R is the radius of the particle, $\dot{\gamma}$ is the shear rate and σ is the interfacial surface tension. With this equation it is possible to predict the critical radius of the system when the forces of shear and surface tension are balanced and the capillary number is in the order of unity.

Using Equation (4) and considering a fixed viscosity of 4.0 cp, measured for experiment number 3 using Ultra-Turrax (Table 1), we are able to obtain response surfaces of surface tension for each dispersion system. Figure 7 shows the surfaces obtained for Ultra-Turrax and mechanical dispersions with cowles impeller.

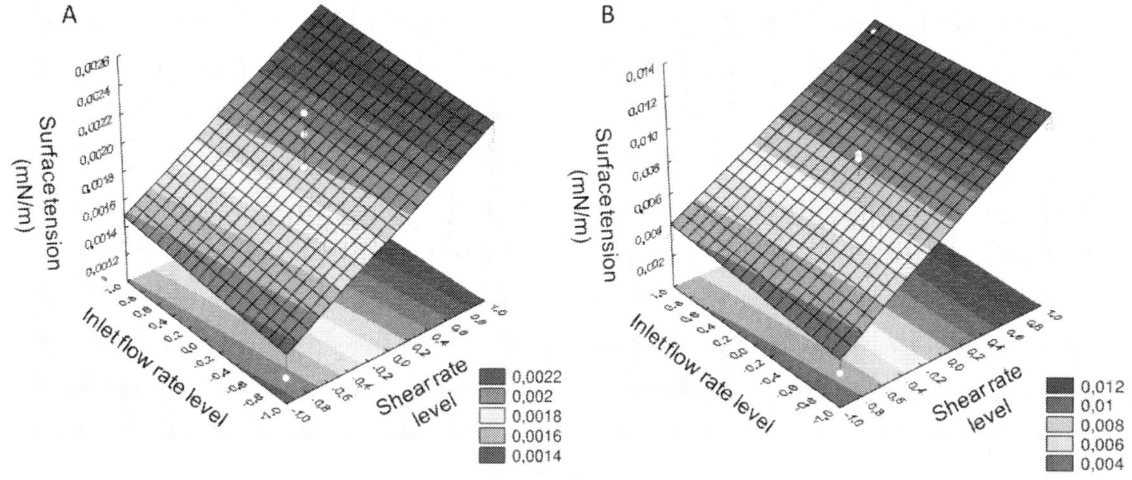

Figure 7 Response surfaces of surface tension for dispersions in (A) Ultra-Turrax and (B) mechanical disperser with cowls impeller. Total phospholipid concentration of 64 mM.

The superficial tension observed in Figure 7 shows a similar tendency for both disperser systems, indicating prevalent effect of shear rate on the superficial tension. The inlet flow rate of lipid dispersions does not influence the response significantly. The low values of superficial tension determined are due the higher lipid concentration (64 mM) used in this study, compared to values reported in the literature.

Conclusions

The investigated effects of the shear rate and flow rate on the size, polidispersity and zeta potencial of liposomes, allow modulating these properties in a useful range for drug delivery applications. The capability for size reduction of liposomes was dependent of the shear rate levels, which were complementary for the ultra-turrax and cowls impeller dispersors. Smaller egg-phosphatidylcholine liposomes may be obtained in the range of 250 nm to 550 nm with relative narrow polydispersity (0.30 to 0.58), through the higher shear rate levels provided by the ultra-turrax, compared to cowls impeller. These results contribute for the production of small liposomes using scalable processes.

References

1. Barnadas-Rodríguez, R., & Sabés, M. (2001). Factors Involved In The Production Of Liposomes With A High-Pressure Homogenizer. *International Journal Of Pharmaceutics*, *213*, Pp. 175-186.

2. Barros Neto, B. B., Scarminio, I. S., & Bruns, R. E. (2003). *Como Fazer Experimentos – Pesquisa E Desenvolvimento Na Ciência E Na Indústria*. Campinas:

3. Bernard, A.; Guedeau-Boudeville, M.; Marchi-Artzner V.; Gulik-Krzywicki, T.; Di Meglio, J.; Jullien, L. (2005). Shear-induced permeation and fusion of lipid vesicles. *Journal of Colloid and Interface Science, 287*, Pp. 298-306.

4. Diat, O., Roux, D., Frédéric, N. (1993). Effect Of Shear On A Liotropic Lamellar Phase. *Journal de Physique II*, 3, Pp. 1427-1452.

5. Gregoriadis, G. (1993). Liposome *Technology*. Boca Raton: Crc Press.

6. Lasic, D. (1993). *Liposomes: From Physics To Applications.* Amsterdam: Elsevier Science Publishers B.V.

7. Liu, D. Z., Chen, W. Y.; Tasi, L. M.; Yang, S. P. (2000). Microcalorimetric And Shear Studies On The Effect Of Cholesterol On The Physical Stability Of Lipid Vesicles. *Colloids and Surface A*, 172, Pp. 57-67

8. Mayhew, E., Conroy, S., Kink, J., Lazo, R., Nikolopoulus, G., Siciliano, A., Et Al. (1987). High-Pressure Continuous-Flow System For Drug Entrapment In Liposomes. *Methods Enzymol*, Pp. 64-77.

9. Nagata, S. (1975). *Mixing Principles And Applications*. New York:

10. Richardson, E. S., Pitt, W. G., Woodbury, D. J. (2007). The Role of Cavitation in Liposome Formation. Biophysical Journal, 93, Pp. 4100-4107.

11. Steffe, J. F. (1992). *Rheological Methods In Food Process Engineering*. Michigan:

12. Talsma, H., Özer, A. Y., Van Bloois, L., & Crommelin, D. J. (1989). The Size Reduction Of Liposomes With A High Pressure Homogenizer (Microfluidizer®). Characterization Of Prepared Dispersions And Comparison Withj Conventional Methods. *Drug Development And Industrial Pharmacy*, *15*, Pp. 197-207.

13. Thompson, A. K., & Singh, H. (2006). Preparation Of Liposomes From Milk Fat Globule Membrane Phospholipids Using A Microfluidizer. *J. Dairy Sci.*, *89*, Pp. 410-419.

14. Van der Linden, E., Dröge, J. H. M. (1993). Deformability Of Lamellar Droplets. *Physica A, 193*, Pp. 439-447.

A Study of Green (Environmentally-Benign) Transfer Films through the Tribo-Comminution of Powder Pellets: Experiments and Modeling

E.Y.A. Wornyoh, Mechanical Engineering Department, Carnegie Mellon University, Pittsburgh, PA

P.S. Dougherty, Mechanical Engineering, Carnegie Mellon University, Pittsburgh, PA

C.F. Higgs III, Mechanical Engineering, Carnegie Mellon University, Pittsburgh, PA

Compacted pellets were formed from titanium dioxide ($TiO2$), molybdenum disulfide ($MoS2$), and tungsten disulfide ($WS2$) powders. Using a pellet-on-disk with slider pad tribometer, a series of tribo-comminution tests was conducted to ascertain the friction coefficient of a metal slider pad on a rotating metal disk covered with the powder transfer film. The purpose of the tests was to intentionally transfer third-body particulates to a tungsten carbide (WC) disk where its lubrication characteristics could be studied as it is depleted by a WC pad. It was shown that the pellet actually acted as a self-repairing, self-replenishing, oil-free lubrication mechanism. Results from the experiment were validated by developing an asperity-based fractional coverage (AFC) model, where the fractional coverage represents the fraction of the disk surface asperities that is covered by the lubricious transfer film. The AFC model was used to predict both (1) the friction coefficient at the pad/disk interface and (2) the wear of the powder pellet through comminution against the spinning disk. The disk surface topography was taken from an atomic force microscopy (AFM) image of a 50 microns by 50 microns patch. Using the AFC model, the wear rate (i.e., rate of pellet comminution) of the pellet and the coefficient of friction at the pellet/disk and slider pad/disk interfaces can be determined as a function of the pellet load, slider pad load, disk speed, and material properties. Results from the AFC model reasonably predict qualitatively and quantitatively the experimental behavior.

See more of this Session: Comminution - Experiments, Theory & Modeling
See more of this Group/Topical: Particle Technology Forum

A Novel Method for Stabilizing Particle Size of Milled Powders

Kwek J W[1], Ng W K[1], Tan R B H[1,2]

[1] *Institute of Chemical and Engineering Sciences (ICES), A*STAR, 1 Pesek Road, Jurong Island, Singapore, Singapore 627833*

[2] *Department of Chemical and Biomolecular Engineering, National University of Singapore (NUS), 4 Engineering Drive 4, Singapore, Singapore 117576*

Introduction

Milling or grinding forms an integral part in pharmaceutical secondary manufacturing. However, occurrence of either particle size increases or reduction during storage had been reported [1]. Changes in the particle size can affect, for instance, the powder homogeneity, dissolution rates, and therapeutic effect in dry powder inhalers. The current practice in the pharmaceutical industry is to store milled powders for durations between 2 weeks to 3 months and the duration can vary from material to material. To the best of the authors' knowledge, no techniques are currently available in the market to shorten or avoid the stabilization time. On the other hand, it is known that fluidized bed agglomeration can be used to reduce dust particles as well as stabilize mixtures of various particulate components [2]. Hence, fluidization after ball milling will be used to stabilize the particle size of ball-milled adipic acid (1, 6-hexanedioic acid) powders stored over a period of 7 days under various conditions. An alternative method of stabilizing the particle size by subjecting the milled adipic acid to low temperature and thawing treatment before storing over a period of 7 days under similar conditions will also be tested.

Materials and Methods

Adipic acid (1, 6-hexandioic acid) is selected because it is a single polymorph [3] and is used as an acidulent in effervescent tablets or lubricant in tablets. 99% pure adipic acid from Sigma-Aldrich was recrystallized in ultra-pure water using a batch cooling method [4]. The crystals were vacuum filtered, subsequently dried for 24 hours in an oven set at 55°C and then stored for another 24 hours in a dessicator using phosphorus pentoxide.

A planetary ball mill (Fritsch Pulverisette 5, Germany) with stainless steel milling jars and balls of diameter 10mm was used to mill the recrystallized adipic acid crystals. The mass ratio of the balls to sample was kept at 50:1. The rotation speed was set at 400rpm with a milling time of 30 minutes for all samples. After milling, the sample was stored in an electrical dessicator overnight.

About 36g of milled adipic acid was then subject to fluidization using dry air in a column of diameter 50mm (Fig. 1) for various durations (3 hours and 4 hours) at 30L/min (+/- 5L/min). The fluidized milled adipic acid was sieved using the sonic sifter (L3P Sonic Sifter Separator, USA) and sieved fraction between 250 and 500μm was recovered to be stored for 7 days in dessicators having relative humidities of 25% and 75% respectively. Particle size measurements using the laser diffraction method (Malvern Mastersizer MS-2000, UK) with n-hexane as the dispersion medium and a pump rate of 800rpm were done on samples before and after storage. All particle size measurements were done in triplicate. The particle morphology was examined by high resolution scanning electron microscopy (SEM, JSM-6700F, JEOL, Japan) operating between 2kV to 5kV. Prior to surface analysis, the samples were sputter coated with gold for 1min by a sputter coater (Cressington Sputter Coater 208HR, UK).

Fig. 1. Fluidization column.

An alternative method of stabilizing the particle size over time was also tested. The milled adipic acid was first stored at around 4°C overnight before thawing for approximately 4 hours. The cold stored milled adipic acid was sieved and sieve fraction 250 μm to 500μm was stored for 7 days at 25% and 75%RH respectively. Particle size measurements and surface analysis by SEM were carried out before and after storage.

Results and Discussion

The effects of fluidization and low temperature and thawing treatment on the particle size stability are illustrated in the following figure.

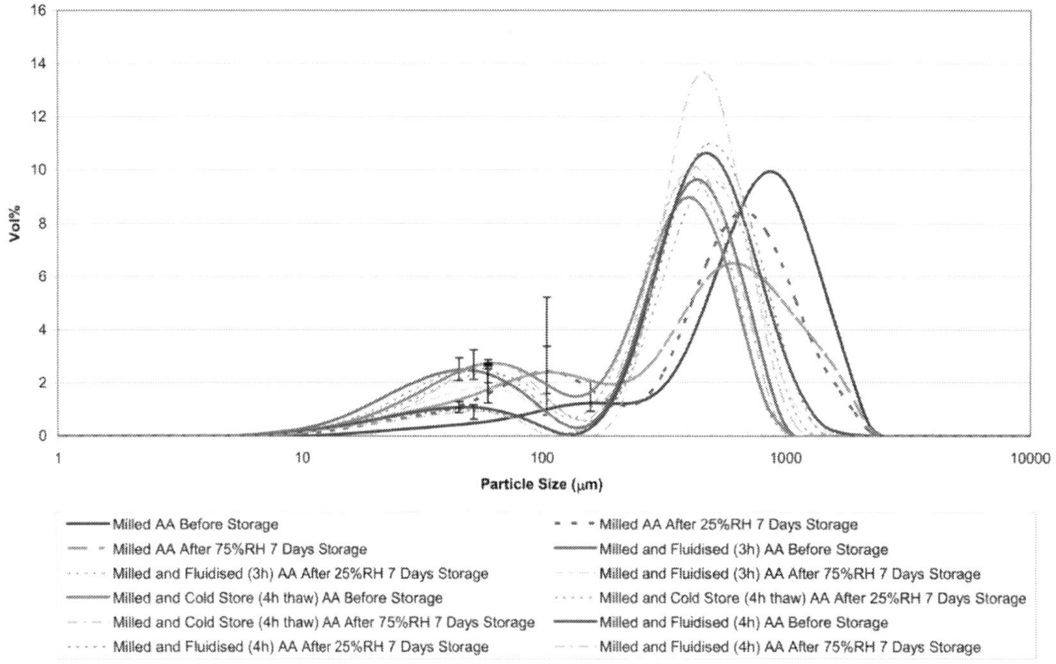

Fig. 2. Effects of fluidization (3 hours and 4 hours) and low temperature treatment (followed by 4hour thawing) on the particle size stability of milled adipic acid.

It can be seen that when the milled adipic acid was treated with fluidization or low temperature storage, there was significantly less particle size shift in the secondary peak as compared to milled adipic acid with no prior treatment after storage for 7 days under various conditions. As fluidization duration increased to 4 hours, the intensity of the secondary peak reduced significantly and particle size shift upon storage for 7 days become even less apparent.

It is observed from SEM images of the surfaces (Fig. 3) that the primary particles appear to be more packed and have less observable voids when they are subject to fluidization. Similarly, milled particles that are treated at low temperature with 4 hour thaw appeared to be more compact than milled particles without treatment.

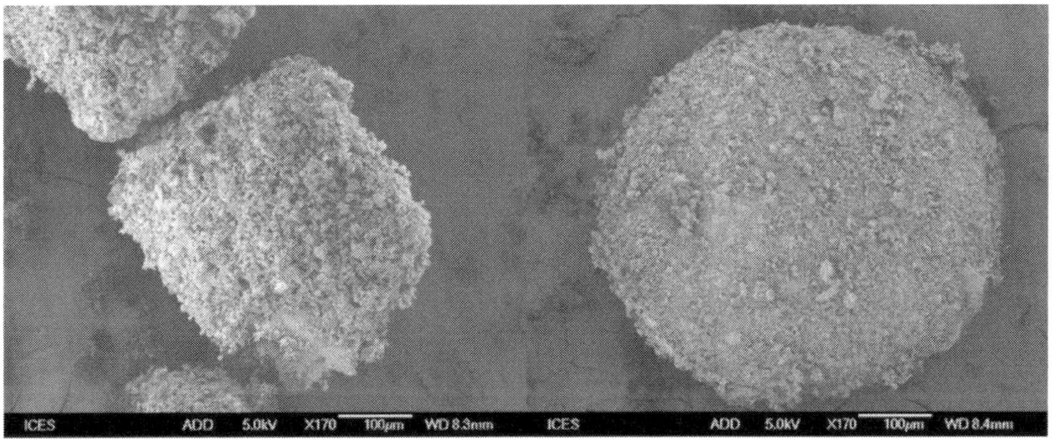

(a) (b)

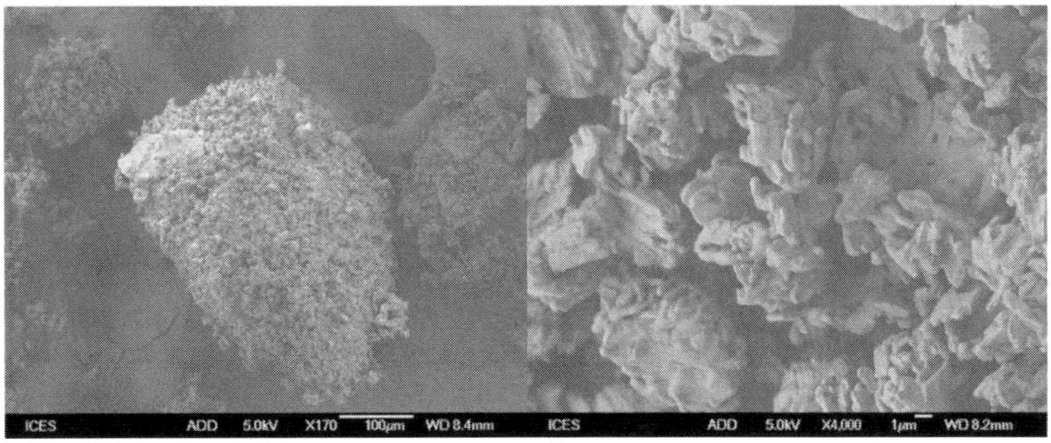

(c) (d)

(e) (f)

Fig. 3. SEM surface morphologies for milled particles [(a) and (d)], milled and fluidized (3h) particles [(b) and (e)], and milled and low temperature treated (4h thaw) particles [(c) and (f)] before storage at 170x and 4000x magnifications respectively.

It is postulated that particle-particle and particle-wall collisions during fluidization aid in removing the loose fines on the bulk primary particle surfaces as well as compacting those remaining on the surfaces as evident in the SEM images. The quantity of loose fines was reduced after fluidization, which was shown by the lower secondary peak of fine particle size fraction when fluidization time increased. As a result, agglomerates which are more stable during storage are formed.

Conclusions

Fluidization after milling could be a potential method to stabilize the particle size of milled materials stored over time under various conditions. Stable agglomerates were formed during fluidization and particle size stability increases with fluidization time. Low temperature storage followed by a 4-hour thaw could also be an alternative method to stabilize the particle size.

References

1. Ng WK, Kwek JW, Tan RBH. Anomalous particle size shift during post-milling storage. *Pharm. Res.* 2008;25:1175-1185.

2. Weber S, Briens C, Berruti F, Chan E, Gray M. Agglomerate stability in fluidized beds of glass beads and silica sand. *Powder Technol.* 2006;165:115-127.

3. Chow KY, Go J, Mehdizadeh M, Grant DJW. Modification of adipic acid crystals: influence of growth in the presence of fatty acid additives on crystal properties. *Int. J. Pharm.* 1984;20:3-24.

4. Keel TR, Thompson C, Davies MC, Tendler SJB, Roberts CJ. AFM studies of the crystallization and habit modification of an excipient material, adipic acid. *Int. J. Pharm.* 2004;280:185-198.

Deagglomeration and Mixing of Nanopowders Using RESS Based Methods

Daniel To, New Jersey Center for Engineered Particulates, New Jersey Institute of Technology, Newark, NJ

Sameer Dalvi, Chemical Engineering, New Jersey Institute of Technology, Clifton, NJ

Sankaran Sundaresan, Department of Chemical Engineering, Princeton University, Princeton, NJ

Rajesh Dave, Otto H. York Department of Chemical, Biological and Pharmaceutical Engineering, New Jersey Institute of Technology, Newark, NJ

Nano-sized materials are gaining popularity because of their unique properties. The high interfacial area can result in improved interactions and increased reaction rates over micron-sized materials, however these desired properties are often not realized due to the agglomeration of the individual nano-sized constituents limiting the available surface area. A common method to overcome this limitation is by mixing two nano-powders together on the particulate scale, which ensures a high interfacial surface area between the two constituents. This is achieved by first breaking up the individual constituent powders and then mix them together. This method is most commonly performed in liquid solvents, however removal of the suspending fluid can be expensive in terms of both time and resources, and potentially lead to drying based segregation. Additionally these solvents can be harmful to the operators and the environment. Therefore the objective of this work was to use dry mixing methods, which are environmentally friendly, to improve the quality of mixing of alumina (dp = 13 nm) and silica (dp = 16 nm) nano-powders using a RESS based mixing method.

The powders were mixed using Rapid Expansion of High Pressure and Supercritical Suspensions process, where nanopowders suspended in high pressure carbon dioxide were simultaneously deagglomerated and mixed upon expansion through a micron-sized capillary nozzle. Agglomerate sizes were determined using a Scanning Mobility Particle Spectrometer (SMPS), by TSI, Inc, which uses differential mobility analysis, an online technique, as well as collection by diffusion and image analysis performed with field emission scanning electron microscopy (SEM). The mixing quality was determined by electron microscopy in conjunction with Energy Dispersive X-Ray Spectroscopy (SEM-EDS). It was shown that mixing quality was dependent on a variety of factors including deagglomeration efficiency, nozzle diameter and premixing.

See more of this Session: Comminution - Experiments, Theory & Modeling
See more of this Group/Topical: Particle Technology Forum

Shearing Induced Breakage of Nanoparticle Agglomerates in a Supercriticle Fluid - A Simulation Study

Yangyang Shen, Chemical and Biochemical Engineering, Rutgers University, Piscataway, NJ

M. Silvina Tomassone, Chemical and Biochemical Engineering, Rutgers University, Piscataway, NJ

A major challenge in making and utilizing materials based on nanoparticles and nanocomposites is the tendency for individual nanoparticle constituents to aggregate due to Van der Waal forces and form large fractal structures on the order of tens of microns. The agglomeration of nanoparticles makes them unusable for the synthesis of nanoparticle-based composites, and ultimately destroys those unique properties.

Recent experiments conducted at NJIT group for deagglomeration of nanoparticles using rapid expansion of supercritical suspensions (RESS) with carbon dioxide (CO_2) revealed that this technique is very effective in breaking nanoparticle agglomerates, and can result in simultaneous mixing when two or more nanoparticle constituents are used in the suspension. To better understand the deagglomeration mechanisms in RESS, it is important to shed light on the interaction forces between nanoparticles in supercritical CO_2, which determine the dominant mechanisms for deagglomeration when the fluid expands rapidly. To the best of our knowledge, there have been no reported simulation studies of breakup forces for nanoscale silica agglomerates in the presence of supercritical CO_2. In this work we perform molecular simulations of silica nanoparticle agglomerates in supercritical CO_2 after exposing them to shear forces. We considered different types of aggregates (with different topologies and strengths) and studied how the agglomerate structure affects their final strength.

Firstly, we used Grand Canonical Monte Carlo to validate the parameters for fluid-fluid and solid-fluid interactions with experimental studies, and then performed molecular dynamics (MD) simulations to investigate the behavior of nanoparticle aggregates under a range of shear rates, i.e. from $10^5 s^{-1}$ (low) to $10^6 s^{-1}$ (high). A group at Princeton studied this process theoretically, and obtained that the flow of the supercritical fluid inside the nozzle in the RESS device generates shear forces of the same order. We could quantitatively confirm that deagglomeration occurs under the influence of those values of the shear force.

See more of this Session: Comminution - Experiments, Theory & Modeling
See more of this Group/Topical: Particle Technology Forum

Breakage of Colloidal Aggregates in Shear Flow

Yogesh M. Harshe, Chemistry and applied biosciences, Institute for chemical and bioengineering, ETH Zurich, Zurich, Switzerland

Marco Lattuada, Institute for Chemical and Bioengineering, Department of Chemistry and Applied Biosciences, ETH Zurich, Zurich, 8093, Switzerland

Massimo Morbidelli, Department of Chemistry and Applied Biosciences, Institute for Chemical and Bioengineering, Zurich, Switzerland

Abstract:

The aggregation, breakage, and restructuring of colloidal clusters have been studied thoroughly in literature through both modeling and experiments. The cluster behaviour in different flows and scaling laws for cluster size on flow parameters for clusters with great variety of cluster geometries has been reported by many researchers. In due of these extensive efforts in modeling, the fundamental understanding of cluster breakage is still incomplete due to the fact that none of the efforts accounted both hydrodynamic particle-particle interactions in detail. Surprisingly, even after the experimental evidence of bending moment presented by Pantina and Furst[1], there are only a few simulation studies where the tangential contact forces are correctly included. In addition, almost no contribution has been published where rigorous simulations have been performed to investigate the dependence of the time required for the on-set of cluster breakage and fragment size distribution on cluster size, structure, shear rate, and inter-particle interactions. In order to correctly predict the fate of a colloidal cluster in a fluid flow both hydrodynamic and inter-particle interactions should be accounted correctly.

In the present work, we have used the Stokesian dynamics method developed by Brady and Bossis[2], which allows computation of the full hydrodynamic interactions between particles. The inter-particle interactions were introduced through the DLVO theory. To account for the contact forces the tangential force model suggested by Becker and Briesen[3] was implemented. Simulations were performed for various cluster sizes with different fractal dimensions, generated using a combination of different Monte-Carlo methods, under different flow conditions such as gravity, simple shear, and elongational flow. The developed model was first used to investigate the dependence of time required for the on-set of cluster breakage on the cluster geometry (structure and size) and primary particle size for various Peclet number values. Precious information on the fragment size distribution has been also obtained.

Keywords:

Aggregation, breakage, Stokesian dynamics, tangential forces

References:

[1] Pantina, J.P., and Furst, E.M., Langmuir, 22, 5282, 2006

[2] Brady, J.F., and Bossis, G., J, Ann. Rev. Fluid Mechanics, 20, 111, 1988

[3] Becker, V., and Brisen, H., Phy. Rev. E, 78(6), 061404, 2008

See more of this Session: Dynamics and Modeling of Particles, Crystals and Agglomerate Formation
See more of this Group/Topical: Particle Technology Forum

Restructuring Phenomena in the Aggregation of a Colloidal Latex in a Turbulent Stirred Tank

Matthäus U. Bäbler, Department of Mechanical and Process Engineering, Institute of Process Engineering, ETH Zurich, Zurich, Switzerland

Miroslav Soos, Department of Chemistry and Applied Biosciences, Institute for Chemical and Bioengineering, ETH Zurich, Zurich, Switzerland

Marco Mazzotti, Department of Mechanical and Process Engineering, Institute of Process Engineering, ETH Zurich, Zurich, Switzerland

Massimo Morbidelli, Department of Chemistry and Applied Biosciences, Institute for Chemical and Bioengineering, ETH Zurich, Zurich, Switzerland

Aggregation of colloidal particles in a flowing suspension is accompanied by two important mechanical effects, i.e., breakage and restructuring of the formed clusters caused by hydrodynamic stresses acting on theme [1-3]. Restructuring is thereby understood as the densification of the clusters as they are processed in an agitated vessel. In this work, we investigate the restructuring of colloidal clusters in a turbulent stirred tank both experimentally and theoretically through a population balance equation (PBE) model. In the experiments, a fully destabilized polystyrene latex with a primary particle size of d_p=810 nm is aggregated in a stirred tank reactor. The evolution of the cluster mass distribution (CMD) is monitored through small angle static light scattering from which we obtain two moments of the CMD, namely a mean radius of gyration and the zero angle intensity that represents a weighted mean cluster mass. The fractal dimension of the initial clusters is inferred from fitting the initial aggregation kinetics to our PBE model. The latter includes a detailed aggregation rate model [4] that accounts for the specific hydrodynamic and colloidal interactions between colliding clusters. From this we found $d_f \approx 2.0$. As the clusters grow larger the hydrodynamic stresses acting on theme increase and eventually breakup and restructuring set in. As a consequence of this, clusters at steady state assume a relatively compact structure. The fractal dimension at steady state is estimated from microscopic images adopting an image analysis routine and an empirical correlation between the fractal dimension and the perimeter fractal dimension of the projection [5]. From this we found $d_f \approx 2.6$. To gain further insight into this densification process, i.e., the transition from $d_f \approx$ 2.0 to 2.6, a PBE model is formulated that combines the previous aggregation rate model with a comprehensive breakage rate model [6]. The evolution of the fractal dimension is then obtained by fitting the PBE model to the experimental data. From this we found a sigmoidal evolution of d_f with time, which can be understood that the clusters have to grow to a certain size before restructuring sets in [7]. However, whether restructuring is a result of breakage, i.e., formation of dense fragments, or if aggregates compactify as coherent entities cannot be inferred from our data.

[1] Soos M., Sefcik J., Morbidelli M., *Chem. Eng. Sci.* **61**, 2349 (2006)

[2] Selomulya C., Bushell G., Amal R., Waite T. D., *Langmuir* **18**, 1974 (2002)

[3] Selomulya C., Bushell G., Amal R., Waite T. D., *Chem. Eng. Sci.* **58**, 327 (2003)

[4] Bäbler M.U., *AIChE J.* **54**, 1748 (2008)

[5] Lee C., Kramer T. A., *Adv. Colloid Interface Sci.* **112**, 49 (2004)

[6] Bäbler M.U., Morbidelli M., Baldyga J., *J. Fluid Mech.* **612**, 261 (2008)

[7] Becker V., Schlauch E., Behr M., Briesen H., *J. Colloid Interface Sci.* submitted (2009)

See more of this Session: Dynamics and Modeling of Particles, Crystals and Agglomerate Formation
See more of this Group/Topical: Particle Technology Forum

An Aggregation Model for Flash NanoPrecipitation

Chungyin Cheng, Chemical and Biological Engineering, Iowa State University, Ames, IA

R. Dennis Vigil, Chemical & Biological Engineering, Iowa State University, Ames, IA

Rodney O. Fox, Chemical and Biological Engineering, Iowa State University, Ames, IA

Flash NanoPrecipitation is a novel approach for producing functional nanoparticles stabilized by amphiphilic block copolymers. A microscale multi-inlet vortex (MIV) reactor is utilized for this process because of its mixing performance and flexibility. During this process, the particle size distribution is frozen and stabilized by the amphiphilic di-block copolymer on the organic particle surface. That is, the particle growth is kinetically arrested and thus a narrow particle size distribution can be attained. To model the precipitation process, a bivariate population balance equation (PBE) has been formulated to account for co-precipitation of the organic and the block copolymer. Aggregation rate kernels have been derived to account for the major different aggregation events: free coupling, unimer insertion, and aggregate fusion. The resulting PBE is solved both by direct integration and by using the quadrature method of moments.

See more of this Session: Dynamics and Modeling of Particles, Crystals and Agglomerate Formation
See more of this Group/Topical: Particle Technology Forum

Analysis of Nucleation Rate Measurements in Precipitation of Silver Nanoparticles

Roberto Irizarry, Electronic Technologies, DuPont, Raleigh, NC

The nucleation rate of silver particles produced by the reduction of a silver-EDA complex with ascorbic acid is studied. The experimental system consists of a rapid mixing stopped flow reactor with a high resolution diode array for full spectra collection (UV-visible) as a function of time. The nucleation rate of silver particle is studied experimentally using plasmonic calculations. The experimental data is analyzed in terms of homogeneous and heterogeneous nucleation theories. To study the possibility of mechanism switch during the particle formation process, the model and parameter identification is made by solving a hybrid dynamic optimization. In this framework, mechanism switching times are determined as a part of the solution of the optimization problem. The optimized model for nucleation is then used in a population balance model to simulate particle growth.

See more of this Session: Dynamics and Modeling of Particles, Crystals and Agglomerate Formation
See more of this Group/Topical: Particle Technology Forum

Breakage Behavior of NaCl Crystals in a Stirred Vessel

Sheena Reeves[1], **Brian White[1], Stephen Castellane[2] and Priscilla Hill[1]**
(1) Dave C. Swalm School of Chemical Engineering, Mississippi State University, Mississippi State, MS
(2) Mississippi Department of Environmental Equality, Jackson, MS

The presence of the small particles resulting from breakage during crystallization may cause unnecessary filter plugging downstream and, often, changes the crystal size distribution. The affect of a particle's shape and size on particle size distribution resulting from breakage can be attributed to two phenomenons – attrition and fragmentation. It is well known that crystals can break due to impact with the crystallizer impeller (1, 2). Some research has indicated that almost all of this breakage is by attrition rather than fragmentation (2, 3). However, it is not clear if this is a factor of crystal size or of other operating parameters such as agitation rate (4).

This breakage was investigated using cubic NaCl crystals, which are usually considered to be more prone to attrition than fragmentation. Studies were performed with the crystals suspended in a saturated solution and a nonsolvent. Due to the highly regular shape of the NaCl crystals, it is more apparent when there is fragmentation rather than attrition. Particle size distributions and weight distributions for each experiment was recorded and quantified. A wide range of initial particle sizes were used. One set was with crystals grown in the lab where the initial size range was ~3500-7500 µm, while other experiments were performed with commercially available crystals in smaller size ranges such as ~350 – 700 µm. In all cases experiments were performed across a range of agitation rates, residence times, and magma densities.

Results are presented which show that there may be significant fragmentation for the larger crystals. Figure 1 depicts the fraction of several particle sizes resulting from breakage over time in saturated solution and the nonsolvent, acetonitrile, respectively. Each size range is based on the value of the resulting major axis. This research indicates that for some cases it is necessary to account for fragmentation as well as attrition. This is demonstrated using population balance modeling.

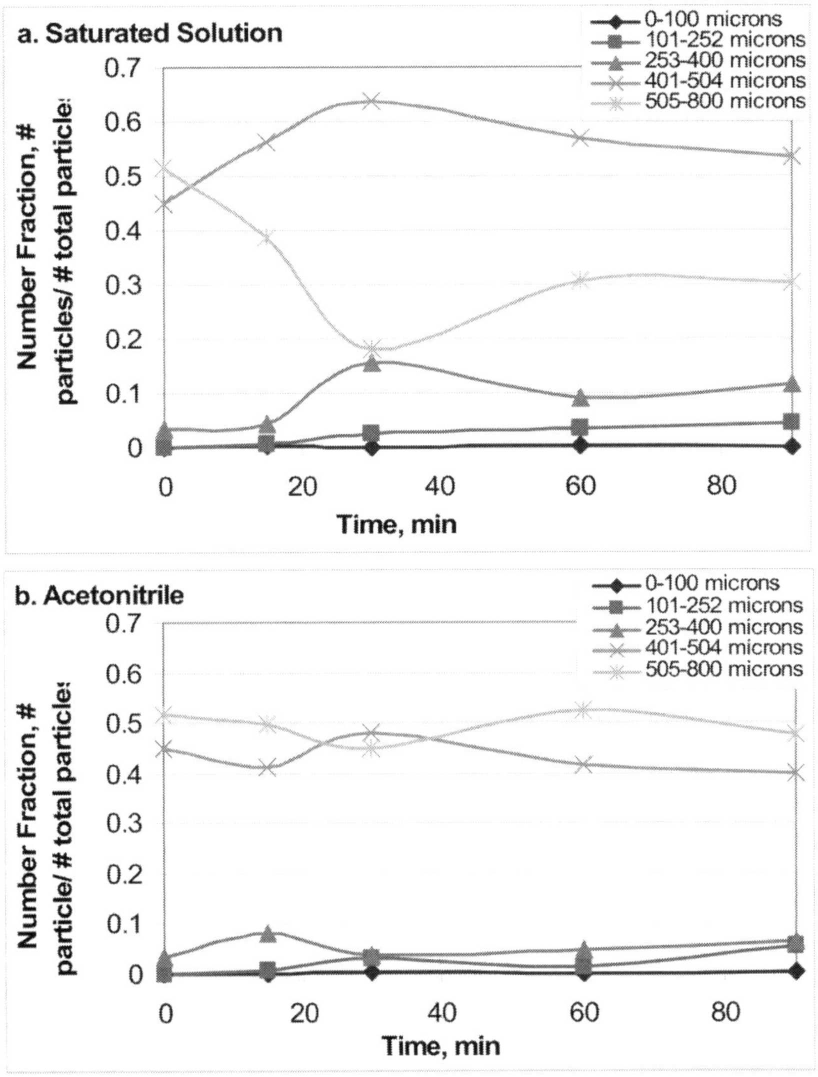

Figure 1: The graphs show the number Fraction associated with various major axis size ranges with respect to time in a) Saturated Solution and in b) Acetonitrile.

REFERENCES

1. Gahn, C., and A. Mersmann, "Brittle fracture in crystallization processes Part A. Attrition and abrasion of brittle solids." *Chem. Eng. Sci.,* **54,** 1273-1282 (1999).
2. Bravi, M., Di Cave, S., Mazzarotta, B., and N. Verdone, "Relating the attrition behavior of crystals in a stirred vessel to their mechanical properties." *Chem. Eng. J.*, Vol. **94**: 223-229 (2003).
3. Nienow, A. W., and R. Conti, "Particle abrasion at high solids concentration in stirred vessels." *Chem. Eng. Sci.*, **33**, 1077-1086 (1978).
4. Hill, P., "Statistics of Multiple Particle Breakage Accounting for Particle Shape." *AIChE J.*, Vol. **50** (5): 937-952 (2004).

Simulating the Shape Evolution of Crystals with Monte Carlo Methods

Andreas Voigt, Process Systems Engineering, University of Magdeburg, Magdeburg, Germany

Kai Sundmacher, Max Planck Institute for Dynamics of Complex Technical Systems, Magdeburg, Germany

Crystallization from a solution is an important process to produce particulate materials in industry. Often not only the size of the particulate products but also the specific shape of the crystals need to be tailored in order to meet the demands of the applications like pharmaceuticals, catalysis or electronics. From the molecular perspective the evolution of a crystal shape is influenced by several phenomena. Diffusion of the molecules or basic building blocks with several molecules transports the material needed to growth crystals from a starting nucleus. The attachment of these primary units on and into the crystal may be different at different crystal faces depending on the corresponding energy barrier to incorporate these units. After the attachment the units might still be able to change their position on the surfaces of the crystal. The incorporation at steps might be less favorable than the incorporation into holes of an already build surface. The dynamical picture of the shape evolution therefore will depend on a number of material and process conditions. Here we will present a Monte Carlo based method where the aforementioned phenomena are incorporated and different scenarios of a shape evolution of crystals from solution are presented. In experiments on barium sulfate we have seen very interesting crystal shapes of different kinds [1]. The simulation of such systems will provide a deeper insight into the shape evolution and will help to improve our understanding of the crystallization process as a whole. [1] C. Steyer and K. Sundmacher: Impact of feeding policy and ion excess on particle shape in semi-batch precipitation of barium sulfate, J of Crystal Growth (2009), doi:10.1016/j.jcrysgro.2009.02.046 .

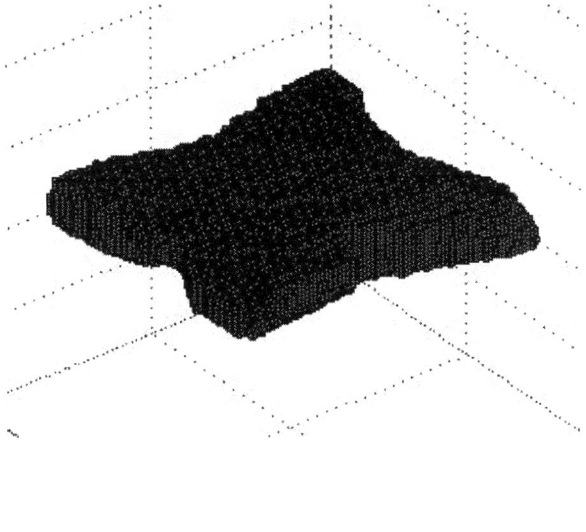

See more of this Session: Dynamics and Modeling of Particles, Crystals and Agglomerate Formation
See more of this Group/Topical: Particle Technology Forum

Two-nozzle flame synthesis of NO_x storage reduction catalysts

Robert Büchel[a,b] Reto Strobel[a1], F. Krumeich[a], Alfons Baiker[b,] and Sotiris E. Pratsinis[a,]

[a]*Particle Technology Laboratory, Department of Mechanical and Process Engineering, ETH Zurich, Sonneggstrasse 3, CH-8092 Zürich Switzerland.*

[b]*Institute for Chemical and Bioengineering, Department of Chemistry and Applied Biosciences, ETH Zurich, Wolfgang-Pauli-Strasse 10, CH-8093 Zurich, Switzerland.*

[1]*Currently with Satisloh Photonics AG, CH-8812 Horgen Switzerland*

Abstract

Catalysts for NO_x storage-reduction (NSR) were made selectively with Pt preferentially on the Al- or the Ba-components without altering significantly the Al_2O_3 or $BaCO_3$ crystal size, Al/Ba weight ratio, specific surface area, porosity, and Pt dispersion using a two-nozzle flame spray pyrolysis (FSP) unit. The NO_x storage performance at 300 °C was best for Pt located near Al_2O_3 as it facilitates the oxidation of NO to NO_2 during the fuel lean period but the reduction rate during the subsequent short fuel rich period was much slower resulting in incomplete regeneration. This contributed to a gradual decrease of the NO_x conversion at increasing cycling. In contrast, Pt on $BaCO_3$ resulted in an initially lower NO_x storage rate but during ten storage-reduction cycles a stable NO_x conversion of about 50% was reached. When using NO_2 instead of NO or higher NO_x oxidation-reduction temperatures (e.g. 350 °C) the Pt location did not affect the NSR performance of the $Pt/Ba/Al_2O_3$ catalysts.

Introduction

Anthropogenic NO_x is formed typically during fuel combustion contributing to smog formation and acid rain [1]. New catalysts are needed to meet stricter emission limits, especially to remove NO_x under oxygen rich conditions of lean fuel and direct injection engines. Among these catalysts, the ones for NO_x storage-reduction (NSR) can trap exhaust NO_x on an alkali- or alkaline earth metal (typically Ba or K) in the form of metal-nitrates [2] without requiring an additional reducing agent, as NH_3 or urea in selective catalytic reduction (SCR). During a subsequent fuel rich period the stored metal-nitrates are reduced to harmless N_2 over a noble metal (typically Pt or Pd) [3]. The storage and the noble metal are supported on a thermally stable carrier material, usually Al_2O_3 or CeO_2 [4]. Pt-free BaO/Al_2O_3 mainly stored NO_2, whereby releasing one NO for three NO_2 stored [5]. In the presence of Pt, NO is oxidized to NO_2 resulting in higher storage capacity and faster NO_x uptake [6], involving NO_2 spillover from the Pt to the Ba sites as proposed from simulations [7] and experimental investigations [8]. Already in 1995 the proximity of Pt to the storage component of NSR catalysts was proposed to strongly influence the performance [9] and was recently investigated by comparing ternary $Pt/Ba/Al_2O_3$ catalysts with binary mixtures of Pt/Al_2O_3 and Ba/Al_2O_3 [10]. For Pt in close contact with Ba a 5-times higher isotopic exchange rate between ^{15}NO was measured, indicative of NO_x forward and reverse spillover from Pt to Ba [10].

In contrast to storage, the reduction (regeneration) is less understood mainly due to parallel occurring phenomena such as the desorption of gaseous NO_x at high temperature [11] and the reaction of the stored nitrate with CO_2 [12, 13]. Nova et al. [14] showed that the reduction of nitrates occurs by Pt-catalyzed surface reactions involving (reverse) spillover processes and concluded that a close proximity of Ba and Pt would be necessary to increase the Ba-nitrate decomposition.

Here, we investigated the importance of Pt being located close to Ba or to Al on NSR catalyst performance. The catalysts were prepared by a two-nozzle flame spray pyrolysis unit as already applied for NSR catalysts [15] with high storage capacity [16]. By this procedure, Pt can be deposited preferentially on Al_2O_3 and/or $BaCO_3$ without altering their structural properties.

Results

With a two nozzle flame spray pyrolysis unit $Pt/Ba/Al_2O_3$ catalysts with controlled deposition of Pt clusters on $BaCO_3$ or Al_2O_3 were prepared for studying the significance of Pt proximity to either Ba or Al in NO_x storage-reduction catalysts. In **Fig. 1** catalysts with Pt preferentially on the Al_2O_3 are shown in A) and with preferential deposition on $BaCO_3$ are shown in B).

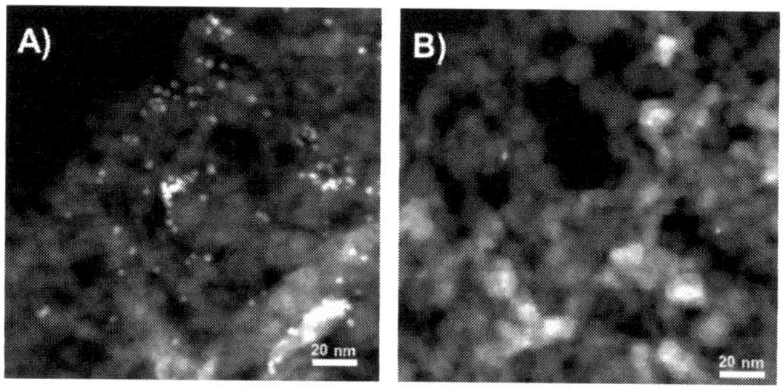

Figure 1: $Pt/Ba/Al_2O_3$ catalysts with Pt preferentially on Al_2O_3 in A) and Pt on $BaCO_3$ in B).

In **Fig.2** the beneficial role of a close interaction of Pt and Ba for NO_x uptake could be confirmed. However, Pt on Al_2O_3 exhibited a better NO oxidation activity, which was the limiting step for the overall NO storage process at low temperatures (300 °C). During reduction, Pt on Ba showed much better activity than Pt on Al_2O_3. This can be attributed to the importance of reverse spill-over and the promotional effect of Ba on the Pt reduction activity. Storage of NO and NO_2 until saturation revealed that the storage capacity is not limited by Ba but by the loss of Pt oxidation activity during catalyst restructuring. We concluded that the spill over is not the limiting step. [17]

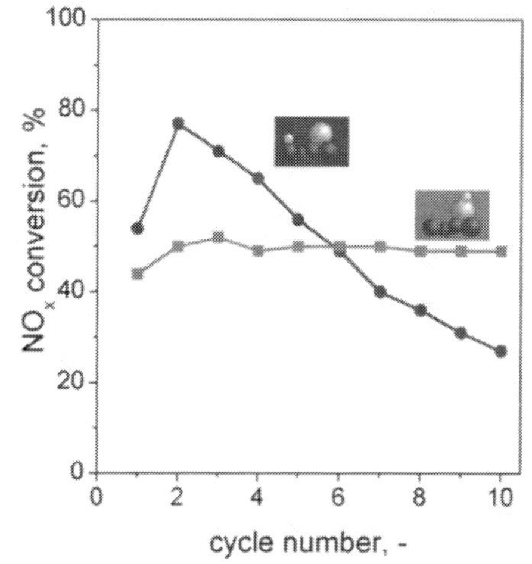

Figure 2: NO_x conversion for the different Pt location at 300 °C.

Acknowledgements

We kindly acknowledge financial support by ETH Zürich (TH-09 06-2) and thank the Electron Microscopy Center of ETH Zurich (EMEZ) for providing the necessary infrastructure. The contribution of platinum chemicals by Johnson Matthey PLC is greatly appreciated.

References

[1] J. Zeldovich, Acta Physicochimica URSS 21 (1946) 577.

[2] N. Takahashi, H. Shinjoh, T. Iijima, T. Suzuki, K. Yamazaki, K. Yokota, H. Suzuki, N. Miyoshi, S. Matsumoto, T. Tanizawa, T. Tanaka, S. Tateishi, K. Kasahara, Catal. Today 27 (1996) 63-69.

[3] W.S. Epling, L.E. Campbell, A. Yezerets, N.W. Currier, J.E. Parks, Cat. Rev. - Sci. Eng. 46 (2004) 163-245.

[4] A. Casapu, J.D. Grunwaldt, M. Maciejewski, F. Krumeich, A. Baiker, M. Wittrock, S. Eckhoff, Appl. Catal., B 78 (2008) 288-300.

[5] N.W. Cant, M.J. Patterson, Catal. Today 73 (2002) 271-278.

[6] B. Subramaniam, A. Varma, Chem. Eng. Commun. 20 (1983) 81-91.

[7] L. Olsson, H. Persson, E. Fridell, M. Skoglundh, B. Andersson, J. Phys. Chem. B 105 (2001) 6895-6906.

[8] Z.Q. Liu, J.A. Anderson, J. Catal. 224 (2004) 18-27.

[9] N. Miyoshi, S. Matsumoto, K. Katoh, T. Tanaka, J. Harada, N. Takahashi, K. Yokota, M. Sugiura, K. Kasahara, SAE Technical Paper 950809 (1995)

[10] N.W. Cant, I.O.Y. Liu, M.J. Patterson, J. Catal. 243 (2006) 309-317.

[11] K.S. Kabin, R.L. Muncrief, M.P. Harold, Catal. Today 96 (2004) 79-89.

[12] S. Balcon, C. Potvin, L. Salin, J.F. Tempere, G. Djega-Mariadassou, Catal. Lett. 60 (1999) 39-43.

[13] A. Amberntsson, H. Persson, P. Engstrom, B. Kasemo, Appl. Catal. B 31 (2001) 27-38.
[14] I. Nova, L. Lietti, L. Castoldi, E. Tronconi, P. Forzatti, J. Catal. 239 (2006) 244-254.
[15]R. Strobel, L. Madler, M. Piacentini, M. Maciejewski, A. Baiker, S.E. Pratsinis, Chem. Mater. 18 (2006) 2532-2537.
[16] M. Piacentini, R. Strobel, M. Maciejewski, S.E. Pratsinis, A. Baiker, J. Catal. 243 (2006) 43-56.
[17] R. Büchel, R. Strobel, F. Krumeich, A. Baiker, S.E. Pratsinis, J. Catal. 261 (2009) 201-207.

Controlled Manufacturing of Nanostructured Catalyst Particles Using Atomic Layer Deposition

J. Ruud van Ommen, DelftChemTech, Delft University of Technology, Delft, Netherlands

Emmanuel Skupien, DelftChemTech, Delft University of Technology, Delft, Netherlands

Guido Mul, DelftChemTech, Delft University of Technology, Delft, Netherlands

Core-shell nanoparticles have high potential in heterogeneous catalysis. The interaction between host material and the coating gives additional degrees of freedom in the design of catalysts. We have applied the Atomic Layer Deposition (ALD) to provide particles with an ultra-thin, uniform layer. This technique is different from CVD in that the chemistry is split into two half-reactions; the different reactant gases are fed to the sample consecutively rather than simultaneously. The number of repetitions of the cycle determines the thickness of the coating, resulting in a well-controlled deposited layer at the atomic level. The quality of the layer is determined by the surface coverage in the adsorption step.

Ferguson et al. (2000) showed that ALD to particles is best carried out when these particles are fluidized, ensuring good gas-solids contacting. We recently showed that ALD of fluidized particles can also be carried out at atmospheric pressure (Beetstra, 2009) which simplifies the fluidization of the particles and facilitates scaling up of the process. In this paper, we will show that in this way, we are able to create catalyst nanoparticles in a controllable and scalable way.

References

Ferguson, J.D., Weimer, A.W., George S.M. (2000), 'Atomic layer deposition of ultrathin and comformal Al2O3 films on BN particles', Thin Solid Films 371, 95-104.

Beetstra, R., Lafont, U., Nijenhuis, J., Kelder, E.M., van Ommen, J.R. (2009), Atmospheric Pressure Process to Coat Particles using Atomic Layer Deposition, Chemical Vapour Deposition, in press.

See more of this Session: Nanostructured Particles for Catalysis
See more of this Group/Topical: Particle Technology Forum

Supercritical Fluid Aided Synthesis of Nanostructured Catalyst

Sermin G. Sunol, University of South Florida, Tampa, FL

Haitoi Li, Chemical Engineering, University of South Florida, Tampa, FL

Aydin Sunol, Department of Chemical Engineering, University of South Florida, Tampa, FL

This presentation deals with synthesis of aerogel catalysts where the support is prepared using templates and sol-gel method followed with supercritical drying. The aerogel support is subsequently impregnated/doped with nano size metal/metaloxide particles using supercritical fluid carriers. The support is titanium dioxide while a portfolio of particles, including Iron and Tungsten, were incorporated. The catalyst activity for phenol degradation and detailed characterization using XPS, BET, Pore size distribution, FTIR, SEM, EDS, and TEM was performed.

See more of this Session: Nanostructured Particles for Catalysis
See more of this Group/Topical: Particle Technology Forum

Formation of Carbon Structures On Stepped and Chiral Surfaces

Perla B. Balbuena, Chemical Engineering and Materials Science and Engineering, Texas A&M University, College Station, TX

Juan C. Burgos, Chemical Engineering and Materials Science and Engineering, Texas A&M University, College Station, TX

Gustavo E. Ramirez-Caballero, Chemical Engineering and Materials Science and Engineering, Texas A&M University, College Station, TX

Diego A. Gomez-Gualdron, Chemical Engineering and Materials Science and Engineering, Texas A&M University, College Station, TX

Single-wall carbon nanotubes grow over catalytic nanoparticles via chemical vapor deposition methods, where a C-containing precursor gas is decomposed over the catalytic surface at ~ 1000 K. A few synthesis methods are able to produce nanotubes in a relative narrow range of chiralities; however the mechanism for such selectivity is not well understood. Our hypothesis is that the surface pattern is able to induce chirality of the nascent tube, and viceversa, under certain conditions carbon adsorption may yield surface reconstruction thus generating chiral surfaces. We present a density functional theory study of adsorption of single carbon atoms and carbon chains on stepped transition metal surfaces, using the projector-augmented wave method and the Perdew-Burke-Ernzerhof exchange-correlation functional, within the generalized gradient approximation. We discuss a series of interesting carbon structures formed over the (211) Co surface which evolve as the C concentration on the surface increases. At low concentration of C we observe the formation of zig-zag chains where each of the atoms in the zigzag are alternatively bonded to a step and to a terrace site. After more carbon is added to the system, perfect graphene structures are formed.1 Interestingly, our findings are in excellent agreement to recently published experimental results.2,3 We use ab initio molecular dynamics to investigate the effect of temperature on the stability of the crabon structures, and we extend the study using chiral surfaces, to analyze if carbon atmospheres on chiral Co surfaces may also lead to the formation of alternative structures. Preliminary calculations suggest that horizontally aligned nanotubes could be also the result of surface patterning.

References

(1) Ramirez-Caballero, G. E.; Burgos, J. C.; Balbuena, P. B. Growth of carbon structures on stepped (211) cobalt surfaces J. Phys. Chem. C 2009, in press.

(2) Zhu, H.; Suenaga, K.; Hashimoto, A.; Urita, K.; Hata, K.; Iijima, S. Atomic-resolution imaging of the nucleation points of single-wall carbon nanotubes Small 2005, 1, 1180.

(3) Zhu, H.; Suenaga, K.; Wei, J.; Wang, K.; Wu, D. A strategy to control the chirality of single-wall carbon nanotubes J. of Crystal Growth 2008, 310, 5473.

See more of this Session: Nanostructured Particles for Catalysis
See more of this Group/Topical: Particle Technology Forum

Size-Selected Pt Nanoparticles Synthesized Via Micelle Encapsulation: Effect of Pretreatment and Oxidation State On the Activity for Methanol Decomposition and Oxidation

Jason R. Croy, University of Central Florida, Orlando, FL

Simon Mostafa, University of Central Florida, Orlando, FL

Helge Heinrich, University of Central Florida, Orlando, FL

Beatriz Roldan-Cuenya, University of Central Florida, Orlando, FL

This article presents the effect of pretreatment conditions on the oxidation state and subsequent activity for methanol (MeOH) decomposition and oxidation of size-selected, micelle-synthesized, Pt nanoparticles (NPs) supported on nanocrystalline ZrO_2. Exposure to an oxygen-rich atmosphere was observed to be the most effective pretreatment for producing clean, thermally stable, and active NP catalysts from polymeric reverse micelles. The Pt-oxide species formed on our samples upon annealing in O_2 were found to be highly stable, and could not be fully reduced under the reaction conditions. This effect is attributed to a strong Pt/ZrO_2 interaction and the possible stabilization of interfacial Pt-O-Zr species. Nevertheless, the initial presence of Pt-oxide species in our samples was found to have little influence on the conversion or selectivity in MeOH decomposition reactions due to their tendency to reduce (with the exception of interfacial Pt-O-Zr sites) in the presence of MeOH and its decomposition products (i.e. CO). During MeOH oxidation reactions, the oxides of Pt appear to be very stable and as active as metallic Pt. Our inability to reduce these oxidized Pt species in the course of the latter reactions is attributed to their possible participation in Mars-van Krevelen-type processes, in which bound-oxygen (NP shell) may be replenished with oxygen from the gas phase.

See more of this Session: Nanostructured Particles for Catalysis
See more of this Group/Topical: Particle Technology Forum

Base Effect On Hydrogen Production From Biomass Over Transition Metal Oxide Nanoparticles

Zhong He, Chemical, Biological and Pharmaceutical Engineering, New Jersey Institute of Technology, Newark, NJ

Xianqin Wang, Chemical, Biological and Pharmaceutical Engineering, New Jersey Institute of Technology, Newark, NJ

In this work, a series of transition metal oxide nanoparticles with or without base metal or metal oxdes are prepared with different methods. These samples are characterized with Brunauer-Emmett-Teller (BET), chemisorption, temperature programmed reduction (TPR), temperature programmed desorption (TPD) and synchrotron-based X-ray diffraction (XRD) and X-ray absorption (XAS) techniques. The hydrogen production from various biomass are tested over these samples in a fixed bed continuous flow reactor, the products are analyzed with HPGC and HPLC. Comparing the results from the samples with and without base metals, we find that the base metal or metal oxides on the sample have significant effect over the reaction temperature and reactivity. The roles of the base on hydrogen production from biomass will be discussed in detail in this talk.

See more of this Session: Nanostructured Particles for Catalysis
See more of this Group/Topical: Particle Technology Forum

A Systematic Approach to DEM Material Model Calibration

David Curry, DEM Solutions Ltd., Edinburgh, United Kingdom

John Favier, DEM Solutions Ltd., Edinburgh, United Kingdom

Richard D. LaRoche, DEM Solutions USA Inc., Lebanon, NH

Processing and handling of particulate material is a significant factor in the cost of production in many industries. Some of the ways in which these costs can be reduced is by increasing production efficiency, reducing product waste, or increasing processing through-put. In each of these cases, new designs and/or operational techniques are required. By achieving a small percentage improvement in performance, a large amount of savings may be gained.

To assess the benefits of new designs, testing must be performed in either an experimental or "virtual" mode. By using virtual testing to minimize the number of design options and hence the number or prototypes and test points required for engineering decisions, huge savings may again be made. Discrete Element Modeling (DEM) is such a tool for modeling particulate flows and processes.

One challenge with DEM is to have the virtual experiments match the physical experiments within an acceptable level of accuracy. To achieve this, DEM requires the input of material properties and interaction parameters; e.g., coefficients of friction, shear modulus, etc. The current methods for determining such parameters are based on physical or virtual tests. In many cases, experimentally-derived material properties will not reproduce the desired material behavior in a DEM simulation to the level of accuracy that is required. In addition, simple tests may not be able to provide the insight into which parameters significantly influence the desired result.

The purpose of simulating a particulate process with DEM is to achieve an accurate representation of the bulk behavior of the material. This is done by defining material properties that affect the material behavior at the particle scale which consequently create the overall bulk behavior. When the physical particle size is small relative to the volume of bulk material, further assumptions and scaling are needed to make the problem computationally tractable. Since direct experimental measurement of these material properties is not straightforward, and particle scaling reduces the precision of these properties, extracting the correct material properties from one experiment or simulation and applying it to another is a great challenge.

In many cases, the direct measurement of particle properties from a bulk process is not possible. Likewise, measuring individual particle properties and applying them to a simulation of a bulk material process may not yield accurate results either. Much of this is because a DEM simulation will usually consist of uniform particles, (with the exception of a possible size distribution), that are intended to approximate the actual particle shapes. For bulk flows, perfectly matching the sizes and shapes of all the particles is not possible and further increases the approximation errors of the simulation. In addition, parameters that can be measured may not have a direct correlation to a variable that is needed in the simulation.

The goal of this approach is to devise a system that includes basic experiments and an optimization procedure that calculates the DEM material properties necessary for simulations to produce results within a suitable accuracy range. To achieve this goal, four major program tasks have been defined:

1. Develop experiments suitable for characterizing bulk material for subsequent DEM analysis.

2. Develop advanced contact models, particle shapes, and particle scaling rules for the EDEM software package that have the proper physics to model the materials of interest.

3. Develop an optimization algorithm that can take the experimental data as an input and select the

parameters for the contact model to best match the simulation with the experiments.

4. Apply the selected experiments to a system to validate the model.

See more of this Session: Poster Session: Particle Technology Forum
See more of this Group/Topical: Particle Technology Forum

Spray Flame Synthesis as a Possible Rout to High Activity Au/TiO2 Catalyst for CO Oxidation

Sachit Chopra, Department of Chemical and Materials Engineering, University of Cincinnati, Cincinnati, OH

Gregory Beaucage, Department of Chemical and Materials Engineering, Cincinnati, OH

Supported gold catalysts were synthesized on titanium dioxide (TiO_2) using spray flame pyrolysis (SFP) and deposition precipitation (DP) methods. The activity of catalyst prepared by DP method is significantly higher then those prepared by SFP for the same gold loading. The gold particle size and size distribution is significantly smaller for the DP prepared catalysts compared to the SFP catalyst. The shape of deposited gold particles from DP method seem to be raftlike two dimensional (2D: height below two atomic-layers) surface layers of gold on titania. In contrast the gold particles for flame made Au/ TiO_2 are mostly spherical three dimensional (3D) nano-crystals embedded on the TiO_2 surface. In addition to the gold particles the TiO_2 support produced by the spray flame pyrolysis method has different crystal structure and nano structure compared to the Degussa, P25 TiO_2. We relate the differences in the catalytic activity to the shape and size characteristics of the catalysts obtained by the two methods, and investigate the way forward to obtaining higher activity Au/TiO_2 catalyst using SFP.

See more of this Session: Poster Session: Particle Technology Forum
See more of this Group/Topical: Particle Technology Forum

Analytical Solutions to Fragmenting Flows

Narayana P. Rayapati, Department of Mechanical Engineering, Tennessee Technological University, Cookeville, TN

Mahesh V. Panchagnula, Department of Mechanical Engineering, Tennessee Technological University, Cookeville, TN

John Peddieson, Department of Mechanical Engineering, Tennessee Technological University, Cookeville, TN

A closed form analytical solution is developed to predict the temporal and spatial evolution of the particle size distribution in an axisymmetric particulate jet with accompanying fragmentation. Here, the particulate phase is conveyed by a carrier (continuous) phase and the fragmentation process along each stream line is modeled using population balance equations. The two dimensional nature of the particle size evolution equation of a jet is transformed into a one dimensional equation using a von Mises-like transformation, for which the solutions are in a form similar to the Reid's solution to the batch grinding equation. The analytical solution is implemented and studied for the situation of a pulsating jet as this is considered to be a model for an internal combustion engine injector.

See more of this Session: Poster Session: Particle Technology Forum
See more of this Group/Topical: Particle Technology Forum

Rotating Fluidized Bed in a Static Geometry (RFB-SG) : Determination of a Range of Stable Operating Conditions

Rahul P. Ekatpure, Laboratory for Chemical Technology, Ghent University, Ghent, Belgium

Vaishali U. Suryawanshi, Laboratory for Chemical Technology, Ghent University, Ghent, Belgium

Geraldine J. Heynderickx, Laboratory for Chemical Technology, Ghent University, Ghent, Belgium

Axel de Broqueville, Grez-Doiceau, Grez-Doiceau, Belgium

Guy B. Marin, Laboratory for Chemical Technology, Ghent University, Ghent, Belgium

In a Rotating Fluidized Bed with a Static Geometry (RFB-SG) type of reactor, the centrifugal force is balanced by the radial drag force [1]. This leads to an axisymmetric rotation of solid particles which are fluidized. RFB reactors offer several advantages over conventional gravitational fluidized bed reactors (FBR): more uniform fluidization and temperature, higher slip velocity and, hence, higher heat and mass transfer at the particle scale [2, 3]. The present contribution reports on experimental results obtained in a versatile cold-flow experimental set-up [4].

A schematic representation of the set-up and a snapshot of a stable operation are shown in Figures 1 and 2. The rotational motion of the solid particles in a RFB-SG is realized by the tangential feeding of gas, in this case air, under a given angle at the reactor periphery, via regularly positioned inlet slots. The width of the latter, l_O, is varied from 2 mm to 6 mm. They define a circle with a diameter, D_R, of 0.54 m. Polyethylene particles with average an diameter, ranging from 0.3 mm to 3 mm are used. The amount of solids in the reactor is varied in the range of 0.5 kg to 7 kg. The air mass flow rate is varied from 0.4 to 1.2 kg/s. The reactor is operated with a horizontal axis of rotation. A stable rotating bed is obtained over the investigated operating range.

Dependent variables, such as, pressure drop, maximum solids capacity of the reactor, solid particles velocity, both radial and tangential, and solids volume fraction are measured as a function of the above mentioned independent variables. The maximum solids capacity of the reactor, increases with increasing air mass flow rate. The former is determined at the condition at which the solid particles start to be entrained via the reactor exhaust, D_E, from the uppermost position of the reactor. At this condition, the centrifugal force acting on the solid particles is balanced by the gravitational force. As the air mass flow rate is increased, the centrifugal force acting on the solid particles increases. This allows an increased capacity of solids in the reactor, until a new equilibrium of centrifugal and gravitational force acting on solid particles is established. These observations and explanations can be quantified. The corresponding equations allow to design a RFB-SG for industrial operations.

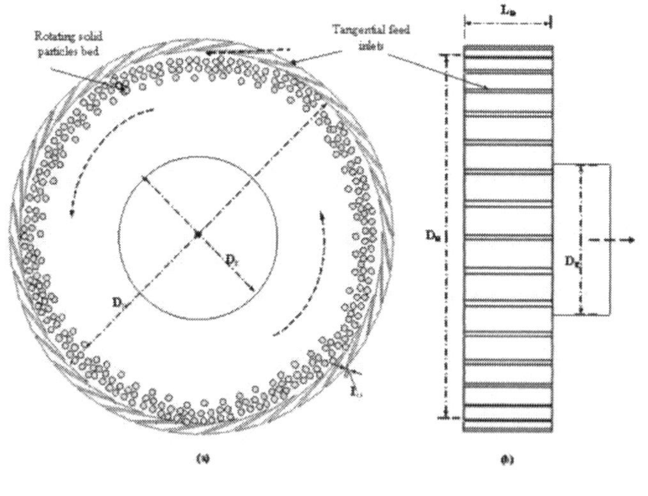

Figure 1: Schematic of the RFB-SG reactor with tangential feed inlets, reactor diameter (D_R), exhaust diameter (D_E), axial length of the reactor (L_R) and inlet opening thickness (l_O). (a) Front view (b) Side view

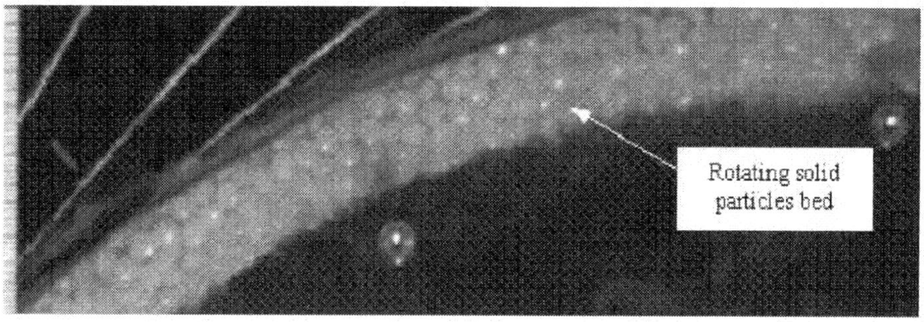

Figure 2: Snapshot of a stable rotating solid particles bed

References:

[1] J. De Wilde, A. Habibi, G. B. Marin, G. J. Heynderickx, R. P. Ekatpure and A. de Broqueville, Experimental Investigation of Rotating Fluidized Bed in a Static Geometry, *AIChE Annual Meeting, Salt Lake City, Utah, November 4-9, 2007*

[2] J. De Wilde and A. de Broqueville, "Rotating Fluidized Beds in a Static Geometry: Experimental Proof of Concept", *AIChE Journal*, Vol. 53, 793-810, 2007

[3] A. de Broqueville and J. De Wilde, "Numerical investigation of gas-solid heat transfer in rotating fluidized beds in a static geometry", *Chemical Engineering Science*, Vol. 64, 6, 1232-1248, 2009

[4] R. P. Ekatpure, G. J. Heynderickx, A. de Broqueville and G.B. Marin, "Experimental Investigation of a Rotating fluidized Bed in a Static Geometry", *2nd European Process Intensification Conference*, Venice, Italy, June 14-17, 2009

See more of this Session: Poster Session: Particle Technology Forum
See more of this Group/Topical: Particle Technology Forum

Shear-Induced Gelation Phase Diagram for Repulsive Particles

Delong Xie, Institute for Chemical and Bioengineering, Department of Chemistry and Applied Biosciences, ETH, Zurich, Switzerland

Hua Wu, Institute for Chemical and Bioengineering, Department of Chemistry and Applied Biosciences, ETH, Zurich, Switzerland

Leonie Braun, BASF SE, Ludwigshafen, Germany

Massimo Morbidelli, Institute for Chemical and Bioengineering, Department of Chemistry and Applied Biosciences, ETH, Zurich, Switzerland

Pure shear-induced gelation of repulsive colloidal particles has been investigated in a microchannel without adding any electrolytes. The main attention of this work is to identify the boundary (criticality) for the fluid-like to solid-like transition with respect to particle size, particle volume fraction and shear rate. It is found that at the criticality, the effective particle volume fraction forming the gel network is substantially smaller than the initial particle volume fraction, i.e., only part of the particles participates in the gel network. Such effective particle volume fraction at the criticality increases as either the shear rate or the particle size increases. From the structure factor of the clusters forming the gel network determined by the static light scattering, it is surprising to have found that at the criticality, structure of the clusters is insensitive to the shear rate and primary particle size, with the fractal dimension of about 2.4. Finally, a phase diagram for liquid-like to solid-like transition has been constructed, which unifies the effects of the shear rate, particle volume fraction and particle size on the criticality of the transition.

See more of this Session: Poster Session: Particle Technology Forum
See more of this Group/Topical: Particle Technology Forum

Nanoparticle Aerosol Gels From Counterflow Diffusion Flames

Yangchuan Xing, Dept. of Chemical and Biological Engineering, Missouri University of Science and Technology, Rolla, MO

This work was to study the formation of nanoparticle aerosol gels in a counterflow diffusion flame.

See more of this Session: Poster Session: Particle Technology Forum
See more of this Group/Topical: Particle Technology Forum

Simulations of Nanoparticle-Polymer Systems to Study the Factors Affecting the Nanoparticle Dispersion

Ileana C. Carpen, Department of Chemical Engineering, Tennessee Tech University, Cookeville, TN

Deepika R. Gollamandala, Department of Chemical Engineering, Tennessee Tech University, Cookeville, TN

Addition of nanosized particles to a polymer solution/matrix can lead to new and improved properties over conventional composites using larger filler particles. The quality of a composite material depends upon the degree of dispersion of the filler particles. The nanoparticles have a tendency to aggregate which negates any advantages associated with the nanosize. The available experimental procedures to quantify the nanoparticle dispersions are expensive and laborious [1]. Hence, we propose to solve these problems by studying the nanoparticle-polymer systems using computer simulations to find the factors that affect the nanoparticle dispersions in polymer solutions. In this work, we investigated systems of nanoparticles and polymer using Brownian Dynamic (BD)simulations. The shape of the nanoparticles was considered to be spherical and the polymer architecture was modeled as bead rod chains. The degree of dispersion and agglomeration in the system were characterized by mean square displacement. The effects of various individual factors, including nanoparticle polymer interactions, polymer chain length, and relative volume fractions, on the dispersion of nanoparticles in the polymer matrix were analyzed and presented.

Reference:

1. J. B. Putman, "Measuring Surface Roughness to Calculate Filler Dispersion in a Polymer Sample", US patent reference 5974167.

See more of this Session: Poster Session: Particle Technology Forum
See more of this Group/Topical: Particle Technology Forum

Identification of Clusters in a Circulating Fluidized Bed of Geldart Group B Particles by Multi-Resolution Analysis

Lii-ping Leu, Chemical Engineering, National Taiwan University, Taipei, Taiwan

Henry Lin, Chemical Engineering, National Taiwan University, Taipei, Taiwan

In this study, a wavelet -threshold criterion was developed to distinguish the cluster and the void phases from the transient solids holdup/concentration signals measured in a 108 mm i.d. x 5.75 m high circulating fluidized bed (CFB) with glass beads (particle size = 115 micron, density = 2462 kg/m3). An optimal threshold was quantitatively identified by the multi-resolution analysis (MRA) of wavelet transform. In addition, the existence time fraction of clusters, average cluster duration time, cluster frequency and average solids holdup in clusters at different radial and axial positions were determined by the threshold under the fast fluidization flow regime. The results also described the dynamic properties of clusters and flow patterns in the CFBs.

See more of this Session: Poster Session: Particle Technology Forum
See more of this Group/Topical: Particle Technology Forum

Simultaneous Particle Size Reduction and Surface Modification to Enhance Flowability of APIs

Xi Han, Otto H. York Department of Chemical, Biological and Pharmaceutical Engineering, New Jersey Institute of Technology, Newark, NJ

Yuhua Chen, Mechanical Engineering, New Jersey Institute of Technology, Newark, NJ

Ajit Mujumdar, Otto H. York Department of Chemical, Biological and Pharmaceutical Engineering, New Jersey Institute of Technology, Newark, NJ

Rajesh Dave, Otto H. York Department of Chemical, Biological and Pharmaceutical Engineering, New Jersey Institute of Technology, Newark, NJ

In this work simultaneous size reduction and surface coating is considered in order to increase the dissolution of drugs while maintaining their flowability. This is accomplished by coating nano-sized, silica guest particles on the surface of API particles, while they are reduced in size through Fluid Energy Mill (FEM). Two types of APIs are used: acetaminophen and ibuprofen. Two types of nano-silica are applied. One is hydrophilic silica M5-P, another one is hydrophobic silica R-972 pharma. Both are pharma grade excipients. Particle size distributions before and after the FEM process are investigated. The results show that with the increase of grinding pressure, the mean particle size decreases. For ibuprofen, before the process the mean particle size is about 100 microns, after the process the mean particle size decreases to about 10 microns. Angle of repose in various conditions is tested, indicating acceptable flow for such fine particles. Ibuprofen with 1.0 wt % of silica M5-P coated show the best flowability. SEM images confirmed the extent of surface modification with silica.

Another objective of this work is to produce high content API with good dispersion and flowability, which may be suitable for inhalation application. Hydrophobic biodegradable material (such as L-leucine) is applied instead of using nano-sized particles to modify the particle surface.

Proposed approach results in simultaneous particle size reduction and surface modification. The particle size is reduced and at the same time the flowability of powders is improved. Powders in size range 1-10 microns are produced with high API content (>95 %), having good dispersion and flowability without use of nano-particles.

See more of this Session: Poster Session: Particle Technology Forum
See more of this Group/Topical: Particle Technology Forum

Numerical Simulation of Spatially Distributed Population Balances

Michael Fricke, PCP, MPI Magdeburg, Magdeburg, Germany

Alexander Zinser, Process Systems Engineering, University of Magdeburg, Magdeburg, Germany

Christian Borchert, Physical and Chemical Process Engineering, Max Planck Institute for Dynamics of Complex Technical Systems, Magdeburg, Germany

Kai Sundmacher, Max Planck Institute for Dynamics of Complex Technical Systems, Magdeburg, Germany

Population balance models have widely been used to model various crystallization systems, for instance reactive crystallization in emulsion droplets [1,2,3]. However, these models do not consider spatial variations of solute concentrations and spatial particle distributions. Recently, we have proposed a model for emulsion-assisted precipitation which overcomes this drawback [4]. Besides mass balances, a (1+1)-dimensional population balance equation (PBE) is used. The first coordinate is particle size, the second is position along the radial direction of a spherical emulsion droplet.

For the solution of the (1+1)-dimensional PBE the Finite Volume Method has been employed. Higher order schemes were implemented to evaluate the fluxes at the cell boundaries. It is well known that such schemes produce spurious oscillations - especially when the solution shows steep gradients. With a Koren flux limiter function these oscillations were suppressed. Then a total variation diminishing (TVD) numerical scheme is obtained. From the literature it is known that many flux limiter functions are available guaranteeing TVD. This contribution compares different flux limiter methods for selected test cases.

Instead of utilizing flux limiter functions, oscillations can also be suppressed with Essentially Non Oscillating (ENO) schemes. Instead of approximating the flux at the cell boundary with a fixed set of neighboring grid points (stencil), the flux at a given cell boundary is evaluated on different stencils. From these stencils the one is selected on which the fluctuation of the state function is minimal. A further development of this method are Weighted Essentially Non-Oscillating (WENO) schemes. With this method, the full set of stencils is used but with different weights according to their magnitude of fluctuation. Our poster compares the WENO schemes against flux limiter methods with respect to accuracy and computational effort for some test cases and the 2-D emulsion crystallization PBE model.

[1] Niemann, B., Rauscher, F., Adityawarman, D. Voigt, A. Sundmacher, K., "Microemulsion-assisted precipitation of particles: Experimental and model-based process analysis", Chemical Engineering and Processing 45 (2006), 917-935.

[2] Weiss, C. Hennig. T. Kümmel, R., Tschernjaew, J., "Modeling Mass Transfer and Crystallization in Disperse Systems", Chemical Engineering Technology 23 (2006), 485-488.

[3] Feltham, D. L., Garside, J., "A mathematical model of crystallization in an emulsion", The Journal of Chemical Physics 122 (2005), 174910.

[4] Fricke, M., Sundmacher, K., "Model-Based Analysis of an Emulsion-Assisted Precipitation Process", Industrial and Engineering Chemistry Research, submitted.

See more of this Session: Poster Session: Particle Technology Forum
See more of this Group/Topical: Particle Technology Forum

Yield Behavior of Dense Assemblies of Frictional Particles

Sebastian Chialvo, Department of Chemical Engineering, Princeton University, Princeton, NJ

Jin Sun, Department of Chemical Engineering, Princeton University, Princeton, NJ

Sankaran Sundaresan, Department of Chemical Engineering, Princeton University, Princeton, NJ

A static granular assembly under shear yields when the shear stress reaches a critical value. Although the yield behavior of granular assemblies has been studied extensively and many yield functions have been proposed [e.g., see 1-3], the links to particle-level properties and direct measurements of microstructure evolution during yielding have not been fully established. In this poster presentation, we examine such connections between yield and the particle properties and microstructure evolution directly obtained from discrete element method (DEM) simulations.

To probe these connections, we have performed DEM simulations of dense granular assemblies subjected to triaxial compression and extension as well as shear deformation. We apply axisymmetric stress increments to initially isotropic assemblies under constant confining pressures using a stress-controlled procedure. The kinematics information at the continuum level is then obtained through statistical averaging. The assemblies are deemed to yield to flow when the continuous strain following a small stress increment exceeds 0.1. The simulations also afford detailed information about internal variables such as the fabric tensor and the coordination number of particles involved in force chains, which characterize the evolution of the microstructure of the assembly. Such simulations have been repeated at several different confining pressures and particle friction coefficients. These results are then analyzed to first identify the plastic deformation and then correlate the yielding behavior to microstructure evolution and particle properties.

The dependence of the shear stress to pressure ratio on shear strain reveals strain-hardening. This stress ratio is independent of pressure but increases with particle friction coefficient. The strain beyond which the assemblies yield to flow also increases with particle friction coefficient. The yield stress data from the three loading paths can all be fitted nicely using a Lade-Duncan type yield function [2], and to a lesser accuracy with a Drucker-Prager type yield function [1]. More importantly, we found that the strain hardening behavior is positively correlated to the anisotropy evolution, which is characterized by the second invariant of the fabric tensor. The center of the yield surface also translates in stress space as particle friction coefficient increases, which is consistent with experimental observations [4]. These results are then used to formulate a yield function in terms of particle friction coefficient, the coordination number and the fabric tensor. This yield function, coupled with associated flow rule and evolution equations for the microstructural quantities, describes quasi-static flow of dense granular assemblies.

References:

[1] D. C. Drucker and W. Prager. Soil mechanics and plastic analysis of limit design. Quarterly of Applied mathematics, 10:157-165, 1952.

[2] P. V. Lade and J. M. Duncan. Elastoplastic stress-strain theory for cohesionless soil. Journal of geotechnology engineering, 101:1037-1053, 1975.

[3] A. S. J. Suiker and N. A. Fleck. Frictional collapse of granular assemblies. Journal of Applied Mechanics, 71(3):350-358, 05 2004.

[4] P. V. Lade and M. K. Kim. Single hardening constitutive model for frictional materials: III. comparison with experimental data. Computers and Geotechnics, 6:31-47, 1988.

See more of this Session: Poster Session: Particle Technology Forum

See more of this Group/Topical: Particle Technology Forum

Granular Flow and Segregation in a Bladed Mixer

Brenda Remy, Dept. of Chemical and Biochemical Engineering, Rutgers University, Piscataway, NJ

Johannes G. Khinast, Institute for Process and Particle Engineering, Graz University of Technology, Graz, Austria

Benjamin Glasser, Dept. of Chemical and Biochemical Engineering, Rutgers University, Piscataway, NJ

Bladed mixers are commonly used in a variety of areas, ranging from the bulk chemical to the food and pharmaceutical industries. The appeal of these mixers comes from their ability to handle a wide variety of solids ranging from free flowing to cohesive powders and even pastes. These unit operations often involve the use of components with particle size distributions. Particle size differences have been shown to cause segregation in many granular systems, a phenomenon which could jeopardize product quality in industrial scenarios. Despite the common used of polydisperse materials, the effect of particle size distributions on the flow behavior observed in bladed mixers is not fully understood.

In this work, we study the effect of polydispersity on flow and segregation in bladed mixers via particle dynamic simulations. Segregation patterns and resulting velocity profiles are compared to those observed experimentally. In bladed mixers, a 3D recirculation zone develops in-front of the blades which enhances radial and vertical mixing. We find that the intensity of the recirculation zone increases when the number of particle sizes is increased. This suggests that convective mixing in the radial and vertical directions is enhanced with a higher amount of polydispersity. Additionally particle diffusivities increase when the number of particle sizes is increased. This suggests that polydispersity enhances diffusive mixing. The observed increase in convective and diffusive mixing leads to a lower degree of size segregation for the cases with more types of particle sizes.

See more of this Session: Poster Session: Particle Technology Forum
See more of this Group/Topical: Particle Technology Forum

Simple Model for Designing Gas-Phase Coating of Nanoparticles

Beat Buesser, Particle Technology Laboratory, Institute of Process Engineering, Department of Mechanical and Process Engineering, ETH Zurich, Zurich, Switzerland

Sotiris E. Pratsinis, Particle Technology Laboratory, Institute of Process Engineering, Department of Mechanical and Process Engineering, ETH Zurich, Zurich, Switzerland

Nanoparticles are coated to make them compatible to their host matrices. With such core-shell particles, desirable properties of core (e.g. dielectric, scattering or opacifying performance) particles are conserved while modifying their surface by the shell material. For example, rutile TiO_2 particles are made by the "chloride" or "sulfate" process and are coated with silica, alumina and other oxides by wet impregnation processes to prevent the photocatalytic reaction of pigmentary TiO_2 with the organic host matrix in paints (Clark, 1975).

Aerosol coating is attractive as it contributes to gas-phase synthesis of functionalized nanoparticles in one-step facilitating their economic manufacture. Here aerosol coating of TiO_2 core particles with SiO_2 shells is investigated by a combination of particle dynamic models and computational fluid dynamics, accounting for SiO_2 monomer generation and growth by coagulation and sintering. The evolution of the SiO_2 coating particle population is simulated distinguishing monomers and particles with the bimodal model of Jeong and Choi (2003). At industrial scale manufacture of nanoparticles, such aerosols can encounter high concentration dynamics resulting in much faster growth kinetics and even gelation (Buesser *et al.*, 2009).

Figure 1 shows the modeled pathways of the coating process. The uncoated TiO_2 core particles, N_c, (white) enter the reactor at the bottom where they have been produced by flame spray pyrolysis (Mädler *et al.*, 2002). Coating starts by injecting vapor of coating shell precursor, C, HMDSO (hexamethyldisiloxane) through a torus ring after the TiO_2 particles have reached their optimal primary particle size that no longer changes by sintering (Teleki *et al.*, 2009). That way TiO_2 primary particles and aggregates have attained a self-preserving distribution that can be represented by a single size facilitating the simulations. Coating monomers (blue), N_{f1}, are generated by gas-phase reaction of HMDSO and form smooth coating shells, V_s, on the core particles by coagulation. Further downstream, coating monomers grow by coagulation and sintering to coating particles, either aggregates or agglomerates, N_{f2}, which create rough coating films, V_r, by coagulation with core particles. Rough coating shells sinter to smooth ones depending on temperature and primary particle size (Teleki *et al.*, 2005).

The aerosol coating model is combined with computational fluid dynamics to account for the non-premixed gas-phase reaction of the coating shell precursor and the flow and temperature fields defining the important mixing of the core and coating particle aerosol inside of the coating reactor (Teleki *et al.*, 2009). The influence of TiO_2 particle size and concentration, process temperature and cooling rate, degree of mixing and HMDSO precursor concentration on SiO_2 coating thickness and texture and process efficiency are investigated. The simulations are compared to experimental data of nanometer thin SiO_2-coated flame-made rutile TiO_2 particles (Teleki *et al.* 2008).

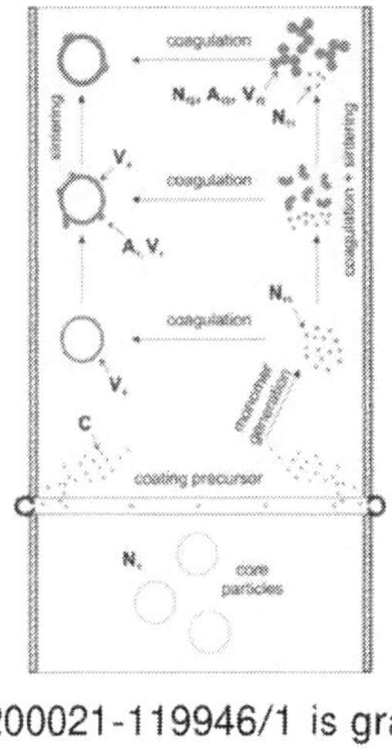

Figure 1 Schematic of the modeled pathways of aerosol coating processes.

Financial support from the Swiss National Science Foundation (SNF) grant # 200021-119946/1 is gratefully acknowledged.

Buesser, B., Heine, M.C., Pratsinis, S.E (2009), "Coagulation of highly concentrated aerosols" *J. Aerosol Sci.*, **40, 89-100**

Clark, H. B., Ed. (1975). Titanium dioxide pigments. Treatise on Coatings, 3, Pigments Part I. New York, Marcel Dekker

Jeong, J. I. & Choi, M. (2003). A simple bimodal model for the evolution of non-spherical particles undergoing nucleation, coagulation and coalescence. *J. Aerosol Sci.* **34**: 965-976

Mädler, L., Stark, W. J. & Pratsinis, S. E. (2002). Rapid synthesis of stable ZnO quantum dots. *J. Appl. Phys.* **92**: 6537-6540

Teleki, A., Pratsinis, S. E., Wegner, K., Jossen R. & Krumeich, F. (2005), "Flame-coating of titania particles with silica" *J. Mater. Res.*, **20**, 1336-1347

Teleki, A., Heine, M. C., Krumeich, F., Akhtar, M. K. & Pratsinis, S. E. (2008). *In-situ* coating of flame-made TiO_2 particles by nanothin SiO_2 films. *Langmuir* **24**: 12553-12558

Teleki, A., Buesser, B., Heine, M. C., Krumeich, F., Akhtar, M. K. & Pratsinis, S. E. (2009), "Role of gas-aerosol mixing during in situ coating of flame-made titania particles" *Ind. Eng. Chem. Res.*, **48**, 85-92

See more of this Session: Poster Session: Particle Technology Forum
See more of this Group/Topical: Particle Technology Forum

Lattice Boltzmann Method: An Efficient Technique for Solving Population Balance Equations

Aniruddha Majudmer, Nanyang Technological University, Singapore, Singapore

Vinay Kariwala, Division of Chemical and Biomolecular Engineering, Nanyang Technological University, Singapore, Singapore

Arvind Rajendran, School of Chemical and Biomedical Engineering, Nanyang Technological University, Singapore, Singapore

Santosh Ansumali, Nanyang Technological University, Singapore, Singapore

Crystallization is widely used in chemical, pharmaceutical, food and semiconductor industries as a method of separation and purification. In industrial crystallization processes, the control of crystal shape and size is of key importance as these properties not only affect the product qualities, e.g. bioavailability, but also the efficacy of the downstream processes. Recent experimental results have demonstrated that model-based control offers significantly improved performance in comparison to model-free approaches [1]. These model-based online control schemes require fast and accurate numerical techniques to solve the model equations.

Population balance equations (PBEs) are widely used to model the dynamics of crystallization and other particulate systems [2]. In a batch crystallization process, these equations have to be coupled with mass balance equations to take into account the mass transfer from solution to solid phase. PBEs are hyperbolic partial differential equations which do not have an analytical solution except for few limited cases and hence need to be solved numerically. The accurate numerical simulation of the crystallization process can be challenging as the crystal size distribution (CSD) can be very sharp and various mechanisms may be present which affect CSD simultaneously. Over the years, researchers have developed various numerical schemes. The available methods show performance limitations as the dimensions of the PBE increases. This motivates the need to develop reliable and computationally efficient numerical methods to solve PBEs for application in online model based control.

In this work, we introduce lattice Boltzmann method (LBM) as an alternate tool for solving PBEs. In recent years, LBM has attracted significant attention from the fluid dynamics community. In this method, fictitious particles resembling groups of molecules are considered in a lattice with finite set of velocities [3, 4]. These particles collide at the lattice nodes and propagate in such a way that the macroscopic behavior of the system is recovered in the long-time limit. In principle, with an appropriate choice of mesoscale equilibrium distribution, any macroscale dynamics other than hydrodynamics can be simulated by lattice Boltzmann type of method. In spite of the ability of LBM to provide fast, accurate and easily implementable numerical scheme, to the best of our knowledge, it has not been used to solve PBEs.

In the current work, we consider crystallization with nucleation and growth. The governing PBE for a crystallization process with size independent growth rate is analogous to the advection equation, which can be handled efficiently by LBM [4, 5]. We further apply LBM to solve PBEs representing a crystallization process with size dependent growth rate. This is done by taking advantage of the functional form used to describe the size dependence of growth rate and proposing a coordinate transformation technique which simplifies the problem to a size independent one. Nucleation as a boundary condition in crystallization is handled in the same manner as open flow boundary condition in a fluid flow problem. Few benchmark problems drawn from literature are used to show the reliability of the solution by analyzing convergence and accuracy. It is found that for 1D problems, solution provided by LBM has the same level of accuracy and computation time as the well established HR method [6]. For multidimensional problems, computation time required by LBM to maintain the same level of accuracy is much lower than that of HR method.

References

[1] Z. K. Nagy, M. Fujiwara, and R. D. Braatz. Modelling and control of combined cooling and antisolvent crystallization processes. J. Process Control, 18(9):856-864, 2008.

[2] D. Ramkrishna. Population balances: Theory and applications to particulate systems in engineering. Acadamic Press, San Diego, CA, 2000.

[3] S. Succi. Lattice Boltzmann equation for fluid dynamics and beyond. Oxford University Press, New York., 2001.

[4] I. V. Karlin, S. Ansumali, C. E. Frouzakis, and S. S. Chikatamarla. Elements of the lattice Boltzmann method. I: Linear advection equation. Commun. Comput. Phys., 1(4):616-655, 2006.

[5] A. Majumder, W. P. Yudistiawan, V. Kariwala, S. Ansumali, and A. Rajendran. Lattice Boltzmann method for solving population balance equations. In proceedings of International Conference on Chemical Engineering 2008, Dhaka, Bangladesh, 2008.

[6] R. Gunawan, I. Fusman, and R. D. Braatz. High resolution algorithms for multidimensional population balance equations. AIChE J., 50(11):2738-2749, 2004.

See more of this Session: Poster Session: Particle Technology Forum
See more of this Group/Topical: Particle Technology Forum

Formation of ZnSe and ZnTe Nanocrystals by Ultrasonic-Assisted Aerosol Spray Pyrolysis

Dae-Jin Kim, Department of Chemical and Biomolecular Engineering, Sogang University, Seoul, South Korea

Jun-Woo Kim, Department of Chemical and Biomolecular Engineering, Sogang University, Seoul, South Korea

Chang-Hwa Jo, Department of Chemical and Biomolecular Engineering, Sogang University, Seoul, South Korea

Kee-Kahb Koo, Department of Chemical and Biomolecular Engineering, Sogang University, Seoul, South Korea

Colloidal monodisperse ZnSe nanospheres consisted of primary ZnSe nanocrystals and one-dimensional (1-D) ZnTe nanocrystals such as nanorods, nanowires, and nanobundles were prepared by aerosol spray pyrolysis with ultrasonic nebulization. TEM image shows that average diameter of ZnSe nanospheres is 41 nm and primary ZnSe nanocrystals have a size of 2-3 nm. XRD pattern of ZnSe nanoparticles shows that the crystal structure is zinc blende. The shape evolution of ZnTe nanocrystals from nanodots to nanowires was achieved by controlling the reaction temperature ranging from 300 to 500 oC. 1-D ZnTe nanocrystals with zinc blende structure were formed from the linear aggregation by oriented attachment of ZnTe primary nanocrystals due to insufficient surface capping of surfactant molecules and strong dipole-dipole interaction of nanocrystals, followed by reorganization of linear ZnTe aggregates at high reaction temperatures.

See more of this Session: Poster Session: Particle Technology Forum
See more of this Group/Topical: Particle Technology Forum

Co-Solvent Effect On Cooing Crystallization of DADNE

Jin-Hwan Ahn, Department of Chemical and Biomolecular Engineering, Sogang university, Seoul, South Korea

Jae-Kyeong Kim, Department of Chemical and Biomolecular Engineering, Sogang university, Seoul, South Korea

Jun-Woo Kim, Department of Chemical and Biomolecular Engineering, Sogang university, Seoul, South Korea

Hyoun-Soo Kim, Agency for Defense Development, Daejeon 305-600, Korea

Kee-Kahb Koo, Department of Chemical and Biomolecular Engineering, Sogang university, Seoul, South Korea

The ability to control the size and shape of energetic materials is essential in preparing crystals which have enough mechanical strength and high purity in military formulation. In the present study, yellow crystals of 1,1-diamino-2,2-dinitroethylene (DADNE) were precipitated by cooling of saturated solution under various operating conditions including cooling rate, stirring rate, solvent, and co-solvent composition. X-ray diffraction analysis of the DADNE crystals showed highly crystalline nature of gamma form. SEM images of DADNE crystals showed shapes of bundled needle, platelet and rhombic. The particle sizes were ranged from 100 to 500 micrometer.

See more of this Session: Poster Session: Particle Technology Forum
See more of this Group/Topical: Particle Technology Forum

Manufacturing Polymer Nano Particles by Using Supercritical Fluids

Jun Hyuk Yim, Department of Chemical and Biomolecular Engineering, Sogang University, Seoul, South Korea

Ki-Pung Yoo, Department of Chemical and Biomolecular Engineering, Sogang University, Seoul, South Korea

Jong Sung Lim, Department of Chemical and Biomolecular Engineering, Sogang University, Seoul, South Korea

The aerosol solvent extraction system (ASES), which is one of the supercritical anti-solvent system (SAS), has been studied to produce nano-particles of polymers such as poly(l-lactic acid) (PLLA) and polystyrene (PS), etc. Dichloromethane (DCM, CH2Cl2) and supercritical carbon dioxide were adopted as a solvent and as an anti-solvent material, respectively. The effect of the process parameters such as temperature, pressure and solution concentration ,etc., on particle size has been investigated. The morphology and size of the particles re-crystallized by ASES process were observed by using a scanning electron microscope (SEM) image. The mean particle size and its distribution of processed particles were measured by using a laser diffraction particle size analyzer (PSA). The particle size increased with increasing temperature and pressure. Also, higher solution concentration led to lager particle size and broader particle size distribution

See more of this Session: Poster Session: Particle Technology Forum
See more of this Group/Topical: Particle Technology Forum

Soft-Agglomerate Breakage by Discrete Element Method (DEM) Simulations

Michael O. John, Particle Technology Laboratory, Institute of Process Engineering, Department of Mechanical and Process Engineering, ETH Zurich, Zurich, Switzerland

Max L. Eggersdorfer, Particle Technology Laboratory, Institute of Process Engineering, Department of Mechanical and Process Engineering, ETH Zurich, Zurich, Switzerland

Sotiris E. Pratsinis, Particle Technology Laboratory, Institute of Process Engineering, Department of Mechanical and Process Engineering, ETH Zurich, Zurich, Switzerland

Nanoparticle product properties are depending on primary particle size distribution and the structure of agglomerates formed by them. The fluid flow [1] as well as particle-particle interactions determine the structure of agglomerates and their size distribution [2]. Furthermore, agglomerates of nanoparticles occupy far more volume than their equivalent sphere volume and can influence fluid-particle interactions in aerosol reactors leading to agglomerate restructuring and even gelation [3]. Many force models for weak (e.g. van-der-Waals) interparticle forces inside agglomerates in DEM simulations have been proposed. The objective of this work is to investigate the impact of process variables in such force models focusing on breakup duration and mechanism of agglomerates in shear flows.

1. P.T. Spicer, S.E. Pratsinis, J.A. Raper, R. Amal, G. Bushell, G.M.H. Meesters "Effect of Shear Schedule on particle Size and Structure during Flocculation in Stirred Tanks" Powder Technol. 97, 26-34 (1998).

2. Flesch, J.C., P.T. Spicer, and S.E. Pratsinis, Laminar and turbulent shear-induced flocculation of fractal aggregates. AIChE Journal, 1999. 45(5): p. 1114-1124.

3. M.C. Heine, S.E. Pratsinis, "Brownian coagulation at high concentrations", Langmuir, 23, 9882-9890 (2007).

See more of this Session: Poster Session: Particle Technology Forum
See more of this Group/Topical: Particle Technology Forum

High Shear Wet Granulation Using Self-Emulsifying System: Granule Shape and Size as a Function of Some Important Process Variables

Mauro Cavinato, Department of Chemical Engineering, University of Padua, Padova, Italy

Erica Franceschinis, Department of Pharmaceutical Science, University of Padua, Padova, Italy

Serena Cavallari, Department of Pharmaceutical Science, University of Padua, Padova, Italy

Nicola Realdon, Department of Pharmaceutical Science, University of Padua, Padova, Italy

Andrea Santomaso, Department of Chemical Engineering, University of Padua, Padova, Italy

Controlling granule shape may be desirable for many reasons; among these are for example the flow properties. A spherical shape possesses a minimum surface area to volume ratio resulting in reduced cohesive forces and mechanical interlocking thereby resulting in an improved flowability of the bulk powder. Obtaining spheronisated shape is a desired prerequisite also when a subsequent coating or drug layering of the granules is necessary.

The advantage of wet granulation process in high shear mixer is that mixing, massing and granulation are performed in few minutes in the same equipment. However the high shear wet granulation (HSWG) process does not always warrant spheronisated granules. The process needs to be controlled with care as the granulation progresses so rapidly that a usable granule can be transformed very quickly into an unusable system.

The focus of this paper is therefore on the influence of process parameters on granule shape. Two different systems have been considered and results compared. Both systems include a fixed mixture of Microcrystalline Cellulose and Monohydrate Lactose as powder mass but two different binder were considered. In the first case water was considered whereas in the second case an oil-in-water emulsion was used as liquid binder in order to produce a solid self-emulsifying systems. All experiments were performed in the same vertical, bottom driven high shear granulator. The results obtained through DOE plus data statistical analysis showed that some of the process variables studied (impeller speed, granulation time, amount of initial powder and of binder) were able to induce different granule shapes. In particular a clear dependence on the binder quantity and on its nature has been observed and two main behaviours were individuated. Final structure of granules depended on structural changes which occurred during the granulation process and these were determined by the rheology of the wet bulk. Granules must be strong enough to resist the forces in the granulator, and deformable enough to generate spherical granules. Using an excess of liquid both binders produced a loss of roundness. Different surface characteristics were also observed. In particular using water, pellets resulted smaller, more rough ad with irregular shape. Composition of the second system containing an oil-in-water emulsion produced longer, more flat and smooth aggregates. These differences have been quantified using various shape factors and explained in terms of MCC granulation mechanisms where phenomena such as powder wettability, binder adsorption/desorption, intra- and inter-particle lubrication play a major role. Measures of granules strength were used in addition to corroborate our conclusions.

See more of this Session: Poster Session: Particle Technology Forum
See more of this Group/Topical: Particle Technology Forum

Analysis of the Surface Shielding of Fractal Like Particles

Flurin Haenseler, Particle Technology Laboratory, Institute of Process Engineering, Department of Mechanical and Process Engineering, ETH Zurich, Zurich, Switzerland

Max L. Eggersdorfer, Particle Technology Laboratory, Institute of Process Engineering, Department of Mechanical and Process Engineering, ETH Zurich, Zurich, Switzerland

Sotiris E. Pratsinis, Particle Technology Laboratory, Institute of Process Engineering, Department of Mechanical and Process Engineering, ETH Zurich, Zurich, Switzerland

Surface characteristics of nanoparticles play a key role in their performance and can effectively influence their transport characteristics. Furthermore, selective deposition of desired materials can increase the functionality of tailor-made particles or prevent toxic implications retaining however their performance (e.g. opacity of TiO_2 particles without undesirable photocatalytic activity by their coating with nanothin layers of SiO_2 [1,2]). Furthermore, the superparamagnetic activity of Fe_2O_3 nanoparticles can be retained by similar coatings that allow their ready incorporation in polymer matrices for actuators or biomedical applications [3]. Also Heine and Pratsinis [4] simulated the growth of silica agglomerates with fractal dimension $D_f=1.8$ and showed that even for low initial concentrations, agglomerates upon growth can effectively occupy far more volume than a mass equivalent sphere so their coagulation can lead to gelation within 3-4 seconds. This work investigates deposition on generated agglomerates in the free-molecular regime through analysis of the surface shielding [5]. Fractals are generated by diffusion limited agglomeration (DLA) and cluster-cluster agglomeration and compared to literature prior to proposing quantitative relationships between agglomerate structure and growth.

1. Teleki, A., M.C. Heine, F. Krumeich, M.K. Akhtar, S.E. Pratsinis, In-situ coating of flame-made TiO2 particles by nanosized SiO2 films, Langmuir, 24,12553-8 (2008).

2. A. Teleki, B. Buesser, M.C. Heine, F. Krumeich, M.K. Akhtar, S.E. Pratsinis, The role of gas-aerosol mixing during in-situ coating of flame-made titania particles, Ind. Eng. Chem. Res., 48, 85-92 (2009).

3. A. Teleki, M. Suter, P. Kidambi, O. Ergeneman, F. Krumeich, B.J. Nelson, S.E. Pratsinis, Hermetically-coated superparamagnetic Fe2O3 particles with SiO2 nanofilms, Chem. Mater., in press (2009).

4. Heine, M.C. and Pratsinis, S.E., Brownian coagulation at high concentration. Langmuir, 23(19), 9882-9890 (2007).

5. Xiong, Y. and Pratsinis, S.E., Formation of agglomerate particles by coagulation and sintering. 1. A 2-dimensional solution of the population balance equation. J. Aerosol. Sci., 24(3), 283-300 (1993).

See more of this Session: Poster Session: Particle Technology Forum
See more of this Group/Topical: Particle Technology Forum

Restructuring and Fragmentation of Agglomerates in Shear Flows

Max L. Eggersdorfer, Particle Technology Laboratory, Institute of Process Engineering, Department of Mechanical and Process Engineering, ETH Zurich, Zurich, Switzerland

Sotiris E. Pratsinis, Particle Technology Laboratory, Institute of Process Engineering, Department of Mechanical and Process Engineering, ETH Zurich, Zurich, Switzerland

The nanoparticle product performance is strongly affected by the primary particle size distribution. However the morphology and degree of particle agglomeration can influence drastically the product properties also. At the initial reactor stages particles grow by Brownian motion while later on and in particular at larger sizes particles may grow by flow-induced coagulation also. As agglomerates become comparable to the size of fluid eddies, the effect of fragmentation increases. Finally a balance between coagulation and fragmentation is reached.

Spicer et al. [1] used light scattering to monitor size and structural dynamics of polystyrene-alum flocs in a stirred tank. They showed that for the production of fast settling particles for particle removal, e.g. in wastewater treatment, it is beneficial to apply cycled-shear-induced flocculation. Flesch and Pratsinis [2] developed a sectional population balance model to simultaneously trace the coagulation and fragmentation in turbulent shear. They accounted for the agglomerate fractal structure using appropriate kinetic expressions. Good agreement of the evolution of the mass mean agglomerate diameter was found between flocculation experiments of polystyrene-Al(OH)3/ water systems in a stirred tank and the model results.

The stability of nanoclusters made by particle-particle interactions is investigated by simulations focusing on the forces holding together such nanostructured particles made in aerosol reactors. Agglomerates are formed by cluster-cluster agglomeration, the dominant mechanism in flame reactors, and are exposed to shear flow fields. The attainment of asymptotic or self-preserving-like structures and distributions is explored.

1. Spicer, P.T., S.E. Pratsinis, J. Raper, R. Amal, G. Bushell, and G. Meesters, Effect of shear schedule on particle size, density, and structure during flocculation in stirred tanks. Powder Technology, 1998. 97(1): p. 26-34.

2. Flesch, J.C., P.T. Spicer, and S.E. Pratsinis, Laminar and turbulent shear-induced flocculation of fractal aggregates. AIChE Journal, 1999. 45(5): p. 1114-1124.

See more of this Session: Poster Session: Particle Technology Forum
See more of this Group/Topical: Particle Technology Forum

Process Simulation and Economics of a Ferrite-Based Water Splitting Cycle to Produce Renewable H2

Melinda M. Channel, Chemical & Biological Engineering, University of Colorado, Boulder, CO

Jonathan Scheffe, Chemical & Biological Engineering, University of Colorado, Boulder, CO

Alan W. Weimer, Chemical & Biological Engineering, University of Colorado, Boulder, CO

Using concentrated solar energy from a heliostat field, metal ferrite materials ($NiFe_2O_4$) can be reduced according to the reaction $NiFe_2O_4 \rightarrow 1.2\ FeO + 0.4\ Fe_2O_3 + NiO + 0.3\ O_2$ at 1,400°C. The resulting solid solution is then used to split water at 1,000°C via an exothermic oxidation reaction $1.2\ FeO + 0.4\ Fe_2O_3 + NiO + 0.6\ H_2O \rightarrow NiFe_2O_4 + 0.6\ H_2$, producing high purity renewable hydrogen. This process has been designed and modeled using AspenPlus™ to produced 100,000 kg H2/day with reduction conversions of 35%, 70%, and 100%. Once the base case AspenPlus™ models were developed, the process was optimized to produce electricity by generating steam in the heat removal steps of the cycle. The models were then used to size and cost all capital equipment and the solar heliostat field to determine the total capital investment (TCI) of the process. The TCI and fixed/variable operating costs were then entered into the Department of Energy's (DOE) H2A economic analysis program to establish the allowable purchase price of metal ferrite for a given hydrogen selling price to meet the DOE cost targets through 2017. The results of this analysis are very promising with solar to receiver annual average efficiencies of 45% and thermal cycle efficiencies as high as 50%. The targeted hydrogen selling price of $4/kg was met with reasonable allowable ferrite capital costs. The economic analysis indicates that the development of solar-thermal water splitting processes using ferrite materials in a redox cycle is a potentially economical pathway to renewable hydrogen.

See more of this Session: Poster Session: Particle Technology Forum
See more of this Group/Topical: Particle Technology Forum

Fluid-Particle Drag in Inertial Polydisperse Gas-Solid Suspensions

William Holloway, Chemical Engineering, Princeton University, Princeton, NJ

Xiaolong Yin, Chemical Engineering, Princeton University, Princeton, NJ

Sankaran Sundaresan, Chemical Engineering, Princeton University, Princeton, NJ

Suspension flows in large devices are generally analyzed using continuum models that treat the particle and fluid phases as interpenetrating continua. These continuum hydrodynamic models consist of balance equations for particle and fluid phase mass, momentum, and energy associated with the fluctuating motion. In the momentum balance equation for the particle phase, the dominant forces acting on the particles are those due to gravity and fluid-particle drag; as a result, a significant amount of research has been performed in the literature, with the aim of developing constitutive relations for the fluid-particle drag, e.g. see Koch et al. [1] and references cited therein. Most of the research performed to date has been devoted to developing constitutive models for monodisperse suspensions (i.e., particles of the same size and/or density). However, polydispersity is inevitable in industrial scale devices, and it can alter the flow characteristics appreciably [2]. The effect of polydispersity on the fluid-particle drag force in fixed beds has also been studied extensively in the literature; for example, see recent publications by van der Hoef et al. [3] and Beetstra et al. [4], who employed a computational approach to formulating constitutive models. In polydisperse suspensions, particles of different sizes (and/or densities) will, in general, have different local average velocities and this relative velocity between particles of different types can impact the magnitude and direction of the net fluid-particle drag force. More general constitutive models for the gas-particle drag force that permit relative motion (in a local average sense) between particles of different types, valid in the Stokes flow limit, were presented recently by Yin & Sundaresan [5, 6].

In this study we extend the fluid-particle drag relation presented by Yin & Sundaresan by taking into account the effect of moderate fluid inertia. Towards this end, we have performed simulations of fluid flow through bidisperse particle assemblies in periodic domains using the lattice Boltzmann simulation method [7], which has been used in a number of previous studies [1, 3-6], to obtain computational data for formulating constitutive models for the drag force. We find that a fluid-particle drag force model that combines the bidisperse suspension drag law in the Stokes flow limit proposed by Yin & Sundaresan [6] and the moderate inertia correction for polydisperse fixed bed drag proposed by Beetstra et al. [4] adequately captures our simulation data.

In this poster, we will present the polydisperse fluid-particle drag force model that allows for the presence of moderate fluid inertia and relative motion (in a local average sense) between particles of different types, and discuss the relative importance of the fluid-mediated particle-particle interactions and that resulting from direct particle-particle collisions at different levels of fluid inertia.

References:

[1] Koch, DL & Hill, R. (2001). Inertial effects in suspension and porous-media flows. Annu. Rev. Fluid Mech., 33: 619-647.

[2] Hoffman, AC, Janssen, LPBM & Prins, J. (1993). Particle segregation in fluidized binary mixtures. Chem. Eng. Sci., 48: 1583-1592.

[3] van der Hoef, MA, Beetstra, R & Kuipers, JAM. (2005). Lattice-Boltzmann simulations of low-Reynolds-number flow past mono- and bidisperse arrays of spheres: results for permeability. J. Fluid Mech., 528: 233-254.

[4] Beetstra, R, van der Hoef, MA & Kuipers JAM. (2007). Drag Force of Intermediate Reynolds Number

Flow Past Mono- and Bidisperse Arrays of Spheres. AIChE J., 53: 489-501.

[5] Yin, X & Sundaresan, S. (2009). Drag Law for Bidisperse Gas-Solid Suspensions Containing Equally Sized Spheres. Ind. Eng. Chem. Res., 48: 227-241

[6] Yin, X & Sundaresan, S. (2009). Fluid-particle drag in low-Reynolds-number polydipserse gas-solid suspensions. AIChE J., (in press)

[7] Ladd, A.J.C. & Verberg, R. (2001). Lattice-Boltzmann simulation of particle-fluid suspensions. J. Stat. Phys., 104: 1191-1251.

See more of this Session: Poster Session: Particle Technology Forum
See more of this Group/Topical: Particle Technology Forum

Nano-Scale Atomic Layer Deposited Films On High Surface Area Particles for Water Splitting Cycles

Victoria J. Aston, Department of Chemical and Biological Engineering, University of Colorado, Boulder, CO

Christopher Perkins, Sundrop Fuels, Louisville, CO

Alan W. Weimer, Department of Chemical and Biological Engineering, University of Colorado, Boulder, CO

Iron oxide (Fe_3O_4) can be reduced to iron metal using synthesis gas (CO, H_2, and CO_2) at moderate temperatures (700 - 1200°C). Iron metal is then used to split water at lower temperatures (300 - 600°C), producing a stream of pure, uncontaminated hydrogen and regenerating the Fe_3O_4 for syngas reduction again. However, both the reduction and oxidation reactions are controlled by diffusion through the reacted layer, making the reaction slow when bulk iron is used. The higher temperature reduction step also results in sintering of the powder, decreasing the surface area for diffusion and greatly slowing the reaction rate. This limits the number of reduction and oxidation cycles that bulk iron can undergo. By coating high surface area substrates with nano-thin films of iron using Atomic Layer Deposition (ALD), the reaction can be completed more quickly than with bulk iron powder for the same mass of iron. The thin films and high surface area substrates also mitigate the effects of sintering. In this study the kinetics of the syngas reduction and water oxidation of these iron thin films on different substrate materials (SiO_2, Al_2O_3, and ZrO_2) have been studied and compared with the kinetics of the reactions when using bulk iron powder.

See more of this Session: Poster Session: Particle Technology Forum
See more of this Group/Topical: Particle Technology Forum

Hydrogel Nanoparticles by Inverse-Miniemulsion Polymerization and Supercritical CO2 Drying

Melinda G. Hemingway, Chemical Engineering, Auburn University, Auburn, AL

Ram B. Gupta, Chemical Engineering, Auburn University, Auburn, AL

Hydrogel nanoparticles can be successfully produced by polymerization in inverse miniemulsions, for use in a variety of applications including diagnosis, drug delivery, separation, and absorption. Unfortunately, conventional drying techniques result in particle agglomeration due to the sticky nature of the wet hydrogel particles. This work utilizes supercritical CO2 drying to obtain free flowing dry hydrogel nanoparticles. Using inverse miniemulsion technique (cyclohexane continuous phase, water dispersed phase, and a non-ionic surfactant), nanoparticles (500 nm diameter) of polyacrylamide are produced. The polymerized miniemulsion is injected into supercritical CO2 which results in rapid removal of cyclohexane, water, and surfactant. The morphology, particle size and size distribution of the nanoparticles are determined using dynamic light scattering and scanning electron microscopy. The effect of supercritical CO2 flow on the drying efficiency is examined and compared with via freeze-drying and spray-drying techniques. Dry nanoparticles obtained from of supercritical CO2 drying show much lower agglomeration.

See more of this Session: Poster Session: Particle Technology Forum
See more of this Group/Topical: Particle Technology Forum

A Comparative Economic Analysis of Cooling, Antisolvent, and Joint Cooling-Antisolvent Recrystallization Methods for Pharmaceutical Compounds

David Widenski, Cain Department of Chemical Engineering, Louisiana State University, Baton Rouge, LA

Jose Romagnoli, Chemical Engineering, Louisiana State University, Baton Rouge, LA

Crystallization is a widely used industrial process for the production of pharmaceuticals. The traditional method for crystallizing pharmaceuticals has been to use cooling crystallization. Recently, there has been extensive research on using antisolvent crystallization to produce pharmaceuticals. However, to the author's knowledge there has been no analysis to see if antisolvent crystallization is economically advantageous over cooling.

A first look at a typical pharmaceutical compound solubility curve reveals that the compound's solubility is more affected by solution composition than temperature. However, the yield for antisolvent recrystallization can be much worse than for cooling. Antisolvent recrystallization yield is dependent on both the operating temperature and solution composition, and can vary significantly from cooling. One way to improve the yield and still utilize antisolvent crystallization is to combine it with cooling. This combined technique has begun to be recently studied and is called joint cooling-antisolvent crystallization.

A comparative analysis of recrystallization yield will be carried out for cooling, antisolvent, and joint cooling-antisolvent crystallization methods. This analysis will be coupled with energy and material costs to determine the profitability of each method. This study will determine the optimal operating points for all three crystallization methods, and determine what range of conditions each method is most profitable. For cooling crystallization, the optimal solution composition will be determined. The optimum operating temperature, initial, and final compositions will be determined for antisolvent crystallization. For combined cooling-antisolvent crystallization, the optimum initial and final temperatures and compositions will be determined. The optimum values for all three methods will be compared. For this study the ternary systems evaluated will be acetone-water-paracetamol, isopropanol-water-paracetamol, and ethanol-water-acetylsalicylic acid.

There has been much interest in using other methods than cooling for use in recrystallizing pharmaceutical compounds. This study will investigate several different systems and evaluate what the most profitable crystallization method is.

See more of this Session: Poster Session: Particle Technology Forum
See more of this Group/Topical: Particle Technology Forum

Rheological Behavior of Dense Granular Matter

Mehrdad Kheiripour Langroudi, Department of Chemical Engineering, The Graduate Center, The City University of New York, New York, NY

Gabriel I. Tardos, Department of Chemical Engineering, The City College of the City University of New York, New York, NY

The main objective of this research is to study the rheological behavior of granular matter in the Quasi-Static (solid-like) and Intermediate (liquid-like) regimes of flow using different flow-geometries. We used a Couette cell where powder is sheared between two concentric cylinders and modified it by providing continuous powder feed and discharge that enabled a small axial flow. This allowed a slight decrease of bed solid fraction (measured by a capacitance probe) and shifting of the flow from the quasi-static to intermediate regime. Furthermore, by measuring the normal stresses inside the shearing zone and defining the coefficient of apparent friction as the ratio of shear to normal stresses, we suggest a visco-plastic constitutive law for dense granular materials. We studied the effects of size and its distribution, shape, stiffness and surface roughness of the particles on the rheology of bulk [1,2].

We also used in this work a classical Jenike cell and a slightly modified version of it that allows faster shearing. Wall-friction coefficients were measured by adding one or several normal force (stress) transducers on the shearing wall and applying known stresses to the free surface of the powder. Using this experimental technique we were able to compare the pressure (force) applied to the top surface of the powder with stresses measured on the opposite side of the cell as sensed by the normal stress transducer. The experimental results were compared with a DEM theoretical model. We found that stresses are transmitted through thin layers without significant attenuation if the material is in continuous shear but that large fluctuations are introduced especially for rigid materials such as glass and steel. Particles of different size, shape, roughness and stiffness were tested in layers of different thicknesses and the response of the transmitted stress was recorded and analyzed.

Several other geometries of flow were also analyzed by measuring stresses, stress fluctuations and porosity. The studied geometries were, centrifugal flow in a "spheronizer" and "roping" flow in a high-shear mixer.

Reference

1.M. Kheiripour Langroudi, S. Turek, A. Ouazzi, and G. Tardos; Submitted to Powder Technology, 2009

2.M. Kheiripour Langroudi, G. Tardos, J. Michaels, and Paul Mort; Accepted to be published in American Institute of Physics Journal, 2009

See more of this Session: Poster Session: Particle Technology Forum
See more of this Group/Topical: Particle Technology Forum

Solar Water Splitting Using Cobalt Ferrites Deposited On Al_2O_3 Substrates Via Atomic Layer Deposition

Jonathan R. Scheffe, Department of Chemical and Biological Engineering, University of Colorado, Boulder, CO

Jianhua Li, Department of Chemical and Biological Engineering, University of Colorado, Boulder, CO

Alan W. Weimer, Department of Chemical and Biological Engineering, University of Colorado, Boulder, CO

We are reporting the results of an alternative ferrite cycle in which the ferrite ($CoFe_2O_4$) is deposited on a porous Al_2O_3 support via atomic layer deposition (ALD). Rather than a traditional cycle, in which the ferrite is deposited on an inert support or is unsupported, the ferrite reacts with Al_2O_3 during thermal reduction to form hercynite ($FeAl_2O_4$). This is advantages because the reaction between the ferrite and Al_2O_3 is favored at lower temperatures than the reduction of Fe^{3+} to Fe^{2+} on an inert support. Subsequent water oxidation is then achievable at 1000 °C, similar to other ferrite cycles. This redox cycle is shown in the two step reaction below:

$CoFe_2O_4 + Al_2O_3 +$ solar thermal energy $\Rightarrow CoO + 2FeAl_2O_4 + 0.5\ O_2$ (1)

$CoO + 2FeAl_2O_4 + H_2O \Rightarrow CoFe_2O_4 + Al_2O_3 + 2H_2$ (2)

In the first step, solar thermal energy is used to reduce the ferrite, and oxygen is evolved. In the second step, the reduced material is reacted with water to generate hydrogen and is reoxidized. Therefore, the only inputs are solar energy and water, and the only outputs are H_2 and O_2.

Thermal reduction was attempted at temperatures ranging from 1200 °C to 1500 °C, and results indicate that reduction occurs at lower temperatures than traditional ferrite cycles. It was observed that hercynite ($FeAl_2O_4$) was formed after reduction, due to a reaction between FeO and the Al_2O_3 substrate.

Subsequent water oxidation was achievable at 1000 °C and the reaction was shown to be repeatable over 8 cycles. Experimental results are compared to thermodynamic modeling performed using FactSage and results agree well with thermodynamic predictions, as seen in figure 1.

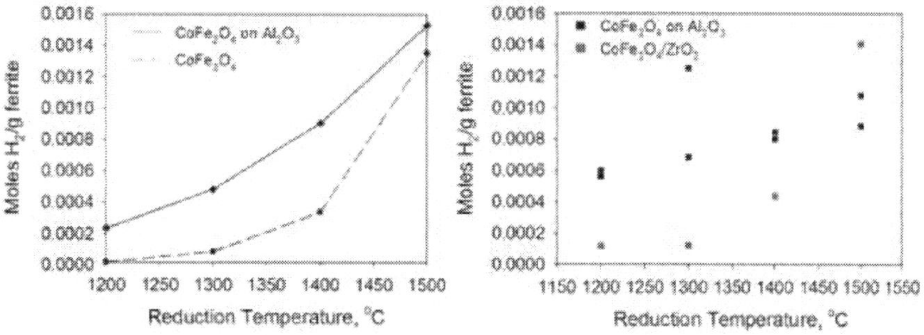

Figure 1. Left) Thermodynamic modeling results predict more H_2 generation for $CoFe_2O_4$ on Al_2O_3 over reduction temperatures from 1200 to 1500 °C. Right) Experimental results predict more H_2 generation for $CoFe_2O_4$ on Al_2O_3 over reduction temperatures from 1200 to 1400 °C

H_2 conversions varied between 18% and 25 % when cycling was performed at 1200 °C reduction and 1000 °C water oxidation. Crystallinity changes were measured before and after reduction using powder x-ray diffractometry, and conversions were calculated via in situ mass spectrometry. Film morphology and composition were measured via high resolution transmission electron microscopy and induced coupled plasma-atomic emission spectroscopy, respectively.

See more of this Session: Poster Session: Particle Technology Forum

See more of this Group/Topical: Particle Technology Forum

Spray Flame Pyrolysis Verses Deposition Precipitation as Synthesis Routes for Au/TiO_2 Catalyst for CO Oxidation

Sachit Chopra, Department of Chemical and Materials Engineering, University of Cincinnati, Cincinnati, OH

Gregory Beaucage, Department of Chemical and Materials Engineering, Cincinnati, OH

Supported gold catalysts were synthesized on titanium dioxide (TiO_2) using spray flame pyrolysis (SFP) and deposition precipitation (DP) methods. The activity of catalyst prepared by DP method is significantly higher then those prepared by SFP for the same gold loading. The gold particle size and size distribution is significantly smaller for the DP prepared catalysts compared to the SFP catalyst. The shape of deposited gold particles from DP method seem to be raftlike two dimensional (2D: height below two atomic-layers) surface layers of gold on titania. In contrast the gold particles for flame made Au/ TiO_2 are mostly spherical three dimensional (3D) nano-crystals embedded on the TiO_2 surface. In addition to the gold particles the TiO_2 support produced by the spray flame pyrolysis method has different crystal structure and nano structure compared to the Degussa, P25 TiO_2. We relate the differences in the catalytic activity to the shape and size characteristics of the catalysts obtained by the two methods, and investigate the way forward to obtaining higher activity Au/TiO_2 catalyst using SFP.

See more of this Session: Functional Nanoparticles and Nanocoatings On Particles
See more of this Group/Topical: Particle Technology Forum

Surface Decoration of Cobalt Nanoparticles On Silica Colloids

Bijith Mankidy, Chemical & Biomedical Engineering, University of South Florida, Tampa, FL

John Wolan, Chemical & Biomedical Engineering, University of South Florida, Tampa, FL

Babu Joseph, Chemical & Biomedical Engineering, University of South Florida, Tampa, FL

Vinay K. Gupta, Chemical & Biomedical Engineering, University of South Florida, Tampa, FL

The growing concerns of pollution and depletion of natural oil resources have made it inevitable to produce clean and effective synthetic fuels to meet future demands. Oxide supported metal nanoparticles have shown enhanced catalytic properties for reactions such as Fischer-Tropsch reaction to produce synthetic fuels. Conventional methods of preparation that rely on reduction of cobalt precursor to metallic cobalt at high temperatures (~350-700oC) and pose challenges due to catalyst deactivation from sintering of pores of the support, agglomeration of cobalt nanoparticles, and reaction with the silica support to cobalt silicates. We have focused on synthesis of novel composite colloids such as surface modified silica supports that are decorated with nanoparticles of cobalt. Cobalt nanoparticles of different sizes (~5-15 nm) with extended stability in solution are synthesized by two methods: (A) inverse micelle technique and (B) thermal decomposition of a cobalt pre-cursor. These cobalt nanoparticles can then be easily immobilized onto the surfaces of silica micro-particles that are surface modified with both small ligands and polymer chains. The interaction of cobalt nanoparticles with the modified silica depends upon parameters such as the functional groups of the small molecule ligands and polymer chains as well as the cobalt precursor and solvent medium that are used. Our approach not only avoids the limitations of conventional impregnation techniques but also provides several benefits such as increased availability of monodisperse cobalt nanoparticles on the surface of the silica supports, ease of preparation, and manipulation of cobalt loading. In this presentation, the various methods of preparation will be discussed along with the characterization of the novel nanomaterial using FTIR spectroscopy, X-ray diffraction, dynamic light scattering, transmission electron microscopy and temperature programmed reduction.

See more of this Session: Functional Nanoparticles and Nanocoatings On Particles
See more of this Group/Topical: Particle Technology Forum

Large-Scale Preparation of Highly Dispersed Pt Nano-Particles On Mesporous Silica Gel by ALD

Jianhua Li, Chemical and Biological Engineering, University of Colorado, Boulder, CO

David King, Chemical and Biological Engineering, University of Colorado, Boulder, CO

Alan Weimer, Chemical and Biological Engineering, University of Colorado, Boulder, CO

Nano-particulate platinum has been well-dispersed onto large quantities of micron-sized mesoporous silica gel using Atomic Layer Deposition in a Fluidized Bed Reactor (i.e. ALD-FBR). Transmission Electron Microscopy (TEM) cross-section investigations showed that the Pt nanoparticles are homogeneously distributed throughout primary 30-75μm silica gel particles , including the inner surface comprising 5-7nm pores. A narrow particles size distribution is maintained across the entire substrate. The Pt catalyst loading level is tightly controlled by the number of ALD coating cycles. An extremely low Pt loading with 0.03mg Pt/cm2 and 0.23mg Pt/cm2 was prepared using 3 and 10 ALD coating cycles respectively. The Pt particles size is approximately 1.3nm for 3 ALD coating cycles. For 10 ALD coating cycles, the Pt particle size is measured to be about 2.4 nm. The sample with a 0.03 mg Pt/cm2 loading shows the highest metal dispersion of 90%, i.e. the highest values ever seen in the literature measured by using the H2 chemical absorption method. Metal dispersion is 46% for a 0.23 mg Pt/cm2loading. Thermal stability studies indicate that there is an initial agglomeration of the Pt clusters during the first 4 hours' of heat treatment at 450°C. The particle size increased from 1.3 nm to about 2nm for 3 cycles Pt coated silica gel. After that there is no noticeable change of the particle size during the following heat treatment process. This study shows that the ALD-FBR method is a rational procedure for the low-cost, simple, and scalable preparation of highly dispersed supported catalysts. The mesoporous substrate not only helps to improve the surface activity of the catalyst, but also stabilizes the Pt nano-particles due to its extremely high surface area and tortuous structure.

See more of this Session: Functional Nanoparticles and Nanocoatings On Particles
See more of this Group/Topical: Particle Technology Forum

New Precursors Leading to Single Crystalline WO3 Nanospheres

Suman Pokhrel, University of Bremen IWT Foundation Institute of Materials Science, Bremen, Germany

Marco Schowalter, Physics, Institute of Solid State Physics, Bremen, Germany

Andreas Rosenauer, Institute of Solid State Physics, Bremen, Germany

Lutz Mädler, Department of Production Engineering, University of Bremen IWT Foundation Institute of Materials Science, Bremen, Germany

Widely applicable Flame Spray Pyrolysis (FSP) route has been employed for the syntheses of spherical, single crystalline WO3 nanoparticles using tungsten precursors with the metal in usual and unusual oxidation states. The tungsten oxides have been characterized by powder X-ray diffraction and the patterns have been refined using Rietveld method (space group P21/n, monoclinic system; a = 7.300, b = 7.530, c = 7.680, β = 90.90deg, Z = 8, unit cell volume 422.1 A^3). The size of the particles obtained from different precursor solutions were in the range of 8-10 nm reasonably agreeing with the particle sizes determined from BET and microscopic analysis. The spherical morphology of the particles and well-developed lattice fringes with spacing of ~ 3.94 Å was determined by low and high resolution transmission electron microscopy (HRTEM) respectively for WO3 (obtained from different precursors). Line indexing of SAED patterns of all WO3 nanomaterials was isostructural with identical monoclinic phases. The precursors used to synthesize these nanopowders are inexpensive and easily available. The synthesis of WO3 nanostrutures using different oxidation states of W in the precursors is unusual but fascinating because it allows an elegant combination of high combustible organic solvent inspired morphosynthesis with FSP thermochemistry.

See more of this Session: Functional Nanoparticles and Nanocoatings On Particles
See more of this Group/Topical: Particle Technology Forum

Characteristics of Self-Assembled Metal Oxide Monolayers On Zeolite Surfaces

Pei-Yoong Koh, Chemical & Biomolecular Engineering, Georgia Institute of Technology, Atlanta, GA

Jason Ward, Chemical & Biomolecular Engineering, Georgia Institute of Technology, Atlanta, GA

Amyn Teja, Chemical & Biomolecular Engineering, Georgia Institute of Technology, Atlanta, GA

William Koros, Chemical & Biomolecular Engineering, Georgia Institute of Technology, Atlanta, GA

Metal oxide -- zeolite nanocomposites are employed as catalysts and adsorbents in organic syntheses, and in waste-water treatment. For example, ruthenium oxide deposited on zeolite-Y is an efficient catalyst for the photochemical reduction of water to hydrogen; and iron oxide - zeolite-Y composites exhibit high adsorption capacity for Cu^{2+}, Cr^{3+}, and Zn^{2+} which are commonly found in wastewaters from metal-plating and tanning. Recently, gas separation membranes have been developed by incorporating magnesium oxide -- zeolite composites in polymeric membranes. The presence of MgO nanowhiskers has been shown to significantly improve interfacial adhesion between the zeolite and the polymer, and results in enhanced separation efficiency. Here, we report on the self-assembly of metal oxide nanostructures such as MgO nanowhiskers on zeolite surfaces via hydrothermal treatment starting from aqueous solutions containing magnesium chloride and ammonia. The effect of temperature, pH, time, and reactant concentration on the size and morphology of the deposited MgO have been investigated. Adsorption isotherms were measured to deduce the nature of MgO bonding on the zeolite surface and to determine optimal concentrations, pH, and residence times for monolayer formation. Mechanisms of nanowhisker formation and deposition are postulated.

See more of this Session: Functional Nanoparticles and Nanocoatings On Particles
See more of this Group/Topical: Particle Technology Forum

Seed-Mediated Growth of Au Nanorods with Controllable Sizes

Juncheng Liu, Department of Chemical Engineering, Auburn University, Auburn, AL

Joshua W. Morgan, Chemical Engineering Department, Auburn University, Auburn, AL

Christopher B. Roberts, Auburn University, Auburn University, AL

The size and shape controlled synthesis of nanostructures is of contemporary interest because their properties at this scale are highly size and shape dependent. Gold nanorods exhibit transverse and longitudinal surface plasmon resonances that correspond to electron oscillations perpendicular and parallel to the rod length direction, respectively. The longitudinal surface plasmon wavelengths (LSPWs) of Au nanorods are desired in the spectral range of 650-900 nm. Light irradiation in this region can penetrate deeper in tissues and cause less photodamage than UV-visible irradiation. As such, development of methodology to tailor both scattering and absorption of Au nanorods with different LSPWs is of ultimate importance for practical application in biomedical imaging and therapeutic systems.

In this presentation, we report the seed-mediated growth of Au nanorods with controllable size by synthesizing Au seeds of designed size through careful control of kinetics. The reduction of the Au salt by $NaBH_4$ in the presence of cetyltrimethyl ammonium bromide (CTAB) within the aqueous phase was demonstrated to be a slow reaction. This allows for an ability to manipulate Au nanoparticle sizes through control of the reaction/growth duration (2, 8, 24, 48, 72, 146 hours were selected in this work). These differently sized Au nanoparticles were subsequently used as seeds for secondary growth where the addition of Au salt, $AgNO_3$, and weak reducing agent, ascorbic acid was required. Smaller Au nanoparticle seeds obtained from short reduction/growth times resulted in Au nanorods with both longer LSPWs and higher aspect ratio, and the larger Au nanoparticle seeds, those obtained from longer reduction/growth times, resulted in Au nanorods with shorter LSPW and lower aspect ratio, illustrating the ability to manipulate size and aspect ratio of the Au nanorods through careful control of the original Au seeds. Moreover, the effect of Au salt, $AgNO_3$, and ascorbic acid concentration used during the seed mediated growth process, as well as the growth temperature, on the LSPWs and aspect ratio of the resultant Au nanorods were systematically investigated and will be discussed in this presentation.

See more of this Session: Functional Nanoparticles and Nanocoatings On Particles
See more of this Group/Topical: Particle Technology Forum

Latex/Vesicle Templated Synthesis of Hollow Inorganic Nanoparticles

Spyros Monastiriotis, Chemical Engineering, The Graduate Center and The City College of the City University of New York, New York, NY

Alex Couzis, Chem. Eng., City College and Graduate Center of the City University of New York, NY, NY

Hollow inorganic nanospheres exhibit catalytic activities and properties from their solid counterparts with the advantages of high surface area and low density. This research focuses on the development of facile scalable methods for the fabrication of porous hollow inorganic spheres using equilibrium catanionic vesicles and amidine functionalized polystyrene latex as templates. Catanionic vesicles [Cetyltrimethylammonium Bromide (CTAB), Sodium Perfluorooctanate (FC7)] and amidine functionalized polystyrene latex with the same size ($\approx$ 90nm) and concentration were the tools in order to study the behavior of acid catalyzed hydrolysis of Tetramethoxysilane (TMOS) and Dihydroxybis(ammonium lactato)titanium(IV) (TALH) as precursors for the formation of the inorganic shell. The vesicle size was tuned by the composition ratio of the two surfactants CTAB and FC7 and characterized by using Cryo-Transmission Electron Microscopy (Cryo-TEM), Small Angle Neutron Scattering (SANS), and Dynamic Light Scattering. By tuning pH, precursor's concentration and template population we can control shell thickness of the inorganic hollow particle and also minimize the gel formation. Multiple inorganic layers were fabricated by surface functionalization of the previous inorganic deposition on the vesicular template. Silica and Titania hollow particles were calcined and characterized with Cryogenic Transmission Electron Microscopy (Cryo-TEM), Transmission Electron Microscopy (TEM), Scanning Electron Microscopy (SEM), Scanning Transmission Electron Microscopy (STEM), Atomic Force Microscopy (AFM), Dynamic Light Scattering (DLS)

See more of this Session: Functional Nanoparticles and Nanocoatings On Particles II
See more of this Group/Topical: Particle Technology Forum

Functionalized Alumina Particles as pH-Responsive Drug Carriers

Bradley R. Gordon, Department of Chemical and Biomolecular Engineering, University of Maryland, College Park, MD

Charles E. Luckett, Department of Chemistry and Biochemistry, University of Maryland, College Park, MD

Daniel D. Lim, Department of Chemistry and Biochemistry, University of Maryland, College Park, MD

Sheryl H. Ehrman, Department of Chemical and Biomolecular Engineering, University of Maryland, College Park, MD

Douglas S. English, Department of Chemistry and Biochemistry, University of Maryland, College Park, MD

Porous alumina particles were functionalized and made pH-responsive using three silane coupling agents (SCAs) with different carbon chain lengths and terminal groups. Release of the anionic dye sulforhodamine 101 (SR101) and the cationic anti-cancer drug Doxorubicin (DOX) from the particles was performed at 37°C and at pH's 5.0 and 7.4 to mimic physiological conditions. The isoelectric points of the functionalized particles were analyzed and are expected to play a crucial role in the pH-response and release characteristics. Charge interactions between the diffusing species and the internal surface of the particles were found to dominate the overall transport for both SR101 and DOX. The effects of silane chain length and layer formation on the release characteristics were also explored and are further discussed in this work.

See more of this Session: Functional Nanoparticles and Nanocoatings On Particles II
See more of this Group/Topical: Particle Technology Forum

Coating Ultra-Thin Microporous/Mesorporous Films On Particle Surface

Xinhua Liang, Department of Chemical and Biological Engineering, University of Colorado, Boulder, CO

Miao Yu, Chamical and Biological Engineering, University of Colorado at Boulder, Boulder, CO

Alan W. Weimer, Department of Chemical and Biological Engineering, University of Colorado, Boulder, CO

Porous thin films/coatings have wide applications in separations, corrosion protective coatings, gas storage, controlled drug delivery, and catalyst encapsulation for controlling reaction selectivity and maintaining catalytic activity. Ultra-thin microporous/mesorporous films/coatings are especially important for some applications. Inorganic films/coatings with few defects, such as zeolite membranes and mesorporous films, have been prepared by hydrothermal synthesis and sol-gel method. While these films/coatings have well defined uniform porous structures, it is difficult to control their thickness with the precision of angstroms. In order to create a porous structure, organic templates must be removed generally by calcination at high temperature. In this paper, we demonstrate a general method to create porous inorganic films/coatings with precisely controlled thickness down to several angstroms from dense organic/inorganic hybrid aluminum alkoxide (alucone) films grown by molecular layer deposition (MLD). The organic constituents in the film can be removed either by mild water etching at room temperature, or calcination in air at elevated temperatures. Because of the layer by layer growth process for MLD, the deposited polymer films have regular structure and the oxidation of MLD polymer films produces uniform interconnected highly porous structures with high surface area. The porous structure has both micropores and mesorpores with BET surface area as high as 1452 m2/g. We expect this porous thin film/coating growth technique, which offers both precise control of the thickness down to several angstroms and well defined porous structure, to have wide applications in catalyst functionalization, controlled drug delivery and nano-filtration.

See more of this Session: Functional Nanoparticles and Nanocoatings On Particles II
See more of this Group/Topical: Particle Technology Forum

The Impact of ATRP Initiator Spacer Length On Grafting Poly(Methyl Methacrylate) From Silica Nanoparticles

David L. Green, Department of Chemical Engineering, University of Virginia, Charlottesville, VA

Chinlun Huang, Merck and Company, Philadelphia

The focus of this work is to quantify the impact of the carbon spacer length (CSL) of immobilized alkoxysilanes initiators on grafting poly(methyl methacrylate) (PMMA) from the surfaces of monodisperse silica nanoparticles. PMMA was grafted using surface-initiated atom transfer radical polymerization (SIATRP), a facile technique to produce well-controlled polymer brushes of high graft densities; the polymerizations were carried out in environmentally friendly 4:1 (v/v) methanol-water solutions at room temperature. Monoethoxysilane initiators of 3, 11, and 15 carbon spacer lengths were synthesized and characterized with 1H NMR and 13C NMR. The initiators were then used to modify the surfaces of monodisperse silica nanoparticles in methyl isobutyl ketone, producing dense initiator monolayers with site densities between 1.80 - 3.64 initiators/nm2. PMMA was subsequently grafted from the functionalized nanoparticles using both CuCl and CuBr catalysts. It was found that polymerizations performed with CuBr were uncontrolled, whereas those with CuCl were controlled. PMMA graft densities ranged between 0.10 - 0.43 polymers/nm2, which increased with the initiator carbon spacer length (CSL). Interestingly, longer CSLs make nanoparticle surfaces hydrophobic, causing nanoparticle aggregation in methanol-water solutions. Our results indicate that surface hydrophobicity is correlated to increases in PMMA graft density through the adsorption of hydrophobic MMA monomers on initiators with longer CSLs. Thus, to augment PMMA graft densities, a subtle balance must be struck between enabling particle stability and increasing MMA adsorption in methanol-water solutions.

See more of this Session: Functional Nanoparticles and Nanocoatings On Particles II
See more of this Group/Topical: Particle Technology Forum

Facile Preparation of Highly-Scattering Metal-Nanoparticle Coated Polystyrene Latex Beads

Jung-Hyun Lee, School of Chemical and Biomolecular Engineering, Georgia Institute of Technology, Atlanta, GA

Mahmoud A. Mahmoud, School of Chemistry and Biochemistry, Georgia Institute of Technology, Atlanta, GA

Valerie Sitterle, Georgia Tech Research Institute, Atlanta, GA

Jeffrey Sitterle, Georgia Tech Research Institute, Atlanta, GA

Carson Meredith, School of Chemical and Biomolecular Engineering, Georgia Institute of Technology, Atlanta, GA

Noble-metal coated polymer composite microspheres and nanoshells have attracted intense interest due to applications in electronics, photonics, medical imaging, drug delivery, surface-enhanced Raman scattering (SERS) and catalysis. By incorporating nanoparticles (NPs) exhibiting preferential scattering characteristics on a polymeric bead surface, the composite microsphere can be used as a scattering contrast agent for medical imaging, and other optical and electronic devices. Here we report a new technique for the facile preparation of highly-scattering metal NP-coated PS latex beads. Different sizes (30, 60 and 80 nm), chemistries (Au and Ag), and shapes (sphere and cube) of NPs were successfully incorporated on commercially available PS beads using this technique. Homogeneous and dense metal coatings were obtained via solvent-controlled swelling and heteroaggregation of polymer capped-metal NPs with the PS beads by the addition and removal of an organic solvent. The organic solvent controlled the heteroaggregation of colloidal particles as well as the swelling/shrinking mechanism of PS beads in this strategy. No prior functionalization of the bead was needed. The fabricated composite beads were stable during long-term water storage. The morphology and coverage of the metal coating on the beads, and thus optical properties, can be effectively controlled by the proper choice of the concentration of solvent and metal NPs. Complete and close packed metal coating on the PS bead can be obtained with smaller and spherical metal AuNPs, while metal surface coverage reached to a certain limitation with larger or cubic metal NPs. The resulting metal coated-PS microspheres exhibited highly enhanced scattering and tunable optical characteristics, which were investigated with the dark-field microscopy and UV-vis absorption spectra. Their performances as SERS substrates were evaluated with Raman Spectroscopy

See more of this Session: Functional Nanoparticles and Nanocoatings On Particles II
See more of this Group/Topical: Particle Technology Forum

Preparation of Quantum Dot-Embedded Polymeric Nanoparticles in a Micromixer

Yanjie Zhang, Chemical & Biological Engineering, Iowa State University, Ames, IA

Aaron R. Clapp, Dept. of Chemical & Biological Engineering, Iowa State University, Ames, IA

We describe a process for stabilizing luminescent quantum dots within a block co-polymer particle. The method uses a staged micromixer which initially forms nanocrystal-homopolymer aggregates in the first mixer, which are then fed into the second mixer where they are stabilized by a suitable block co-polymer. The precipitation is controlled by the nanocrystal and polymer concentrations, flow rate, resident time, and solvent ratio. Under turbulent mixing conditions, the generated particles have a low polydispersity index and relatively small size (~100-500 nm diameter). Below the turbulence threshold, polydispersity increases dramatically. The loading of quantum dots and ultimate composite particle size depends strongly on the efficiency of aggregation obtained within the first mixer and can be controlled by adjusting the homopolymer concentration as well as the capping ligands present on the nanocrystal surface. Careful selection of the block co-polymer introduced in the second mixer provides long-term colloidal stability in aqueous solutions and preserves the luminescence properties of the quantum dots. This method is a rapid (< 1 minute) and effective means for stabilizing quantum dots within a polymeric particle for applications including cell labeling, drug delivery, and particle image velocimetry.

See more of this Session: Functional Nanoparticles and Nanocoatings On Particles II
See more of this Group/Topical: Particle Technology Forum

Effective Parameter Estimation within a Multi-Dimensional and Multi-Scale Population Balance Model Framework

Rohit Ramachandran, Department of Chemical Engineering, Massachusetts Institute of Technology, Cambridge, MA

Paul I. Barton, Department of Chemical Engineering, Massachusetts Institute of Technology (M.I.T.), Cambridge, MA

Multi-dimensional and multi-scale population balance models are often required for the modelling and analysis of particulate systems that are distributed with respect to two or more of their internal coordinates (e.g. granulation) [1]. The population balance model is essentially a meso-scale framework that utilizes microscopic information (e.g. material properties, kinetic and thermodynamic parameters) in the form of kernels, to predict macroscopic particle traits (e.g. size distribution of crystals, porosity distribution of granules etc.). These kernels can be mechanistic (i.e., formulated based on the physics and chemistry of the process) or empirical (i.e., formulated based on power-law expressions with unknown fitting parameters). Given that most particulate processes are complex and intricate, most of the kernels in the literature are of the latter form. An important part of the formulation of the model of a given process is the estimation of the values for model parameters. The estimation of these model parameters is crucial to the applicability of the formulated model for process design, control and optimization. Parameter estimation can sometimes be straightforward, for instance when the process and formulated model are relatively simple and sufficient data is available. However, these conditions are not always met and this can result in difficulties in determining optimal parameter values. The focus of this study is on systems described by population balances and the nature of these systems is such that conventional optimization routines often fail when attempting to determine the optimal set of parameters. In many instances, due to the complexity of these model structures, optimization techniques are not used and parameters are estimated based on trial and error, often resulting in sub-optimal parameter values. The multiple dimensions, nonlinearity, non-convexity, hyperbolic and integral terms associated with the population balance partial-differential equation are potential causes amongst others, for convergence failure [2].

In this work, via two-dimensional crystallization and three-dimensional granulation applications, we aim to investigate and circumvent these failures systematically. A multi-dimensional population balance model is formulated for each application within which an objective function is defined. The objective function is formulated as a nonlinear least squares problem, minimizing the difference between experimental data and simulated model output. Results show that derivative-free methods, such as the Nelder-Mead simplex method, fail to converge to an optimal solution. A similar result was obtained with gradient-based methods such as BFGS, quasi-Newton, Newton, Gauss-Newton, Levenberg-Marquardt and SQP, and with a stochastic genetic algorithm. It was hypothesized that two main issues could contribute to the convergence failure: 1) gradients were calculated based on finite differences; as a result of improper step size determination, the numerical error could be prohibitive resulting in inaccurate derivative information and 2) optimal parameters may not be identifiable. To circumvent these issues, this work addresses the calculation of derivative information based on automatic differentiation to ensure increased accuracy. Issues such as parameter identifiability are also dealt with by analysing an accurate Fisher information matrix. Given the computational burden in calculating accurate Jacobians and Hessians, compounded by the discontinuities introduced into the objective function as a result of crystal/granule nucleation, a global optimization strategy may be warranted and this work addresses that accordingly. Overall, by systematically assessing the problem formulation and mechanisms, results will show that improvements in convergence can be achieved by utilizing appropriate optimization techniques, thus leading to more successful and optimal parameter estimation.

References

1. Iveson, S. M., 2002. Limitations of one-dimensional population balance models of wet granulation processes. Powder Technology 124, 219-229.

2. Wang, F. Y. and Cameron, I. T., 2007. A multi-form modelling approach to the dynamics and control of drum granulation processes. Powder Technology 179, 2-11.

See more of this Session: Population Balance Modeling for Particle Formation Processes: Nucleation, Aggregation and Breakage Kernels
See more of this Group/Topical: Particle Technology Forum

Particle heteroaggregation: Analysis by a property discrete population balance model

S. Rollié[1], K. Sundmacher[1,2]

[1] Max Planck Institute for Dynamics of Complex Technical Systems, Sandtorstr. 1, 39106 Magdeburg

[2] Process Systems Engineering, Otto-von-Guericke University Magdeburg, Universitätsplatz 2, 39106 Magdeburg

In important technical processes (crystallization, polymerization, coating processes) hetero-aggregation between unlike particles plays a decisive role. If the particle species are of different size and zeta potential, formation of morphologically different heteroaggregates (clusters) can be observed. The cluster composition is influenced by adjusting the mixing ratio of the particle species. By using fluorescently labeled particles in a binary particle mixture of polystyrene (PS, 2µm) and melamine formaldehyde particles (MF-RhB, 366nm, labeled with Rhodamine-B) it is possible to analyze the heteroaggregation process by flow cytometry, which provides a multidimensional resolution of cluster composition [1].

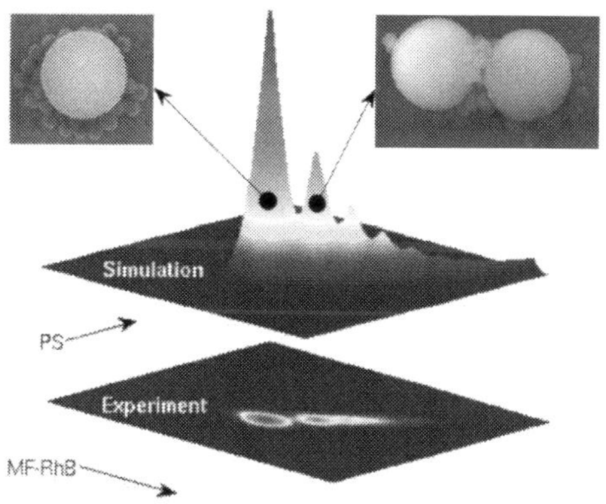

Fig. 1: Simulated and experimental cluster distributions (insets: FE-SEM of heteroaggregates)

The dynamics of the cluster formation process and the tuning of cluster composition via the particle mixing ratio by means of discrete bivariate population balance simulations are presented. The two internal coordinates represent the particle species and characterize the cluster composition. The property space is semi-heuristically reduced to the physically relevant coordinates for improved computational handling [2]. Aggregation rates are determined by two different kernel concepts, which are based on deterministic models under consideration of Brownian motion and DLVO theory. The kernel for 'patchy-particles' accounts for the heterogeneous surface structure of aggregated clusters [3], while a 'charge-balance' kernel calculates the interaction potential between two clusters on the basis of a mean electrostatic charge [2]. A comparison with experimental data shows the applicability of both concepts. To improve the deterministic prediction of aggregation kinetics and consolidate the multi-scale approach, interaction potentials will be measured by atomic force microscopy and compared with DLVO models.

References:

[1] S. Rollié, K. Sundmacher; Langmuir, 2008, 24, 13348-13358.

[2] S. Rollié, H. Briesen, K. Sundmacher, J. Colloid Interface Sci., 2009, 366, 551-564.

[3] A. Moncho-Jorda, G. Odriozola, M. Tirado-Miranda, A. Schmitt, R. Hidalgo- Álvarez; Phys. Rev. E, 2003, 011404.

Equilibrium Aggregation-Breakage: Connecting Molecular Simulations and Rate Kernels

R. Dennis Vigil, Chemical & Biological Engineering, Iowa State University, Ames, IA

A characteristic feature of particulate systems that evolve due to competition between aggregation and breakage is that they sometimes produce non-trivial steady-state particle size distributions. If such solutions satisfy detailed balance conditions, then they are equilibrium solutions. The conditions that must be satisfied by aggregation and fragmentation rate kernels in order for equilibrium solutions to be produced are elaborated, and it is shown that the rate kernels are uniquely determined by the aggregation and breakage rate constants for the reactions involving monomers. Consequently, for equilibrium systems there is a significant reduction in the amount of information needed in order to infer the general form for aggregation or breakage kernels. Lastly, by making use of a statistical mechanical approach, it is also shown how macroscopic rate kernels can be constructed for equilibrium systems by using information obtained from molecular simulations.

See more of this Session: Population Balance Modeling for Particle Formation Processes: Nucleation, Aggregation and Breakage Kernels
See more of this Group/Topical: Particle Technology Forum

Modelling and Simulation of Extensional Flow-Units in Emulsion Formation

Cristhian P. Almeida-Rivera, Process Science, Unilever R&D Vlaardingen, Vlaardingen, Netherlands

Peter Bongers, Process Science Deparment, Unilever R&D Vlaardingen, Vlaardingen, Netherlands

Every day consumers are in direct contact with emulsions and dispersions. Food products, detergents and paints, for instance, do not form homogeneous mixtures but are composed of at least one component in the form of solids or droplets. Within food emulsions the product attributes relevant to the consumers are distinguishable and particular of each product category. Thus, margarine should be spreadable, mayonnaise spoonable and ice-cream creamy. All attributes of an emulsion products (e.g. mouthfeel, colour, aroma, stability, etc) are the result of the close marriage between product composition, microstructure and processing. Due to the inherent business importance of emulsion products, the improvement of current processes and synthesis of novel processing routes are key activities within FMCG companies.

Current technologies for the production of food emulsions include static mixers, stirred tanks, rotator-stator devices, and high-pressure and ultrasonic homogenisers, among others. The governing droplet break-up and coalescence mechanism in each device is highly dependent on its geometry, operation policy and the actual physico-chemical properties of the colloidal system. For instance, stirred tanks are basically driven by simple shear in the laminar regime, whereas turbulent inertial mechanism is mainly responsible for droplet break-up in high-pressure homogenisers.

Despite the sound expertise generated over the last decades regarding droplet disruption and coalescence mechanisms, the effort has been exclusively channelled to current available units. Thus, relatively small attention has been paid to explore novel unit designs. The aim of this contribution is to broaden the spectrum of current emulsification devices, by addressing the design, modelling, and simulation aspects of extensional-flow units. This type of apparatus is characterised by the dominant extensional flow of material in the turbulent regime, and from a mechanical perspective, it is realized in a converging element with high extensional stresses.

The strong feature of our model approach resides in the fully mechanistic description of the governing phenomena. Namely, a population balance equation is formulated and solved to account for the disappearance and appearance of droplets at each size class. Moreover, coalescence mechanism is included to account for the instability of newly created droplets. Additionally, our model estimates the prevailing break-up mechanism at each size class as a function of droplet diameter, the acting forces on the droplet and the exposure time in the high energy zone.

To validate our modelling approach we aim to obtain an oil-in-water emulsion with a dispersed phase volume fraction close to 0.74 and Sauter diameter ranging 4-5 ƒÝm. The initial pre-emulsion is represented by a log-normal distribution of droplets. The results obtained by simulation show that at a given set of operational conditions and pre-emulsion properties the product obtained is within the desired and narrow specifications space.

A sensitivity analysis suggests that the breakage mechanism changes with the flowrate. Regardless the disperse phase volume fraction, large throughputs lead to finer dispersions, while coarser pre-emulsions lead to coarser product. The simulated data show an acceptable degree of agreement with experimental pilot plant data, stressing the practical relevance of extensional-flow units in emulsion formation.

See more of this Session: Population Balance Modeling for Particle Formation Processes: Nucleation, Aggregation and Breakage Kernels
See more of this Group/Topical: Particle Technology Forum

Predicting the Effect of Homogenization Pressure On Emulsion Drop Size Distributions

Neha B. Raikar, Chemical Engineering, University of Massachusetts, Amherst, MA

Surita R. Bhatia, Chemical Engineering, University of Massachusetts Amherst, Amherst, MA

Michael F. Malone, College of Engineering, University of Massachusetts, Amherst, MA

Michael H. Henson, Department of Chemical Engineering, University of Massachusetts Amherst, Amherst, MA

Emulsions constitute numerous natural as well as processed products with a wide variety of applications like foods, pharmaceuticals, cosmetics etc. We have developed a population balance equation (PBE) model to predict drop size distributions obtained from high pressure homogenization. Starting with the simple case of dilute oil-in-water emulsions, we identified the need for two breakage mechanisms to adequately describe the experimental model system (Raikar et al., 2009). We then advanced our PBE model to incorporate multiple daughter drop formation, which tremendously improved the model predictions over the case of binary breakage. We have also added coalescence to make the model applicable to more industrially relevant emulsion systems with high disperse phase volume fractions.

The quality of the final emulsion is strongly affected by the type and concentration of the ingredients used as well as the method used for preparation. We have tested the extensibility of our model for changes in formulation variables as well as processing conditions. The model could capture the effect of changes in formulation variables but failed to describe the effect of the operating pressure. The operating pressure was incorporated through the energy dissipation rate for the homogenizing valve. Our current description lacks extensibility for changes in pressure, and the present contribution is focused on identifying deficiencies in the energy dissipation rate function and identifying alternative functional forms that allow improved predictions. Computational fluid dynamics (CFD) simulations are used to understand the variation of the energy dissipation rate across the homogenizing value and to relate the local dissipation rate to the local pressure and to the applied pressure. The CFD results are used to construct alternative energy dissipation rate functions that provide improved drop size distribution predictions across a range of applied pressures.

Raikar N.B., Bhatia S.R., Malone M.F. and Henson M.A, "Experimental Studies and Population Balance Equation Models for Breakage Prediction of Emulsion Drop Size Distributions", Chemical Engineering Science, 64 , 2433-2447 (2009)

See more of this Session: Population Balance Modeling for Particle Formation Processes: Nucleation, Aggregation and Breakage Kernels
See more of this Group/Topical: Particle Technology Forum

Mechanistic Modeling of the Breakage Function in Eukaryotic Cell Populations

Martin A. Hjortsø, Chemical Engineering, Louisiana State University, Baton Rouge, LA

Michael Benton, Chemical Engineering, Louisiana State University, Baton Rouge, LA

All eukaryotes share a similar mechanism governing progress of the cell through the cell cycle. The mechanism is based on a chemical oscillator and cell division occurs when triggered by the state of this oscillator.

A model of this oscillator has been formulated and solved numerically using Monte Carlo simulations. From simulations of a large number of cell cycles, one easily obtains the distribution of states of the cells in the population as well as the distribution of states of dividing cells. Knowing these two distributions, the breakage function can be calculated.

We will describe the cell cycle model and present results for the breakage function for various growth conditions and for different distribution of daughter cell sizes.

See more of this Session: Population Balance Modeling for Particle Formation Processes: Nucleation, Aggregation and Breakage Kernels
See more of this Group/Topical: Particle Technology Forum

Membrane Encapsulation of Catalysts by Fluidized Bed Film Coating

Maxx Capece, Chemical Engineering, New Jersey Institute of Technology, Newark, NJ

Ajit Mujumdar, Chemical Engineering, New Jersey Institute of Technology, Newark, NJ

Rajesh Davé, Chemical Engineering, New Jersey Institute of Technology, Newark, NJ

This study describes the application of fluidized bed film coating as an efficient method to coat porous membranes onto catalytic particles. These composite materials described as membrane encapsulated catalysts have the potential to increase reaction yield and selectivity by controlling the diffusion of products and reactants to and from the catalytic core. Membrane coated catalysts were synthesized by spray coating spherical zeolite particles with an alumina suspension in hydroxylpropyl cellulose (HPC) using a conventional fluidized bed (Mini-Glatt). Coated particles were subsequently calcined in a furnace to remove the HPC leaving a porous alumina membrane. The fluidized bed coating process was evaluated by its ability to coat uniformly as well as to control the coating thickness of the membrane. Scanning electron microscopy (SEM), Near Infrared Spectroscopy (NIR), and surface area and pore size distribution analysis (BET) were used as methods to characterize the coated catalyst. Results show that membrane coatings were uniformly coated and free of voids upon optimization of fluidization parameters for the range of membrane thicknesses (6.2-22.1 micrometers) tested in this study. Furthermore, membrane coating thickness was shown to adhere very closely to a theoretical thickness equation of Dewettinck (1997). The coating thickness estimated from the experimental techniques was in fairly good agreement with the theoretically predicted thickness of 22.1 micrometers and deviated by only ±0.83 micrometers. Deviation was smaller for thinner membrane thicknesses. The alumina membrane was also shown to adhere well to the catalyst after removal of HPC and its thickness remained unchanged after calcining. BET data shows that zeolite pores remained unchanged after the coating process. NIR absorption spectra was shown to correlate coating thickness with absorbance at 1186nm ($R2=.987$) as a method to replace image analysis for fast non-destructive characterization.

See more of this Session: Applications of Engineered Structured Particulates
See more of this Group/Topical: Particle Technology Forum

Mesoporous silica particles for formulation of poorly water soluble ingredients

Leonard Chia[1], Shoucang Shen[1], Wai Kiong Ng[1], Reginald B. H. Tan[1,2]*

[1] *Institute of Chemical and Engineering Sciences, A*STAR (Agency for Science, Technology and Research), 1 Pesek Road, Jurong Island, Singapore 627833.*
[2] *Department of Chemical and Biomolecular Engineering, The National University of Singapore, 4 Engineering Drive 4, Singapore 117576.*

** corresponding author: Tel: +65 6796 3847; Fax: +65 6316 6183; E-mail: leonard_chia@ices.a-star.edu.sg*

Introduction

Drug insolubility is one of the most challenging problems in the development of pharmaceutical products. Currently, over a third of drugs listed in US Pharmacopoeia and half of new chemical entities (NCEs) are classified as poorly water soluble or insoluble. Poor solubility in the digestive fluid will normally result in low bio-availability which in turn leads to irregular and incomplete absorption. The lack of success in commercialisation of such NCEs has been attributed to their poor solubility properties. As such, much effort has been put in to increase the dissolution rate of such poorly water soluble drugs. There are various formulation methods to improve dissolutions rates, including particle size mediation [1,2], co-crystallisation [3], inclusion in cyclodextrins [4,5], dispersions, nanocrystalline [6] or amorphous forms of the active pharmaceutical ingredients (APIs). One approach is to produce the drugs in an amorphous form and thus increase its dissolution rate. However, the greatest drawback of working with amorphous drugs is that they are generally thermodynamically unstable and may revert back to their more stable crystalline form during storage and transport. In our work, we have overcome these difficulties by using ordered mesoporous silica particles, such as SBA-15, which possess high surface areas and pore volumes to attain a rapid dissolution rate property. A poorly water soluble drug, such as ibuprofen (IBU) or fenofibrate (FEN) is typically co-spray dried with mesoporous SBA-15 to produce a solid dispersion. The amorphous state of the drug is achieved by the quick drying of the spray drying method and is confined within the pore channels of the mesoporous structure. Dissolutions tests show that the dissolution rate of the formulated drug is enhanced and would improve its absorption in the digestive tract and thus its efficacy. As the drug is contained within the nanospaces of the mesoporous structures and enclosed by thick pore walls, re-crystallisation of the drug is prevented, even under stress test conditions. The co-spray drying process could offer much potential in the formulation of other such poorly water soluble drugs which may otherwise find much difficulty in commercialisation.

Results and Discussions

Formulation methods and physical state of IBU

X-Ray diffraction (XRD) tests have been performed on the co-spray dried ibuprofen with SBA-15 with different ratios of IBU:SBA-15. In Figure 1a, when the IBU:SBA-15 ratio is not more than 50:50 (wt:wt), no XRD peaks were found. This indication is that the IBU stored in the mesoporous pore channels of the SBA-15 is in the amorphous form. An investigation into crystallisation has revealed that crystallisation can only occur in channels with a diameter to molecular size ratio above 20 [7]. In our study, the pore diameter of the SBA-15 is approximately 6 nm, and with the IBU molecule estimated have dimensions of 1.0 x 0.6 nm [8], their pore diameter to molecular size ratio is well below 20 at 6~8. However, the pore channels that limit the crystallisation also limit the drug loading of IBU molecules. It was found that above a 50:50 (wt:wt) IBU:SBA-15 loading ratio, 75:25 (wt:wt) IBU:SBA-15 loading ratio, the IBU that is not loaded into the pore channels will reside on the surface of the SBA-15, resulting in some diffraction peaks being observed in the XRD pattern (Figure 1c). Nevertheless, the diffraction peaks observed were much lower in intensity than that of a spray-dried pure IBU crystal (Figure 1d), indicating that both amorphous IBU (in the pore channels) and crystalline IBU (on the surface) co-exist together on the SBA-15.

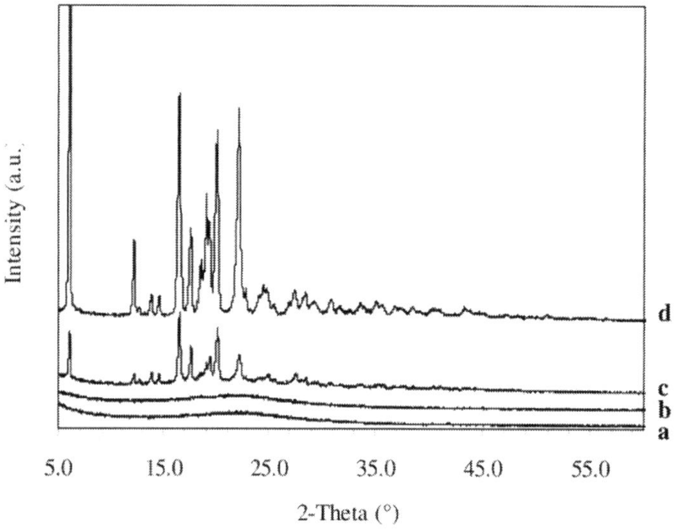

Figure 1. XRD patterns of co-spray dried IBU-SBA-15 with IBU:SBA-15 ratios of: (a) 25:75 (wt:wt), (b) 50:50 (wt:wt), (c) 75:25 (wt:wt) and (d) spray dried IBU crystal (Sigma-Aldrich).

From the DSC curves in Figure 2d, we can notice a strong endothermic peak at 78 °C for a physical mixture of equal amounts of IBU and SBA-15. This peak characterises the melting of the bulk phase of IBU. However, the co-spray dried IBU-SBA-15 formulation with an IBU:SBA-15 ratio of 50:50 (wt:wt), Figure 2b displays no such peak. These results further confirm the amorphous state of IBU when co-spray dried with SBA-15 and is in good agreement with the XRD measurements. When a co-spray dried sample of higher drug loading [IBU:SBA-15 75:25 (wt:wt)], Figure 2c, is tested, the DSC curve shows an endothermic peak at 78 °C, however the endothermic intensity was lower than the physically mixed sample, indicating that only the IBU that exists on the surface of the SBA-15 contributed to the endothermic peak.

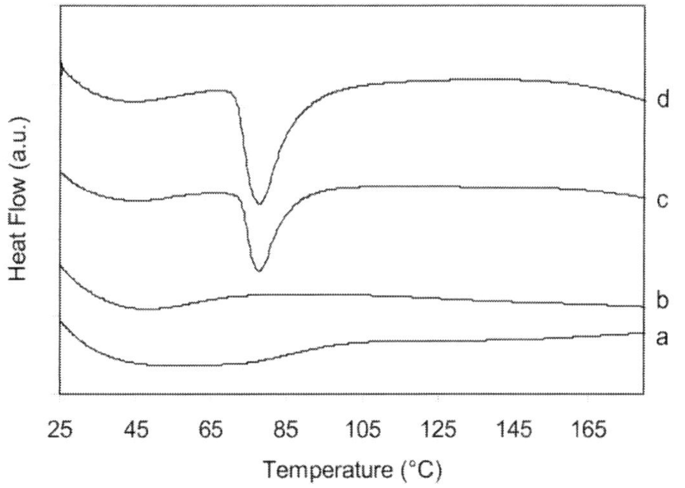

Figure 2. DSC curves of co-spray dried IBU-SBA-15 with IBU:SBA-15 ratios of (a) 25:75 (wt:wt) (b) 50:50 (wt:wt), (c) 75:25 (wt:wt) and (d) physical mixture of IBU and SBA-15 50:50 (wt:wt).

Drug distribution in solid dispersions and morphology

The morphologies of submicron SBA-15 particles and co-spray dried IBU and SBA-15 particles were observed. From the images, it can be seen that the morphology of the co-spray dried IBU-SBA-15 (50:50 wt:wt) (Figure 3B) was not changed as compared to the pure SBA-15 particles (Figure 3A) as the IBU was loaded into the pore channels of the SBA-15. However, when the loading of IBU was increased, (75:25 wt:wt), new nanoparticles with a rod-like morphology were found on the surface of the original SBA-15 submicron particles (Figure 3C). A spray dried sample of pure IBU was also observed under SEM in Figure 3D, the resulting image showed that the IBU particle was much larger than compared to the nanoparticles formed on the SBA-15 in figure 3C The atomisation of the IBU particles have been affected by the SBA-15 submicron particles, leading to a reduced particle size and preventing the formation of large IBU crystals, even though an excess of IBU was used. The small particle size of the IBU in the resulting powder is in good agreement with the XRD patterns (Figure 1c) earlier which showed a reduced intensity compared to pure spray dried IBU crystals.

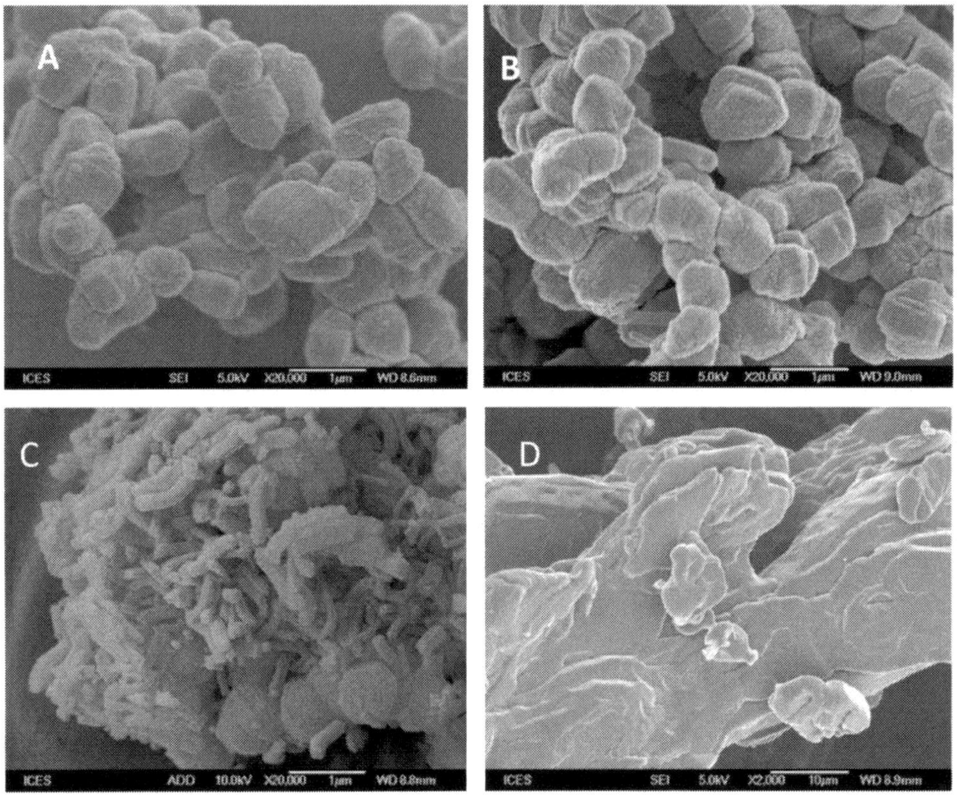

Figure 3. SEM images of (A) submicron particles of SBA-15, (B) co-spray dried IBU-SBA-15 (50:50 wt:wt), (C) co-spray dried IBU-SBA-15 (75:25 wt:wt) and (D) spray dried pure IBU.

Long term stability tests

As the amorphous state of the co-spray-dried IBU and SBA-15 is attributed for its improved dissolution rate, it is important to ensure that the amorphous state is stable during storage. Accelerated stress test conditions of 40 °C /75 %RH was established to investigate the stability of the formulated co-spray dried IBU and SBA-15. It was found that even at such severe stress test conditions, the IBU co-spray dried with SBA-15 exhibited an excellent stability against re-crystallisation. From Figure 4, it can be seen that the XRD pattern for the co-spray dried IBU and SBA-15 exhibited typical amorphous characteristics and no crystal growth was noticed even after 3, 6, 10 and 12 months of the sample being kept under stress test conditions. The broad peak at 15-30 ° of 2θ is a typical diffraction pattern of X-ray amorphous silica. The uniform silica pore walls perfectly separate the fine amorphous particles of IBU and prevent the re-crystallisation of IBU even under severe storage conditions. The small amount of IBU that stays on the surface of the SBA-15 particles during the spray drying process is bound by hydrogen bonds from the hydroxyl groups on the surface of the SBA-15 particle, which prevents the movement and re-crystallisation of the surface IBU. These properties distinguish the use of mesoporous SBA-

15 as a drug carrier in the formulation of poorly water soluble drugs. Whilst amorphous APIs of poorly water soluble drugs can also be obtained by formulation with soluble polymers to form solid dispersions, very few reports investigate their stability during storage.

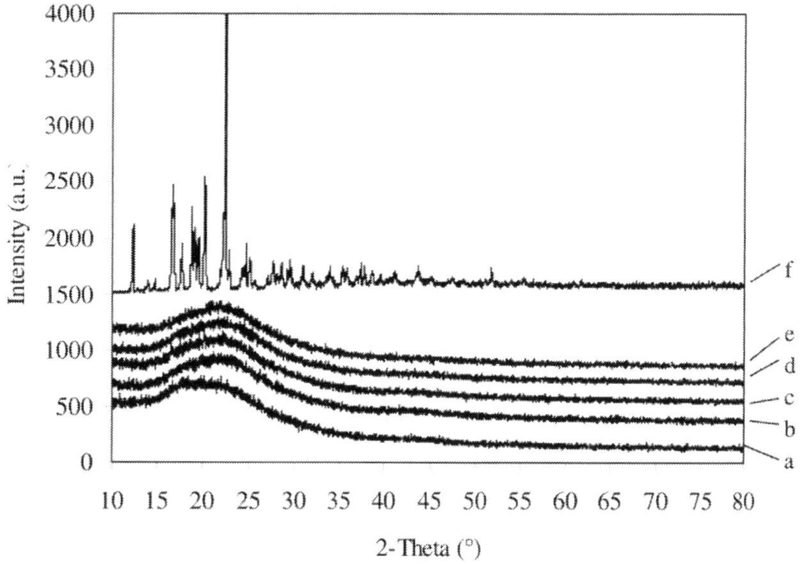

Figure 4. XRD patterns of co-spray-dried IBU with mesoporous SBA-15 after storage at 40°C/75%RH for (a) fresh (b) 3 months, (c) 6 months, (d) 10 months, (e) 12 months; and (f) a control sample of commercial ibuprofen crystal.

Dissolution of IBU from solid dispersion

The dissolution tests comparing between co-spray dried IBU-SBA-15 and pure, untreated IBU crystals are shown in Figure 5. For both types of formulations, tablet and powder forms were used and there was a slight difference in dissolution rates owing that the tablet takes a while to disintegrate. However, the co-spray-dried IBU-SBA15 showed a remarkably increased dissolution rate in comparison to the pure IBU crystals. Within the first 15 minutes, the percentage of IBU dissolved from the IBU-SBA-15 solid dispersion reached 95% whereas the pure crystal had only dissolved by 16%. 100% of the IBU of the co-spray dried solid dispersion was dissolved after 50 minutes, while the pure IBU crystal was only 50% dissolved.

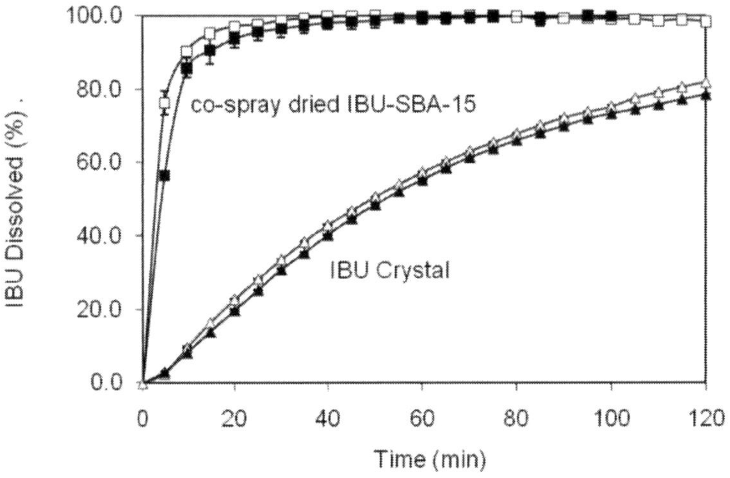

Figure 5. Dissolution profile of IBU from solid dispersion of co-spray dried IBU-SBA-15 with 50:50 (wt:wt) (solid symbol: Tablet form, empty symbol: powder form).

The dissolution profiles of a low (IBU:SBA-15 25:75 wt:wt) and high (IBU:SBA-15 75:25 wt:wt) drug loading of co-spray dried IBU-SBA-15 were compared below in Figure 6. The low drug loaded formulation displayed a rapid dissolution of IBU from the solid dispersion. As shown earlier, this rapid release is attributed by the completely amorphous state of the IBU entrapped within the mesoporous structure of the SBA-15. By comparison, the high drug loaded co-spray dried IBU-SBA-15 formulation showed a slightly delayed dissolution rate. Only 56% of the IBU was dissolved after 10 minutes. Nevertheless, this high drug loaded formulation still managed to achieve 95% IBU dissolution in 40 minutes, which is a much faster dissolution rate compared to the pure IBU crystal. Due to the high drug loading, the IBU on the surface of the SBA-15 was in a crystalline form, albeit nano-sized. The nanoscaling facilitated its ability to achieve complete dissolution within an hour. This is in stark comparison to other studies, such as that by van Speybroeck et al [9], who formulated poorly soluble drugs with mesoporous SBA-15 by solvent impregnation. However, the drug loading was only about 20%. A high drug loading is usually desired and it is an important factor for effective formulation to avoid excess silica in the final product.

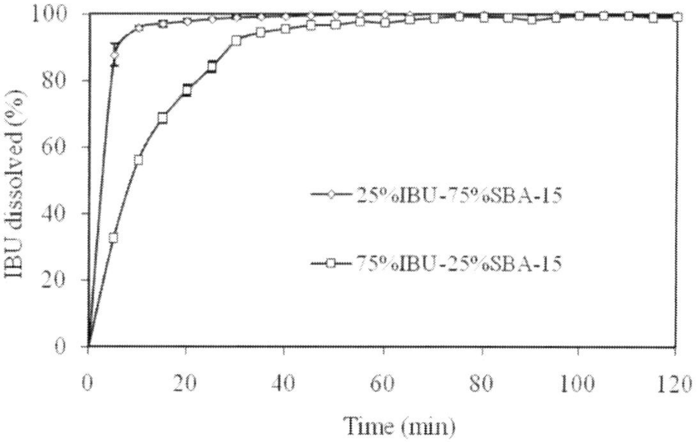

Figure 6. Dissolution profiles of IBU from solid dispersion of co-spray dried IBU-SBA-15 with different drug loadings.

As mentioned earlier, the IBU-SBA-15 formulation was stable even under severe stress test conditions. The XRD patterns shown above in Figure 1 showed that the formulation remained in the amorphous form even after 12 months at the accelerated stress test conditions. As can be expected, the dissolution profile (Figure 7.) of the formulation is not obviously changed as compared with the fresh sample. 90% of the IBU in the formulation was released into the dissolution medium after 15 minutes and 100% dissolution was achieved within 60 min.

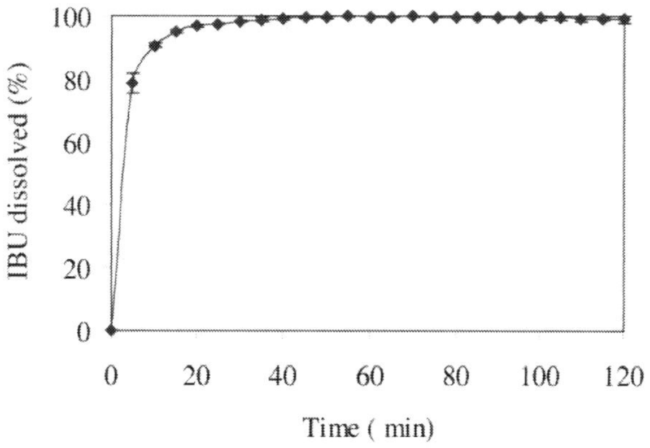

Figure 7. Dissolution profile of IBU from co-spray dried IBU-SBA-15 (50:50 wt:wt) solid dispersion after storage at 40°C/75%RH

Conclusions

It has been found that co-spray drying with mesoporous silica SBA-15 is an effective formulation method for poorly water soluble drugs, such as IBU, to achieve a rapid dissolution rate. The nano-sized mesoporous channels of the the SBA-15 particle helps to entrap the drug particles in its amorphous state. The amorphous state of the IBU contributed to a significant enhancement of the dissolution rate compared to the pure IBU crystal. As the pore walls help to separate the drug particles, preventing re-crystallisation or crystal growth, excellent physical stability is achieved. Even under severe stress test conditions of 40 °C/75 %RH for 12 months, the amorphous state of the IBU-SBA-15 and its corresponding rapid dissolution rate were maintained. The formulation is also capable of achieving high drug loadings of 75:25 (wt:wt) IBU:SBA-15 ratios and still significantly enhancing its dissolution rate. This co-spray drying process could have much potential in the formulation of other poorly water soluble drugs.

References

1. Shukla AJ, Price JC. 1989. Effect of Drug (core) particle-size on the dissolution of theophylline from microshperes made from Low-molecular weight cellulose-acetate propionate. Pharm Res 6: 418-421.

2. Jinno J, Kamada N, Miyake M, Yamada K, Mukai T, Odomi M, Toguchi H, Liversidge GG, Higaki K, Kimura T. 2006. Effect of particle size reduction on dissolution and oral absorption of a poorly water-soluble drug, cilostazol, in beagle dogs. J Control Res 111: 56-64.

3. Shiraki K, Takata N, Takano R, Hayashi Y, Terada K. 2008. Dissolution improvement and the mechanism of the improvement from cocrystallization of poorly water-soluble compounds. Pharm Res 25: 2581-2592.

4. Xiang TX, Anderson BD. 2003. Pharmaceutical formulation for poorly water soluble camptothecin analogues, US Patent No. 6,653,319.

5. Palmieri G F, Antonini I, Martelli S. 1997. Inclusion complexation of fenofibrate with β-cyclodextrin and hydroxypropyl-β-cyclodextrin. Evauation of interactions in solution and solid complex characterization. STP Pharma Sci. 7: 174-181.

6. Wunderlich JC, Schick U, Werry J, Freidenreich J. 1999. Gelatin or collagen hydrolysate containing drug formulation that provides for immediate release of nanoparticle drug compounds, US Patent No. 5,932,245.

7. Sliwinska-Bartkowiak M, Dudziak G, Gras R, Sikorski. R, Radhakrishnan R, Gubbins KE. 2001. Freezing behavior in porous glasses and MCM-41. Colloid Surf. A: Physicochem Eng Aspects 187-188: 523-529.

8. Vallet-Regi M, Rámila A, Real RPD, Pérez-Pariente J. 2001. A new property of MCM-41: drug delivery system. Chem Mater 13: 308-311.

9. Van Speybroeck M, Barillaro V, Thi TD, RANDY Mellaert R, Marten J, Van Humbeeck J, Vermant J, Annaert P, Van Der Mooter G, Augustijn P. 2009. Ordered Mesoporous Silica Material SBA-15: A Broad-Spectrum Formulation Platform for Poorly Soluble Drugs. J Pharm Sci 98: 2648-2658.

Co-Solvent Assisted Spray Pyrolysis for the Preparation of Metal Microparticles

Kai Zhong, Chemical and Biomolecular Engineering, University of Maryland, College Park, MD

Ranjan Pati, Chemical and Biomolecular Engineering, University of Maryland, College Park, MD

Alex Langrock, Chemical and Biomolecular Engineering, University of Maryland, College Park, MD

Howard Glicksman, DuPont Electronic Technologies, Research Triangle Park, NC

Sheryl Ehrman, Chemical and Biomolecular Engineering, University of Maryland, College Park, MD

Spray pyrolysis processes offer many advantages over solution phase routes for powder production, but extension to metals on an industrial scale has been restricted to only a few oxidation resistant systems such as silver and palladium. The requirement of addition of high concentrations of reducing gas to produce metallic particles has limited extension to base metal production. In our approach, a co-solvent is added to the aqueous solution of metal salt precursors, which decomposes in the reactor to produce small amounts of hydrogen, less than the flammability limit in air, reducing inexpensive metal salt precursors to metallic powders. We have constructed a reactor system capable of producing several grams/hour of micron sized metallic copper powder using an ultrasonic droplet generator operating at a frequency of 1.65 MHz. This process operates with nitrogen as a carrier gas, and particles are quenched with additional nitrogen. We have also shown that these metal powders can be produced using ethylene glycol as the co-solvent rather than ethanol as used previously by our group. This represents an additional process safety advance, because of the higher flash point of ethylene glycol relative to ethanol, our original co-solvent choice. The oxide free copper metal particles were made from copper nitrate precursor, and different reaction conditions were investigated to optimize the process. It was found that the amount of co-solvent, carrier and quench gas flow rates, and reaction temperature play an important role in this process. We have successfully made pure copper particles with a narrow size distribution using 30 volume% ratio of ethylene glycol, 5 L/min nitrogen (N2) carrier gas and 10L/min N2 quench gas. The experiment results also indicated that the co-solvent-assisted spray pyrolysis technology is a promising method for the preparation of multi components alloy particles such as Ag/Cu/Co.

See more of this Session: Applications of Engineered Structured Particulates
See more of this Group/Topical: Particle Technology Forum

Adsorption of Warfare Simulants with Ordered Porous Materials

Tyler J. Selbe, Kansas State University, Manhattan, KS

Keith L. Hohn, Chemical Engineering, Kansas State University, Manhattan, KS

Jennifer L. Anthony, Chemical Engineering, Kansas State University, Manhattan, KS

This presentation will report our findings of the adorption of several chemical warfare agent simulants (CWAs) using ordered microporous and mesoporous materials. Zeolites and zeolite-like materials have well-ordered structures and pores. They have varying capacities for molecules based upon size, functional groups, polarity, and intermolecular and intramolecular forces. Zeolites, therefore, can be used as highly specific coatings on sensing devices. CWAs have a diverse range of features that pose threats to humans. CWAs can blister skin, attack the human nervous system, and deplete oxygen supply. The various features and functionalities in CWAs lead to different adsorption capabilities in zeolites. Evaluating the uptake in microporous and mesoporous zeolitic materials can help to quickly identify and isolate structural features in CWAs without the need for human presence in a contaminated area. We will report the uptake of several CWA simulants with microporous and mesoporous materials including: ZSM-5, VPI-7, VPI-9, zeolite 13X, and zeolite 4A.

See more of this Session: Applications of Engineered Structured Particulates
See more of this Group/Topical: Particle Technology Forum

The Synthesis of Magnetic Nano-Photocatalyst Via Atomic Layer Deposition

Yun Zhou, Department of Chemical and Biological Engineering, University of Colorado, Boulder, Boulder, CO

David King, Department of Chemical and Biological Engineering, University of Colorado, Boulder, Boulder, CO

Alan Weimer, Department of Chemical and Biological Engineering, University of Colorado, Boulder, Boulder, CO

Iron nanoparticles were produced in this study via the decomposition of iron oxalate particles. The iron oxalate powder was first ground using a cryogenic milling process. TiO2 thin films were then deposited on the synthesized iron nanoparticles using an in situ Atomic Layer Deposition (ALD) method at 100oC with TiCl4 and H2O2 precursors. However, due to a high surface area, the iron nanoparticles were unstable and spontaneously oxidized when exposed to H2O2 during the TiO2 ALD process, thus reducing the magnetic property of the magnetic core particles. Therefore, prior to TiO2 deposition, Si3N4 films were deposited in-situ to coat and passivate the iron nanoparticles. This passivation provided for asignificant decrease in the the iron oxidation. The photoactivity of this nanocatalyst was tested by decomposing methylene blue under UV irradiation. The magnetic properties were measured using a magnetometer.

See more of this Session: Applications of Engineered Structured Particulates
See more of this Group/Topical: Particle Technology Forum

Large Eddy Simulations of Titania Nanoparticle Evolution in a Turbulent Flame with Detailed Chemistry

Yonduck "James" Sung, Department of Mechanical Engineering, University of Texas at Austin, Austin, TX

Venkatramanan Raman, Department of Aerospace Engineering & Engineering Mechanics, University of Texas at Austin, Austin, TX

Maulik Mehta, Department of Chemical & Biological Engineering, Iowa State University, Ames, IA

R. O. Fox, Department of Chemical & Biological Engineering, Iowa State University, Ames, IA

Sotiris E. Pratsinis, Particle Technology Laboratory, Institute of Process Engineering, Deparment of Mechanical and Process Engineering, ETH Zürich, Zürich, Switzerland

Titania (TiO_2) nanoparticles have widespread practical use in pigments, photocatalysis, and as catalyst supports. Each year, over 3 million tons of titania nanoparticles are produced, for which gas-phase combustion based synthesis has been a dominant method due to its advantage in large-scale production. In this flame-based synthesis, titanium chloride ($TiCl_4$) vapor is oxidized in a pre-existing turbulent jet flame, which then forms solid nanoparticles through interactions with the gas-phase species. The final size distribution of the nanoparticles is dictated by the nonlinear interaction of turbulence, gas-phase flame and titanium chloride chemistry, and titania particle evolution. In order to accurately compute this complex process, it is crucial to account for these interactions using detailed physical models. For this, we present a computational approach based on the large eddy simulation (LES) methodology, where the gas-phase titanium chloride evolution and methane combustion are represented using very detailed chemical kinetics. In particular, the transformation of titanium chloride to the solid primary nucleating particle is represented using 30 species and 51 reactions. The evolution of the titania number density function is tracked using the quadrature method of moments (QMOM) approach. For validation purposes, this detailed computational model is first compared against an experimental setup. The effect of flame configuration on the final product distribution is then discussed.

See more of this Session: Applications of Engineered Structured Particulates
See more of this Group/Topical: Particle Technology Forum

AIChE Thomas Baron Award Lecture (Nashville, TN, November 11, 2009)

Functionalization of Fine Particles Using Atomic and Molecular Layer Deposition

D. M. King[a], X. H. Liang[a], and A. W. Weimer[a]

[a]Department of Chemical and Biological Engineering, University of Colorado, Boulder, Colorado 80309, USA

The functionalization of fine primary particles, including nanoparticles and nanotubes, is easily carried out using atomic or molecular layer deposition (ALD or MLD, respectively) techniques. Particle ALD/MLD can be used to deposit conformal and pinhole-free films of refractory oxides, non-oxides, metals, and hybrid polymer-based materials, amongst others. Fluidized bed reactors are well-suited for large scale operations and can be operated at reduced pressures while using inert gases necessary for standard self-limiting, flow-based ALD/MLD processes. The continuous-flow processing allows for process control using an in-line mass spectrometer downstream from the reactor chamber. Many insulator and semiconductor films, including Al_2O_3, ZnO, TiO_2 and SiO_2 have been successfully deposited on primary particle surfaces in fluidized bed reactors using a variety of precursor types. The applications of functionalized or passivated nanomaterials are innumerable, and the flexibility of this scalable process demonstrates its economic viability.

Introduction

There are many industrial and scientific applications where there is a strong desire to functionalize the surfaces of ultrafine powders to achieve a variety of purposes. Surface functionality can be broken up into two categories, inactive and active. Inactive coatings include inert films that are applied to eliminate the spontaneous interaction between the core powder and its surroundings. One example of an inactive coating is the deposition of insulating Al_2O_3 films on metal nanopowders in order to prevent detrimental combustion reactions that render the metal useless. Another is the deposition of SiO_2 films on TiO_2 nanopowders used in sunscreens and pigments, and are pore-free and thick enough to eliminate the photoactivity of the core powder that prevents the powder from adversely affecting the surrounding organic molecules in the liquid vehicle. Many other industrially-relevant functionalization needs exist, including the incorporation of ceramic films on polymeric particles that can then be fabricated into hybrid organic-inorganic composite materials.

Using self-limiting atomic layer deposition (ALD) growth methods, a wide variety of materials have been deposited with atomic layer control (1-6). ALD involves the sequential dosing of two or more precursors, oftentimes ones used for chemical vapor deposition (CVD), with essential purge steps that prevent unwanted gas-phase reactions between precursors. The generic formula for CVD of a metal oxide can be described as

$$x \cdot M(L)_n + y \cdot H_2O \rightarrow M_xO_y + 2y \cdot (HL) \quad [1]$$

where M represents the metal species, n is the number of ligands, L; and x:y is the metal:oxide ratio. The two main classes of precursors are metal halides and metalorganics. Metal halides emit strong acid vapors as products, whereas metalorganics tend to emit alkanes, alcohols or alkylamines for alkyl-, alkoxide- and amine-based precursors, respectively. Any binary metal oxide CVD reaction can be divided into two half-reactions that occur solely on a particle surface to define ALD:

$$\text{A)} \quad [S]:(OH)^*_y + x \cdot M(L)_n \rightarrow [S]:(O)_y M_x(L)^*_{n-y} + y \cdot (HL), \quad [2]$$

$$\text{B)} \quad [S]:(O)_y M_x(L)^*_y + y \cdot H_2O \rightarrow [S]:(M_xO_y):(OH)^*_{2y} + y \cdot (HL). \quad [3]$$

[S] represents the particle surface and an asterisk represents the active surface species. If each half-reaction is self-limiting and self-terminating, the repetitive application of these reactions in an ABAB... sequence can produce atomic layer controlled metal oxide deposition (2). These types of sequential surface reactions have been studied extensively with Fourier transform infrared (FTIR) vibrational spectroscopy and other in-situ techniques and have been shown to be self-limiting . In addition to ceramic films, ALD methods for metal and polymer films, which have been extensively demonstrated on flat surfaces, have also been transferred to particle-based substrates .

Ferguson et al were the first to demonstrate that ALD techniques could be used to coat ALD films on the surfaces of primary particles (7-15). The particles were pressed into a stainless steel grid as a support for deposition in a viscous flow reactor. Though the particles were sometimes not usable for application-specific testing, this proof of concept work was critical to begin exploring more traditional particle coating unit operations. Shortly thereafter, Wank et al demonstrated that a fluidized bed reactor operating under vacuum could be used to deposit Al_2O_3 films on bulk quantities of micron-sized iron particles using ALD techniques (16-18). Hakim et al studied the fluidization behavior of nanoparticles and was the first to demonstrate that ALD techniques can be used to apply conformal, pinhole-free films to the surfaces of high surface area nanoparticles (19-25). Wilson et al was the first to deposit metal films (tungsten) on metal or polymeric particles using a rotary reactor (26, 27), and Liang et al was the first to deposit ceramic films on polymers using a fluidized bed reactor (28-31). King et al was the first to demonstrate ZnO, TiO_2 and SiO_2 ALD in a fluidized bed reactor (32-36). Even though the initial groundwork for particle ALD research was performed at the University of Colorado (37), many other institutions have since begun independent explorations of the benefits of applying particle ALD techniques to sparingly deposit ceramic films and metal nanoclusters on the surfaces of particles for a variety of applications (38-41). Several particle ALD advances have been made in the deposition of Pt and Pd nanoclusters on carbon nanotubes, carbon aerogels and graphitic and graphene particles (amongst others) for improved catalysis processes and next generation electrode materials for fuel cells and batteries (42-44).

Molecular layer deposition (MLD), also known as alternating vapor deposition polymerization (AVDP) and layer-by-layer growth, was first introduced by several groups in Japan in the 1990s. The MLD method was developed from the earlier gas phase polymer growth technique known as vapor deposition polymerization (VDP). VDP utilizes the co-evaporation of two types of bifunctional monomers under vacuum with each evaporator controlled thermostatically. MLD is an analog of a one-step VDP reaction, and it is based on bifunctional monomer precursors such as A-L-A and B-M-B. In these precursors, "A" and "B" are chemical functional groups and "L" and "M" are organic fragments. In the MLD process, the chemical functionality of the surface species switches in a digital manner and molecules are stacked on substrates one by one in order of preference under vacuum. MLD can deposit conformal and precise ultra-thin polymer films with molecular layer control. The thickness and molecular weight of polymeric thin films can be effectively controlled by the number of reaction cycles. This process is similar to the ALD method except for the anisotropic linear chain molecules. The linear chain molecules play an important role in molecular orientation perpendicular to the substrate. This dry vapor-phase method, which operates under vacuum and does not require solvents or catalysts, is a useful and promising technique for the fabrication of functional ultra-thin polymeric layers (45-47).

A number of studies have been performed on the growth of different polymer MLD films, such as polyamide, polyurea, polyimide, polyimide-polyamide, and poly(p-phenylene terephthalamide). MLD can also be used to deposit hybrid inorganic/organic thin films using suitable precursors, such as TMA and ethylene glycol (EG) for aluminum alkoxide (alucone) hybrid polymer films. The properties of the composite film can be tuned by the assembly of different organic/inorganic compositions. MLD on particles has been demonstrated by alucone MLD on ZrO_2 particles (~30 nm) and $BaTiO_3$ particles (~200 nm) on a small (< 0.1 g) scale in a viscous flow reactor (45, 46). Liang et al was the first to successfully deposit conformal, pinhole-free MLD films on particles using a fluidized bed reactor (48). Challenges for polymer film MLD on large quantities of fine particles include the tendency of particle aggregation during the MLD coating process and the difficulty to deliver the sticky low vapor pressure precursors into the reactor.

Particle Fluidization in a Fluidized Bed Reactor

Significant work has been performed to study the fluidization behavior of fine and ultrafine particles of varying material, composition and morphology in fluidized bed reactors (49-52). It is widely accepted that most larger micron-sized particles (ϕ > 100 μm) can be easily fluidized as individual particles at any pressure, as long the fluidizing medium (here, inert gases) is flowing at the minimum fluidization velocity or faster. The actual superficial gas velocity corresponds to the pressure drop across a bed of a given cross-sectional area. When proceeding downward in particle size, London-van der Waals interparticle forces dominate, which causes nanoparticles to form soft aggregates, they break apart and re-form as new aggregates during a normal fluidization process. In order to overcome these interparticle forces during nanoparticle fluidization, external energy must be imparted to the bed continuously in order to achieve a well-fluidized state (50-52). Operating an ALD fluidized bed reactor, i.e. fluidization in the presence of surface reactions, the additional need to break apart aggregates is to expose all surfaces to reactive gas-phase precursors throughout a precursor exposure or dose (53). It has also been shown that the ALD films themselves will modulate interparticle forces as the particles go from surfaces consisting of the substrate, to those of the film material. This evolution has been observed in a change in the Hamaker Constant of TiO_2 nanoparticles with increasing number of ALD cycles of Al_2O_3 (20).

It is important to have a fundamental understanding of the two-phase fluid flow inherent to particle fluidization processes, as well as an understanding as to how the kinetics and thermodynamics of the surface reactions taking place will affect heat and mass transfer within the reaction vessel (49). These issues can become quite problematic when using new chemistries for particle ALD/MLD in a fluidized bed reactor based on process conditions optimized for coating flat surfaces in a viscous flow reactor. From a surface chemistry perspective, the concept of the sticking coefficient of a precursor, or the number of molecules that participated in surface reactions in a system per number of molecules delivered to the system, is interesting when applied to particle ALD processes in fluidized bed reactors. For low reactivity precursors used in ALD processes on flat surfaces, it may take the administration of 10^3 or 10^4 molecules to the system to have one that reacts, which corresponds to very low precursor utilization. In a fluidized bed reactor however, those 10^3 or 10^4, less one, molecules may have an opportunity to see 10^9 or 10^{10} particles, each of which has many available active surface sites at the onset of the process. At high turnover frequencies characteristic of well-fluidized beds, the entering precursor molecules will be exposed to available nucleation sites prior to escaping out of the top of the bed, and the probability that this number of molecules won't find active sites before exiting the bed becomes very low. This means that full utilization is attainable for *some* amount of time, and oftentimes a very high stochastic surface conversion can be reached. The FBR is ideal for promoting very efficient uptake of precursor molecules while maintaining appropriate reaction conditions throughout the bed (54).

From an engineering perspective, it is more interesting to monitor the gases that exit the reactor, which can be achieved using a residual gas analyzer (RGA) downstream of the reactor (54). The specific surface area of a given powder can be easily measured using the Brunauer-Emmett-Teller technique. With knowledge of the overall surface area of the particle bed, in conjunction with a given precursor dosing rate, in situ mass spectrometry can be a very powerful tool to help close a mass balance around the reactor in order to quantify surface conversion and precursor utilization. During the initial stages of a precursor dose, the gaseous ALD reaction byproduct is the only species that is observed in the effluent stream. After some amount of time, the unreacted precursor exits the bed. This phenomenon is observed in the RGA and evidenced by a characteristic "breakthrough" of the precursor molecule. Breakthrough properties are a function of the reactants, substrate type, operating conditions and fluidized bed parameters. For highly reactive precursors, the breakthrough of the reactant coincides with a sharp decline in reaction product, signifying that a very high stochastic surface conversion was attained. For low reactivity precursors, the flexibility exists to stop administering a precursor once breakthrough occurs, and the stochastic growth rate would then be equivalent to the surface coverage at the point of breakthrough multiplied by the typical growth rate for a fully-saturated system. This notion of "running-lean" typically does not add to the overall process time, as purging step times are greatly reduced due to a lack of residual unreacted precursor in the bed. The downstream RGA technique has been effectively employed to provide real-time feed-forward process control to modulate dosing times based on the breakthrough time for a given system. A real-time understanding of the breakthrough properties of the coating process leads to optimal precursor usage and minimized processing costs.

Sol-gel techniques are the most common for liquid-phase functionalization of particle surfaces (55). Liquid-phase processes yield films that tend to be inconsistent, uncontrollable and may have spotty coverage. Liquid-phase processes also tend to be relatively expensive due to precursor/solution waste and the potential need for separation and drying steps prior to particle usage. Chemical vapor deposition (CVD) is the most common vapor-phase method for coating particles, in which reactive gas precursors are delivered to the reactor system simultaneously (56). CVD films tend to be non-conformal, granular and porous, which reduce the quality of surface passivation. Sol-gel and CVD processes tend to be run thicker to compensate for the inconsistency and film porosity resulting from these techniques. Film porosity is an attribute that directly correlates to a decrease in core material lifetime, regardless of the application. The quality of pinhole-laden films depends on the tortuosity of the pores, which may not be controllable. Sol-gel and CVD coating processes may also tend to agglomerate primary particles. The lack of control of these processes operated at an industrial scale requires even thicker films to account for potential batch-to-batch variations. Neither sol-gel nor CVD processes offer adequate opportunity for precision thickness control and may result in wide variation within and across batches. A conformal thin film deposition method may enhance the performance attributes and extend the usable lifetime for passivated powders used in a variety of applications.

ALD is an ideal technique to deposit conformal, pinhole-free films on substrate surfaces with precision thickness control (57). In contrast to liquid and many other vapor-phase deposition techniques, ALD is not a line-of-sight dependent process. Any exposed particle surface with the ability to support functional groups can be coated, including internal surfaces of porous substrates such as catalyst supports, mesoporous materials, aerogels, etc. The typical process windows for ALD/MLD vs. CVD processes operated in fluidized bed reactors is shown in Figure 1. CVD processes cannot be adequately performed on submicron-sized particles due to the tendency for these particles to agglomerate. For applications that require passivating layers of reasonable thickness, ALD may be an ideal technique to replace CVD operations for systems with large enough throughput due to material savings alone.

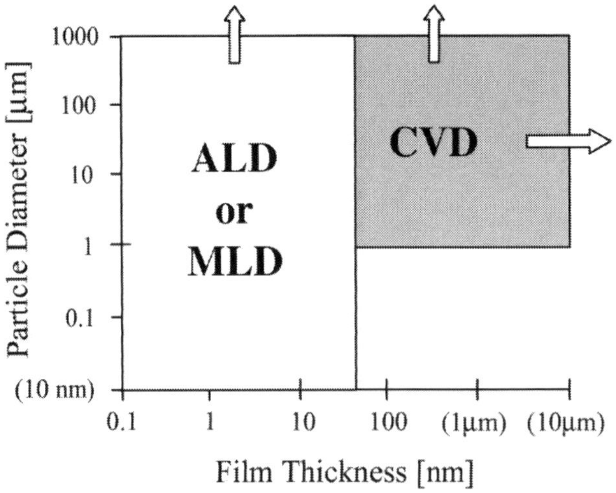

Figure 1. Depiction of the ALD/MLD vs. CVD process windows for particle coating

Experimental Methods

The Fluidized Bed Reactor (FBR).

A custom FBR was used for the gas-phase delivery of reactive ALD and MLD precursors to coat various ceramic or hybrid organic/inorganic films on particle surfaces. The details of this reactor configuration have been described previously (54). Typical ALD operating pressures are on the order of 1 to 10 Torr, and increase with increasing reactor diameter and bed mass to be fluidized. The reaction vessel diameter is sized appropriately based on the desired batch size. Two lab-scale reactors with diameters of 2.5 and 6.3 cm, as well as a 2 L pilot scale system with a 15 cm diameter have been used to functionalize particle surfaces. Larger particle ALD systems are currently being designed. A schematic of a lab-scale reactor configuration is shown in Figure 2. Dosing sequence and times are controlled using LabView® software and pneumatically actuated valves. This flexible system allows for the deposition of multiple film types on a given batch of powder.

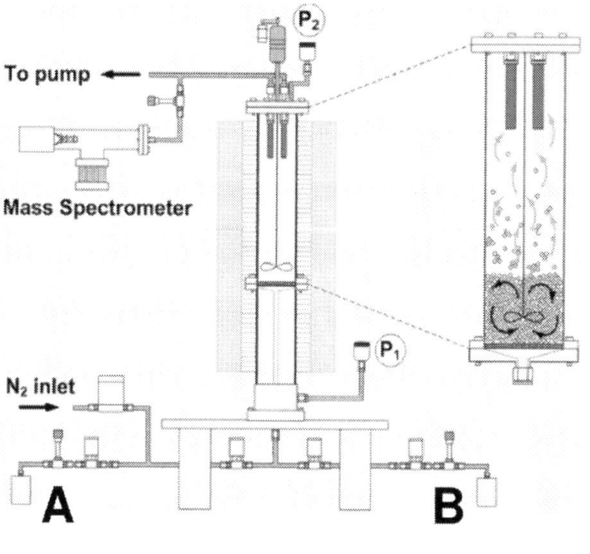

Figure 2. Schematic of the ALD fluidized bed reactor.

Several fluidization curves for 550 nm SiO_2 spheres are shown in Figure 3 for various bed masses. As expected, the pressure drop required to fluidize the bed increases with increasing mass. Also shown here is the relative effectiveness of fluidizing a 50 g bed of ultrafine spherical powder with and without stirring. When the stirrer is in use, each bed is well-fluidized by a superficial gas velocity of 6.0 cm/s, independent of bed mass. The pressure drop across the unstirred bed continually increases, even beyond a gas flow of 10.0 cm/s.

Precursor delivery to the system works well when the operating pressure of the system is less than the vapor pressure of the precursor. Heating the precursor will raise its vapor pressure in accordance with the Antoine equation. If the required vapor pressure cannot be achieved by this strategy, bubblers can be used to deliver the precursor vapor into the fluidized bed reactor. This may also be required for dosing precursors into industrial scale fluidized beds that operate with higher pressure drops, or when performing ALD processes at higher operating pressures. Ideally, the bubbler inlet is controlled using an MFC to allow a calibrated amount of N_2 to be bubbled through the precursor reservoir. More novel precursors with low vapor pressures (e.g. many amine-based precursors), or solid precursors (e.g. metallocenes or some metal chlorides) can be easily used on particles in this flow-based reactor configuration.

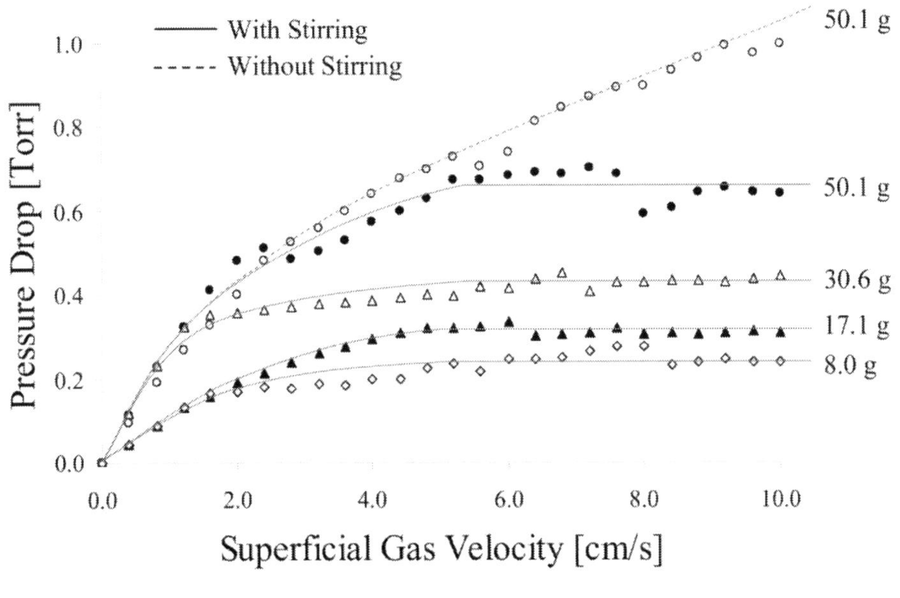

Figure 3. Fluidization curves for stirred beds of 550 nm SiO_2 spheres of varying mass. Top two curves depict the fluidization behavior of stirred and unstirred beds of 50 g of powder.

Carbon nanotube surfactant pre-treatment. Some particle surfaces do not possess adequate surface functionality to support the chemisorption of reactive ALD precursors such as single-walled carbon nanotubes (SWNTs). In these cases, liquid-phase surfactant-based or gas-phase pretreatments (e.g., NO_2 functionalization, described elsewhere) can be employed such that ALD precursors can react on a physisorbed scaffold, which has proven to be sufficient to support a fully-dense ALD layer on particle surfaces (58, 59). For the SWNT case, high-pressure decomposition of carbon monoxide (Pure HiPco) bundles (Carbon Nanotechnologies Incorporated, Houston, TX) containing an iron catalyst were used as a substrate. The as-received SWNTs were gently dispersed in an aqueous vehicle using a low-power ultrasonic bath. Dispersion can be defined as the distribution of nanotube bundles and the splitting of the bundles into significantly smaller tube networks. The ethanol-treated SWNT suspension was stirred over heat until all excess ethanol was evaporated. For the liquid-phase surfactant treatment, an anionic surfactant, sodium dodecylsulfate (SDS) was used at 1.0 wt % in D_2O, which exceeds the critical micelle concentration. A probe sonicator had been used in other studies to enhance the surfactant-based flocculation process, which was shown to be non-detrimental to SWNTs; a low-power ultrasonic bath was used for this study for all liquid-phase pretreated dispersions. Immediately after sonication, SDS-functionalized samples were placed in a centrifuge for decantation; D_2O was only used to enhance the density disparity between phases and ease of separation. The upper 75 to 80% of the supernatant was carefully decanted, leaving a concentrated micelle-suspended nanotube solution.

In situ mass spectrometry. A residual gas analyzer (Stanford Research Systems, Sunnyvale, CA) with an internal quadrupole mass spectrometer was attached to the outlet of the fluidized bed reactor, as shown in Figure 2. Each molecular species may be individually tracked with time throughout the entirety of the ALD process. This allows for the real-time visualization of both reactants and products of each ALD half-reaction. The entire molecules are shown as well as their fragments, which are derived through the ionization of their parent molecule. An example of one full ZnO ALD cycle is shown in Figure 4a-b, with an HRTEM image of the resulting ZnO film coated on TiO_2 nanoparticles in Figure 4c. ZnO was deposited using sequential exposures of diethylzinc (DEZ) and water at 177°C, with ethane being the reaction byproduct. It is clear from Figure 4a that all DEZ entering the reactor was completely utilized until the point at which the signal began to rise, or "breakthrough". This mass spectrometry excerpt was taken during the coating of a 25 g batch of 79 m^2/g TiO_2 nanopowder. The total bed surface area was ~2000 m^2 and required ~33 min per entire ALD cycle, or 1 sec/m^2. in the 6.3 cm diameter reactor. The DEZ dose time for this particular process was 540s (open triangle) and DEZ breakthrough occurred after ~450s (closed triangle). Until breakthrough, the generation rate of ethane decreased with time. The signal increased upon breakthrough due to the confounded nature of the reaction-generated product and the reactant fragmentation that appears at the same m/z value. This signal can be deconvoluted for real-time process control.

The H_2O half-reaction was simpler to monitor using this *in situ* technique as the m/z values of this reactant and the product did not overlap (Figure 4b). The reactant breakthrough (closed triangle) occurred just before 300s and was extremely rapid. The appearance of H_2O coincided with a rapid drop in generated product. Integrating under the cumulative intensity curve yielded a 98.4% surface conversion at the time of breakthrough, relative to ethane generated throughout the entire H_2O dose. There was no discontinuity in the second derivative of the ethane signal at the point the water dose was stopped, which implied that increasing the water dose would not have generated additional ethane and the reaction was essentially complete.

This in situ monitoring technique can be utilized at any scale, and as such, an optimized powder loading and precursor dosing strategy can be put into place to utilize 100% of any precursor, independent of cost, type or sticking coefficient. Even novel, bulky precursors for Pt ALD, which cost hundreds of dollars per gram can be efficiently utilized in a fluidized bed reactor. Highly reactive precursors such as trimethylaluminum may be better suited in a shallow bed, whereas lower reactivity precursors such as silicon chloride may achieve higher conversions at the point of breakthrough in taller beds. ALD processes, which leverage the lower energies found on surfaces, run at lower temperature than their CVD counterparts (typically 300-500°C lower). Efficiently run unit operations for particle ALD drive down the variable operating costs for this technique, making it competitive, if not superior, to current particle coating techniques used in industry today.

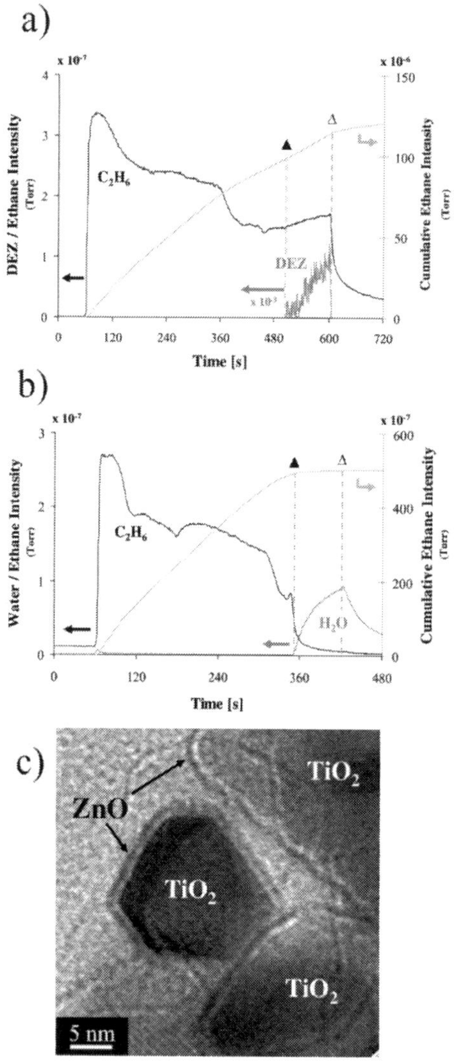

Figure 4. In situ mass spectrometry plots for the a) diethylzinc dosing step and the b) water dosing step during a ZnO ALD process. The actual precursor/ethane signals are on the left axis and the cumulative ethane generated is on the right axis of each. The closed triangles signify the initial breakthrough of each unreacted precursor and the open triangles represent the end of each dose. c) An HRTEM image of the resulting ZnO ALD film on TiO_2 nanoparticles after 10 cycles.

ALD Film Composition Analytical Techniques.Electron microscopy, X-ray photoelectron spectroscopy (XPS, Axis Kratos analytical XPS spectrometer), and Fourier-transform infrared (FTIR) vibrational spectroscopy are other common techniques, amongst others, used to verify film presence or composition on particles at the nanoscale.

Electron microscopy. Transmission electron microscopy (TEM) analysis was performed using a Philips CM10 microscope. The particles were put directly onto Cu grids with a holey carbon overlay film. TEM is a useful technique to observe film/particle interfaces when there is at least a moderate density difference between the materials. Films thinner than ~5 nm may be difficult to observe using standard TEM techniques alone. High resolution transmission electron microscopy (HRTEM) was conducted using a JEOL 2010F 200 kV Schottky field emission TEM. Cross-sectional TEM samples were prepared by cutting thin slices of a cured epoxy resin loaded with the particles of interest. The thickness of the polymer slices was ~60 nm. The deposited films on nanoparticles can be better observed by cross-sectional HRTEM. Scanning transmission electron microscopy (STEM) utilizes similar equipment and provides a z-contrast image, such that materials with a large difference in molecular weight can be readily observed.

FTIR spectroscopy. A Nicolet 750 Magna-IR FTIR spectrometer was used to study the surfaces of particles via surface vibrational changes specific to the core and shell materials. Uncoated and coated particles were pressed into tungsten grids (Buckbee Mears, 0.5 mm thick, 3.9 lines/mm) for this analysis. Bulk vibrational modes of the ALD film material were increasingly visible with increasing number of ALD cycles.

X-Ray Diffraction. A powder X-ray Diffractometer (XRD, PAD5, Scintag, Inc.) was used to study the crystallinity of ALD films. A Cu x-ray source was used, operating at $K\alpha = 1.540562$ Å. Most ALD films grown on curved particle surfaces are amorphous as deposited, except for ZnO, which grows as polycrystalline, independent of temperature. Thermal annealing post-treatments can be used to crystallize ALD films. Powder XRD was also used to measure the crystallite size propagation of ZnO ALD films with the number of cycles.

Inductively coupled plasma atomic emission spectroscopy (ICP-AES). ICP-AES was performed using an Applied Research Laboratories ICP-AES 3410+. ICP-AES can provide the mass-based concentration in parts per million with respect to the substrate. Since the amount of material deposited on the particle substrates by ALD is determined by the surface area of the substrate, the ALD film growth rate, and the number of coating cycles, the metal concentration based on ICP-AES can be highly affected by substrate morphology. Analysis by ICP-AES was achieved by completely digesting the ALD films and the substrate particles in a strong basic or acidic solution. Concentrated NaOH was sufficient for the dissolution of amorphous Al_2O_3 and polycrystalline ZnO ALD films. SiO_2 ALD films and underlying particles were digested using an $HF/HNO_3/HCl$ mixture. Most other films and particles were efficiently dissolved using a concentrated H_2SO_4 solution at 95°C.

ALD Film Performance Evaluation Techniques

Since there are widely varying applications for ultrathin coatings on particles, many techniques are available to test their performance. Once the presence of the films has been verified, thickness dependent behavior can be better understood using proper experimental design techniques. For screening experiments, standard partial- or full-factorial statistical designs of experiments can be performed to study whether film thickness, deposition temperature, etc., affect the responses of interest for a given nanocomposite system. Once relationships have been determined, central composite designs can be used to map the surface response over multiple factors in order to optimize the application-specific performance. Scale-up factors can also be included in this optimization process, including film deposition cost for different reactor configurations and sizes, to understand the value-added performance enhancement at any annual throughput, even at the industrial scale.

Photoactivity mitigation. The photoactivity of both coated and uncoated TiO_2 particles, a well-known UV-active photocatalyst substrate, was measured using a standard methylene blue (MB) dye decomposition technique. A known amount of powder was loaded into the MB solution and irradiated with a 0.8 mW/cm2 UV lamp (λ = 360 nm). A UV-Vis spectrophotometer was used here to detect the change in absorbance at the blue wavelength. The UV absorbance of the TiO_2 particles, with and without Al_2O_3 ALD films remained unchanged, signifying that the elimination of photoactivity was not due to the disruption of the UV irradiation that reached the core TiO_2 surfaces (22).

Oxidation resistance. A thermogravimetric analyzer (TGA) was used to measure the change in mass of coated and uncoated metal particles during oxidation. The powder sample was loaded into an alumina crucible suspended from a microbalance by a Pt wire. Air was fed such that an excess amount of oxygen was available to oxidize the metal powder. The oxidation onset temperature and oxidation rate with temperature are readily visible using this method. It has also been shown that the magnetic moment of iron particles was unchanged after being coated by ALD Al_2O_3 films, which is further evidence that ALD can apply functional coatings on particles that do not detract from the bulk properties (18).

Pinhole-free SiO_2 ALD acid resistance. ICP-AES was also used to test the conformal, pinhole-free nature of SiO_2 ALD films deposited on pigment grade TiO_2 particles, since a TiO_2 particle fully encapsulated by SiO_2 should be stable in H_2SO_4 indefinitely, whereas a pinhole-laden or inconsistent SiO_2 film would allow the underlying particle to be degraded with time.

Results and Discussion

Conformal, Pinhole-Free ALD Films

The ALD technique yields conformal, pinhole-free films on surfaces that possess the proper surface functionality that allows for precursor nucleation. The standard technique to "prove" that a deposition process provides conformal and uniform films that has been commonly employed in scientific papers is to use TEM images. Though this strategy may be serviceable for ALD films deposited on flat surfaces, since a cross-sectional image may show a significant percentage of the surface area that was coated, it is much more difficult to rely solely on TEM images for coatings on particles. Three methods have been developed specifically to study the pinhole-free nature of the particles and test their performance as barrier films.

Oxidation resistance of coated metals. Al_2O_3, the most frequently studied ALD barrier film, was deposited on a variety of metal powders of varying size. The simplest set of experiments utilized micron-sized metal spheres coated with several thicknesses of Al_2O_3. The results from these experiments showed that the particle ALD technique could effectively apply an oxidation-resistant barrier to the surfaces of metal powders, even when the coating thickness was on the order of 10 nanometers on a particle with a 10 micron diameter. For micron-sized iron spheres, both the oxidation resistance with temperature and the magnetic properties were tested. The 10 nm film provided complete passivation of the 10 μm iron spheres up to a temperature of approximately 600°C. Above this temperature, the mismatch between thermal expansion coefficients of the core and shell was too great and the passivating layers were fractured. This problem can be overcome by depositing alternating multilayers of materials to perform thermal expansion matching between the composite film and the core particle. Nevertheless, the magnetic moment of the micron-sized iron spheres coated by various thicknesses of Al_2O_3 remained unchanged, which was anticipated due to the extreme thinness of each shell (25).

A more intensive method was used to produce metal nanoparticles, which cannot be obtained directly without having the ALD reactors contained within an environmental chamber. The pyrophoric nature of metal nanopowders is extremely hazardous, but a method to produce passivated metal nanopowders that retain their metallic properties can be extremely beneficial to a variety of industries. Rather than pursuing metal nanoparticles directly, extensive work has been performed on the in situ decomposition of metal salts followed by ALD coating within a fluidized bed reactor. Several metal oxalate powders have been studied, including those of iron, nickel, cobalt and copper. The resulting particles are micron-sized, although highly porous due to the macroscopic release of CO or CO_2 upon decomposition. Several size reduction processes have been employed, including standard ball milling and cryogenic milling. The latter allowed particles to be as small as 30-50 nm, which provided a suitable test vehicle for a proof of concept that even metal nanopowders could effectively be passivated (25). The temperature dependent oxidation behavior of uncoated Fe nanopowder and the oxidation resistance of Fe nanopowder coated by 30 and 50 cycles of Al_2O_3 ALD is shown in Figure 5. It is clear that the Fe nanopowder began to oxidize almost instantaneously in air, and resulted in a mass gain of 20% at a temperature of 250°C. For the 30 cycle film (~6 nm film thickness), the onset of oxidation was delayed until a temperature of approximately 350°C. The 10 nm film (50 cycles) showed no signs of oxidation when held at a temperature of 400°C for over three hours. The resulting elemental analyses from these experiments showed that the oxygen content of the 50 cycle material matched the anticipated content for Al_2O_3 coated on iron metal, not an oxidized form.

Similar results have been obtained for Al_2O_3 ALD films coated on the other core metals produced from the decomposition of their respective oxalates. Recently, TiO_2 was used and showed similar beneficial properties when coated on metal nanoparticles (36). For TiO_2 coatings on Ni nanoparticles (from nickel oxalate) that had been brought to temperatures that caused the films to fail, an XRD was performed to determine the resulting crystal structure. It was verified that the resulting composite nanoparticle was actually a shell of nickel titanate surrounding a core of nickel oxide, rather than a cracked TiO_2 film that allowed the nickel to oxidize. In this case, the mixed-metal oxide reaction occurred, which is coincidently a very useful technique to form a completely homogeneous mixed-metal oxide material.

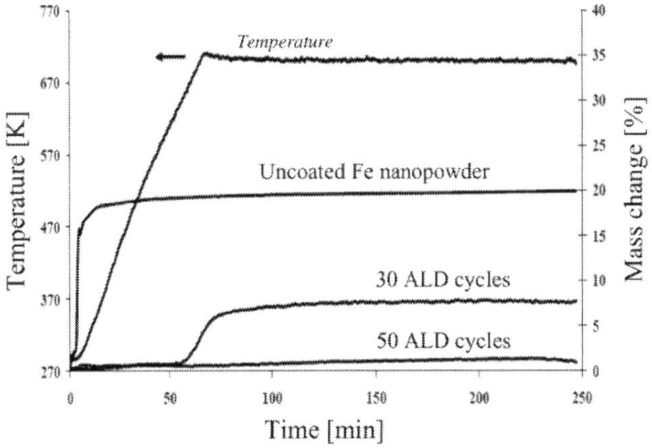

Figure 5. TGA plot of the thermal stability of uncoated and coated Fe nanopowder at elevated temperatures in the presence of air. The uncated Fe begins to oxidize at 50°C, the Fe nanopowder coated with $30Al_2O_3$ ALD cycles begins to oxidize at 350°C, and Fe nanopowder coated with 50 Al_2O_3 ALD cycles is stable in air at 400°C indefinitely.

Chemical resistance of coated ceramics. TiO_2 is the most widely used inorganic pigment material, and can be found in paints, plastics and paper, amongst others (60). The integration of TiO_2 into several of these end-use products can be challenging for two reasons. The first is that TiO_2 is a well-known photocatalyst material that readily degrades surrounding organic molecules. The second is that TiO_2 also readily degrades in H_2SO_4, and as such its usage in external paints in locations with a high content of acid rain can be problematic. In order to solve both of these problems simultaneously, SiO_2 ALD layers were applied to the surfaces of anatase and rutile pigment-grade (~200-300 nm diameter) TiO_2 particles. Trisdimethylamino-silane (TDMAS) and concentrated hydrogen peroxide were employed as precursors at a reactor temperature of 500°C. The growth rate of this chemistry was 0.4 Å/cycle, and 50 cycles were employed to deposit

2.0 nm conformal, pinhole-free films. The uncoated and coated particles were separately placed in 5.0 M H_2SO_4 and each was maintained at 95°C for 24 hours. The ICP-AES results, including an inset HRTEM image are shown in Figure 6. The ICP-AES technique, which has a detection limit of less than one part per million, was used to measure the presence of Ti that escaped into the solution. The uncoated particles were steadily dissolved in the strong acid solution throughout the 24 hour measurement time, whereas the particles coated by 2.0 nm SiO_2 ALD films did not allow more than 1% of the TiO_2 to escape. This set of experiments was the most aggressive strategy to test the claim that the films were not pinhole-free, as 2.0 nm is on the lower edge of practicality (33).

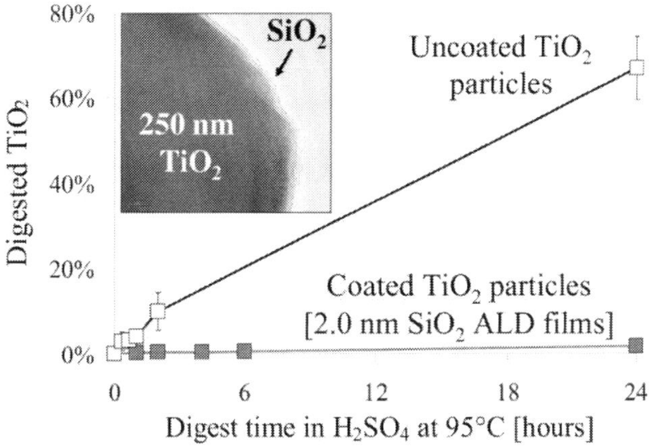

Figure 6. Digest rate of uncoated and SiO_2-coated pigment-grade TiO_2 particles in H_2SO_4 at 95°C for 24 hours. SiO_2 ALD film thickness is 2.0 nm and is fully dense; films are pinhole-free and all TiO_2 nanoparticles are fully encapsulated.

Separately, the photoactivity mitigation was studied for this set of coated particles and it was determined that the 2.0 nm film eliminated more than 98% of the photoactivity of the underlying anatase particles. The photoactivity of the rutile phase of TiO_2, which is the more common pigment material due to its higher refractive index, is much lower than that of the anatase, so again this provided data for the lower bound of the performance of a 2.0 nm film (33).

Non-Ideal Behavior During High Surface Area Particle ALD

Oftentimes the transference of a known ALD process and chemistry that has been well-characterized on flat surfaces is non-trivial. Significant issues can arise when coating large batches of high surface area particles (potentially tens/hundreds of thousands m^2 total, or more) due to the significantly longer dose times to administer a saturating dose of precursor. Some ALD processes utilize temperatures above which a precursor may thermally decompose, as is the case when using diethylzinc (DEZ) for ZnO ALD. DEZ, which begins to decompose at 120°C, is much more cost effective than its more stable counterpart, dimethylzinc. As such, it is more preferable to develop lower temperature industrial processes that utilize DEZ for ZnO ALD on particles (34, 35).

Thermal decomposition. DEZ-based ZnO ALD processes (30 cycles) were carried out at 50 and 100°C using concentrated H_2O_2 as an oxidizer, and at the more common particle ALD temperature of 177°C using DEZ and H_2O as precursors. Each of the films was deposited on 25g batches of TiO_2 nanoparticles using a fluidized bed reactor. Since ZnO and TiO_2 are each white materials, it was initially surprising to see the deviation from a brilliant white color to gray after the ZnO ALD process operated at a temperature of 177°C. The surface area of this sample was significantly lower than expected, which was indicative of hard aggregation that normally does not occur during typical particle ALD processes. This was the initial impetus behind studying low temperature ALD processes for coating particles (35). At lower temperatures, the discoloration issue subsided, but the surface area once again dropped significantly. The predominant causes of hard aggregation in a powder bed are poor mixing (e.g., due to inadequate mechanical agitation) and CVD.

Liquid bridging. In a properly purged system, reactive precursors do not come in contact with one another, thus eliminating an opportunity for CVD to occur. With decreasing temperatures, residual water becomes increasingly difficult to remove. The water removal efficiency from a fully saturated 2000 m^2 nanopowder bed was cut in half when decreasing the operating temperature from 125°C to 100°C. In general, ALD processes can be operated at lower temperatures when using a stronger oxidizing or oxidation agent, such as the migration from H_2O to H_2O_2. Another benefit of transitioning from H_2O to concentrated H_2O_2 (50% in H_2O) at lower temperatures is that the use of the latter removes 50% of the H_2O that is administered to the system, assuming constant exposures. This significantly reduces the required purging times for processes operated at 100°C or below, even down to room temperature. With the use of the in situ mass spectrometer, it was observed that the enhanced removal efficiency of H_2O_2 diminished with increasing temperature, and the removal rates became equivalent at a reactor temperature of 125°C. The removal of water from the reaction vessel is paramount to preventing unwanted gas-phase reactions as is the case with any ALD process, but this becomes even more important when coating high surface area particles using ALD techniques. Liquid bridging, one of the strongest interparticle forces, becomes increasingly problematic in particle coating processes that operate at low temperatures (35).

Elemental analysis. The Zn content of 30 cycle ZnO films at various temperatures is shown in Figure 7. The XPS (surface) and ICP-AES (bulk) also suggested that different high and low temperature mechanisms were present. XPS percentages included the presence of surface carbon and oxygen from being exposed to air between the coating and analysis. Film composition is expressed in terms of Zn content as there is an indiscernible XPS peak shift between free Zn and oxygen-bonded Zn. The Zn content in the film at 177°C increased in the presence of thermal decomposition, and was 42% greater than the Zn content at 100°C. The opposite trend was observed in the ICP-AES data, in that the overall Zn content was significantly greater at 50°C than any other temperature. Liquid bridging from retained H_2O within aggregates explains this observation, as only a bulk measurement technique can quantify ZnO formed by localized CVD within aggregates.

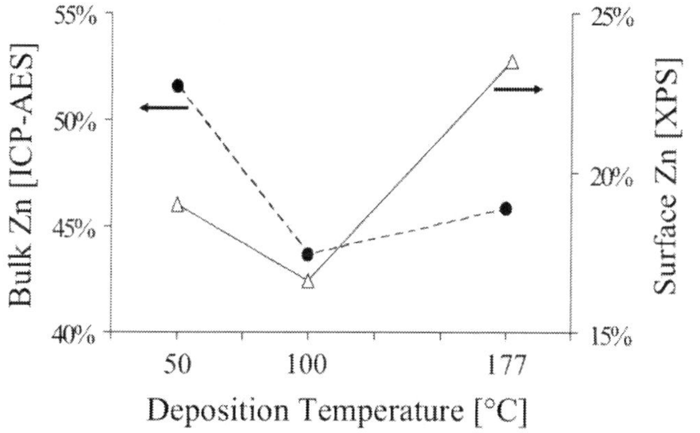

Figure 7. Temperature-dependent Zn content measured in the bulk (left, ICP) vs. the surface (right, XPS) for TiO_2 nanoparticles coated with 30 ZnO ALD cycles.

Flexibility of the Fluidized Bed Reactor Configuration

The efficiency attained through the use of a fluidized bed reactor has been mentioned repeatedly, based on the claim that there exists a finite amount of time during which 100% of the precursor is utilized. The degree of utilization varies widely for individual systems and reactor geometries, but this efficiency opens up the opportunity to use more costly, novel precursors that cannot afford to be wasted. Further, fluidized bed processes are used to produce both small quantities of high-valued pharmaceutical products as well as most of the polyethylene and gasoline/diesel products used throughout the world today.

Classes of ALD precursors. There are several classes of precursors that are commonly used for ALD ranging from inexpensive and corrosive, to solid materials that can only be fed via dilute phase conveying of the sublimated vapor. The FBR system can accommodate any of these precursors, and an engineering decision can be made to balance the precursor utilization with cost, throughput, operating pressures, etc. Three HRTEM images showing examples of the use of varying precursor classes for different films is shown in Figure 8. Si_3N_4 was deposited using TDMAS and NH_3 as precursors at a deposition temperature of 400°C. High surface area TiO_2 nanoparticles (79 m^2/g) were used as the substrate. Both fully saturating and lean-running processes were employed to deposit Si_3N_4 in the FBR. The 2.0 nm film shown in Figure 8a was from the lean-running process, where the precursor utilization rose from 80% up to approximately 97%. Figure 8b shows a BN ALD film that was deposited on ZrO_2 nanoparticles using BCl_3 and NH_3 as precursors (13). The corrosive vapors emitted from metal halide processes can be effectively abated through the use of simple traps in line between the continuous gas flow reactor and the pump(s). This strategy can be implemented at any scale, and can allow for relatively inexpensive precursors (e.g. $TiCl_4$ or $SiCl_4$) to be used in industrial-scale particle ALD processes. More recently, the popularity of metallocene precursors, or those with cyclopentadienyl ligands, especially for the deposition of transition metal oxides or highly catalytic noble metal nanoclusters. Figure 8c shows an iron oxide ALD film deposited on the surfaces of primary ZrO_2 nanoparticles, using ferrocene (or bis-cyclopentadienyl iron) and oxygen (61). An experimental plan was performed that determined the optimal exposure time for ferrocene, which was delivered to the reactor system by diverting the N_2 fluidizing gas through the solid precursor container and into the FBR. The ZrO_2 particles remained in a fluidized state before, during and after the administration of dilute ferrocene to the reactor. With the metallocene processes, which are typically operated with O_2 as the oxidizing agent, CO_2 is emitted and can be monitored using the mass spectrometer. Even with a diluted precursor stream, the evolution of the reaction product was continuous at the onset of a dosing step. Unreacted precursor breakthrough was also observed after a certain amount of time, signifying that the precursor was 100% utilized prior to that point. Ferrocene is relatively inexpensive, but significant gains can be made during the deposition of more costly noble metal nanoclusters such as ruthenium or platinum. Cyclopentadienyl-based precursors can be used to coat particles with 100% efficiency in the fluidized bed reactor.

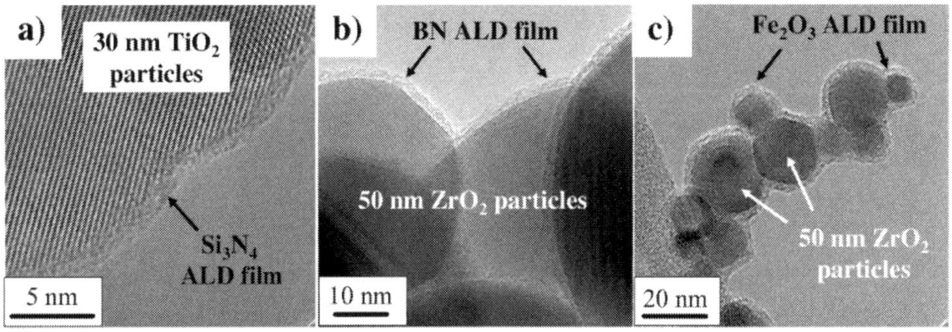

Figure 8. HRTEM images of a) TiO_2 nanoparticles coated with a 2.0 nm silicon nitride ALD film using an amine-based precursor; b) ZrO_2 nanoparticles coated with a 3.0 nm boron nitride ALD film using a chloride precursor; c) ZrO_2 nanoparticles coated with a 5.0 nm iron oxide ALD film using a metallocene precursor.

ALD on Polymeric and Carbon-Based Particles

Another reason for developing lower temperature ALD processes is so that they can be used to deposit ceramic films on the surfaces of primary polymer particles in the fluidized bed reactor. Significant work has been performed on the deposition of Al_2O_3 ALD films on polymer powders.

ALD on HDPE particles. In situ mass spectrometry allows for the effective monitoring of gaseous product evolution and subsequent unreacted precursor breakthrough. For Al_2O_3 ALD with the alternating reaction of TMA and H_2O, the expected reaction product is CH_4, which, however, is also formed directly from the decomposition of TMA in the mass spectrometry. To distinguish the amount of methane formed as a reaction product formed directly from TMA, D_2O was used instead of H_2O. The reaction product from TMA and D_2O was expected to be CH_3D, instead of CH_4. The m/z peaks of interest (i.e. primary and fragmentation peaks) were 28 for N_2, 42 ($-AlCH_3$), 57 ($-Al(CH_3)_2$) and 72 ($Al(CH_3)_3$) for TMA, 18 (-OD) and 20 (D_2O) for D_2O, and 17 for CH_3D. The in situ mass spectrometry results of Al_2O_3 ALD on 16 μm HDPE particles at

77°C, shown in Figure 9a, indicate that the chemistry of TMA and D_2O ALD was self-limiting. When TMA was dosed into the reactor (similar to Equation [2]), an instantaneous increase of the signal of CH_3D byproduct was not seen. This was reasonable, since there was no deuterium present before the very first dose of D_2O into the reaction system. TMA is an extremely reactive precursor and as such, very high surface conversions have been observed at the point of breakthrough. It is clear that all TMA entering the reactor was completely utilized until the time at which the TMA signal increased, which is called the "breakthrough" time. Shortly after the unreacted TMA was observed, the TMA dose was stopped, and N_2 was fed into the reactor to purge any residual reaction product or unused TMA from the system. After the N_2 purge, D_2O was dosed (an analog of equation [3]) and the CH_3D signal increased instantaneously. As the D_2O dose proceeded, the reaction product began to decrease, and the signal of D_2O (m/z = 20) appeared. At this point, the surface reaction was nearly complete and N_2 was fed into the reactor to purge any residual reaction product or unreacted D_2O from the system. After the N_2 purge, TMA was dosed into the reactor to begin the subsequent cycle, and there was an instantaneous increase of the CH_3D byproduct. This was reasonable since particle surface had been saturated with –OD groups. During both precursor doses, it was apparent that the reactions were self-limiting and self-terminating, because the reaction products increased and then decreased while the reactants were still being dosed. If these reactions had not been self-limiting, the products would have continued to be emitted as long as the reactants were dosed. An STEM image of an Al_2O_3 ALD-coated micron-sized HDPE polymer particle, and a zoomed in TEM image of the shell are shown in Figures 9b and 9c, respectively.

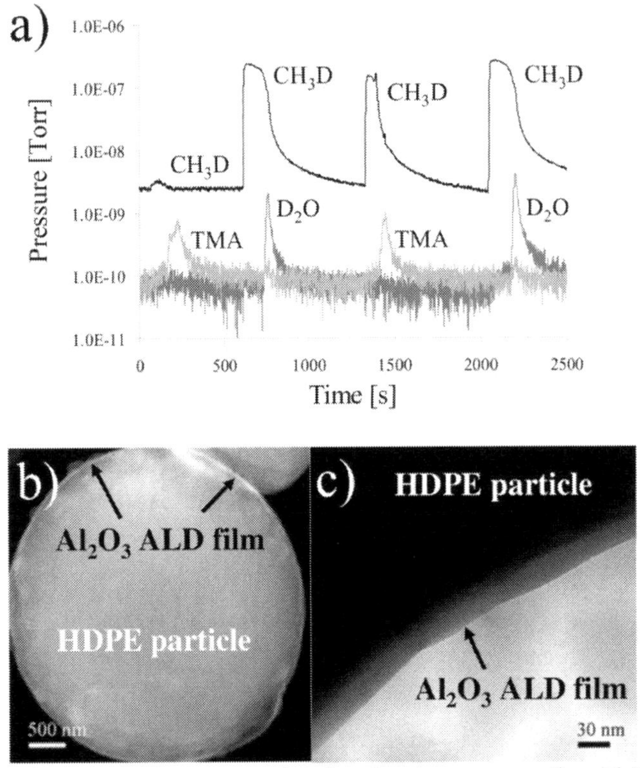

Figure 9. a) In situ mass spectrometry plot showing two consecutive Al_2O_3 ALD cycles on HDPE polymer particles using TMA and D_2O as precursors; b) an STEM image of an Al_2O_3-coated HDPE particle, and c) a TEM image of the same at higher magnification.

ALD on highly-porous polymer particles. A process to synthesize a porous polymer ceramic composite material was demonstrated and the properties of the composite materials have been investigated for a variety of applications. Highly porous polystyrene divinyl benzene (PS-DVB) particles were successfully coated with alumina films at the atomic level in a fluidized bed reactor using a low-temperature particle ALD process. This fabrication technique was considered to be novel since it allowed for the ceramic layers to form on any surface site the gaseous molecules could reach, meaning that Al_2O_3 was successfully deposited along all of the internal pore walls and external surface of the particles. A cross-sectional TEM image of the Al_2O_3 ALD films deposited along the walls of the porous polymer is shown in Figure 10. The dark lines along the edges are Al_2O_3 layers, and the lightest gray areas are the pores. This functionalized nanocomposite system can be utilized for a variety of applications, many of which can be enabling in the field of biotechnology. Novel scaffold materials can be developed that utilize the deposition of biocompatible films, such as Al_2O_3 or TiO_2, onto the surfaces of porous supports.

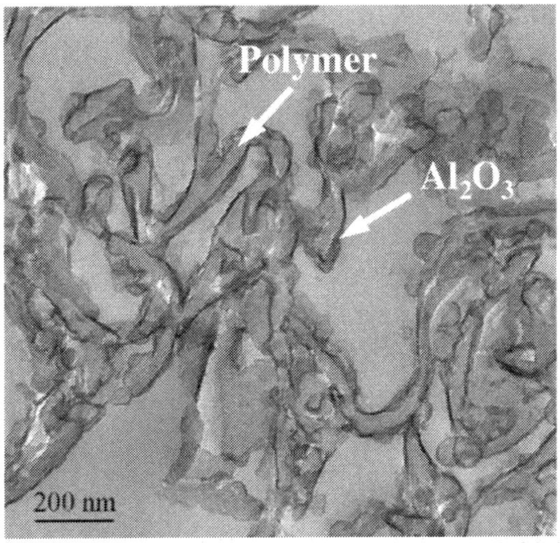

Figure 10. Cross-sectional TEM image of an ALD-coated highly-porous polymer particle. The internal surfaces of the nanoporous network are coated with Al_2O_3.

ALD on carbon nanotubes and diamond powder. Recently, there has been significant emphasis on research to develop ALD coating processes to effectively coat the surfaces of carbonaceous particles, including single- and multi-walled carbon nanotubes (SWNTs and MWNTs, respectively), graphitic carbon, diamond powder and mesoporous carbon particles (43, 58). ALD requires surfaces with proper functionality in order for reactive precursor molecules to chemically bond to the particle surface. Some carbon-based particles, such as SWNTs or highly crystalline graphite/graphene tend to have pristine surfaces that do not support adequate functionality to yield a dense ALD nucleation site pattern. In these cases, liquid pretreatments such as dispersion of materials in ethanol or a surfactant-laden solution, or gas-phase pretreatment such as pre-exposure of NO_2 to the particle bed, may be necessary in order to physisorb a dense pattern of molecules onto the powder surfaces. The physisorbed 'scaffold' can remain on the particle surface during the initial stages of regular precursor growth, and can be a viable strategy to functionalize the surfaces of carbonaceous particles that do not lend themselves to facile coating processes. An HRTEM of an Al_2O_3-coated SWNT and several Al_2O_3-coated MWNTs are shown in Figures 11a and 11b, respectively. Figure 11c depicts a TEM image of of an Al_2O_3-coated diamond particle. The SWNT was coated using a sodium dodecylsulfate liquid-phase pre-treatment strategy. The MWNTs and diamond particles were able to support a dense nucleation site pattern suitable for conformal, pinhole-free growth of Al_2O_3 ALD without any pre-treatment steps.

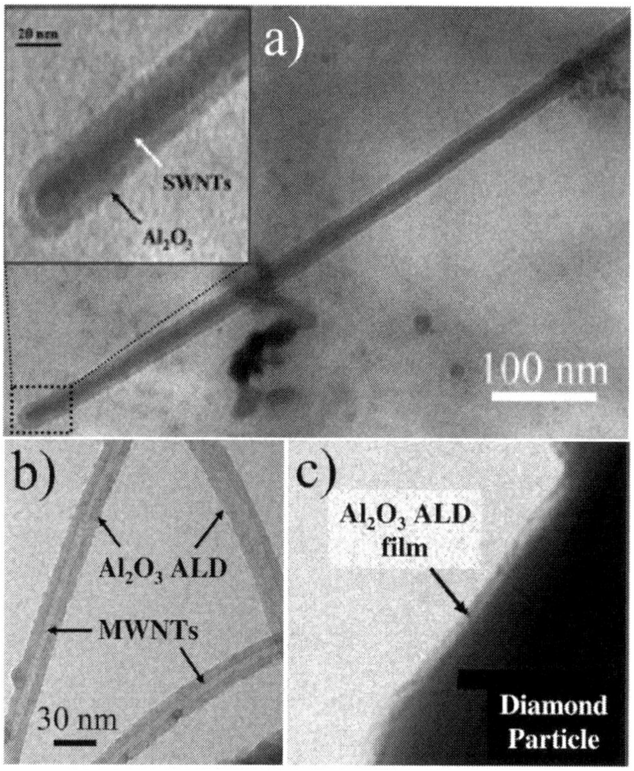

Figure 11. Al_2O_3 ALD coatings on carbonaceous fine and ultrafine particles including a) single-walled carbon nanotubes, b) multi-walled carbon nanotubes and c) micron-sized diamond particles.

Molecular Layer Deposition on Particles

Recently, MLD techniques have been extended from the coating of flat surfaces to the functionalization of nanoparticle surfaces. Dameron et al were the first to investigate the chemistry of aluminum alkoxide (or "alucone") MLD using TMA and ethylene glycol. Alucone MLD was successfully deposited on ZrO_2 particles (<30 nm), and $BaTiO_3$ particles (<200 nm) at small (<0.1 g) scale using a viscous flow reactor (46). Liang et al was the first to demonstrate that the MLD process could be operated on nanoparticles in a fluidized bed reactor at larger scale (48). Hybrid organic/inorganic films have been successfully deposited on a variety of metal oxide particle surfaces, including 250 nm SiO_2 spheres and 160 nm anatase TiO_2.

The MLD process. Hybrid organic/inorganic films were deposited on particles using the sequential administration of TMA and ethylene glycol as precursors at temperatures between 100°C and 160°C. The dosing lines were kept at 80°C to avoid excessive adsorption of the relatively sticky ethylene glycol precursor on the internal walls of the delivery area. The room-temperature vapor pressure of EG (anhydrous, 99.8%, Sigma Aldrich) is very low, and is much lower than that required for the fluidization of reasonable bed masses. Heating the EG liquid raises its vapor pressure in accordance with the Antoine equation. The required operating pressures were achieved using this strategy, however, the flow rate control offered by the needle valve was then insufficient to prevent the pure-component EG vapor stream from flooding into the system. Since EG is a sticky molecule, it can be very difficult to remove any overdosed vapor from the particle surfaces or the reactor walls. A bubbler was employed to dilute the heated EG vapor stream, and allowed for vapor delivery to the reactor in a controllable fashion, and prevented the overdose of the precursor. The bubbler inlet was controlled using a mass-flow controller (MKS Instruments) to allow a calibrated amount of N_2 to be bubbled through the precursor reservoir. Before the reaction, the particles were dried at 120°C under continuous N_2 flow for 3h. A typical coating cycle consisted of the following sequence: TMA dose, N_2 purge, evacuation; EG dose, N_2 purge, evacuation. In this manner, there was no overlap between the two reactants, such that no undesirable CVD side-reactions could occur.

MLD film characterization. Alucone MLD films have been coated on silica and titania nanoparticles (48). In order to determine the composition of MLD films, the uncoated and coated silica particles were analyzed by EDS. For the uncoated silica particles, only O and Si signals were present. In contrast, for the 50 cycles MLD coated silica particles, from which the STEM image in figure 12a was taken, there was a strong peak of Al and the presence of both carbon and oxygen was much more pronounced. This is a direct confirmation of the presence of alucone MLD films, which have a backbone structure of the -Al-O-CH_2-CH_2-O- species. The aluminum concentrations on the titania and silica particles were analyzed by ICP-AES. The concentration of aluminum on the particle surfaces was directly proportional to the number of coating cycles. Since the coating thicknesses were small compared to the nanoparticle radius, it indicated a constant growth rate of alucone hybrid films. In other words, the film thickness and the molecular weight of the polymer can be controlled by the number of reaction cycles. An HRTEM image of the MLD-coated TiO_2 nanoparticle is shown in Figure 12b.

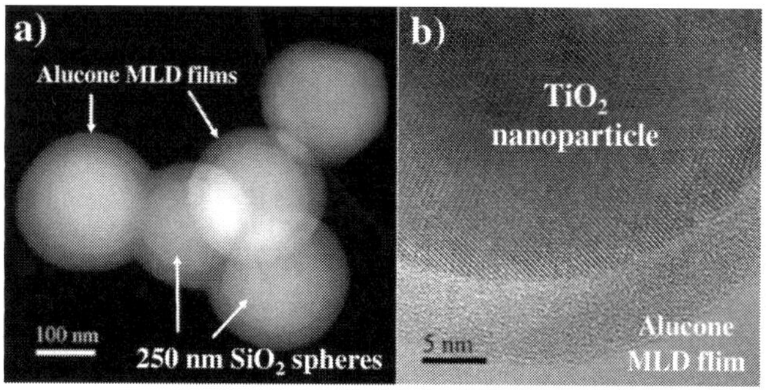

Figure 12. a) STEM image of Alucone MLD films on 250 nm SiO_2 spheres; b) HRTEM image of an Alucone MLD film on a TiO_2 nanoparticle.

Pinhole-free MLD films. It is clear from the images in Figure 12 that the displayed films are conformal as deposited. In order to evaluate the pinhole-free nature of the MLD films, the film thickness dependent photoactivity mitigation of the MLD coatings on anatase TiO_2 was used. A process identical to the SiO_2-coated TiO_2 via ALD was used here as the MLD films were coated on the same anatase TiO_2 substrate. The absorbance of the methylene blue solution remains unchanged if the TiO_2 surfaces are completely passivated, and a slurry of uncoated TiO_2 particles readily degrades color of the solution from blue to clear. The photoactivity results are shown in Figure 13. For increasing MLD layers deposited on TiO_2 surfaces, the color degradation diminished over any given time interval. This shows that the MLD films had completely passivated the TiO_2 particles over the analysis time period. It is interesting to note here that Dameron et al. found that the alucone films do not have superior stability properties, and the structure of the film may change in the presence of water. The photoactivity results were repeated for particles that had been soaked in water for one week, and the performance indeed had diminished due to the rearrangement of molecules within the film, possibly through localized hydrolysis pathways. Studies are currently being performed on other MLD films that are based on other elements that have been deposited at lower temperatures (e.g. Ti and Zn), as well as other routes to synthesize alucone films that may provide enhanced long term stability. Though the current long-term performance does show film instability in aqueous solutions, there are many other conditions where MLD-coated particles will remain stable and pinhole-free.

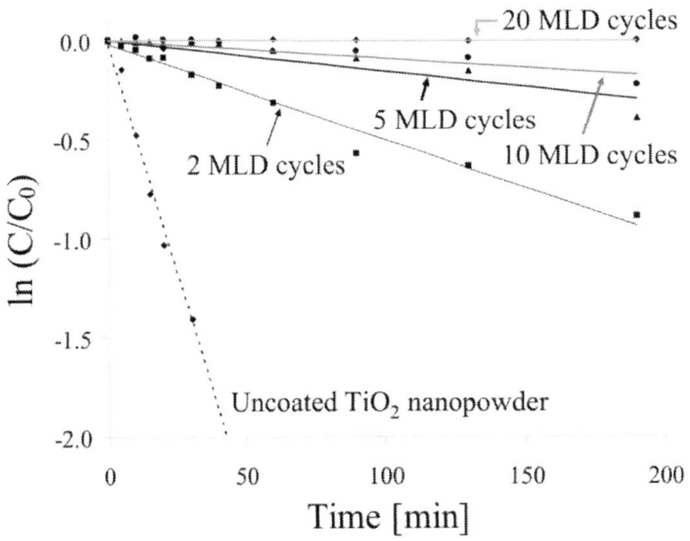

Figure 13. Decreasing photoactivity of TiO_2 nanopowder with increasing MLD coating thickness as evidenced by the decrease in absorbance accompanied by the decomposition of MB in the presence of UV light irradiation.

Multilayered ALD Coatings

The ability to deposit multiple layers of alternating materials on particles without having to expose them to air is a very powerful technique, but ideally the films should be stable. In addition, it has been shown that stable, nanoscale films fabricated using ALD can exhibit quantum confinement (32, 62). To test the stability of a multilayered ALD film, a deposition strategy consisting of a series of ZnO ALD cycles sandwiched between layers of Al_2O_3 ALD was performed on the surfaces of SiO_2 spheres (in situ) using the ALD technique. From the STEM image (Figure 14a), two external films can be discerned due to the differences in atomic mass between Zn and either Al or Si. Each of the visible concentric layers does not appear to have interdiffused into one another. The inner-most Al_2O_3 layer was not visible using the STEM technique, so the particles were cross-sectioned using a diamond knife for HRTEM imaging (Figure 14b). Once again, the inner Al_2O_3 layer was not discernable, but using ICP-AES bulk elemental analysis, the proportionate amount of Al_2O_3 expected in two shells was present. Energy Dispersive Spectroscopy (EDS) mapping was also performed at various locations across the entire particle shown in the STEM image. The EDS probe position and elemental intensity are shown in Figure 14c, with a schematic of probe locations in Figure 14d. It is clear that the EDS intensity plot only contained Al and O when the outermost layer was probed. Zn appeared when the probe was moved toward the center of the particle and over the thickest portion (i.e. the edge) of the ZnO shell. When the probe continued to the visually indiscernible Al_2O_3 layer, the Zn content decreased while the oxygen content increased, which was expected based on the increased net oxygen ratio. Si was only observed when the beam passed through the core SiO_2 sphere. Overall, the ZnO ALD films appear to be quite stable at temperatures up to 550°C, which is a temperature at which significant grain growth can occur. There was no evidence that a solid state reaction between the polycrystalline ZnO and either amorphous Al_2O_3 layer had occurred (32).

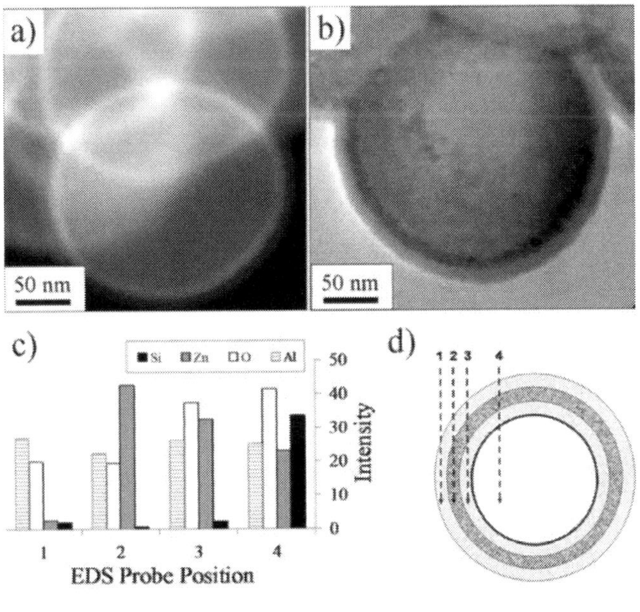

Figure 14. a) STEM image of an ALD Al_2O_3-ZnO-Al_2O_3 multilayered film coated on SiO_2 nanospheres; b) cross-sectional HRTEM image of the same; c) EDS mapping data for various points along the radius of the composite structure in accordance with d) the schematic of the system.

Conclusions

It has been shown that a fluidized bed reactor is a robust unit operation for particle ALD and MLD processing. In situ mass spectrometry is a useful tool that allows for optimization of the process variables at any scale. 100% precursor utilization was shown to be attainable at relatively high surface conversions. Many materials can be readily deposited using this flexible reactor configuration and virtually any particle substrate, including high aspect ratio nanotubes, with the appropriate surface functionality can be coated using this gas-phase adsorption process. Internal walls for porous particle subsrates can also be easily coated. Particle ALD/MLD has been used to deposit many types of ceramic metal oxides and nitrides, and metal nanoclusters that demonstrate superior catalytic activity over conventional techniques, as well as hybrid organic-containing films on particle surfaces. In addition, particle ALD/MLD provides an opportunity to use unconventional precursors that normally would have been cost prohibitive. The quality of the deposited layers is unparalleled from a particle functionalization perspective. Particle ALD processes are anticipated to replace conventional CVD and sol gel liquid solution processes on the basis of precursor cost savings and improved performance (i.e. improved cost/performance benefit). A multitude of applications can be addressed in a competitive fashion using ultrafine particles that have been surface-modified using ALD/MLD techniques in scalable, high-throughput unit operations.

Acknowledgments

The authors thank the many graduate students and postdoctoral fellows under the advisement of Dr. Steven M. George and Dr. Alan W. Weimer at the University of Colorado at Boulder who have contributed to the advancement of the particle ALD technology platform at a fundamental level. The authors thank the many government sponsored projects that have been funded over the years. Any opinions, findings, and conclusions or recommendations expressed in this work are those of the authors and do not necessarily reflect the views of these government organizations.

References

1. A. C. Dillon, A. W. Ott, J. D. Way and S. M. George, Surf. Sci., **322**, 230 (1995).
2. S. George, A. Ott and J. Klaus, Journal of Physical Chemistry, **100**, 13121 (1996).
3. M. Ritala and M. Leskela, Atomic Layer Deposition, Academic Press, San Diego (2002).
4. B. S. Lim, A. Rahtu and R. G. Gordon, Nature Materials, **2**, 749 (2003).
5. C. H. L. Goodman and M. V. Pessa, Journal of Applied Physics, **60**, R65 (1986).
6. T. Suntola, Materials Science Reports, **4**, 261 (1989).
7. J. D. Ferguson, A. W. Weimer and S. M. George, Chemistry of Materials, **16**, 5602 (2004).
8. J. D. Ferguson, K. J. Buechler, A. W. Weimer and S. M. George, Powder Technology, **156**, 154 (2005).

AIChE Thomas Baron Award Lecture (Nashville, TN, November 11, 2009)

9. J. D. Ferguson, E. R. Smith, A. W. Weimer and S. M. George, Journal of the Electrochemical Society, **151**, G528 (2004).
10. J. D. Ferguson, A. W. Weimer and S. M. George, Chemistry of Materials, **12**, 3472 (2000).
11. J. D. Ferguson, A. W. Weimer and S. M. George, Applied Surface Science, **162**, 280 (2000).
12. J. D. Ferguson, A. W. Weimer and S. M. George, Thin Solid Films, **371**, 95 (2000).
13. J. D. Ferguson, A. W. Weimer and S. M. George, Thin Solid Films, **413**, 16 (2002).
14. J. D. Ferguson, A. W. Weimer and S. M. George, Journal of Vacuum Science & Technology A, **23**, 118 (2005).
15. J. D. Ferguson, A. R. Yoder, A. W. Weimer and S. M. George, Applied Surface Science, **226**, 393 (2004).
16. J. R. Wank, S. M. George and A. W. Weimer, Powder Technology, **121**, 195 (2001).
17. J. R. Wank, S. M. George and A. W. Weimer, Powder Technology, **142**, 59 (2004).
18. J. R. Wank, S. M. George and A. W. Weimer, Journal of the American Ceramic Society, **87**, 762 (2004).
19. L. F. Hakim, J. Blackson, S. M. George and A. W. Weimer, Chemical Vapor Deposition, **11**, 420 (2005).
20. L. F. Hakim, J. H. Blackson and A. W. Weimer, Chemical Engineering Science, **62**, 6199 (2007).
21. L. F. Hakim, S. M. George and A. W. Weimer, Nanotechnology, **16**, S375 (2005).
22. L. F. Hakim, D. M. King, Y. Zhou, C. J. Gump, S. M. George and A. W. Weimer, Advanced Functional Materials, **17**, 3175 (2007).
23. L. F. Hakim, J. A. McCormick, G. D. Zhan, A. W. Weimer, P. Li and S. M. George, Journal of the American Ceramic Society, **89**, 3070 (2006).
24. L. F. Hakim, J. L. Portman, M. D. Casper and A. W. Weimer, Powder Technology, **160**, 149 (2005).
25. D. M. King, X. H. Liang, Y. Zhou, C. Carney, P. Li, L. F. Hakim and A. W. Weimer, Powder Technology, **In Press** (2007).
26. C. A. Wilson, D. N. Goldstein, J. A. McCormick, A. W. Weimer and S. M. George, Journal of Vacuum Science & Technology A, **26**, 430 (2008).
27. C. A. Wilson, J. A. McCormick, A. S. Cavanagh, D. N. Goldstein, A. W. Weimer and S. M. George, Thin Solid Films, **516**, 6175 (2008).
28. X. Liang, D. M. King, M. D. Groner, J. H. Blackson, J. D. Harris, S. M. George and A. W. Weimer, Journal of Membrane Science, **322**, 105 (2008).
29. X. H. Liang, S. M. George and A. W. Weimer, Chemistry of Materials, **19**, 5388 (2007).
30. X. H. Liang, L. F. Hakim, G. D. Zhan, J. A. McCormick, S. M. George, A. W. Weimer, J. A. Spencer, K. J. Buechler, J. Blackson, C. J. Wood and J. R. Dorgan, Journal of the American Ceramic Society, **90**, 57 (2007).
31. X. H. Liang, D. M. King, P. Li and A. W. Weimer, Journal of the American Ceramic Society, **92**, 649 (2009).
32. D. M. King, S. I. Johnson, J. H. Li, X. H. Du, X. H. Liang and A. W. Weimer, Nanotechnology, **20** (2009).

AIChE Thomas Baron Award Lecture (Nashville, TN, November 11, 2009)
33. D. M. King, X. H. Liang, B. B. Burton, M. K. Akhtar and A. W. Weimer, Nanotechnology, **19** (2008).
34. D. M. King, X. H. Liang, C. S. Carney, L. F. Hakim, P. Li and A. W. Weimer, Advanced Functional Materials, **18**, 607 (2008).
35. D. M. King, X. H. Liang, P. Li and A. W. Weimer, Thin Solid Films, **516**, 8517 (2008).
36. D. M. King, Y. Zhou, L. F. Hakim, X. H. Liang, P. Li and A. W. Weimer, Industrial & Engineering Chemistry Research, **48**, 352 (2009).
37. S. George, J. Ferguson and A. Weimer, U.S. Patent 6,613,383 (2003).
38. R. E. Winans, S. Vajda, G. E. Ballentine, J. W. Elam, B. D. Lee, M. J. Pelline, S. Seifert, G. Y. Tikhonov and N. A. Tomczyk, Topics in Catalysis, **39**, 145 (2006).
39. S. Vajda, M. J. Pellin, J. P. Greeley, C. L. Marshall, L. A. Curtiss, G. A. Ballentine, J. W. Elam, S. Catillon-Mucherie, P. C. Redfern, F. Mehmood and P. Zapol, Nature Materials, **8**, 213 (2009).
40. S. T. Christensen, J. W. Elam, B. Lee, Z. Feng, M. J. Bedzyk and M. C. Hersam, Chemistry of Materials, **21**, 516 (2009).
41. S. T. Christensen, J. W. Elam, F. A. Rabuffetti, Q. Ma, S. J. Weigand, B. Lee, S. Seifert, P. C. Stair, K. R. Poeppelmeier, M. C. Hersam and M. J. Bedzyk, Small, **5**, 750 (2009).
42. S. Ghosal, T. F. Baumann, J. S. King, S. O. Kucheyev, Y. M. Wang, M. A. Worsley, J. Biener, S. F. Bent and A. V. Hamza, Chemistry of Materials, **21**, 1989 (2009).
43. J. S. King, A. Wittstock, J. Biener, S. O. Kucheyev, Y. M. Wang, T. F. Baumann, S. K. Giri, A. V. Hamza, M. Baeumer and S. F. Bent, Nano Letters, **8**, 2405 (2008).
44. X. R. Jiang, H. Huang, F. B. Prinz and S. F. Bent, Chemistry of Materials, **20**, 3897 (2008).
45. S. M. George, B. Yoon and A. A. Dameron, Accounts of Chemical Research, **42**, 498 (2009).
46. A. A. Dameron, D. Seghete, B. B. Burton, S. D. Davidson, A. S. Cavanagh, J. A. Bertrand and S. M. George, Chemistry of Materials, **20**, 3315 (2008).
47. N. M. Adarnczyk, A. A. Dameron and S. M. George, Langmuir, **24**, 2081 (2008).
48. X. Liang, D. M. King, P. Li, S. M. George and A. W. Weimer, Aiche Journal, **55**, 1030 (2009).
49. D. Kunii and O. Levenspiel, Fluidization Engineering, p. 534, John Wiley and Sons, New York (1969).
50. C. Nam, R. Pfeffer, R. Dave and S. Sundaresan, AICHE J, **50**, 1776 (2004).
51. J. Jung and D. Gidaspow, J NANOPART RES, **4**, 483 (2002).
52. J. Wank, S. George and A. Weimer, Powder Technology, **121**, 195 (2001).
53. L. Hakim, J. Portman, M. Casper and A. Weimer, Powder Technology, **160**, 149 (2005).
54. D. M. King, J. A. Spencer, X. H. Liang, L. F. Hakim and A. W. Weimer, Surface & Coatings Technology, **201**, 9163 (2007).
55. C. J. Brinker and G. W. Scherer, Sol-Gel Science, Academic Press, San Diego (1990).
56. Powell, J. Mater. Res., **12**, 557 (1997).
57. M. Leskela and M. Ritala, THIN SOLID FILMS, **409**, 138 (2002).
58. X. H. Liang, G. D. Zhan, D. M. King, J. A. McCormick, J. Zhang, S. M. George and A. W. Weimer, Diamond and Related Materials, **17**, 185 (2008).

AIChE Thomas Baron Award Lecture (Nashville, TN, November 11, 2009)
59. D. B. Farmer and R. G. Gordon, Nano Letters, **6**, 699 (2006).
60. J. H. Braun, A. Baidins and R. E. Marganski, Progress in Organic Coatings, **20**, 105 (1992).
61. J. R. Scheffe, A. Frances, D. M. King, X. H. Liang, B. A. Branch, A. S. Cavanagh, S. M. George and A. W. Weimer, Thin Solid Films, **517**, 1874 (2009).
62. D. M. King, X. H. Du, A. S. Cavanagh and A. Weimer, Nanotechnology, **19** (2008).

Every Grain of Sand – Highlights of a Career in Particle Technology

Martin Rhodes, Monash University

My work in particle technology started in 1983 when I joined Derek Geldart to study for a PhD at Bradford University. Prior to that I had been teaching in high schools and working in industry. At the time I had no idea what prestigious company I was in at Bradford (Derek Geldart, John Williams, Lado Svarovsky, Terry Allen, Norman Harnby), but I was to benefit greatly from studying and working in that environment. After one year as research assistant I was fortunate to be appointed to a lectureship – my colleagues at Bradford were taking a gamble since at that time I had no publications and was only one year into the PhD.

CFB Model:

One of the topics for my PhD studies was the circulating fluidized bed. We built CFBs with 15cm and 30cm diameters risers and ran a range of Group A powders through them. I recall Derek Geldart suggesting that we incorporate the pressure balance with some ideas that I had had on the CFB. I did this with the help of correlations for entrainment and the result was a fairly crude model for the observed trends in CFB behaviour – variation in pressure and voidage profile with gas velocity and solids circulation rate [1]. I recall being flattered that this model was later inadvertently mimicked by Kunii and Levenspiel [2].

Turbulent fluidization:

Another topic of my PhD studies was turbulent fluidization. The literature at the time suggested that increasing gas velocity in a bubbling bed brought about a rise and fall in the amplitude of pressure fluctuations. The velocity U_c where the peak in amplitude occurred was regarded as the start of the transition to turbulent fluidization and velocity U_k where the amplitude of the pressure fluctuations levelled off at a low value was regarded at the onset of turbulent fluidization. After many hours of observing the transition in 2D and 3D apparatus with range of particles from Geldart Groups A to D, I realized that the so-called transition was actually the onset of dilute transport and exposure of probes to freeboard [3, 4]. This was difficult to see in beds of fine powders but much easier with coarser powders and with the insight from observing the coarser powders one could interpret the fine powder behaviour. Many disagreed with this interpretation and still do. However, it is interesting that over the years there came about a convenient redefinition of the transition – U_c became the *onset* of turbulent fluidization and U_k (originally marking the onset) now receives little mention.

Regime map:

While I was still studying for the PhD, Arthur Squires (at that time with City College New York) visited Derek Geldart at Bradford University. Prior to his visit, Arthur sent us a draft of his Science paper [5]. The paper included a regime map including several question marks over the area of fast fluidization and how it linked with pneumatic transport, turbulent fluidization and conventional bubbling fluidization. I puzzled over this for weeks prior to his visit against the background of our CFB model and my ideas on turbulent fluidization and I came up with what I thought was the answer to all the questions in Arthur's regime diagram and created my own. On Derek Geldart's advice, we decided not to discuss this with him during his visit, but instead I drafted my own paper. Unfortunately, I decided to publish with the IChemE - the process took 2 years [6]. Since, to my mind, I had filled the gaps in our knowledge that had been identified by Arthur Squires, well known in the field, I assumed that the work would be welcomed with open arms by the community. Unfortunately, it passed virtually unnoticed (except Prabir Basu, who due to his English connections, read the IChemE journal. I think this was partly due to delay in publication (the community's interest had waned) and partly due to the journal, which receives little following outside the UK. My only recourse was to promote the ideas at conferences. I found this a hard road. Surprisingly conferences are generally not ideal means of getting ideas across. In my frustration I may have inadvertently adopted a style that may have appeared arrogant. On the whole, however, as I mellowed in my approach, I think people looked forward to and enjoyed the discussions on fast fluidization, slip velocity and choking.

Similar profiles:

In the late 80s I became fascinated by the idea of measuring the flux profile and pursued the use of the non-isokinetic probe originally conceived by van Breugal et al [7] for the FCC riser studies. We confirmed the existence of a similar profiles regime, first noticed by Monceaux and co-workers [8] from Compiegne, France. We studied the onset and end of similar profiles regime – it seems that as particles are added at a greater and greater rate to a flowing gas stream, the velocity profile is initially disturbed, but then settles into a steady shape over a wide range of conditions until a point is reached when the annulus can hold no more solids and grows inwards changing the profile shape [9, 10, 11]. I think this is a really interesting and significant phenomenon that has really not been given the attention that it deserved amongst the fluidization community. We had earlier shown theoretically that under some circumstances voidage profile would be determined by mean voidage – similar voidage profiles [12]. Others later showed this to be true experimentally.

Wall Motion, Solids Mixing, Gas Mixing:

For two years at Bradford I had the pleasure to work in the company of Toshi Hirama. Toshi brought with him his younger son Mak, who attended the local high school. My family got to know Toshi's family well – particularly his elder son Tom, who lived with us for seven months after Toshi had returned to Japan. Toshi was a meticulous experimentalist and with the help of postdoctoral researcher Hideaki Mineo and students Peter Laussman, Gabriel Amos and Shihui Zhou we studied various aspects of the CFB, including the motion of solids at the wall, axial solids mixing and axial gas mixing [13, 14, 15].

CFB combustor – coal

In a collaboration between Bernie Gibbs, renowned fluid bed combustion man from Leeds University, and myself from nearby Bradford University, we designed and built a CFB combustor, which worked with few problems from the start – a great credit to Shan Wang, the postdoctoral researcher on the project. This was my first contact with Shan

Wang and the start of a collaboration which was to continue for 19 years, both in the UK and in Australia. Using the combustor, we produced a model for coal combustion in a CFB and looked at the impact of air staging on NO_x and N_2O production [16, 17]. Then went on to study heat transfer at high temperature using a probe developed by Shan [18, 19, 20, 21]. I moved to Monash University in Australia in 1994 and Shan joined me as research fellow a year or so later.

Use of tomography – structure of dense phase – dilute phase interface:
I had always been interested in what happens at the dilute-dense phase interface in the CFB – the S in the S-shaped profile that at one time defined fast fluidization. We used capacitance tomography, pressure profiles and momentum probe to understand this [22]. We found it was necessary to use caution when applying the early commercial versions capacitance tomography; when calibrating the system we found that it was possible to make quite large hollow objects placed inside the bed disappear – i.e. they were not seen by the imaging technique!

Influence of Interparticle Forces on Fluidization:
My student, Loretta McLaughlin, confirmed that increased interparticle force due to liquid addition can give rise to change in Geldart behaviour from B to A (characterized by $U_{mb} > U_{mf}$) and on to C (characterized by channeling and plug rising) [23]. Quantifying the actual critical interparticle forces at regime change is difficult, whether done by calculation or measurement. However, this lead to further work using magnetic particles in a magnetic field and DEM simulation, where similar transitions in Geldart behaviour were observed and further estimates of the critical interparticle forces at transitions change [24, 25]. At this point I was in the camp that thought that B to A transition was determined solely by the interparticle forces, but later our DEM simulation (with Jai Kant Pandit [26, 27, 28]) was to show that Group A behaviour was possible without the imposition of interparticle force, provided the particles were small enough. So the AB boundary and fluidized bed stability seems to be far more complex that was at first realized by those schools of thought that put it down to only interparticle forces or only particle size.

Scaling Laws Over Wide Range of Scales:
In that late 90s a student called John Sanderson joined my growing team at Monash University. We got along very well together and had a great time over the 4 years he was with me. John managed to get through a tremendous amount of work – meticulously studying the scaling laws over scales from 0.15m to 1.5m – analyzing pressure fluctuations and studying mixing using novel techniques that he developed himself [29, 30, 31]. One night John fell asleep while working on his literature review and listening to Chuck Berry's Johnny B. Goode on the computer. He awoke with the literature set to the tune of Johnny B. Goode. He performed this at the Clean Coal Cooperative Research Centre's annual conference in Adelaide that year. I convinced him that it would go down well with the international fluidization community and so we performed it together at the Fluidization X conference in Beijing in 2001. When I volunteered to take charge of the after dinner entertainment for the 4th World Congress on Particle Technology in Sydney in 2002, with a zero budget, I had no hesitation in inviting John to join me on stage to

deliver Scaling Laws with apologies to Chuck Berry – this time with a good PA system and a dance routine. (see Appendix for the lyrics). During his time at Monash University John also created the short movies "Group E Powders" (the *lesser known* Geldart Group) and "The Igor has Landed", in which student Igor Sidorenko dons a spacesuit and "enters" the pressurized fluidized bed at Monash University for inspection.

Studies using DEM simulation:
Our work using DEM simulation was triggered by a 2-month visit of Masa Horio to Monash University as Visiting Professor. One of his students had written out Tsuji's DEM fluidized bed code and added a liquid bridge force to simulate cohesive behaviour in bed with added liquid. Horio left a copy of the code behind for our use. We took out the liquid bridge force and tidied up the code (which reverted to Tsuji's code) and eventually used it quite effectively as a tool to study various aspects of fluidized bed behaviour that were not easy to study by physical experimentation: solids mixing, the effect of interparticle forces, increased gravity, pulsed fluidization and solids motions near to the wall [25, 32, 33, 34, 35].

DEM Simulation of the Spouted Bed:
I was thankful to Tak Tsuji also for recommending a young PhD graduate to me. Shintaro Takeuchi (Shiggy) worked with me for two years, during which time he gave tremendous help to some of my students, developed a successful 3D DEM code [36, 37] for the spouted bed and programmed a model for the Brazil-Nut effect (see below). Interestingly, Shiggy first attempted to simulate the spouted bed in a 2D code, but was frustrated by not being able to achieve a stable spout. Fortunately at that time, Jim Lim from UBC visited us. He was most pleased with our results, since in his experience, in practice 2D spouted beds do not produce stable spouts – a case of the model inadvertently agreeing with the data! Shiggy seemed to have the gift of being able to put the physics directly into code. Tsuji described him as near genius in CFD. He was also a wizard at origami, or paper engineering as he preferred to call it. During his time with us, he made a 3D version the Penrose Triangle, which is supposed to exist only in two dimensions! He had a great sense of humour; when Shiggy learned that his name written Takeucci and pronounced in Italian would have the same pronunciation as the conventional Takeuchi, he adopted the new spelling. I suggested that the Irish Tacky O'Shay would almost sound correct, but for some reason he decided not to adopt that spelling.

Brazil Nut Effect:
My first experience of the Brazil Nut effect was the segregation demonstrations done by John Williams in his short course lectures. After that I always used the steel ball in the sand demonstration in my talks and lectures to show that you should never take powder behaviour for granted. The demonstration has an impact on all who see it – from the school student to the engineer with 20 years experience working with powders. In the late eighties and early nineties, the research by physicists, such as Knight and co-workers [38, 39], drew our attention to the fact that we had not clear understanding of what caused the effect. I decided to take a look, with the help of some undergraduate students and Kurt Liffman and Guy Metcalfe, colleagues from CSIRO. Our major preliminary observation was that the dense "intruder" seemed to punch its way upward through the particle bed,

rather that be carried up on a tide of smaller particles as the favoured models at the time suggested. The larger and denser the intruder, the faster it rose. We modeled this and published [40]. Soon after, Mobius and co-workers [41] showed experimentally that very low density intruders rise more rapidly again. The plot thickens. To us, this seemed to be a buoyancy effect on the intruder. It was already known that an oscillating particle bed would create an oscillating pressure gradient across the bed. A pressure gradient would create a buoyancy force on the intruder, which would have a relatively greater effect on a low density intruder than on a high density one. We modeled this and published [42]. Although the resulting model was not completely satisfactory, I feel that this could be the way forward. During the experimental work we also observed that interparticle forces (electrostatics and low humidity and liquid forces at high humidity) had a dramatic effect on the rise times for the intruder. This surprised us since the bed particles were 1mm diameter and one would not usually expect interparticle forces to have much effect on such particles – a cautionary tale.

Interparticle forces and powder behaviour:
With the help of Adam Forsyth and Sean Hutton of the Physics Department at Monash University, we also studied the influence of interparticle forces on particle packing, free-flow to slip-stick transition and dynamic angle of repose [43, 44, 45]. To do this we used iron particles in a magnetic field and glass particles in air of controlled humidity. With the glass particles in humid air the transition from free-flowing to slip-stick behaviour was found to be due to the system achieving a certain critical level of cohesive interparticle force and not due to an abrupt rise in this force (as monitored by AFM). In the particle packing work we found that the void fraction was determined only by the ratio of interparticle force to particle weight, regardless of particle size.

Nanoparticle fluidization:
The fact that nanoparticles can be fluidized at all surprised the fluidization community. The mechanism is still not completely clear, but is most likely associated with the formation of stable, low density agglomerates. In recent years Shan Wang and I collaborated with Monash University mechanical engineer Julio Soria to apply laser-based planar imaging to study the agglomerates of nanoparticles in the splash zone above the nanoparticle fluidized bed. This enabled us to make reasonable predictions of the size of agglomerates in the fluidized bed [46, 47, 48]. Another interesting feature of the nanoparticle bed is its ability to expand considerably at velocities lower than minimum fluidization. This puzzled me until I realized that tiny spouts carried particles from within the bed and deposited them as a low density layer on top of the bed – effectively causing the packed bed to expand. This was confirmed in a paper by Ammendola and Chirone [49] presented at Fluidization XII in Canada – they used visual observation of a bed containing layers of different coloured particles, which clearly demonstrated the effect.

Particle Curtain Heat Exchanger:
Pursuing the "curtain heat exchanger" concept of Owen Potter, we built a cold model in which air was caused to horizontally through a curtain of particles over a range of air velocities, particle flowrates and curtain thicknesses. We studied the hydrodynamics and heat transfer from gas to particles experimentally and theoretically [50, 51, 52], made

some interesting observations and struck up a useful collaboration with researchers Madoc Sheehan and Andrew Lee at James Cook University in far north Queensland. Perhaps rather unfairly, I gave Madoc and Andrew the parameters for our experiments and asked them to model the particle curtain trajectories without having seen our experimental results. To their great credit, they rose to the challenge and their CFD model produced results which showed surprisingly good agreement with our data. The most interesting results were those at zero air flow. Depending on its initial thickness, the curtain was found to converge or diverge. We believed that this was the result of competition between the particle-particle collisions, causing the curtain thickness to increase, and the pressure gradient within the curtain, trying to compress the curtain. Based on this idea, there would be an equilibrium curtain void fraction which the curtain would tend towards by contracting or expanding. We modeled this with encouraging results. Although the CFD model was able to predict the converging-diverging effect on the curtain, we ran out of time and were unable to use the CFD model to confirm our physical model ideas (for example, by comparing predicted pressure forces and particle collision forces).

Education:
Bradford University used to run a number of specialist short courses on powder technology, primarily through the IChemE. Derek Geldart suggested that we need an introductory course. We called it Getting Started in Powder Technology" and I was asked to be director. There turned out to be a tremendous market for this course, which was successful financially and in terms of getting the message across to industry. With Bradford contributions from Lado Svarovsky, Derek Geldart, Norman Harnby, John Williams, Nayland Stanley-Wood and Arthur Hawkins, by directing the course, I learned a lot about Particle Technology and how to teach it - especially from John. During my time at Bradford hundreds of engineers and scientists from Europe were enlightened in the area of particle technology – it was always wonderful to see the faces when they suddenly realized that many of their problems were common across different industries and had simple solutions. I was aware of the need for a powder technology textbook and encouraged my fellow lecturers on the short course to convert their lecture notes into chapters for a book. I was naïve to think that 12 authors could work together to produce a textbook. The result [53] was more a handbook than a textbook, but still probably better and more comprehensive than what was around at the time. So, as soon as I moved to Monash University, I set about writing my own textbook, based on my experience teaching particle technology to students at Bradford and Monash University and the short course in the UK and Australia. I was keen to include as many worked examples and exercises as possible and this took some time. However, I was pleased with the result when "Introduction to Particle Technology" was published in 1998 [54] together with a solution manual for the exercises [55]. Having realized that there were many phenomena in particle technology that could not be described in words or with diagrams – I put together the CDROM "Laboratory Demonstrations in Particle Technology" [56] using some of Derek Geldart's fluidization videos, pneumatic transport footage courtesy of NEU Engineering, new material on mixing, segregation, explosions, cyclones, hopper flow and prepared with help of students and technical staff at Monash University. The whole thing put together by CSIRO external student Alfi Zakhari (with help from student

Belinda Mathers on the 2nd edition). We made it available to educators worldwide who have found it a helpful teaching tool for staff and students.

Although I was pleased with the first textbook, I felt that it was lacking in certain areas: colloids and fine particles, hydraulic transport, weak in granulation and health effects. So I added these with the help of contributors Jennifer Curtis (Florida), Karen Hapgood (Monash) and George Franks (Melbourne), added more exercises, test yourself questions and case studies with help from Derek Geldart, Peter Wypych (Wollongong), Peter Arnold (Wollongong), Lyn Bates (AJAX) and others (not wishing to be named) to produce a second edition [57]. The Solution Manual, which was available only from Monash University for the first edition, is now available online via Wiley [58]. Wiley promised that they would be able to make the CDROM "Laboratory Demonstrations in Particle Technology" available online. However, as this paper goes to press, this has yet to come to pass.

People:
I have been fortunate to work with some really bright students and postdoctoral researchers, to whom I am very thankful - my apologies for any omissions. Special thanks go to Shan Wang, who was my right-hand man for many years – a true professional researcher of high intellect, who was always willing to tackle something different.

Shan Wang	Art Yew Looi	Ali Akhavan	Alfi Zakhari
Toshi Hirama	Belinda Mathrs	Farhana Rahman	Ouyang Shan
Hideaki Mineo	Loretta McLaughlin	Sean Hutton	Ming Mao
Gabriel Amos	Greg Davies	Igor Sidorenko	Marit Sollaart
Shihui Zhou	John Sanderson	Jai Kant Pandit	
Peter Laussman	Adam Forsyth	Kanni Muniandi	
Cheng Hong	Shintaro Takeuchi	Chrestella Wardjiman	

Fluidized Bed Scaling - Literature Review
To the tune of Johnny B Goode – Chuck Berry
John Sanderson, Monash University

Scale-up ain't that easy or so it would seem,
You gotta build a lotta models in between.
In my opinion that's a pretty good reason for
Developing a groovy little scaling law.
So I'll tell you what's been written and what's been said
For scaling up a process in a fluidized bed

Chorus:
Blow, blow! Blow plenum blow!
Flow, flow! Flow fluid flow!
Go, go! Go powder go!
Grow, grow! Grow bubble grow!
Show, show! Show me the scaling law!

Glicksman's law started theoretically.
Based on di-men-sional similarity.
Horio's law was based instead
On a model of the bubbles in the fluid bed.
But it doesn't matter how their derivations went,
Their laws are mathematically equivalent.

Chorus

Deterministic chaos was the next thing tried.
To pressure fluctuations it was surely applied.
Lot of engineers were struggling to see
What the hell was Kol-mo-go-rov entropy?
And if the chaos theory was to be believed
Information conservation had to be achieved.

Chorus
(musical interlude)

Engineers were saying these laws were a farce.
Matching up the systems was a pain in the arse.
Chaos was OK was far as it went,
But no-one really knew what the answers meant.
Which pretty much brings us right back to the start:
Building scale models is still the state of the art.

Chorus

References

1. Rhodes, M.J. and D. Geldart, (1987), "A Model for the Circulating Fluidized Bed", Powder Technology, 53, pp. 155-162.

2. Kunii, D. and O. Levenspeil, "Entrainment of Solids from Fluidized Beds, I. Hold-Up of Solids in the Freeboard, II. Operation of Fast Fluidized Beds", Powder Technology, 61 (1990) 193 – 206.

3. Rhodes, M.J. and D. Geldart, (1986), "Transition to Turbulence?", Presented at the FLUIDIZATION V Engineering Foundation Conference, Helsinoor, Denmark, May 18-23, 1986. Published in FLUIDIZATION V, eds. K. Ostergaard and A.Sorenson, Engineering Foundation, New York, 1986, pp.281-288.

4. Rhodes, M.J., (1996) "What is Turbulent Fluidization?", Powder Technology, 88, 1, pp. 3-14.

5. Squires, A.M., Mooson Kwauk, and A. A. Avidan, "Fluid Beds: At Last, Challenging Two Entrenched Practices", Science, 20 December 1985, 230: 1329-1337.

6. Rhodes, M.J. (1989), "The Upward Flow of Gas/Solid Suspensions, Part 2: A Practical Quantitative Regime Diagram for the Upward Flow of Gas-Solid Suspensions", Chem. Eng. Res. Des., 67, pp. 30-37.

7. van Breugel, J. W., Stein, J. J. and de Vries, R. J., 1969b, "Isokinetic sampling in a dense gas solids stream", Patent 3C. Proc. Inst. Mech. Engng 184, 18-21.

8. Monceaux, L., Azzi, M., Molodtsof, Y. and Large, J.-F., 1986, "Overall and local characterization of flow regimes in a circulating fluidized bed", in Circulating Fluidized Bed Technology (Edited by P. Basu), p. 185. Pergamon Press, Oxford.

9. Rhodes, M.J. and P. Laussmann, (1992) "A Simple Non-Isokinetic Sampling Probe for Dense Suspensions", Powder Technology, 70, (2), pp. 141-151.

10. Rhodes, M.J., X.S. Wang, H. Cheng, T. Hirama and B.M. Gibbs, (1992) "Similar Profiles of Solids Flux in Circulating Bed Risers", Chem. Eng. Sci. , Vol. 47, No. 7, pp. 1635-1643.

11. Wang, X.S., M.J. Rhodes and B.M. Gibbs, (1995) "Influence of Temperature on Solids Flux Profiles in the Riser of a Circulating Fluidized Bed", Chem. Eng. Sci., Vol. 50, No. 15, pp. 2441-2447.

12. Rhodes, M.J., S. Zhou and H. Benkreira, (1992) "Flow of Dilute Gas-Particle Suspensions", AIChE J., Vol.38, No.12, pp.1913-1915.

13. Rhodes, M.J., H. Mineo and T. Hirama, (1992) "Particle Motion at the Wall of a Circulating Fluidized Bed", Powder Technology, 70, (3), pp.207-214.

14. Rhodes, M.J., S. Zhou, T. Hirama and H. Cheng, (1991) "Effects of Operating Conditions on Longitudinal Solids Mixing in a Circulating Fluidized Bed Riser", AIChE Journal, 37 (10), pp. 1450-1458.

15. Amos, G.L., M.J. Rhodes, (1993), "Gas Mixing in Gas-Solids Risers", Chem. Eng. Sci. Vol. 48, No. 5, pp. 943-949.

16. Wang, X.S., N.A. Akhtar, B.M. Gibbs and M.J. Rhodes, (1993), " Modelling of Circulating Fluidized Bed Combustion of Coal", Fuel, Vol. 73, No. 7, pp. 1120-1127.

17. Wang, X.S., M.J. Rhodes and B.M. Gibbs, (1994), "Impact of Air Staging on the Fate of NO and N_2O in a CFB Combustor", Combustion and Flame, Vol. 99, pp. 508-515.

18. Wang, X.S., M.J. Rhodes, B.M. Gibbs, and D. Geldart, (1997) "Heat transfer in dilute gas-particle suspensions", Chem. Eng. Sci., 52, 20, pp. 3617 - 3621.

19. Wang, X.S., M.J. Rhodes, B.M. Gibbs, M. Ahmad and D. Geldart, (1996) "Thermal Entrance Effect for Heat Transfer in Up-flowing Gas-Particle Suspensions", Ind. Eng. Chem. Res., 35, pp. 4781-4787.

20. Wang, X.S., M.J. Rhodes, and B.M. Gibbs and D. Geldart, (1996) "A correlation for suspension-to-wall heat transfer coefficient in circulating fluidized beds", Ind. Eng. Chem. Res., Vol. 35, No. 10, pp. 3822.

21. Wang, X.S., M.J. Rhodes, and B.M. Gibbs and D. Geldart, (1996) "Convective Wall-Suspension Heat Transfer in Circulating Fluidized Bed Risers", AIChEJ , Vol. 42, No. 8, p. 2112.

22. Rhodes, M.J., M. Sollaart and X.S. Wang, (1998) "Flow structure in a fast fluidized bed", Powder Technology, Vol. 99, 2, pp. 194-200.

23. McLaughlin, L.J. and M.J. Rhodes, (2001)"Prediction of gas fluidized bed behaviour in the presence of liquid bridge forces". Powder Technology, 114, 213-223.

24. Rhodes, M.J., X.S. Wang, A.J. Forsyth, K.S. Gan and S. Phadtajaphan, (2001) "Use of a magnetic fluidized bed in studying Geldart Group B to A transition", Chem. Eng. Sci., 56, 18, pp. 5429–5436.

25. Rhodes, M.J., X.S. Wang, M. Nguyen, P. Stewart and K. Liffman, (2001) "Onset of cohesive behaviour in gas fluidized beds: a numerical study using DEM simulation", Chem. Eng. Sci. 56, 14, pp. 4433-4438.

26. Pandit, J.K., X.S. Wang and M.J. Rhodes, (2006) "On Geldart Group A behaviour in fluidized beds with and without cohesive interparticle forces: A DEM study", Powder Technology, Volume 164, Issue 3, Pages 130-138.

27. Pandit, J.K., X.S. Wang and M.J. Rhodes (2005), "Study of Geldart's Group A behaviour using the discrete element method simulation", Powder Technology, Vol. 160, Issue 1, pp. 7-14.

28. Pandit, J.K., X.S. Wang and M.J. Rhodes, "A DEM study of bubble formation in Group B fluidized beds with and without cohesive inter-particle forces", Chemical Engineering Science, Vol. 62, Issue 1-2, 2007, pp. 159-166.

29. Sanderson, J.P. and M.J. Rhodes, (2003) "Hydrodynamic similarity of solids motion and mixing in bubbling fluidized beds" AIChEJ, Vol 49, No. No 9, pp. 2317-2327.

30. Sanderson, P.J., K. Seng Lim, I. Sidorenko and M. J. Rhodes, (2004), "Hydrodynamic Similarity in Bubbling Fluidized Beds: The Importance of the Solid-to-Gas Density Ratio", Ind. Eng. Chem. Res., Vol.43, No. 18, pp.5466-5473.

31. Sanderson, J.P. and M.J. Rhodes, (2005) "Bubbling fluidized bed scaling laws: Evaluation at Large Scales", AIChEJ, Vol 51, No. 10, October 2005, 2686-2694.

32. Rhodes, M.J. X.S. Wang, M. Nguyen, P. Stewart, K. Liffman, (2001) "Study of mixing in gas-fluidized beds using a DEM model", Chem. Eng. Sci., Vol 56, 8, pp 2859-2866.

33. Wang, X.S. and M.J. Rhodes, (2003) "Determination of particle residence time at the walls of gas fluidized beds by discrete element method simulation", Chemical Engineering Science, 58, pp387-395.

34. Wang, X.S and M.J. Rhodes, (2004) "Numerical study of gas fluidization under increased 'gravity' ", Advanced Powder Technology, Vol. 15, No. 6, pp. 629-638.

35. Wang, X.S and M.J. Rhodes, (2005) "Pulsed fluidization - A DEM study of a fascinating phenomenon", Powder Technology, Vol. 159, Issue 3, pp.142-149.

36. Takeuchi, S., X. S. Wang and M.J. Rhodes, (2005) "Discrete element study of particle circulation in a 3-D spouted bed", Chemical Engineering Science, Volume 60, Issue 5, pp 1267-1276.

37. Takeuchi, S., X.S. Wang, M.J. Rhodes, (2004) "Discrete element simulation of a flat-bottomed spouted bed in the 3-D cylindrical coordinate system", Chemical Engineering Science, Volume 59, Issue 17, pp 3495-3504.

38. Knight, J.B., Jaeger, H.M. and Nagel, S., "Vibration-Induced Separation in Granular Media: The Convection Connection", Phys. Rev. Lett., 70, 3728-3731, (1993).

39. Knight, J.B., Ehrichs, E.E., Kuperman, V.Y., Flint, J.K., Jaeger, H.M. and Nagel, S.R., "Experimental Study of Granular Convection", Phys. Rev. E, 54, 5726-5738, (1996).

40. Liffman, K., K. Muniandy, M.J. Rhodes, D. Gutteridge and G. Metcalfe (2001) "A segregation mechanism in a vertically shaken bed", Granular Matter Vol 3, Issue 4, 205-214.

41. Mobius, E., Lauderdale, B.E., Nagel, S.R. and Jaeger, H.M., "Size Separation of Granular Particles", Nature, 414, 270, (2001).

42. Rhodes M., S. Takeuchi, K. Liffman and K. Muniandy. Role of interstitial gas in the Brazil Nut Effect. Granular Matter (2003), 5, 107-114.

43. Forsyth, A.F., S. Hutton, M.J. Rhodes and C.F. Osborne, (2001) "Effect of applied interparticle force on the static and dynamic angles of repose of spherical granular material", Physics Review E, *Vol. 63*, pp. 031302-1 to 031302-5.

44. Forsyth, A.F., S. Hutton, M.J. Rhodes and C.F. Osborne, (2001) "Effects of interparticle force on the packing fraction of spherical granular material", Physical Review Letters, Vol. 87, 24, pp. 244301-1 to 244301-3.

45. Forsyth, A.F., S. Hutton and M.J. Rhodes, "Effect of cohesive interparticle force on the flow characteristics of granular material", Powder Technology, 126 (2002) 150-154.

46. Wang, X.S., F. Rahman and M.J. Rhodes, "Nanoparticle fluidization and Geldart's classification", Chemical Engineering Science, 62 (2007), pp. 3455 - 3461.

47. Wang, X.S., V. Palero, J. Soria and M.J. Rhodes, (2006) Laser-Based Planar Imaging of Nano-Particle Fluidization: Part II - Mechanistic analysis of nanoparticle aggregation, Chemical Engineering Science, 61, 8040-8049.

48. Wang, X.S., V. Palero, J. Soria and M.J. Rhodes, (2006) "Laser-based planar imaging of nano-particle fluidization: Part I—determination of aggregate size and shape", Chemical Engineering Science, 61, pp. 5476– 5486.

49. Ammendola, P. and Chirone, R. "Role of Sound Vibration during Aeration of Nano-Sized Powders", paper 43, The 12th International Conference on Fluidization - New Horizons in Fluidization Engineering, May 13-17, 2007 - Vancouver, Canada. Eds: Franco Berruti, Xiaotao (Tony) Bi, Todd Pugsley.

50. Wardjiman, C., A. Lee, M. Sheehan and M.J. Rhodes, "Shape of a particle curtain falling in stagnant air", Powder Technology, 192, 3, (2009), pp 384-388.

51. Wardjiman, C. and M.J. Rhodes, "Heat transfer in a particle curtain falling through a horizontally-flowing gas stream," Powder Technology, 191, 3, (2008), pp 249-253.

52. Wardjiman, C., A. Lee, M. Sheehan and M.J. Rhodes, "Behaviour of a curtain of particles falling through a horizontally-flowing gas stream", Powder Technology, 188 (2008), pp 110-118.

53. Rhodes, M.J., (ed.), (1990), "Principles of Powder Technology", John Wiley & Sons, Chichester, England, 439 pages.

54. Rhodes, M.J. (1998), "Introduction to Particle Technology", John Wiley & Sons, Chichester, England, 320 pages.

55. Rhodes, M.J. (1999) "Solution manual for exercises in the textbook, Introduction to Particle Technology", Department of Chemical Engineering, Monash University, 185 pages.

56. Rhodes, M.J., B. Mathers and A. Zakhari (2002) CD ROM: "Laboratory demonstrations in particle technology", 2nd Edition, Department of Chemical Engineering, Monash University.

57. Rhodes, M.J. (2008), "Introduction to Particle Technology – 2nd Edition", John Wiley & Sons, Chichester, England, 450 pages.

58. Rhodes, M.J. (2008) "Solution manual for exercises in the textbook, Introduction to Particle Technology, 2nd Edition", John Wiley & Sons, Chichester, England.

Effect of Sample Concentration On the Determination of the Anisotropy Constant of Magnetic Nanoparticles

Victoria L. Calero-DdelC, Chemical Engineering, University of Puerto Rico, Mayagüez Campus, Mayagüez, PR

Carlos Rinaldi, Chemical Engineering, University of Puerto Rico, Mayagüez Campus, Mayagüez, PR

Techniques such as equilibrium (DC) and dynamic (AC) magnetic measurements have been used to determine the anisotropy constant of a variety of magnetic nanoparticles, obtaining values that are often an order of magnitude higher than the corresponding bulk material. Unfortunately, the effect of particle-particle interactions is often neglected, hence the reported values are an effective collective parameter, rather than an intrinsic property. In this study DC and AC magnetic measurements are used to determine the anisotropy constant of magnetic nanoparticles fixed in a cross-linked polymer matrix. Nanoparticle concentrations of 0.1%, 1%, 6% and 33% (w/w) were used in order to determine the effect of concentration, and hence interactions, on the value determined for the anisotropy constant. The effect of interaction on determining the magnetocrystalline anisotropy constant was accounted for by using the Néel and Vogel-Fulcher relaxation models, the latter with an effective interaction temperature determined from independent measurements. A decrease in the anisotropy constant determined from these measurements was observed with decreasing particle concentration from ~6100 kerg/cm^3 to ~500 kerg/cm^3 using the Vogel-Fulcher model and from ~9100 kerg/cm^3 to ~1100 kerg/cm^3 using the Néel model. The values obtained for the most dilute samples using the Vogel-Fulcher model are of comparable magnitude to bulk values for magnetite. More importantly, the value of the characteristic time τ_0 obtained from AC susceptibility decreased from 10^{-32} to 10^{-9} s with decreasing particle concentration. As τ_0 is expected to be of the order of 10^{-9} s, these results indicate that the value of magnetic anisotropy determined for the dilute sample represents an intrinsic rather than effective property. These measurements and analysis illustrate the importance of particle concentration/interactions in determining the intrinsic magnetic properties of nanoparticles, particularly the anisotropy constant.

See more of this Session: Magnetic Particle Synthesis and Properties
See more of this Group/Topical: Particle Technology Forum

Magnetic Gelation

Marco Furlan, Institute for Chemical and Bioengineering, Department of Chemistry and Applied Biosciences, ETH Zurich, Zurich, Switzerland

Marco Lattuada, Institute for Chemical and Bioengineering, Department of Chemistry and Applied Biosciences, ETH Zurich, Zurich, Switzerland

Massimo Morbidelli, Institute for Chemical and Bioengineering, Department of Chemistry and Applied Biosciences, ETH Zurich, Zurich, Switzerland

Magnetic nanocomposites, based on magnetite nanocrystals and polymer matrix, are of great interest for many fields (e.g. material science, medicine, diagnostic). A possible application in material science is for example the production of porous material with controlled pore size and pore size distribution, which can be used for magnetic separation or applications in chromatography.

For this work, we introduce a novel technique to create porous materials with controlled structure that we call magnetic gelation. First of all, magnetic colloids, composed of magnetite nanocrystals dispersed into a polystyrene matrix, and stabilized by SDS, have been prepared via miniemulsion polymerization. Due to the superparamagnetic behavior of the nanocrystals embedded in the polymer matrix, the magnetic colloids, which can be prepared in a size range from 70 to 300 nm, develop strong and reversible dipolar interactions only in the presence of an external magnetic field.

We have then studied the gelation behavior of dispersions of these nanoparticles in presence and absence of an external magnetic field. The nanoparticels, which are stabilized by means of electrostatic interactions, are partially destabilized through the addition of controlled amount of electrolytes. In the absence of magnetic fields, the particles self-assemble into random fractal clusters, which eventually percolate to form a colloidal gel, at particle volume fraction of a few percents. Instead, when an external magnetic field is applied, the particles align themselves in columnar structures in the direction of the field. If the external magnetic field is applied when the gelation process has already begun, both particles and clusters are aligned in the direction of the magnetic field. By tuning the time at which the field is applied, different structures and characteristic pore sizes in the gel are obtained. To be able to study the final structure of the obtained materials, their structure is fixed by a procedure called reactive gelation [1]. The reactive gelation process consists in swelling the polymer particles with a small amount of monomer and crosslinker and let them react after the gelation process to transform the weak physical bonds keeping the particles together in the gel matrix into much stronger covalent bonds. The structures so obtained have been characterized by means of scanning electron microscopy.

[1] Marti N., Quattrini F., Butté A., Morbidelli M., MACROMOLECULAR MATERIALS AND ENGINEERING, 290, 221-229 (2005)

See more of this Session: Magnetic Particle Synthesis and Properties
See more of this Group/Topical: Particle Technology Forum

Gold-Coated Iron-Oxide Nanoparticles for Optical Imaging and Hyperthermia of Cancer

Martin O'Toole, Chemical Engineering, University of Louisville, Louisville, KY

Jianting Wang, Chemical Engineering, University of Louisville, Louisville, KY

Krishna Kanth Sanapala, Chemical Engineering, University of Louisville, Louisville, KY

Jacek Jasinski, Chemical Engineering, University of Louisville, Louisville, KY

Kyung A. Kang, Chemical Engineering, University of Louisville, Louisville, KY

Fe_3O_4 nanoparticles are effective heat guides for minimally invasive cancer hyperthermia using an alternating electromagnetic (AEM) field. Conjugation of cancer-specific targeting molecules to the nano-sized particles allows cancer-specific treatment, with minimal damage to normal tissue.

Nano-scale, gold surfaces have a plasmon field that can be beneficially used to manipulate the fluorescence intensity of fluorophores for molecular imaging. When the fluorophore is at an appropriate distance from a nano-scale gold surface, the fluorescence intensity can be greatly enhanced. This distance for enhancement (in our previous studies, 5 - 10 nm) can be established by placing a spacer between the fluorophore and the gold surface. This fluorescence enhancement can increase the signal-to-noise ratio, and therefore the sensitivity of optical imaging.

Here, we are combining the magnetic properties of Fe_3O_4 nanoparticles and the plasmon field of nano-gold surfaces into a single nano-entity. When a fluorophore is linked to this entity, the resulting product is a highly fluorescent contrast agent for optical detection of a tumor and heat guide to the tumor. In order to achieve this goal, gold-coated Fe_3O_4 nanoparticles were synthesized. Fluorophores were then conjugated onto the particle via bio-compatible spacers. The entire entity was designed to have a total size of 10 - 50 nm for efficient cell uptake. The level of heat generation under an AEM field was correlated to the Fe_3O_4 core size and gold shell thickness. The fluorescence intensity of the entity was studied with changes in the particle size and gold shell thickness.

The resulting single nano-entity can be used for both highly sensitive optical detection and hyperthermic treatment of cancer. The particles can also be used, without further modification, as contrast agents for MRI. Recently gold nanoparticles have been found to provide contrast for x-rays, and the entity may be used for x-ray/CT imaging as well.

See more of this Session: Magnetic Particle Synthesis and Properties
See more of this Group/Topical: Particle Technology Forum

Multistage Enrichment of Particles Using Temperature-Responsive Polymers

Jiamin Wu, Chemical and Petroleum Engineering, University of Pittsburgh, Pittsburgh, PA

Di Gao, Chemical and Petroleum Engineering, University of Pittsburgh, Pittsburgh, PA

Enrichment of cells is an important process for sample preparations, which are the foundations for effective technology validation as well as for meaningful biological and clinical research. A conventional method for cell enrichment is magnetic particle separation technology, which allows for the simple, rapid and efficient enrichment of cells. A well known problem that prevents the magnetic particle separation technology from achieving high enrichment factors is the non-specific interactions between target species and the magnetic particles. To circumvent this problem we develop an enrichment technology by reversibly capturing and releasing target species with poly (N-isopropylacrylamide) (PNIPAM) coupled magnetic particles. This is realized by manipulating hydrophobic interactions between the temperature-responsive PNIPAM molecules covalently grafted both on the magnetic particles and the target species. At 37 °C, the PNIPAMs exhibit a hydrophobic conformation, and the target species are captured by the magnetic particles through hydrophobic interactions. While at 4 °C, the PNIPAMs are trigged to their hydrophilic state, and the target species are released from the magnetic particles. We design a multistage enrichment process by introducing multiple capture-and-release cycles to the magnetic separation procedure. Experiments demonstrated the multistage enrichment of modeled particles, an analogy to target cells and showed that the cumulative enrichment factor increases as the number of stage increases.

See more of this Session: Magnetic Particle Synthesis and Properties
See more of this Group/Topical: Particle Technology Forum

Effect of Graft Molecular Weight On the Colloidal Stability of Iron Oxide Nanoparticles with Grafted Poly(ethylene glycol)

Carola Barrera, Chemical Engineering, University of Puerto Rico, Mayaguez Campus, Mayaguez, PR

Adriana Herrera, Chemical Engineering, University of Puerto Rico, Mayaguez Campus, Mayaguez, PR

Nayla Bezares, Chemical Engineering, University of Puerto Rico, Mayaguez Campus, Mayaguez, PR

Carlos Rinaldi, Chemical Engineering, University of Puerto Rico, Mayaguez Campus, Mayaguez, PR

We studied the stability of colloidal suspensions of iron oxide nanoparticles with grafted poly(ethylene glycol) of various molecular weights in aqueous media, buffers, and cell culture media. The particles consist of iron oxide cores with narrow particle size distribution and coated with individual polymer coronas. Particle surface charge and hydrodynamic size were studied as a function of pH and ionic strength using zeta potential measurements and dynamic light scattering. DLVO theory was used to model particle-particle interactions, which were correlated to aggregation state as determined from experimental measurements.

See more of this Session: Magnetic Particle Synthesis and Properties
See more of this Group/Topical: Particle Technology Forum

Multifunctional Iron Oxide Nanoparticles Coated by Novel Biocompatible Zwitterionic Polymer

Lei Zhang, Department of Chemical Engineering, University of Washington, Seattle, WA

Shaoyi Jiang, Chemical Engineering, University of Washington, Seattle, WA

A novel multifunctional magnetic nanoparticles(MNPs) coated by the biocompatible zwitterionic polymer, poly(carboxybetaine methacrylate)(pCBMA) were prepared by a self-assembling method. The pCBMA coating not only stabilized the MNPs in complex mediums, but also provided abundant functional groups to the MNPs. The prepared pCBMA-MNPs had a hydrodynamic particle size of about 130 nm and a strong saturated magnetization of 110 emu/g Fe, which is 1.6 times higher than that of the commercial product Feridex. In vitro stability studies showed the pCBMA-MNPs could achieve long-term stability in commonly use medium like PBS and also in high ionic strength solution like 10% NaCl solution for at least 6 monthes. Moreover, comparing to the dextran coated MNPs, pCBMA-MNPs presented better stability in 100% human blood serum incubated at 37 C. Macrophage cells uptake studies revealed that the uptake ratio of pCBMA-MNPs was much lower than the dextran-MNPs. Cell targeting studies showed that, after conjugated with targeting ligand, the cell uptake rated was notably increased.

See more of this Session: Magnetic Particle Synthesis and Properties
See more of this Group/Topical: Particle Technology Forum

Controlled Surface Modification of Superparamagnetic Iron Oxide Nanoparticles (SPIONs)

Esther Amstad, Torben Gillich, Idalia Bilecka, Marcus Textor, Erik Reimhult

ETH Zurich, Switzerland

Abstract

The number of applications for iron oxide nanoparticles especially in the biomedical field has been rapidly increasing during the last few years. Most of these applications require stable nanoparticles under physiologic conditions, with close control over the hydrodynamic diameter and interfacial chemistry. The latter is a prerequisite for efficient targeting of such nanoparticles. Irreversibly binding low molecular weight dispersants have potential to stabilize iron oxide nanoparticles such that all these stringent requirements are met. Due to the importance of anchor groups for nanoparticle stabilization, we compared stability of iron oxide nanoparticles coated with low molecular weight poly(ethylene glycol) (PEG) based dispersants which were anchored to the nanoparticle surface through a set of catechol derivatives. NitroDOPA and nitrodopamine were shown to irreversibly bind PEG to the iron oxide surfaces at high dispersant packing densities and thus lead to unsurpassed iron oxide nanoparticle stability under physiologic conditions and heating up to 90 °C. Irreversible binding of these anchors allowed functionalization of SPIONs with optimized saturation magnetization by co-adsorption of biotinylated and non-biotinylated PEG-nitroDOPA. Fast and specific binding of such-functionalized nanoparticles towards neutravidin presenting surfaces indicated their potential e.g. as targeted magnetic resonance contrast agents for *in vivo* applications.

Introduction

Oxide nanoparticles can be used for many different applications such as for diagnostics[1] and therapeutics[2]. Especially iron oxide nanoparticles have recently been a heavily investigated subject. Due to their magnetic properties and good biocompatibility[3], iron oxide nanoparticles are often used e.g. as magnetic resonance (MR) contrast agents, for hyperthermia treatment, for cell separation purposes or for guided drug delivery[4,5]. Successful application of these nanoparticles crucially depends on nanoparticle stability, size and surface chemistry[6]. Especially for biomedical application, where nanoparticles need to be stable under physiologic conditions, agglomeration is a major, still not satisfactorily resolved obstacle. Iron oxide nanoparticles are typically stabilized with high molecular weight dispersants such as dextran. However, the rather weak affinity of dextran to iron oxide nanoparticles leads to reversible adsorption[6]. Furthermore, typically multiple iron oxide cores are embedded within one dextran molecule[7]. This results in a poorly defined hydrodynamic diameter. Reversibly adsorbed dispersants lead to poor control over the interfacial chemistry, which is a severe obstacle to create targeted nanoparticles. To enhance nanoparticle stability, dextran has been crosslinked after its adsorption on iron oxide nanoparticles[8]. However, also for this approach, control over the hydrodynamic diameter and interfacial chemistry is still limited.

Alternatively, iron oxide nanoparticles have been stabilized with low molecular weight dispersants which consist of an anchor group covalently linked to a biocompatible polymer such as PEG through a simple grafting to approach[9-11]. Such dispersants adsorb on the iron oxide nanoparticle surface in a well defined way which allows close control over the interfacial chemistry and hydrodynamic diameter. The latter is solely determined by the iron oxide core diameter and the dispersant molecular weight, configuration (such as linear or dendritic structures), packing density and solubility, all parameters which can be tuned. Therefore, different functionalities such as fluorophores or targeting moieties can be either presented at the interface which leads to efficient targeting[9] or buried in the dispersant shell. However, stabilization of oxide nanoparticles with low molecular weight dispersants relies on anchors which have a high affinity, low probability of desorption and which pack dispersants densely onto the nanoparticle surface. In the past, dopamine and DOPA, two catechol derivatives, the latter is an amino acid abundantly present in the mussel adhesive protein Mytilus edulis[12], have been used as high affinity anchor groups[10,13-15]. However, their performance is debated. Dopamine has been reported to lead to iron oxide core dissolution accompanied by an iron catalyzed dopamine oxidation which results in poorly binding semiquinones and quinones[16]. Because of the high importance of anchor groups for nanoparticle stability and functionality, we have investigated eight different catechol derived anchor groups selected to shed light into the chemical factors important for optimal binding affinities of these anchors.

Experimental

Materials:

Fe(ac)$_2$ (batch 517933, Lot 03901JJ), mimosine, dopamine, diethylether, 4-(2-hydroxyethyl)-1-piperazineethanesulfonic acid (HEPES), sodium tetraborate decahydrate (borate buffer), Tris-(hydroxymethyl)aminomethane (Tris) and N,N-Dimethylformamide (DMF) were purchased from Sigma-Aldrich, DOPA, hydroxydopamine and benzylalcohol from Acros and PEG(5)-NHS, PEG(1.5)-NHS and biotin-PEG(3.4)-NHS from Jemkem, NaCl from J. T. Baker and ethanol from Scharlau, 1,2-Dipalmitoyl-*sn*-Glycero-3-Phosphoethanolamine-N-(Cap Biotinyl) (biotin-PE) and 1-palmitoyl-2-oleoyl-sn-glycero-3-phosphocholine (POPC) from Avanti Polar Lipids and neutravidin from Invitrogen.

Anchor groups:

Nitrodopamine and nitroDOPA were synthesised according to[17]. Starting from kojic acid hydroxypyridine was synthesised by replacing the aliphatic alcohol by chlorine according to[18] followed by a removal of chlorine[19]. The aromatic alcohol was protected with benzyl[18] before the aromatic oxygen was substituted with ethylenediamine[19]. Hydroxypyrone was synthesised starting from kojic acid. The aliphatic alcohol was replaced by chlorine[18] and subsequently by azides[20] before it was converted to kojic amine[21].

Dispersants:

PEG(5)-NHS, biotin-PEG(3.4)-NHS and PEG(1.5)-NHS respectively were coupled to the different anchor groups according to a slightly modified protocol from Zurcher et al.[22]. Briefly, PEG-NHS (1 mol equ) was dissolved in borate buffer (pH=8.4) (20 ml) before the respective anchor group (0.8 mol equ) was added. After 24 h reaction at room temperature under constant magnetic stirring, the solution was acidified with HCl. It was extracted with chloroform before it was precipitated in diethylether and vacuum dried.

Iron oxide nanoparticles:

Iron oxide nanoparticles were synthesized according to Bilecka et al.[23]. Briefly, $Fe(ac)_2$ (173 mg) was dissolved in benzylalcohol (5 ml) and heated in argon atmosphere for 3 or 30 min respectively at 180°C in a microwave (Discover S-class, CEM, NC, USA). These nanoparticles were washed once with ethanol (10 ml) before they were re-dispersed in ethanol (10 ml) resulting in an iron oxide nanoparticle concentration of approximately 10 mg/ml. Washed iron oxide nanoparticles were kept at 4°C for 2-6 h before these particles (1 mg) were added to DMF or EtOH respectively (1 ml) containing dispersants (unless stated otherwise 8 mg for particles which have been heated 3 min and 6 mg for particles which have been heated for 30 min). This suspension was kept at 50°C for 24 h under constant mechanical stirring at 500 rpm (Thermomixer comfort, Vaudaux-Eppendorf, Switzerland). To functionalize nanoparticles, as-synthesized nanoparticles which had been heated for 30 min (1 mg) were added to biotin-PEG(3.4)-nitroDOPA (3.8 mg corresponding to 30 mol%) which was dissolved in EtOH (1 ml) and kept for 30 min at 50°C under constant mechanical stirring at 500 rpm before they were back-filled with PEG(1.5)-nitroDOPA (2.2 mg) and incubated for further 23.5 h at 50°C under constant mechanical stirring at 500 rpm. Excessive dispersants were removed by 24 h dialysis against Millipore water (R=18.2 Ω, TAC<6 ppb) using dialysis membranes with a cut-off of 14-16 kDa (Spectra/Por dialysis membrane, spectrum labs, Netherlands) before so-stabilized iron oxide nanoparticles were freeze-dried (freeze dryer ALPHA 1-2 / LDplus, Kuhner LabEquip, Switzerland).

Dynamic light scattering (DLS)

DLS experiments were run on a Zetasizer Nano ZS (Malvern, UK) in the 173° backscattering mode. Because intensity weighted diameters scale with r^6 and thus large particles are predominant, volume weighted data which scale with r^3 [24] were analyzed using the multiple narrow modes evaluation incorporated into the Malvern software. Temperature dependent DLS experiments have been performed in 10 mM HEPES containing 150 mM NaCl (pH=7.4) at an iron oxide nanoparticle concentration of 200 µg/ml. Solutions were filtered with 200 nm Minisart syringe filters (Sartorius, Germany) prior to DLS analysis. These suspensions were heated from 25 to 90°C in 5°C steps with 10 min equilibration time before the respective DLS experiment. The dispersant binding reversibility was tested on iron oxide nanoparticles (500 µg/ml) in 10 mM HEPES solutions containing 150 mM NaCl at room temperature. Solutions were repeatedly filtered using Microcon filters with a cut-off of 30 kDa (Millipore, MA, USA) by centrifuging these solutions at 13 rpm for 20 min (MiniSpin, Vaudaux Eppendorf, Switzerland). After filtration, the filter cake was re-suspended in HEPES containing 150 mM NaCl (0.5 ml) and equilibrated for 24 h at room temperature before the next DLS experiment was performed.

Thermogravimetry analysis (TGA)/differential scanning calorimetry (DSC)/Mass spectrometry (MS):

Stabilized iron oxide nanoparticles (5 mg) were re-dispersed in Millipore water (500 µl) for 24 h before they were filtered with Microcon centrifuge filters with a cut-off of 30 kDa. The filter cake re-dispersed in Millipore water (0.5 ml) was freeze-dried. Alternatively, re-dispersed iron oxide nanoparticles were dialyzed against Millipore water for 24 h using dialysis membranes with a cut-off of 25 kDa before these nanoparticles were freeze-dried. So-prepared nanoparticles (3-5 mg per run) were analyzed on a NETZSCH STA 449 C Jupiter, which for some experiments was coupled to a NETZSCH QMS 403 Aeolos MS (NETZSCH Geraetebau GmbH, Austria). Samples were heated from 35°C to 600 °C at 10°C/min using a flow rate of 47.4 sccm Ar and 12.6 sccm O_2.

Transmission electron microscopy (TEM)

Transmission electron micrographs were taken on a Philips CM12 microscope operated at 100 kV and on a HRTEM Philips CM30 operated at 300 kV. Iron oxide nanoparticles suspended in Millipore water were dried on a 10 nm carbon coated 400 mesh Cu-grid.

Vibrating sample magnetomter (VSM)

Vibrating sample magnetometer measurements were performed on a MicroMag^{TM} equipment (Princeton Measurements Corperation, NJ, USA). Freeze dried PEG(5)-nitroDOPA stabilized and unstabilized iron oxide nanoparticles (2-3 mg) were analyzed per run. At least 2 independent identical samples were measured. The saturation magnetization was normalized assuming magnetite (Fe_3O_4) iron oxide cores. Furthermore, dispersant:iron oxide core ratios measured in TGA were assumed.

Vesicle preparation:

Liposomes were produced by extrusion following a protocol adapted from MacDonald et al[25]. A mixture of 95 wt% POPC and 5 wt% 1,2-Dipalmitoyl-*sn*-Glycero-3-Phosphoethanolamine-N-(Cap Biotinyl) (biotin-PE) dissolved in chloroform was dried for 1 h under a constant N_2 stream before lipids were rehydrated in 10 mM Tris containing 150 mM NaCl (pH=7.4) to a concentration of 5 mg/ml during 1 h. Vesicles were extruded 31 times through double polycarbonate membranes with a pore size of 100 nm (Avestin).

Quartz crystal microbalance with dissipation monitoring (QCM-D)

QCM-D experiments[26] were performed on a Q-Sense E4 equipment (Q-Sense AB) in 10 mM Tris containing 150 mM NaCl (pH=7.4) at RT. Vesicles (100 µg/ml) were adsorbed on SiO_2 coated QCM-D crystals (Q-Sense AB), resulting in supported lipid bilayer (SLB) formation[27,28]. Neutravidin (50 µg/ml) was immobilized on the biotinylated SLBs before biotinylated or PEG(1.5)-nitroDOPA stabilized iron oxide nanoparticles re-suspended in water at a stock solution concentration of 100 mg/ml and diluted in 10 mM Tris containing 150 mM NaCl to a concentration of 50 µg/ml, were injected into the QCM-D chamber.

Results and Discussion

Iron oxide nanoparticle stabilization

Iron oxide cores were synthesized by a microwave assisted non-aqueous sol-gel route[23]. $Fe(ac)_2$ dissolved in Benzylalcohol was heated to 180°C and kept at this temperature for 3 and 30 min, resulting in single domain iron oxide cores with a diameter of 5.7±0.6 and 14.2±1.2 nm respectively (Figure 1). These nanoparticles were washed once with EtOH and subsequently stabilized with different dispersants depicted in Figure 2.

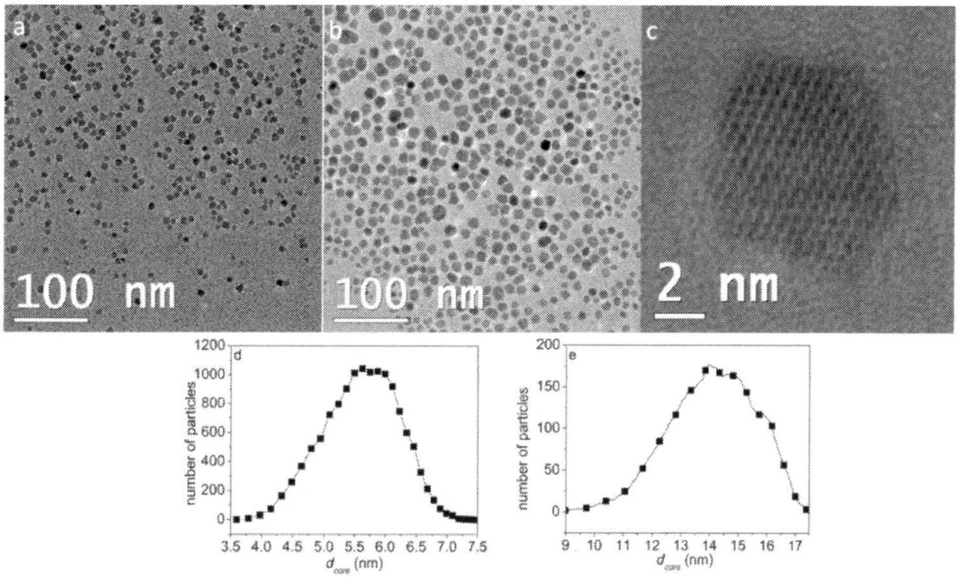

Figure 1: TEM micrographs of PEG-nitroDOPA stabilized iron oxide nanoparticles. Overview of iron oxide nanoparticles kept at 180°C for a) 3 min and b) 30 min respectively. C) HRTEM of iron oxide nanoparticles kept at 180 °C for 3 min. Particle size distributions of iron oxide nanoparticles kept for d) 3 min and e) 30 min at 180 °C. Statistics was done on 13000 particles have for d) and on 7000 for e).

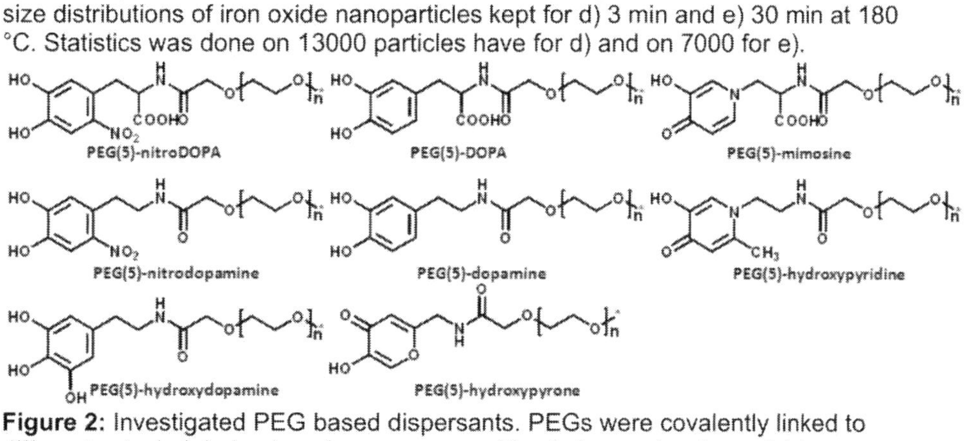

Figure 2: Investigated PEG based dispersants. PEGs were covalently linked to different catechol derived anchor groups resulting in low molecular weight dispersants. n≈114 for PEG(5), n≈77 for PEG(3.4) and n≈34 for PEG(1.5).

Nanoparticle stability was primarily investigated with dynamic light scattering (DLS) by analyzing volume weighted hydrodynamic diameters and count rates. Iron oxide nanoparticles stabilized with PEG(5)-hydroxypyrone did not even pass 200 nm cut-off filters. This was in contrast to particles stabilized with the other investigated dispersants and implied that PEG(5)-hydroxypyrone stabilized iron oxide nanoparticles readily agglomerated upon dilution in HEPES containing 150 mM NaCl. Consequently, DLS was only performed on nanoparticles which passed 200 nm cut-off filters. Because the scattering intensity scales with r^6, count rates are very sensitive to nanoparticle agglomeration. Thus, an increase in count rate indicates an increase in particle size, which here corresponds to nanoparticle agglomeration. Iron oxide nanoparticles stabilized with the well known PEG(5)-dopamine and PEG(5)-

hydroxydopamine already started to agglomerate at room temperature if dispersed in HEPES containing 150 mM NaCl (Figure 3 a, b). Furthermore, agglomeration of such-stabilized nanoparticles became even more pronounced at elevated temperatures, which led us to use volume weighted hydrodynamic diameters at elevated temperatures as a probe to distinguish the affinity of the different dispersants. Iron oxide nanoparticles stabilized with PEG(5)-hydroxypyridine, PEG(5)-mimosine and PEG(5)-DOPA respectively started to agglomerate between 50 and 80°C while iron oxide nanoparticles stabilized with PEG(5)-nitroDOPA and PEG(5)-nitrodopamine remained stable even after repeatedly heating them up to 90°C.

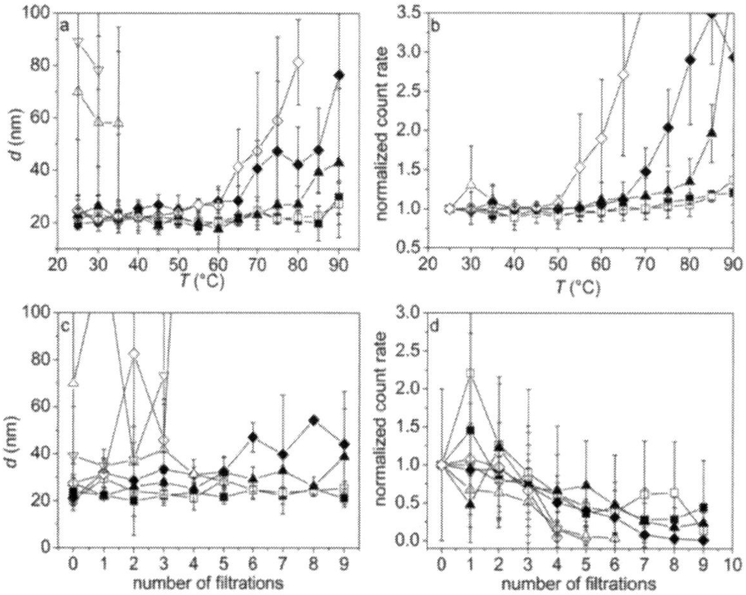

Figure 3: DLS of 5.7 nm iron oxide cores stabilized with different dispersants. Volume weighted diameters of iron oxide nanoparticles stabilized with PEG(5)-nitroDOPA (-■-), PEG(5)-nitrodopamine (-□-), PEG(5)-DOPA (-▲-), PEG(5)-dopamine (-△-), PEG(5)-mimosine (-●-), PEG(5)-hydroxypyridine (-○-) and PEG(5)-hydroxydopamine (-▽-) dispersed in HEPES containing 150 mM NaCl. Temperature dependence of the a) volume weighted hydrodynamic diameter and b) normalized derived count rate. To assess the binding reversibility of these dispersants, c) the hydrodynamic diameter and d) the normalized derived count rate of stabilized nanoparticles is shown as a function of the number of times particles were filtered, which is equivalent to the number of times free dispersants were removed.

For reversibly adsorbing dispersants, there is equilibrium between free and on iron oxide nanoparticle adsorbed molecules. To gain further insights into the dispersant binding reversibility, stabilized iron oxide nanoparticles were successively filtered and re-dispersed in HEPES containing 150 mM NaCl. Between every filtration, particles were equilibrated at RT for 24 h. While the hydrodynamic diameter of iron oxide nanoparticles stabilized with PEG(5)-nitroDOPA and PEG(5)-nitrodopamine respectively remained unchanged, even after 9 filtrations, it rapidly increased especially for particles stabilized with PEG(5)-dopamine and PEG(5)-hydroxydopamine (Figure 3c and d). The derived count rates decreased with the number of filtrations for all nanoparticles. This, and the high standard deviations, can be assigned to losses due to repeated sample transfers between centrifuge filters and DLS cuvettes. However, the derived count rates of particles stabilized with PEG(5)-dopamine and PEG(5)-hydroxydopamine decreased markedly faster compared to PEG(5)-nitroDOPA and PEG(5)-nitrodopamine stabilized nanoparticles. This is indicative of nanoparticle agglomeration resulting in sedimentation and thus in a decreased count rate with increasing number of filtrations.

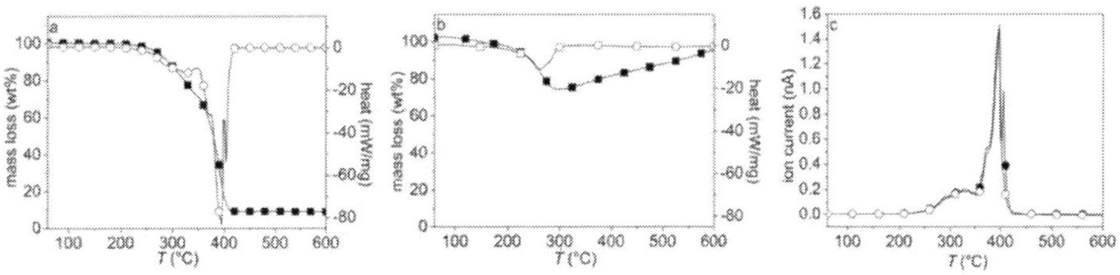

Figure 4: Quantification of dispersants adsorbed on iron oxide nanoparticles. TGA (-■-) and DSC (-□-) of 5.7 nm core diameter iron oxide nanoparticles a) stabilized with PEG(5)-nitroDOPA and b) unstabilized. c) The burned products of PEG(5)-nitroDOPA stabilized iron oxide nanoparticles were *in situ* analyzed by MS and could be assigned to H_2O (-○-) and CO_2 (-●-).

Furthermore, simultaneous thermogravimetry analysis (TGA) and differential scanning calorimetry (DSC) were performed on stabilized and unstabilized iron oxide nanoparticles respectively (Figure 4 a and b). Burned fragments were further *in situ* analyzed by a mass spectrometer (MS) which was coupled to the TGA/DSC. Decomposition products could be assigned to equal amounts of H_2O and CO_2 without any other measureable residues. Especially the absence of nitrogen containing fragments indicated that all DMF, in which these nanoparticles were stabilized, could be removed from the nanoparticle surface and dispersant shell.

Table1: Packing densities of dispersants adsorbed on iron oxide cores with a diameter of 5.7 nm measured with TGA and the resulting ethylene glycol (EG) density.

dispersant	dispersant/nm^2	EG/nm^2
PEG(5)-nitroDOPA	2.8±0.3	296±27
PEG(5)-nitrodopamine	2.6±0.2	290±40
PEG(5)-mimosine	2.5±0.9	247±37
PEG(5)-DOPA	2.4±0.9	210±50
PEG(5)-dopamine	2.3±0.5	254±25
PEG(5)-hydroxydopamine	1.7±0.6	254±26
PEG(5)-hydroxypyridine	1.8±0.6	197±16
PEG(5)-hydroxypyrone	1.8±0.2	244±13

Iron oxide nanoparticles with a core diameter of 5.7 nm and 14.2 nm contained 88 wt% and 80 wt% PEG(5)-nitroDOPA respectively while physisorbed impurities accounted for 14 wt% and 6 wt% of uncoated iron oxide nanoparticles with a core diameter of 5.7 and 14.2 nm respectively. Based on these TGA results, dispersant packing densities were calculated for iron oxide cores with a diameter of 5.7 nm assuming an iron oxide core size distribution shown in Fig. 1d, a Fe_3O_4 density of 5.18 g/cm^3 [29] and a dispersant molecular weight of 5.2 kDa for PEG(5)-anchors. The resulting dispersant packing densities greatly exceeded the previously highest reported PEG(5)-packing densities on flat (0.3 PEG(5)DOPA$_3$/nm^2 if adsorbed on TiO_2 surfaces)[30] and highly curved surfaces (1.3 PEG(5)-hydroxydopamine adsorbed on iron oxide nanoparticles)[9]. Interestingly, the dispersant packing densities scaled with iron oxide nanoparticle stability at elevated temperatures (cf. Table 1 and Figure 3). This close correlation between high temperature nanoparticle stability, dispersant binding reversibility and packing density indicates that these anchor groups not only determine binding strength of dispersants but also the packing density. It further stresses the importance of the anchor group for nanoparticle stability.

While iron oxide nanoparticles with a core diameter of 5.7 nm stabilized with PEG(5)-nitroDOPA resulted in a volume weighted hydrodynamic diameter around 20 nm, the hydrodynamic diameter of PEG(1.5)-nitroDOPA stabilized nanoparticles was approximately 13 nm (Figure 5a). However, no significant difference in nanoparticle stability was measured between iron oxide nanoparticles stabilized with PEG(1.5)-nitroDOPA and PEG(5)-nitroDOPA, for which the temperature dependence of the count rates was the same within experimental errors (Figure 5b). The lower molecular weight of PEG(1.5)-nitroDOPA compared to PEG(5)-nitroDOPA resulted in a thinner dispersant layer surrounding the iron oxide cores. However, PEG(1.5)-nitroDOPA still resulted in a dispersant layer which was thick enough to prevent nanoparticle agglomeration even under physiologic conditions and elevated temperatures.

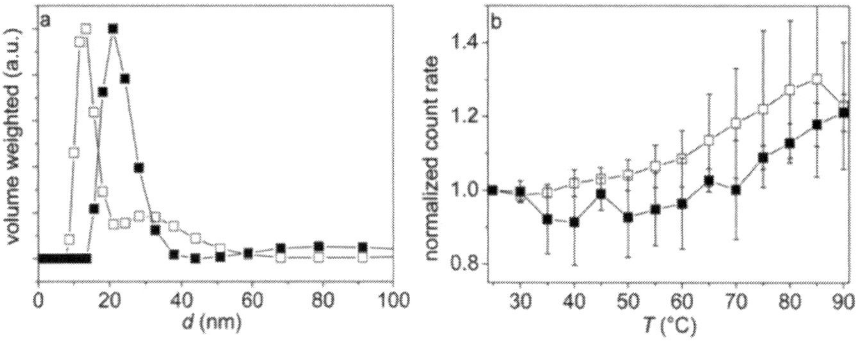

Figure 5: Influence of M_w(PEG). A) Volume weighted hydrodynamic diameters measured at RT and b) temperature dependent normalized count rates of 5.7 nm core diameter iron oxide nanoparticles stabilized with PEG(1.5)-nitroDOPA (-□-) and PEG(5)-nitroDOPA (-■-) respectively.

Magnetic properties

A high saturation magnetization is needed for good contrast at low nanoparticle concentrations if used as MR contrast agents[31] and reduces the magnetic field strength needed for application of the nanoparticles to separate cells or guide drug delivery vehicles[32]. Therefore, saturation magnetizations of PEG(5)-nitroDOPA stabilized and unstabilized nanoparticles were compared for nanoparticles with d=5.7

and 14.2 nm respectively (Figure 6). The saturation magnetization of 5.7 and 14.2 nm iron oxide core nanoparticles was 20 emu/g_{Fe} and 61 emu/g_{Fe} respectively where the latter was in good agreement with reported values for magnetite nanoparticles of comparable size[33]. However, it decreased by 30 and 33 %, respectively, upon stabilization. This decrease in saturation magnetization upon steric stabilization was described in literature and was assigned to a non-magnetic layer[33]. Because of the considerably higher saturation magnetization of 14.2 nm iron oxide cores compared to 5.7 nm cores, which leads to a higher relaxivity r_2[31] and thus better contrast in T_2 and T_2* weighted MR images, the 14.2 nm core nanoparticles were chosen for further functionalization.

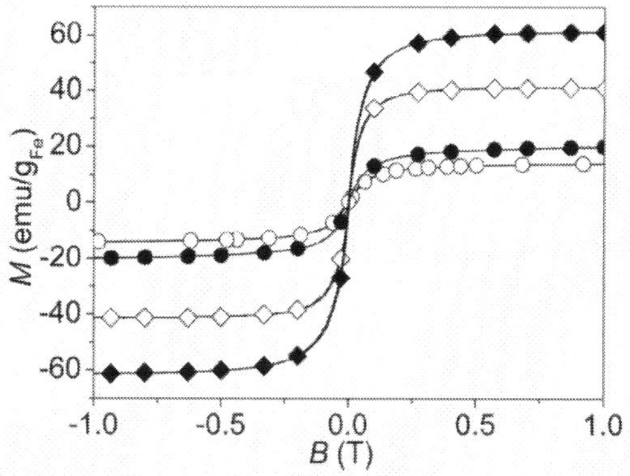

Figure 6: Hysteresis of iron oxide nanoparticles. VSM measurements of unstabilized (filled symbols) and with PEG(5)-nitroDOPA stabilized (open symbols) 5.7 nm (-●-, -○-) and 14.2 nm (-◆-, -◇-) iron oxide cores.

SPION functionalization

To show feasibility for efficient targeting, 14.2 nm core diameter iron oxide nanoparticles were stabilized with a mixture of biotin-PEG(3.4)-nitroDOPA and PEG(1.5)-nitroDOPA. PEG(1.5)-nitroDOPA rather than PEG(5)-nitroDOPA was chosen to back-fill particles due to its lower molecular weight compared to biotin-PEG(3.4)-nitroDOPA, which rendered biotin sites immobilized on the iron oxide nanoparticles more accessible for binding. Binding specificity of biotinylated nanoparticles was tested with quartz crystal microbalance with dissipation monitoring (QCM-D) measurements. Neutravidin was immobilized on a SiO_2 coated quartz crystal through biotinylated supported lipid bilayers, mimicking targeting of membrane receptors. This surface was then exposed to biotinylated and non-biotinylated iron oxide nanoparticles. While biotinylated iron oxide nanoparticles bound quickly and specifically, non-biotinylated PEG(1.5)-nitroDOPA stabilized iron oxide nanoparticles did not bind (Figure 7). Negligible non-specific particle adsorption to neutravidin demonstrated the resultant "stealth" property of the PEG-nitroDOPA stabilized particles that was expected from the high stability under physiologic conditions measured by DLS.

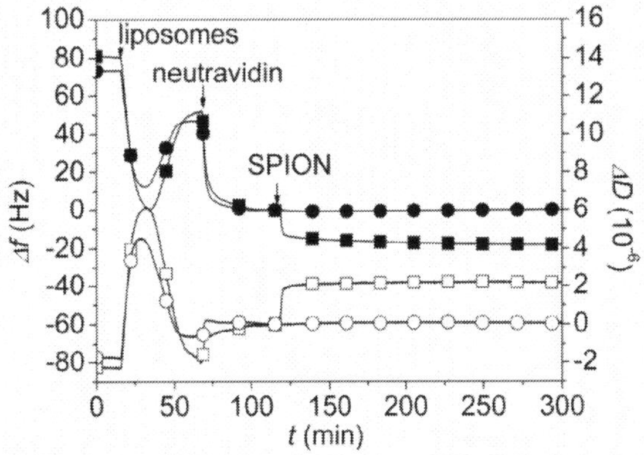

Figure 7: Binding specificity of biotin functionalized iron oxide nanoparticles was measured with QCM-D. Neutravidin was immobilized on SiO_2 coated QCM-D crystals through biotinylated supported lipid bilayers. Fast changes in frequency (filled symbols) and dissipation (open symbols) indicated strong and specific binding of biotin-PEG(3.4)-nitroDOPA/PEG(1.5)-nitroDOPA stabilized iron oxide nanoparticles (-■-, -□-) in contrast to close to no adsorption for PEG(1.5)-nitroDOPA stabilized nanoparticles (-●-, -○-).

Conclusions

The anchor group of low molecular weight dispersants has been proven to be crucial for creating functional, ultra-stable nanoparticles especially under physiologic conditions. We found that

1. the well known catechols such as dopamine and DOPA adsorbed reversibly on iron oxide nanoparticles which leads to nanoparticle agglomeration if dispersed under physiologic conditions at temperatures even below body temperature.
2. nitroDOPA and nitrodopamine anchored PEG for practical purposes irreversibly to the iron oxide nanoparticle surfaces. This irreversible binding of dispersants to the particle surface not only leads to ultra-stable nanoparticles up to 90 °C under physiologic buffer conditions; it also resulted in a previously unsurpassed dispersant packing density.
3. The saturation magnetization increased with increasing iron oxide core diameter. Because of the marked reduction in saturation magnetization of iron oxide nanoparticles upon stabilization, individually stabilized iron oxide nanoparticles should be at least 14.2 nm in diameter for applications such as MR contrast agents, for ferrying cells or other magnetic separation assays.
4. Irreversible binding of nitroDOPA and nitrodopamine towards iron oxide surfaces allowed not only tuning the hydrodynamic diameter by controlling the iron oxide core diameter and the dispersant molecular weight, it also enabled us to efficiently functionalize these nanoparticles. Simple co-adsorption of biotinylated and methoxy terminated dispersants leads to strong and specific binding of such-functionalized nanoparticles to neutravidin presenting surfaces. Furthermore, sequential addition of different molar ratios of functionalized and unfunctionalized dispersants allowed close control over the surface density of functional units.

The experimentally easily performed co-adsorption of differently functionalized dispersants further opens up the possibility to create multifunctional, ultrastable nanoparticles where functionalities and their surface densities can be controlled and tuned according to individual needs by adjusting the molar ratios of dispersants during nanoparticle stabilization. Keys to successful creation of such multifunctional nanoparticles are anchors such as nitroDOPA and nitrodopamine which irreversibly graft dispersants to the particle surface at high dispersant packing densities. Thus these anchors pave the way for many applications especially in the biomedical field, where under physiologic conditions ultra-stable, stealth nanoparticles with the option for controlled and custom tailored surface modification and functionalization are in high demand.

Acknowledgements

The authors would like to thank Prof. Markus Niederberger for providing the infrastructure for the iron oxide nanoparticle synthesis, Dr. Lucio Isa for help with TEM image processing and evaluation and Dr. Andreas Gehring and Claudio D'Addario for their help with VSM measurements. Furthermore, we are indebted to NETZSCH-Geraetebau GmbH which kindly performed some TGA-MS experiments for us. Prof. L. Gauckler (ETH Zurich) and EMEZ (ETH Zurich) and are acknowledged for providing TGA and electron microscopy facilities respectively. COST Action No. D43 "Multifunctional Superparamagnetic Iron Oxide Nanoparticles for Targeted Magnetic Resonance Imaging" and ETH Zurich are acknowledged for their financial support.

References

1 Lee, J.H. *et al.*, Artificially engineered magnetic nanoparticles for ultra-sensitive molecular imaging. *Nature Medicine* 13 (1), 95-99 (2007).

2 Namdeo, M. *et al.*, Magnetic Nanoparticles for Drug Delivery Applications. *J Nanosci Nanotechno* 8 (7), 3247-3271 (2008).

3 Weissleder, R. *et al.*, Superparamagnetic Iron-Oxide - Pharmacokinetics and Toxicity. *American Journal of Roentgenology* 152 (1), 167-173 (1989).

4 Sun, C., Lee, J.S.H., & Zhang, M.Q., Magnetic nanoparticles in MR imaging and drug delivery. *Advanced Drug Delivery Reviews* 60 (11), 1252-1265 (2008).

5 Lin, M.M., Kim, D.K., El Haj, A.J., & Dobson, J., Development of Superparamagnetic Iron Oxide Nanoparticles (SPIONS) for Translation to Clinical Applications. *Ieee Transactions on Nanobioscience* 7 (4), 298-305 (2008).

6 McCarthy, J.R. & Weissleder, R., Multifunctional magnetic nanoparticles for targeted imaging and therapy. *Advanced Drug Delivery Reviews* 60 (11), 1241-1251 (2008).

7 Bautista, M.C. *et al.*, Comparative study of ferrofluids based on dextran-coated iron oxide and metal nanoparticles for contrast agents in magnetic resonance imaging. *Nanotechnology* 15 (4), S154-S159 (2004).

8 Josephson, L., Tung, C.H., Moore, A., & Weissleder, R., High-efficiency intracellular magnetic labeling with novel superparamagnetic-tat peptide conjugates. *Bioconjugate Chem* 10 (2), 186-191 (1999).

9 Amstad, E. *et al.*, Surface Functionalization of Single Superparamagnetic Iron Oxide Nanoparticles for Targeted Magnetic Resonance Imaging. *Small* 5 (11), 1334-1342 (2009).

10 Xu, C.J. *et al.*, Dopamine as a robust anchor to immobilize functional molecules on

the iron oxide shell of magnetic nanoparticles. *Journal of the American Chemical Society* 126 (32), 9938-9939 (2004).

11 Xie, J., Xu, C., Kohler, N., Hou, Y., & Sun, S., Controlled PEGylation of monodisperse Fe3O4 nanoparticles for reduced non-specific uptake by macrophage cells. *Advanced Materials* 19 (20), 3163-+ (2007).

12 Waite, J.H. & Tanzer, M.L., Polyphenolic Substance of Mytilus-Edulis - Novel Adhesive Containing L-Dopa and Hydroxyproline. *Science* 212 (4498), 1038-1040 (1981).

13 Xie, J. *et al.*, Linking hydrophilic macromolecules to monodisperse magnetite (Fe3O4) nanoparticles via trichloro-s-triazine. *Chemistry of Materials* 18 (23), 5401-5403 (2006).

14 Gu, H.W., Yang, Z.M., Gao, J.H., Chang, C.K., & Xu, B., Heterodimers of nanoparticles: Formation at a liquid-liquid interface and particle-specific surface modification by functional molecules. *Journal of the American Chemical Society* 127 (1), 34-35 (2005).

15 Gao, J.H. *et al.*, Multifunctional yolk-shell nanoparticles: A potential MRI contrast and anticancer agent. *Journal of the American Chemical Society* 130 (35), 11828-11833 (2008).

16 Shultz, M.D., Reveles, J.U., Khanna, S.N., & Carpenter, E.E., Reactive nature of dopamine as a surface functionalization agent in iron oxide nanoparticles. *Journal of the American Chemical Society* 129 (9), 2482-2487 (2007).

17 Napolitano, A., Dischia, M., Costantini, C., & Prota, G., A New Oxidation Pathway of the Neurotoxin 6-Aminodopamine - Isolation and Characterization of a Dimer with a Tetrahydro[3,4a]Iminoethanophenoxazine Ring-System. *Tetrahedron* 48 (39), 8515-8522 (1992).

18 Ma, Y.M., Luo, W., Quinn, P.J., Liu, Z.D., & Hider, R.C., Design, synthesis, physicochemical properties, and evaluation of novel iron chelators with fluorescent sensors. *Journal of Medicinal Chemistry* 47 (25), 6349-6362 (2004).

19 Liu, Z.D. *et al.*, Synthesis of 2-amido-3-hydroxypyridin-4(1H)-ones: Novel iron chelators with enhanced pFe(3+) values. *Bioorganic & Medicinal Chemistry* 9 (3), 563-573 (2001).

20 Durden, J.A., Stansburry, H.A., & Catlette, W. Correspondence-The Reaction of Azidoorganic Compounds with Acetylenes (1964), Vol. 9, pp. 480-480.

21 Atkinson, J.G. *et al.*, Kojic Amine - Novel Gamma-Aminobutyric Acid Analog. *Journal of Medicinal Chemistry* 22 (1), 99-106 (1979).

22 Zurcher, S. *et al.*, Biomimetic surface modifications based on the cyanobacterial iron chelator anachelin. *Journal of the American Chemical Society* 128 (4), 1064-1065 (2006).

23 Bilecka, I., Djerdj, I., & Niederberger, M., One-minute synthesis of crystalline binary and ternary metal oxide nanoparticles. *Chemical Communications* (7), 886-888 (2008).

24 Mie, G., Articles on the optical characteristics of turbid tubes, especially colloidal metal solutions. *Annalen Der Physik* 25 (3), 377-445 (1908).

25 Macdonald, R.C. *et al.*, SMALL-VOLUME EXTRUSION APPARATUS FOR PREPARATION OF LARGE, UNILAMELLAR VESICLES. *Biochimica Et Biophysica Acta* 1061 (2), 297-303 (1991).

26 Rodahl, M., Hook, F., Krozer, A., Brzezinski, P., & Kasemo, B., Quartz-Crystal Microbalance Setup for Frequency and Q-Factor Measurements in Gaseous and Liquid Environments. *Review of Scientific Instruments* 66 (7), 3924-3930 (1995).

27 Keller, C.A. & Kasemo, B., Surface specific kinetics of lipid vesicle adsorption measured with a quartz crystal microbalance. *Biophysical Journal* 75 (3), 1397-1402

(1998).
28 Reimhult, E., Hook, F., & Kasemo, B., Intact vesicle adsorption and supported biomembrane formation from vesicles in solution: Influence of surface chemistry, vesicle size, temperature, and osmotic pressure. *Langmuir* 19 (5), 1681-1691 (2003).
29 Linde, D.R. & Frederikse, H.P.R., *Handbook of Chemistry and Physics, 75th edition*, 75th edition ed. (CRC, 1995).
30 Dalsin, J.L. *et al.*, Protein resistance of titanium oxide surfaces modified by biologically inspired mPEG-DOPA. *Langmuir* 21 (2), 640-646 (2005).
31 Josephson, L., Lewis, J., Jacobs, P., Hahn, P.F., & Stark, D.D., The Effects of Iron-Oxides on Proton Relaxivity. *Magnetic Resonance Imaging* 6 (6), 647-653 (1988).
32 Qiang, Y. *et al.*, Iron/iron oxide core-shell nanoclusters for biomedical applications. *J Nanopart Res* 8 (3-4), 489-496 (2006).
33 Vidal-Vidal, J., Rivas, J., & Lopez-Quintela, M.A., Synthesis of monodisperse maghemite nanoparticles by the microemulsion method. *Colloids and Surfaces a-Physicochemical and Engineering Aspects* 288 (1-3), 44-51 (2006).

Numerical and Experimental Analysis of Heat Transfer in a Continuous Spouted Bed with Particle Bottom Feed

Cezar A. Rosa, Chemical and Biological Engineering, Illinois Institute of Technology, Chicago, IL

José T. Freire, Chemical Engineering, Federal university of São Carlos, São Carlos - SP, Brazil

Hamid Arastoopour, Department of Chemical and Biological Engineering, Illinois Institute of Technology, Chicago, IL

Spouted beds are largely used due to their high heat and mass transfer rates and good mixing of solids. Specially, spouted beds are applied to deal with large particles (dp > 1mm), where fluidized beds are not efficient. In the literature there are several experimental and simulation studies using batch spouted beds. However, little information can be found regarding continuous spouted beds. The continuous operation has some important advantages such as higher production capacity and lower operating costs. The objective of this work is to perform numerical simulations and experimental analysis of heat transfer in a continuous spouted bed with particle bottom feed. To perform this analysis, experiments using glass beds as a particulate phase were carried out. The air inlet velocity and temperature were set to 20m/s and 80ºC respectively. During the experiments, the air temperature (central channel and annulus) and the particulate phase temperature (inlet and outlet) were measured. The average particle diameter and the particle density were 0.0039m and 2500kg/m3, respectively. The Eulerian-Eulerian multi-fluid modeling approach was applied to predict the gas-solid flow behavior, and the energy balance of each phase was solved to account for the heat transfer. The commercially available FLUENT 6.3 Computational Fluid Dynamics (CFD) code was used to perform the numerical simulations. The effective thermal conductivity of the particulate phase was considered to be a function of the particle concentration and it was added using a compiled executable code by means of a User Defined Function. The simulation results of air temperature (central channel and annulus) and solids temperature (inlet and outlet) showed a good agreement with the experimental data.

See more of this Session: Gas/Solid Mixing and Heat/Mass Transfers in Fluidized Beds
See more of this Group/Topical: Particle Technology Forum

The Pressure Dependence of Atomic Layer Deposition Processes On Particles in a Fluidized Bed Reactor

Joseph A. Spencer, ALD NanoSolutions, Inc., Broomfield, CO

David M. King, Department of Chemical Engineering, University of Colorado, Boulder, CO

J. Ruud Van Ommen, Delft University of Technology, Delft, Netherlands

Alan W. Weimer, Department of Chemical Engineering, University of Colorado, Boulder, CO

Standard atomic layer deposition (ALD) processes are operated under vacuum conditions to allow for the effective chemisorption of precursors while minimizing the physisorption of unwanted contaminants. Particle ALD processes can be performed in a variety of reactor configurations, but the most common is the fluidized bed reactor. There is a desire to increase the operating pressure in order to significantly decrease the processing time to functionalize particle surfaces at the pilot and industrial scales without sacrificing film quality. It is well known that the steric hindrance of adsorbed precursors can be overcome at elevated pressures, causing tighter packing and increased film growth per cycle. Reaction progress was measured using an in situ mass spectrometer downstream from the fluidized bed reactor. A comprehensive evaluation of the pressure-dependent attributes of ALD films was performed. Implications on process time and scale-up are discussed for lab-scale and pilot-scale fluidized beds, and these results are compared to ALD processes performed in alternate reactor configurations.

See more of this Session: Gas/Solid Mixing and Heat/Mass Transfers in Fluidized Beds
See more of this Group/Topical: Particle Technology Forum

2009 AIChE Annual Meeting,
Nashville, TN, USA, November 8-13, 2009

CFD prediction of uranium tetrafluoride particle fluorination in fluidized bed pilot

N'dri A. Konan[1,2], Hervé Neau[1,2], Olivier Simonin[1,2], Marc Dupoizat[3] and Thomas Le Goaziou[3]

[1] Université de Toulouse; INPT, UPS;
Institut de Mécanique des Fluides de Toulouse;
Allée du Professeur Camille Soula, 31400 Toulouse, France
[2] CNRS; IMFT; 31400 Toulouse, France
konan@imft.fr, neau@imft.fr, simonin@imft.fr
[3] Areva NC BU-Chimie
BP 44, 26701 Pierrelatte, France
marc.dupoizat@areva.com, thomas.le-goaziou@areva.com

Abstract: The paper deals with the mechanisms controlling the reaction rates in the framework of the uranium hexafluoride conversion within a fluidized bed reactor. Results show that the reactive gas diffusion around the reactive particles with respect to their sizes does not influence the chemical reaction. The thermal boundary conditions for gas and dispersed phase applied in combination with the radiation effect between particles (formulated using the Rosseland approximation) successfully account for the heat transfer modes within the bed. Additionally, promising simulations with a initial polydispersion of the reactive particles are performed.

Keywords: Fluidized bed, uranium fluorination, shrinking core model, heat transfer

1 Introduction

The nuclear energy efficiently ensures energy independence needed to support the development and facilitates a broader accessibility of the domestic energy to the populations. The manufacturing of nuclear reactor fuels primarily requires the gaseous uranium hexafluoride (UF_6) which will usually undergo enrichment in isotope 235 of uranium. Among the several available processes for the UF_6 production, the oxidation of uranium tetrafluoride (UF_4) with fluorine (F_2) in fluidized bed turns out to be the best way (Janov & LePage (1981)). The efficiency of such a system operating at higher temperatures relies on the excellent heat transfers between phases and with wall, and mass transfers that continuously arise between gas and particles through chemical reactions. Beyond the requirement of the accuracy of the description and the modeling of hydrodynamics that ensure the well mixing of the bed (see e.g. Fan et al. (2004)), the simulation of the gas/particle chemical reactions within the bed is inherent to the mechanisms controlling the reaction rate.

The paper mainly reports both the modeling and the simulations of such mechanisms within a fluidized bed: the diffusion of the reactive gas around the particles and the heat transfers occurring in the reactor. The bed considered herein is comprised of reactive uranium (UF_4) and inert alumina particles (Al_2O_3). Only the following the exothermic fluorination reaction of the UF_4 particles with the fluorine (F_2) is assumed

$$UF_4 + F_2 \longrightarrow UF_6, \quad \text{with} \quad \Delta H = 234.4 kJ.mol^{-1} \tag{1}$$

2 Modeling of the reactive flow

2.1 Hydrodynamics of the bed

2.1.1 Mass balance equation

Considering the reactive flow between the phases within the reactor, the mass balance is governed by

$$\frac{\partial\left(\alpha_k\rho_k\right)}{\partial t}+\frac{\partial\left(\alpha_k\rho_k U_{k,i}\right)}{\partial x_i}=\Gamma_k \tag{2}$$

where ρ_k and α_k denote the density and the volume fraction of the *kth* phase, respectively. $U_{k,i}$ represents the *ith* component of the mean velocity of the phase k. Γ_k accounts for the interphase mass transfer rate and satisfies

$$\sum_k \Gamma_k=0 \tag{3}$$

In the frame of the study, as the alumina particles do not react, their mass transfer is null ($\Gamma_{Al_2O_3}=0$). Accordingly, the mass transfer that continuously arises between gas mixture and the uranium particles is modeled according to (3) by

$$\Gamma_g=-\Gamma_{UF_4} \tag{4}$$

2.1.2 Momentum balance equation

Assuming the gravity as the only external force acting on each phase k, the mean momentum equation is written

$$\alpha_k\rho_k\frac{\partial U_{k,i}}{\partial t}+\alpha_k\rho_k U_{k,j}\frac{\partial U_{k,i}}{\partial x_j}=-\alpha_k\frac{\partial P_g}{\partial x_i}+\alpha_k\rho_k g_i+\left[U_{\sigma,i}-U_{k,i}\right]\Gamma_k+\frac{\partial}{\partial x_j}\left[-\alpha_k\rho_k R_{k,ij}+\Theta_{k,ij}\right]+\sum_{m\neq k}S_{k\rightarrow m,i} \tag{5}$$

in which P_g is the mean pressure of the gas phase while g_i denotes the *ith* component of gravity.
The interfacial momentum transfer rate between the gas and the particles is accounted for in the last term with the drag force model

$$S_{g\rightarrow p,i}=-S_{p\rightarrow g,i}=\alpha_p\rho_p\frac{V_{rp,i}}{\tau_{gp}^F} \tag{6}$$

$V_{rp,i}=U_{p,i}-U_{g,i}-V_{dp,i}$ represents the *ith* component of the mean relative velocity between the particles and the surrounding gas, while the fluid-particle drift velocity $V_{dp,i}$ accounts for the dispersion effects due to the particles transport by turbulent fluid motion and, following Simonin et al. (1993) and Gobin et al. (2003), is approximated by

$$V_{dp,i}=-D_{gp,ij}\left[\frac{1}{\alpha_p}\frac{\partial\alpha_p}{\partial x_j}-\frac{1}{\alpha_g}\frac{\partial\alpha_g}{\partial x_j}\right],\quad \text{with}\quad D_{gp,ij}=\frac{1}{3}q_{gp}\tau_{gp}^t\delta_{ij} \tag{7}$$

where q_{gp} and τ_{gp}^t are the fluid-particle velocity correlation and the eddy-particle interaction time.
The characteristic time τ_{gp}^F in (5) of the particle entrainment by the fluid motion is approximated by

$$\frac{1}{\tau_{gp}^F}=\frac{3}{4}\frac{\rho_g}{\rho_p}\frac{C_D\left(\langle Re_p\rangle_p\right)}{d_p}\langle\mid\vec{v}_r\mid\rangle_p,\quad\text{with}\quad C_D\left(\langle Re_p\rangle_p\right)=\begin{cases}C_{D,\text{Wen \& Yu}} & \text{if }\alpha_g\geq 0.7\\ \min\left(C_{D,\text{Wen \& Wu}},C_{D,\text{Ergun}}\right) & \text{otherwise}\end{cases} \tag{8}$$

$\vec{v}_r$ and $Re_p=\alpha_g\langle\mid\vec{v}_r\mid\rangle_p d_p/\nu_g$ refer to the relative velocity vector and the mean particle Reynolds number, respectively. $C_{D,\text{Wen \& Wu}}$ and $C_{D,\text{Ergun}}$ are the drag correlation of Wen & Yu (1966) and Ergun (1952), respectively.

$R_{k,ij}$ in (5) represents the gas turbulent Reynolds stresses for $k=g$ and the particle kinetic stresses for $k=p$, respectively, and is modeled according to Boussinesq-like approximation. Closure of these modeling is achieved by solving the $k-\varepsilon_g$ transport equations, modified by additional terms accounting for the particle influence on the fluid turbulence, and transport equations for each particle agitation kinetic energy q_p^2 (granular temperature) and for the fluid-particle velocity fluctuation covariance q_{gp} (see Simonin et al. (1993) and Gobin et al. (2003)).

$U_{\sigma,i}$ represents the *ith* component of the mean velocity of the mass crossing the interface between gas and reactive particle, as the reaction proceeds. This velocity is assumed to be equal to particle velocity (i.e. $U_{\sigma,i} = U_{UF_4,i}$).

Tensor $\Theta_{k,ij}$ in (5) represents the laminar viscous stress for the gas ($k = g$) and the collisional stress for the dispersed phases ($k = p$). $S_{k\to p,i}$ represents the collision transfer term between k and p particle species in the momentum balance equation. Details of the closure assumptions for n-fluid modeling of reactive flows with several particle classes can be found in Boelle et al. (1995) and Ciais et al. (2007).

2.2 Thermal exchange in the bed

The temperature of the phases within the bed is continuously modified as the chemical reactions proceed. Assuming that the heat exchanged by contact during the interparticle collisions is negligible, the main heat exchange modes in the bed are summarized by

- the convection/diffusion heat transfer ($\Pi_{g\to p}$) between the gaseous phase and the particles occurring with a characteristic time scale τ_{gp}^T such that

$$\Pi_{g\to p} = -\Pi_{p\to g} = \frac{\alpha_p \rho_p C_{p_p}}{\tau_{gp}^T}(T_p - T_g), \quad \text{with} \quad \frac{1}{\tau_{gp}^T} = \frac{6\lambda_g}{\rho_p C_{p_p}} \frac{\langle Nu \rangle_p}{d_p^2} \tag{9}$$

  where λ_g is the thermal conductivity of the gaseous phase. $\langle Nu \rangle_p = 2 + 0.55 Re_p^{1/2} Pr^{1/3}$ represents the Nusselt number of the particle phase while $Pr = \rho_g \nu_g C_{p_g}/\lambda_g$ denotes the Prandtl number. C_{p_k} is the specific heat of the *kth* phase in the reactor.

- the enthalpy exchanged (H_σ) from reactive particles towards the gas due to the mass flux crossing the particle surfaces because of the mass transfer induced by the chemical reactions. With assumption that the reaction enthalpy is directly released in the surrounding gas, the mean enthalpy of the matter crossing the interface due to reaction is set to the enthalpy of the reactive particles ($H_\sigma = H_{UF_4}$).

- the radiative heat transfer between particles of the dense phases. Assuming that the radiation between the particles in the bed takes place in the frame of the Rosseland approximation through a diffusion mechanism, Konan et al. (2009) write the radiative flux in the alumina particle enthalpy equation as proportional to the temperature gradient with a radiative thermal diffusion coefficient given by

$$K_p^r = \frac{32\sigma}{9\alpha_p} \frac{d_p T_p^3}{\rho_p C_{p_p}} \tag{10}$$

  in which σ is denotes the Stefan-Boltzmann constant and T_p the temperature of the particles.

Then in the absence the influence of external heat sources, the distribution of the enthalpy of each phase in the bed satisfies the transport equation

$$\alpha_k \rho_k \frac{\partial H_k}{\partial t} + \alpha_k \rho_k U_{k,j} \frac{\partial H_k}{\partial x_j} = \frac{\partial}{\partial x_j}\left(\alpha_k \rho_k K_k \frac{\partial H_k}{\partial x_j}\right) + [H_\sigma - H_k]\Gamma_k + \sum_{m \neq k} \Pi_{k\to m} \tag{11}$$

In the last term, $\Pi_{p\to q} = 0$ for both particle phases p and q since the heat exchanged during the interparticle collisions is assumed to be negligible. The first term on the right-hand side is a transport term written in the frame of the gradient approximation with an effective phase diffusivity coefficient K_k

- for the particle phase:

$$K_p = K_p^t + K_p^r \tag{12}$$

  K_p^t and K_p^r are the contributions due to the transport of the enthalpy by the velocity fluctuations and the radiative heat transfer between the particles, respectively.

- for the gas:

$$K_g = K_g^t + K_g^l \tag{13}$$

where K_g^t and K_g^l are the contributions to the transport of the enthalpy due to the gas turbulence and the laminar diffusivity, respectively.

The heat exchange at the wall is due to the laminar heat gas conduction effect and to the radiation exchange between the particles and the wall. Furthermore, Simonin & Konan (2010) found that the main contribution of the heat exchanged by the near-wall computing cell with the thermal boundary layer is dominated by the particle heat transfer. Relying on such analysis, the thermal wall-function approach proposed by Simonin & Konan (2010) to avoid the detailed computation of the thermal boundary layer, leads to the following practical boundary conditions for the gas and the particle phases

$$\alpha_p \rho_p K_p \left(\frac{\partial H_p}{\partial y}\right)_{|y=Y} = \alpha_g \rho_g C_{p_g} K_g \left(\frac{\partial T_g}{\partial y}\right)_{|y=0} + \varepsilon_p \sigma T^4_{p|y=0} - \varepsilon_w \sigma T_w^4 = h_{p\to w}\left(T_{p|_{y=Y}} - T_w\right) \tag{14}$$

$$\alpha_g \rho_g K_g \left(\frac{\partial H_g}{\partial y}\right)_{|y=Y} = 0 \tag{15}$$

T_g and T_w are the temperatures of the gas and the wall, respectively. The subscript Y refers to the first grid point from the wall.
The heat transfer coefficient at the wall $h_{p\to w}$ ($W.m^{-2}.K^{-1}$) which must be modeled depends on many factors, such as the particle properties, the wall emissivity or the bed void fraction. However this heat transfer coefficient may be assessed using experimental data (see Yamada et al. (1995)) or the "wall-function model".

2.3 Evolution of the sizes of reactive particles and gaseous species

2.3.1 Size of the reactive particles

The chemical reaction between reactive uranium particles and the fluorine assumed to be uniformly located at the surface of the particles necessarily induces the decrease of the mean diameter of the particles. The local instantaneous mean diameter is updated according to

$$d_{UF_4} = d_{UF_4,0}\left(\frac{\chi_{UF_4,0}}{\chi_{UF_4}}\right)^{1/3} \tag{16}$$

where $d_{UF_4,0}$ and $\chi_{UF_4,0}$ are the initial mean diameter and the corresponding number of particles per unit mass.
Considering that the chemical reaction does not mofify the particle number per unit volume of the multiphase mixture, the transport equation governing the time-evolution of the mean number of particles per unit mass (χ_{UF_4}) in the beds is written

$$\alpha_{UF_4}\rho_{UF_4}\frac{\partial \chi_{UF_4}}{\partial t} + \alpha_{UF_4}\rho_{UF_4}U_{UF_4,j}\frac{\partial \chi_{UF_4}}{\partial x_j} = \frac{\partial}{\partial x_j}\left(\alpha_{UF_4}\rho_{UF_4}D^t_{UF_4}\frac{\partial \chi_{UF_4}}{\partial x_j}\right) - \chi_{UF_4}\Gamma_{UF_4} \tag{17}$$

in which $D^t_{UF_4}$ denotes the diffusivity coefficient due to the transport by the particle velocity fluctuations (Abbas et al. (2009)).

2.3.2 Balance of species of the gaseous mixture

With the assumption of a mixture of ideal gases within the reactor, the density of the gaseous mixture (ρ_g) and the mass fractions (Y_α) of the αth specie of the gas are governed by

$$\rho_g = \frac{P_{\text{ref}}}{R_g T_g}\left[\frac{Y_{N_2}}{W_{N_2}} + \frac{Y_{F_2}}{W_{F_2}} + \frac{Y_{UF_6}}{W_{UF_6}}\right]^{-1} \tag{18}$$

$$\alpha_g \rho_g \frac{\partial Y_\alpha}{\partial t} + \alpha_g \rho_g U_{g,j}\frac{\partial Y_\alpha}{\partial x_j} = \frac{\partial}{\partial x_j}\left(\alpha_g \rho_g \frac{\nu_g^t}{\sigma_g}\frac{\partial Y_\alpha}{\partial x_j}\right) - Y_\alpha \Gamma_g + \Psi_\alpha \tag{19}$$

ν_g^t and σ_g are the viscosity and the Schmidt turbulent number of the gaseous mixture, respectively. Ψ_{F_2} accounts for the vanishing rate of the fluorine whereas Ψ_{UF_6} represents the apparition rate of uranium hexafluoride. In accordance with the chemical reaction equation (1) these reaction rates satisfy

$$\Psi_{F_2} = \frac{W_{F_2}}{W_{UF_4}}\Gamma_{UF_4}, \quad \Psi_{UF_6} = -\frac{W_{UF_6}}{W_{UF_4}}\Gamma_{UF_4} \tag{20}$$

W_α (with $\alpha = N_2, F_2, UF_6$ and UF_4) in (18) and (20) are the molar mass of the gaseous species and the uranium particles. Note that the reaction rate of nitrogen is null ($\Psi_{N_2} = 0$).

Practically, the transport equation (19) of the gaseous species is solved on the one hand for the fluorine (since it is the only one which directly reacts with UF_4 particles) and on the other hand to the nitrogen because its reaction rate is null. The mass fraction of the uranium hexafluoride is computed according to

$$Y_{UF_6} = 1 - Y_{F_2} - Y_{N_2} \tag{21}$$

2.4 Mass transfer rate of the reactive particles

With the assumption that the UF_4 particles are nonporous and the chemical reaction arises uniformly at the surface of any single particle with an Arrhenius-type law, Randrianarivelo et al. (2007) shows that the Eulerian reaction rate of the uranium particle phase is written

$$\Gamma_{UF_4} = -\frac{6\alpha_{UF_4}}{d_{UF_4}}\frac{W_{UF_4}X_{F_2}^S P_{\text{ref}}}{R_g T_{UF_4}} f_c \exp\left(-\frac{E_a}{R_g T_{UF_4}}\right) \tag{22}$$

E_a the activation energy, R_g the gas constant and f_c the kinetic constant of the reaction. The derivation of the reaction rate (22) assumes that the particle temperature is uniform ($T_{UF_4}^S = T_{UF_4}$). $X_{F_2}^S$ represents the mole fraction of fluorine at the surface of the particle. Following Konan et al. (2009), the mass fraction of fluorine at the surface of the reactive particle ($Y_{F_2}^S$), accounting for the diffusion of fluorine within the nitrogen around the reactive particles, is given by

$$Y_{F_2}^S = Y_{F_2}\left[1 + \frac{d_{UF_4}}{D_{F_2}\,Sh} f_c \exp\left(-\frac{E_a}{R_g T_{UF_4}}\right)\right]^{-1} \tag{23}$$

D_{F_2} represents the diffusion coefficient of fluorine in nitrogen while Sh is the Sherwood number given by the empirical relation of Ranz and Marshall (1952)

$$Sh = 2 + 0.55 Re_p^{1/2} Sc^{1/3} \quad \text{with} \quad Sc = \frac{\nu_g}{D_{F_2}} \tag{24}$$

The characteristic time scale of the chemical reaction in the bed is the one of the reaction of the fluorine which may be written by

$$\tau_{F_2}^{\text{react}} = -\frac{\alpha_g \rho_g Y_{F_2}^S}{\Psi_{F_2}} \tag{25}$$

The reaction time scale is such that

$$\tau_{F_2}^{\text{react}} \propto \frac{1}{f_c}\frac{\alpha_g}{\alpha_{UF_4}} d_{UF_4} \exp\left(\frac{E_a}{R_g T_{UF_4}}\right)\left[1 + X_{F_2}^S\left(\frac{W_{F_2}}{W_{N_2}} - 1\right)\right]^{-1} \tag{26}$$

A priori evaluation of the reaction time (26) is shown in figure 1 for cases with and without accounting for of the fluorine diffusion in the nitrogen around the reactive particles, with respect to the size of the particles at three different temperatures of the bed. The figure firstly shows that the characteristic time of the reaction increases with the particle diameter, but decreases as higher temperatures are considered in the bed. Secondly, it appears in figure 1 that for higher temperatures of the bed and larger reactive particles, the approximated time scale significantly increases when fluorine diffusion is considered. This indicates that the effect of the fluorine diffusion around UF_4 particles would most probably induce a reduction of the reaction rate (22) for the larger particles at the higher temperatures in the bed. This a priori analysis allow to believe that the effect of diffusion may be efficient in the case of the initial polydispersion of the reactive particles.

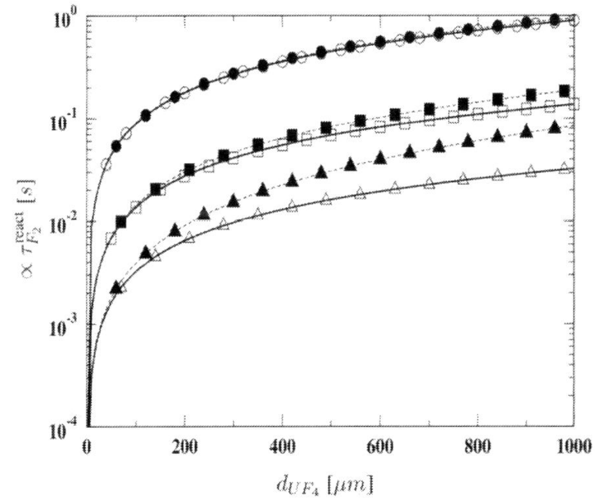

Figure 1: Evolution of the characteristic time scale of the reaction (26) with respect to the mean diameter of the reactive particles at three temperatures within the bed: (○) $400°C$, (□) $500°C$ and (△) $600°C$. Filled and empty symbols refer, respectively, to the chemical reaction gas/particle with and without the external diffusion of fluorine around uranium particles.

3 Computational conditions

Numerical simulations are performed for a laboratory-scale fluidized bed pilot representative of a real reactor. The pilot-plant is typically a column of $1.7\,m$ with diameter about $108\,mm$. The process starts with the prior pre-heating of the bed, during 10 seconds, at $400°C$ using the nitrogen. The second stage of the process consists of the injection of the gaseous mixture of the fluorine (20% in mole fraction) and the nitrogen (80%) within the inlet of the reactor at a lower temperature (about $100°C$) than for the pre-heating stage. The experimental size of the uranium particles spans from $0.5\,\mu m$ to $1600\,\mu m$ while the simulations are carried out with a mean diameter $d_{50} = 250\,\mu m$. Regarding the inert particles, although the size distribution may modify hydrodynamics of the bed, they are assumed to have the same diameter $d_{Al_2O_3} = 150\,\mu m$. The measurements show that a few seconds after the injection of the gaseous mixture, UF_6 is obtained at the outlet of the reactor and rapidly reaches a constant value, corresponding to the 20% mole fraction of F_2 introduced. After three minutes, the amount of UF_6 seamlessly decreases while the leak of the fluorine is observed at the same time at the outlet.

The three-dimensional unsteady simulation of the reactive fluidized bed is performed with NEPTUNE_CFD software using the Eulerian n-fluid modeling approach for fluid-particle turbulent polydispersed reactive flows proposed in this paper and implemented in NEPTUNE_CFD V1.08@Tlse version by IMFT (Institut de Mécanique des Fluides de Toulouse). NEPTUNE_CFD is a computational multiphase flow software developed in the framework of the NEPTUNE project, financially supported by CEA (Commissariat à l'Energie Atomique), EDF (Electricité de France), IRSN (Institut de Radioprotection et de Sûreté Nucléaire) and AREVA-NP. Actually, to account for the requirement imposed by the characteristic transfer time scales that strongly decrease as the reactions proceed, coupled solving of enthalpy and efficient implicit method to account for the source terms were implemented.

Unstructured finite volume method is employed and further details about the numerical resolution methods can be found in Méchitoua et al. (2003). The mesh of the domain (see figure 2) is comprised of around $40,000$ cells such that the mean spacing are about $4\,mm$ in both transverse directions while it is about $19\,mm$ in the streamwise direction. Initial simulation time is $\Delta t = 10^{-3}\,s$. Hydrodynamics simulations not reported here were carried out using meshes comprising two and three times more cells than the mesh presented herein. The initial height of the bed is $0.6\,m$. The inlet velocity of the gas is set to $0.22\,m.s^{-1}$ and $0.12\,m.s^{-1}$ for the pre-heating and the reaction stages, respectively. The initial mass of particles, inside the reactor, is $9\,kg$ of which 10% is uranium tetrafluoride and 90% is the alumina.

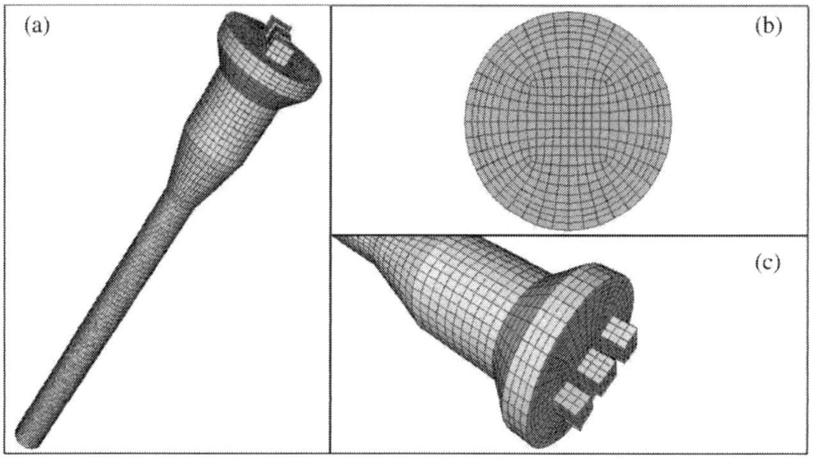

Figure 2: Sketch and mesh of the pilot-plant. (a) Full reactor, (b) bottom-inlet and (c) top of the reactor showing the smokestacks for gas recovery.

During the simulations, locally the gas/particle reaction is not taken into account anymore when the size of the reactive particles is smaller than a given diameter (taken about $1\,\mu m$).

4 Results

4.1 Influence of the external diffusion of the fluorine around the uranium particles

A previous work with the mono-sized initial distribution of uranium particles using the d_{50} (about $250\,\mu\,m$), accounting for the external diffusion of the fluorine concluded that the diffusion of the fluorine within the nitrogen did not limit the reaction rate (Konan et al. (2009)). However, as evidenced by the a priori study concerning the change in the reaction time when both larger reactive particles and higher temperatures are considered (see figure 1), we investigate the effect of the external diffusion of the fluorine with respect to the diameter of the uranium particles. Three diameters are arbitrary chosen: $d^{(1)}_{UF_4} = 500\,\mu\,m$, $d^{(2)}_{UF_4} = 750\,\mu\,m$ and $d^{(1)}_{UF_4} = 1000\,\mu\,m$ to perform initial mono-sized simulations without any heat exchanges. These additional sizes of the uranium particles belong to the measured initial distribution. Plots in figure 3 confirm the previous findings by exhibiting no change in the mole fractions at the outlet of the reactor for the considered diameters. Therefore, even for the larger particles with mean diameter up to four times the d_{50}, globally the diffusion of F_2 in N_2 does not influence the reaction rate. However, as the diffusion is a phenomenon occurring at small scales, its effects may arise for more refined computational meshes. Accordingly more investigations are needed in this way.

4.2 Influence of the bed to wall heat exchange

The efficiency of the thermal transfers is investigated by taking simultaneously into account the radiative heat exchange between the particles (of the dense phase, i.e. the inert alumina particles) and both the convective gas/wall and the radiative particle/wall heat transfers. Indeed, following Konan et al. (2009), the interparticle radiative transfer is not negligible anymore when the thermal exchanges with the wall surface are accounted for. In the framework of the simulations accounting for the bed to wall heat exchange, the thermal wall functions (14) and (15) are applied for the reactor surface defined by $z \leq 0.5\,m$ (in the streamwise direction) while the wall temperature T_w is set to $400°C$ for the considered region.
In a first approximation, the heat transfer coefficient $h_{p\rightarrow w}$ in the thermal wall function (14) is simply assumed to be

UF_4 class #	1	2	3	4	5
Mean diameter (μm)	20	80	150	400	850
Mass ratio	20%	20%	20%	20%	20%

Table 1: Operating conditions for the fictitious initial polydispersion distribution of the uranium particles. The initial total volume fraction of uranium phase is: $\alpha_0 = 4.68 \times 10^{-2}$.

proportional to the volume fraction α_p of the considered dispersed phase such that

$$h_{p \to w} = \kappa \alpha_p \tag{27}$$

κ is a constant coefficient.

The snapshots exhibited in the figure 4: for the adiabatic reactor simulations without interparticle radiation [4-(a)] and the cases with full thermal exchanges [4-(b) to (d)] reveal that the temperature of the particles is uniformly distributed in the bed. This indicates that the reactor operates as a perfect temperature mixer. Furthermore, for the non-adiabatic cases it appears that the particles tend to reach the temperature of the wall (about $400°C$) as the heat exchange coefficient at the wall is increased such that $\kappa = 10^2, 10^3, 10^4$ and 10^5. In figures 4-(b) to (d), we show the snapshots of the alumina temperatures captured at both the height $z = 0.25m$ (in the bed) and the simulation time around $240\,s$ for the four wall-heat exchange coefficients arbitrary chosen in the investigation. However, the chosen values of the constant κ allow getting heat transfer coefficient $h_{p \to w}$ of the same order of those measured in the experimental fluidized bed by Yamada et al. (1995) about $2 - 6 \times 10^3\, W.m^{-2}.K^{-1}$. Figures show that for cases such that $\kappa = 10^4$ and $\kappa = 10^5$, the temperature of the particles in the wall region is almost the same with the wall temperature (the better capture of the wall temperature being for $\kappa = 10^5$) whereas for $\kappa = 10^2$ and $\kappa = 10^3$ the temperatures of the particles are significantly higher than the one of the wall (the worse case being $\kappa = 10^2$). This indicates better heat transfers at the wall for $\kappa = 10^4$ and $\kappa = 10^5$ with a global feedback on the temperature in the reactor expressed by the quasi-uniform temperature of the gas around $400°C$ (the temperature of the wall) whereas it continuously increases at the outlet of the reactor for $\kappa = 10^2$ and $\kappa = 10^3$ like in the adiabatic case (see figure 5). These results indicate the good accounting for of the mechanisms controlling the heat exchanges occurring at the wall and throughout the reactor from the boundary conditions (14), (15) and the modeling of the radiation between particles of the dense phase following the Rosseland approximation's. However, the modeling of the wall-heat exchange coefficient is required.

5 Perspectives

In order to investigate the effect of the initial polydispersion of uranium particles, we performed a pseudo-two-dimensional simulations of the reactor characterized by the two-dimensional reactor with a thin thickness (about two times the mesh spacing in the transverse direction). The particle size distribution used herein is simply a fictitious distribution. Typically, five classes are considered with mean diameter ranging from $20\,\mu m$ to $850\,\mu m$ with the same initial volume fraction (see Table 1). Multi-Class Eulerian approach equations (2), (5) and (17) (including of course the closure modeling of the particle agitation and the covariance fluid/particle, see Gobin et al. (2003) and Ciais et al. (2007)), is used to performed the simulations. Moreover, in accordance with Belt & Simonin (2009) that proposed an extension of the DQMOM formalism of Marchisio & Fox (2005) by accounting for the DQMOM formalism in the derivation of the closure equations, the Eulerian transport equations in the DQMOM framework are the same equations as those in the Multi-class Eulerian approach when there is not coalescence.

Simulations are performed for both the mono-sized initial distribution of uranium particles with the mean diameter set to $d_{50} = 250\,\mu m$ and the fictitious distribution. Figure 6-(a) compares the evolution of the total mass of uranium in the bed between for the mono-sized and the polydispersed cases. Although, both simulations started with the same uranium mass within the reactor, the jump observed before the reaction stage reveals the flight of the small sized particles during the pre-heating stage. The flight that continues during the reaction stage is evidenced by the important diminishing during the first times of the simulations. The total mass of uranium linearly decreases for the initial mono-sized simulations up to the

complete consumption of the uranium within the reactor, whereas the simulations with the initial polydispersion, beyond the flight of the small particles it may be observed at the end of the simulation that the total mass does not linearly decrease anymore.
Figure 6-(b) shows that the temperature linearly increases in both cases and barely changes with the initial polydispersion of uranium.
Regarding the mole fractions, the figure 6-(c) shows a leak of fluorine for initial polydispersion of uranium whereas in mono-sized case, all the fluorine injected within the reactor reacts. Figure 6-(d) compares the amount of the produced UF_6 between the initial poly-dispersed and the mono-dispersed cases. Actually, in the mono-sized simulations the diminishing and the absence of uranium particles within the reactor is expressed by the abrupt stop of the uranium hexafluoride production whereas the amount seamlessly decreases up to zero for the initial poly-dispersed case.

In total it appears that the initial polydispersion of uranium leads to qualitative agreement with measurements in the predictions of the mole fractions at the outlet of the reactor.

6 Conclusion

The conversion of the uranium hexafluoride from the oxidation of the uranium tetrafluoride particles with the fluorine in a fluidized bed reactor is simulated using NEPTUNE_CFD V1.08@Tlse.

The investigation on the external diffusion of the fluorine around the reactive particles with respect to their size demonstrated an non-existent effect on the reaction rate even with larger reactive particles contrary to the expected effect such as suggested by the a priori analysis of the reactive time.

The heat transfers in the reactor has been investigated by considering the radiation between the particles of the dense phases, the convective gas/wall and the radiative particle/wall heat transfers. The radiative particle/wall heat transfers at the wall were parameterized using an heat transfer coefficient proportional to the volume fraction of the particle phase. Results showed a very good efficiency of the mixing inside the bed inducing thus a quasi-uniform temperature of the gas at the outlet of the reactor. However, the heat transfer coefficient at the wall must be accurately modeled. Furthermore, the efficiency of the heat transfers exhibited by the simulations indicates a good accounting for of the main mechanisms controlling the thermal exchanges within the bed.

Additionally, the first simulations using a fictitious initial distribution of the reactive particles demonstrated the good representativeness of the trends of the mole fractions of the fluorine and uranium hexafluoride measured at the outlet of the reactor.

The fully three-dimensional simulations under investigation of the reactive flows within the reactor to predict the trends of the mole fractions of the gaseous species and the temperature is extremely promising because of the ability to account for the main mechanisms controlling the reactive flows and the initial polydispersion using the Multi-class Eulerian approach.

Acknowledgements

This work was granted access to the HPC resources of CINES under the allocation 2009-X2009026012 made by GENCI (Grand Equipement National de Calcul Intensif).

References

Abbas, M., Climent, E., Simonin, O. "*Shear-Induced Self-Diffusion of Inertial Particles in a Viscous Fluid*", Physical Review E, Vol. 79, N°3 (2009).

Belt, R., Simonin, O. "*Quadrature Method Of Moments for The Pdf Modeling Of Droplet Coalescence In Turbulent Two-Phase Flows*". In Proceedings of FEDSM2009, ASME 2009 Fluids Engineering Division Summer Meeting August 2-5, 2009, Vail, Colorado USA (2009).

Boelle, A., Balzer, G., Simonin, O. "*Second-Order Prediction of the Particle-Phase Stress Tensor of Inelastic Spheres in Simple Shear Dense Suspensions*". In Gas-Particle Flows, ASME FED, Vol. 228, pp 9-18.

Ciais, V., Martin, R., Neau, H., Simonin, O. "*3D Multiparticle Simulation of Volcanic Dome Collapse and Pyroclastic Flow: Application to 1984 Event at Merapi Volcano*". 6th ICMF, Leipzig, Germany, paper S7_Mon_D_10 (2007).

Fan, R., Marchisio, D. L., Fox, R. O. "*Application of the direct quadrature method of moments to polydisperse gas-solid fluidized beds*". Powder Technology, 139, pp. 7-20 (2004)

Gobin, A., Neau H., Simonin, O., Llinas, J. R., Reiling, V., Selo, J. L. "*Fluid dynamic numerical simulation of a gas phase polymerization reactor*". International Journal for Numerical Methods in Fluids, 43, pp. 1199-1220 (2003).

Janov, J., LePage, A.H. "*Uranium hexafluoride production in a 150 mm diameter fluidised bed pilot plant*". Chemeca 81, 9th Australasian Conference on Chemical Engineering, Christchurch, New Zealand, Aug. 30 to Sept. 4 (1981).

Konan, N. A., Neau, H., Simonin, O., Dupoizat, M., Le Goaziou, T. "*3D Unsteady Multiphase Simulation of Uranium Tetrafluoride Particle Fluorination in Fluidized Bed Pilot*". In Proceedings of the 20th International Conference of Fluidized Bed Combustion, Xi'an, China, May 18-21. Edited by G. Yue, H. Zhang, C. Zhao and Z. Luo (Springer) (2009).

Marchisio, D. L., Fox, R. O. "*Solution of population balance equations using direct quadrature method of moments*". Aerosol Science 36, pp. 43-73 (2005).

Méchitoua, N., Boucker, M., Laviéville, J., Pigny, S., Serre, G. (2003). International Conference on Supercomputing in Nuclear Applications, SNA'2003, Paris, France (2003).

Randrianarivelo, T. N., Neau, H., Simonin, O., Nicolas, F. "*3D Unsteady Polydispersed Simulation of Uranium Hexafluoride Production in a Fluidized Bed Pilot*". 6th ICMF, Leipzig, Germany, paper S6_Thu_A_46 (2007).

Simonin, O., Deutsch, E., Minier, J.P. "*Eulerian Prediction of the Fluid/Particle Correlated Motion in Turbulent Two-Phase Flows*". Applied Scientific Research, Vol. 51, pp 275-283 (1993).

Simonin, O., Konan, N. A. "*Development of Heat Transfer Wall Boundary Conditions for Gas-Solid Dense Fluidized Bed CFD Simulation*". Submitted to 7th ICMF, Tampa, Florida, USA, May 30 - June, 4 (2010).

Yamada, J., Kurosaki, Y., Satoh, I., Shimada, K. "*Radiative heat exchange between a fluidized bed and heated surface*". Experimental Thermal and Fluid Science, 11, pp. 135-142 (1995).

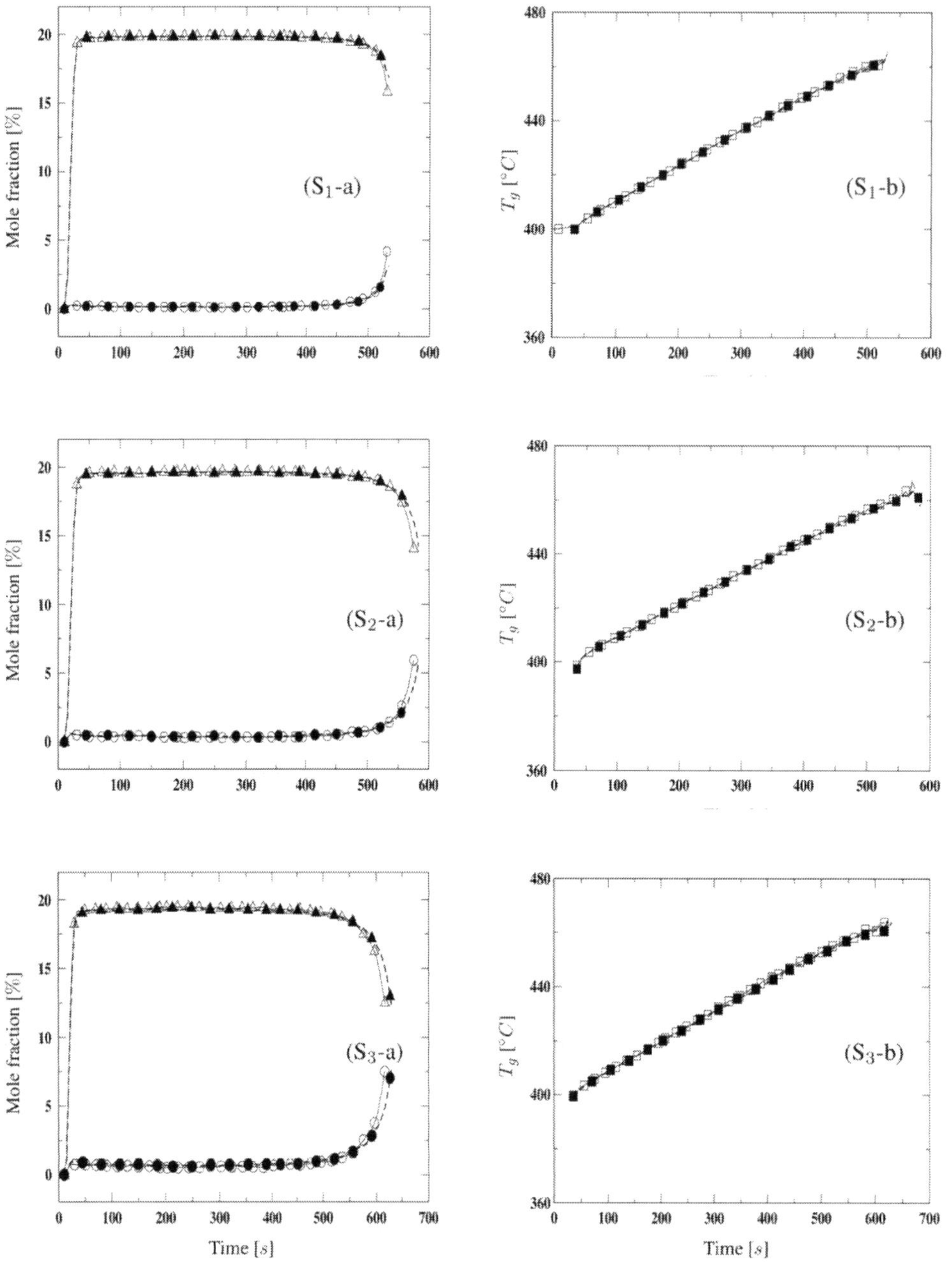

Figure 3: Influence of the fluorine external diffusion around particles for three initial mono-sized distributions of the uranium: (S$_1$-) $d^{(1)}_{UF_4} = 500\,\mu m$, (S$_2$-) $d^{(2)}_{UF_4} = 750\,\mu m$ and (S$_3$-) $d^{(1)}_{UF_4} = 1000\,\mu m$. (a) and (b) refer to the mole fraction and the temperature of the gaseous mixture at the outlet of the reactor. Symbols are: (○) mole fraction of F_2, (△) mole fraction of UF_6 and (□) temperature of the gas. Filled and empty symbols refer to the external diffusion and free-diffusion cases.

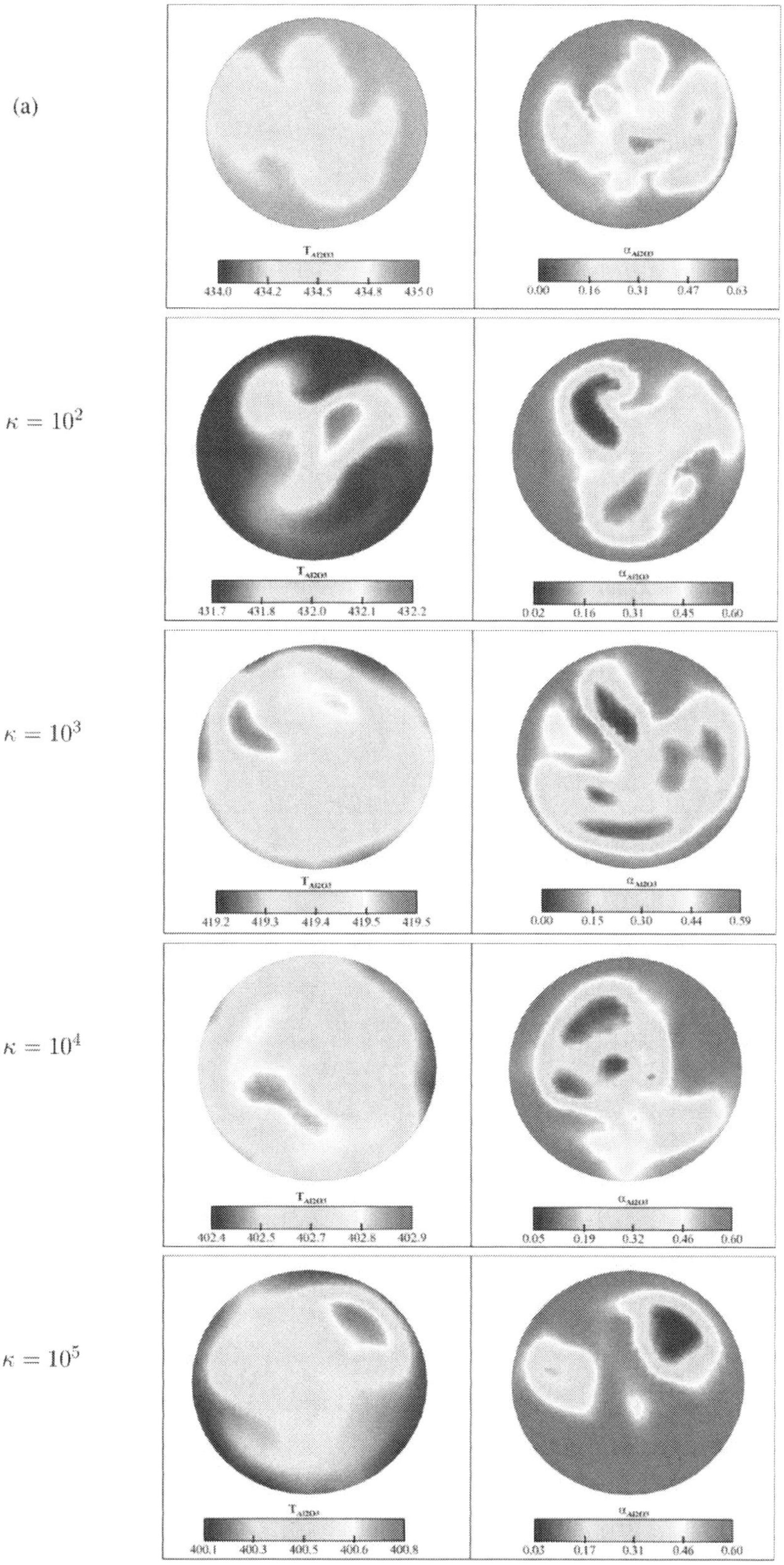

Figure 4: Effect of the heat wall exchange on the instantaneous distribution of the temperature of alumina dense phase within the bed. (Left: temperature and Right: volume fraction). Snapshots are captured at $z = 0.25m$ and $t = 240\ s$: (a) adiabatic simulations, (b) to (e) refer to the simulations with convective gas/wall and both radiative particle/particle and particle/wall heat exchanges. κ is the parameter of the heat transfer coefficient at the wall (27).

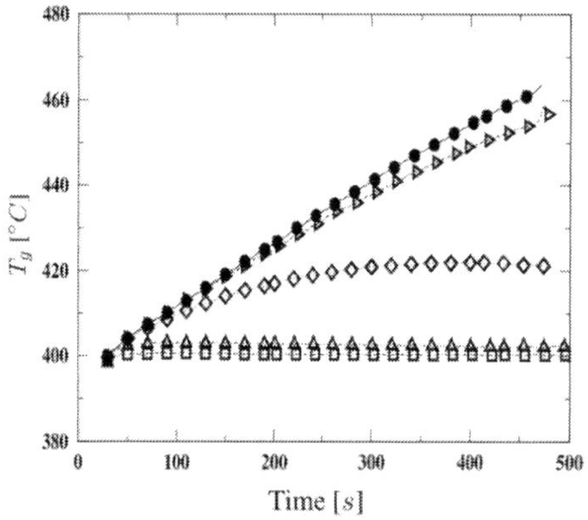

Figure 5: Effect of the heat exchange in the temperature of the gas at the outlet of the reactor. (•) adiabatic simulations while empty symbols refer to heat exchange cases: (▷) $\kappa = 10^2$, (◇) $\kappa = 10^3$, (△) $\kappa = 10^4$ and (□) $\kappa = 10^5$.

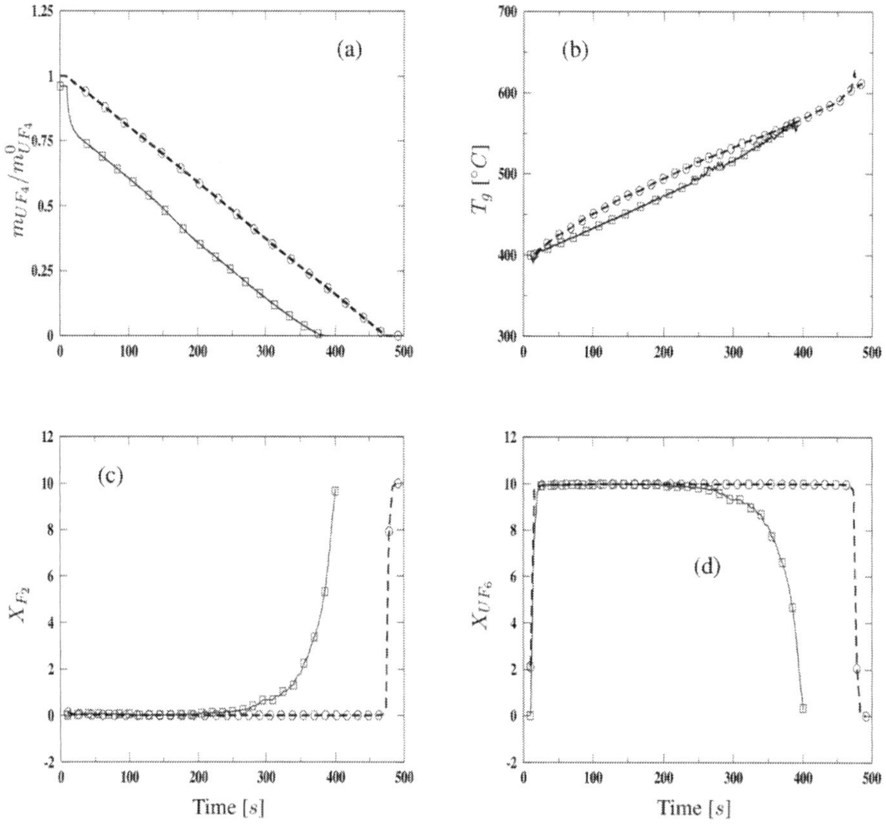

Figure 6: Influence of the initial polydispersion of UF_4 particles in the conversion of the gaseous UF_6. (a) Total mass of UF_4 particles within the reactor, (b) Temperature of the gas, (c) Mole fraction of F_2 and (d) Mole fraction of UF_6 at the outlet. Symbols are: (◯) for the mono-sized and (□) for the polydispersion of the UF_4.

Axial Segregation in Bubbling, Gas-Fluidized Beds with Continuous Size Distributions

Jia Wei Chew, Chemical and Biological Engineering, University of Colorado at Boulder, Boulder, CO

Jeffrey R. Wolz, Chemical and Biological Engineering, University of Colorado at Boulder, Boulder, CO

Christine M. Hrenya, Chemical and Biological Engineering, University of Colorado at Boulder, Boulder, CO

Due to a lack of understanding of polydisperse flows, processes employing particulate flows often operate below design capacity and exhibit undesired flow behavior. Previous work, including experiments, theoretical efforts, and simulations, has largely focused on binary mixtures (two particle species which vary in size and/or density) to characterize and better understand the segregation behavior. Relatively little work has been performed, however, for continuous size distributions, despite their practical importance. Experiments involving Geldart Group B particles with a continuous size distribution have been carried out under atmospheric conditions. The system under consideration is a gas-fluidized bed, in which local segregation measurements are taken. Sand of various widths of Gaussian and lognormal distributions were fluidized, then axial segregation profiles were obtained from frozen-bed sectioning. Experimental results show that (i) the mean particle diameter decreases with increasing height within the bed, (ii) the degree of segregation increases with increasing width of the particle size distribution, and (iii) the shape of the local size distribution (i.e., Gaussian or lognormal) is preserved with respect to the overall distribution, even in the presence of size segregation. These experimental findings agree with discrete particle simulations carried out by Dahl et al. (2003) for granular systems and Dahl and Hrenya (2005) for gas-fluidized systems, and are expected to simplify subsequent continuum modelling efforts. A surprising fourth finding, however, is that the degree of segregation is non-monotonic with respect to the width of lognormal distributions.

References:

Dahl, S.R., Clelland R., and Hrenya C.M. (2003). Three-dimensional, rapid shear flow of particles with continuous size distributions. Powder Technology 138 (1): 7-12.

Dahl, S. R., and Hrenya, C.M. (2005). Size segregation in gas-solid fluidized beds with continuous size distributions. Chemical Engineering Science 60: 6658-6673.

See more of this Session: Gas/Solid Mixing and Heat/Mass Transfers in Fluidized Beds
See more of this Group/Topical: Particle Technology Forum

Mass Transfer in Tapered Fluidized Beds

Duvvuri Subbarao , Makhan Lal Das[1]
Department of Chemical Engineering,
Universiti Teknologi PETRONAS, Bandar Seri Iskandar, 31750, Tronoh, Perak, Malaysia
[1]*Department of Chemical Engineering, Assam College of Engineering, Guwahati, India*

Abstract

Tapered beds can handle fluidization of particles with wide size distribution more effectively. As the area of cross-section increases with height, higher gas velocity at the gas distributor reduces the chances of de-fluidization of larger particles while lower gas velocities at the top ensure less elutriation of fines. Tapered beds can be useful for processes such as combustion of coal, roasting of ores, polymerization, granulation of fines, drying of wet powders in which particle size may change with time. In all these processes gas to particle mass transfer plays a vital role. Mass transfer rates from naphthalene balls to fluidizing air in tapered beds of inert materials are measured and mass transfer coefficients are estimated. Effect of size and density of the mass transferring particle with respect to properties of bed particles is investigated. A model based correlation is developed to explain the observations.

Keywords: Fluidized beds, Tapered fluidized beds, Mass transfer

Introduction

There are many processes where particles need to be processed on a continuous basis such as petroleum refinery reactors with decaying catalysts, or polymerization reactors/agglomerators with particles growing in size, or coal combustors with particles reducing in size. Fluidized beds are widely used as they offer the advantages of good particle mixing (resulting in bed isothermality) and easy handling of particles. Conventionally particles are fluidized in vessels of uniform cross section. These units have some disadvantages such as entrainment of fines and formation of local defluidized zones if the bed material varies widely in size and/or densities of particles. To overcome some of these difficulties, other geometries such as tapered beds are receiving attention in recent times. Tapered fluidized beds are reported to be in use for combustion/gasification of coal, roasting of sulfide ores, incineration of solid wastes, drying and agglomeration etc,. In tapered beds, cross-sectional area of the bed increases with height along the flow and thereby reducing velocity of the fluidizing medium along the height. This reduces the possibility of entrainment of fines at the top bed surface and defluidization of larger particles at the bottom entry point.

Gas to particle mass transfer can restrict rate of reactions with high reaction kinetics such as coal combustion. Reviewing investigations on mass transfer in fluidized bed combustion of coal, LaNauze (1985) observed that mass transfer coefficient is a function of carbon particle size, inert bed particle size, fluidizing velocity and gas properties. To the best of our information, studies on gas to particle mass transfer in tapered fluidized beds are not yet reported. Experimental measurements on gas to particle mass transfer in tapered fluidized beds reported by Das (1999) are presented in this paper.

Experimental Study:

Air supplied by an air cooled twin-lobe blower, with online pressure regulator, was metered through a calibrated orifice meter enters the plenum chamber below the fluidized bed. Two layers of canvas cloth held between flanges were used as gas distributor for tapered beds. Three dimensional tapered fluidized beds were made by rolling hard paper and fixed to the plenum chamber. Transparent plastic sheets were fixed on two slits cut on opposite sides of the wall to facilitate observation of the expanded fluidized bed height. Sintered porous plate distributor was used for cylindrical beds. Dimensions of fluidized beds in which mass transfer was investigated are summarized in Table 1

Table 1. Dimensions of the beds.

Type of Bed	Apex Angle degree	Diameter (cm)		Bed Height (mm)
		Bottom	Top	
Tapered Bed	5.2	2.9	7.4	50.0
Tapered Bed	10.5	2.2	11.4	50.0
Tapered Bed	24	2.3	16.3	33.0
Cylindrical bed	0	4.8	4.8	37.7

Closely sieved silica sand or mustard seed were used as the inert bed material and their properties are summarized in Table 2. Minimum fluidization velocity was obtained from experimental measurements in a cylindrical bed.

Table 2. Properties of inert bed material used.

Material	D_p, cm	ρ_p, g/cm^3	u_{mf}, cm/s	u_t, cm/s
Sand	0.0115	2.6	1.8	52.2
Sand	0.0196	2.6	6.1	88.2
Sand	0.0427	2.6	16.2	191.1
Mustard	0.0725	1.10	18.6	300.4

Sublimation of naphthalene balls was used for mass transfer studies. Naphthalene balls of size ranging from 0.2 cm to 1.56 cm were made from bigger naphthalene balls by crushing them to different sizes and filing them to near spherical shape. The density of naphthalene balls is 1.08 g/cm3 and molecular weight is 128.18.

Experimental method

Known quantity of inert bed material is charged to the bed, fluidized and defluidized to note the initial bed height. The bed is then fluidized at the desired gas flow rate and the average bed height is noted. Bed temperature is noted. A naphthalene ball of known weight is dropped into the bed at time t=0. After a specified period, the air supply was cut off. The test naphthalene ball in the bed is recovered, stored in a air tight sample bottle and weight loss is recorded. Mass transfer rate is estimated from the weight loss during the specified time. Mass transfer coefficient is estimated

$$\frac{\Delta W_{NB}}{M_N \Delta t} = k_g \frac{\pi D_{NB}^2}{4} \frac{p_s}{RT_b}$$

$$D_{NB} = \left(\frac{6}{\pi} \frac{W_{NB}}{\rho_{NB}} \right)^{\frac{1}{3}}$$

The vapor pressure of solid naphthalene is calculated using Antoine Equation. The experiment is repeated in beds of different apex angles (and cylindrical beds as well) using different bed materials and naphthalene balls of different sizes

Results and Discussion:

Table 3 summarizes the experimental conditions at which gas to particle mass transfer measurements using naphthalene particles of different sizes as a function of gas flow rate were performed.

Table.3

Experiment No.	Apex Angle	Bed Material	Dp	Initial Bed Height	Bed Temperature
1	10.5	Mustard	0.0725	5	34
2	10.5	Sand	0.0196	5	34
3	10.5	Sand	0.0427	5	31.5
4	24	Sand	0.0196	5	34.5
5	5.2	Sand	0.0196	5	34.5
6	0	Sand	0.0196	5	35
7	0	Sand	0.0115	5	30.5
8	0	Mustard	0.0725	5	29

As the cross sectional area increases along the bed height, gas velocity decreases with height. To estimate the representative gas velocity, equation for average cross sectional area for flow is obtained through the application of Ergun's equation for flow through tapered beds as

$$A_{avg} = \frac{1}{3}\left(A_{in} + A_{out} + \sqrt{A_{in} A_{out}} \right) \quad (1)$$

and gas velocity is estimated as

$$u_{og} = \frac{Q_g}{A_{avg}} \quad (2)$$

Experimental observations on mass transfer from naphthalene particle to gas in fluidized beds of various apex angles operating with bed particles of 0.0196 cm size is shown in Fig 1. There appears to be no specific dependence with apex angle.

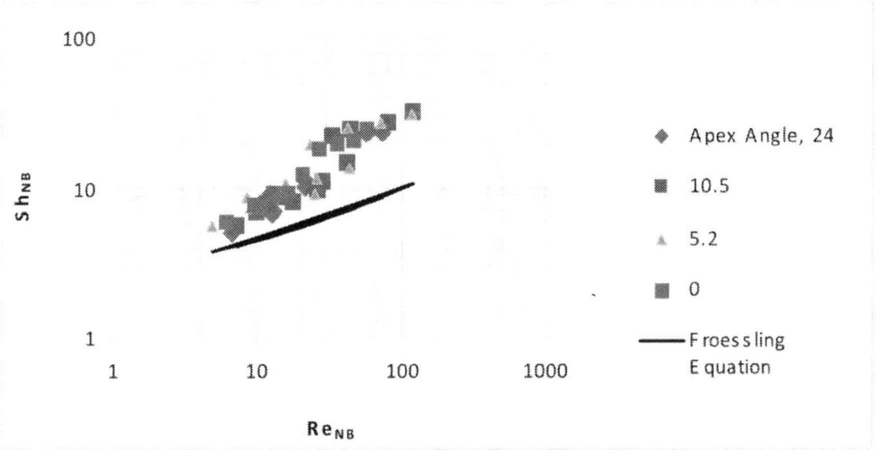

Fig.1. Mass transfer from naphthalene particle to gas in fluidized beds of various apex angles operating with bed particles of 0.0196 cm size.

Experimental observations on mass transfer from naphthalene particle to gas in tapered fluidized beds (apex angle of 10.5) operating with different bed particles is shown in Fig.2. Sherwood numbers appear to be higher for finer bed particles.

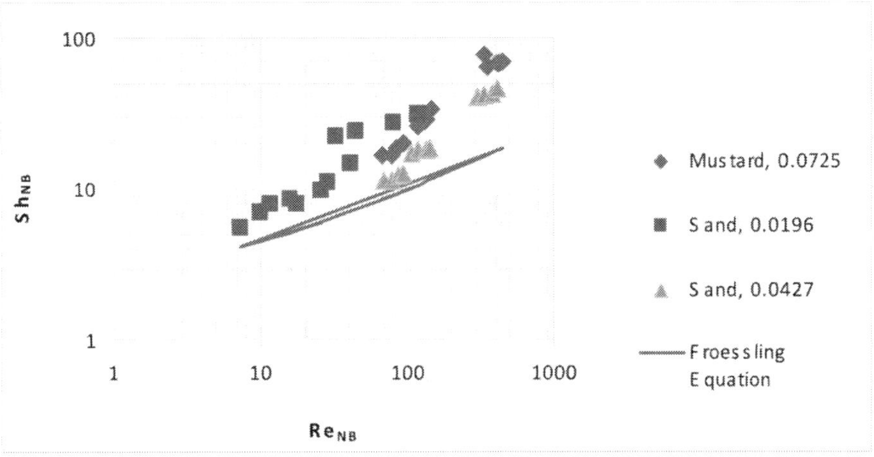

Fig.2. Mass transfer from naphthalene particle to gas in tapered fluidized beds (apex angle of 10.5) operating with different bed particles.

A simple Model

Mass transfer from a single naphthalene particle of size D_{NB} to a flowing gas is generally explained through Higby's penetration theory as

$$k_g = \left(\frac{4D_{AB}}{\pi\tau} \right)^{0.5} \tag{3}$$

Time of exposure can be estimated as

$$\tau = \frac{Path\ length}{Gas\ Velocity} \tag{4}$$

If the same naphthalene particle, immersed in a bed of fluidized particles, is exposed to gas flowing at a superficial gas velocity of u_{og} , the path length is reduced

$$Path\ length = \frac{D_{NB}}{D_{NB}/D_p} \tag{5}$$

And gas velocity is

$$Gas\ velocity = \frac{u_{og}}{\varepsilon} \tag{6}$$

As the naphthalene particle is surrounded by bed particles, effective diffusion coefficient is reduced as

$$Effective\ diffusion\ coefficient = D_{AB}\varepsilon \tag{7}$$

Putting these factors together

$$k_g = \left(\frac{4D_{AB}\varepsilon}{\pi} \frac{u_{og}/\varepsilon}{D_{NB} \, D_p/D_{NB}} \right)^{0.5} = \left(\frac{4D_{AB}u_{og}}{\pi D_p} \right)^{0.5} \tag{8}$$

In terms of dimensionless numbers

$$\frac{k_g D_p}{D_{AB}} = \left(\frac{4D_{AB}u_{og}}{\pi D_p} \right)^{0.5} \frac{D_p}{D_{AB}} = \left(\frac{4}{\pi} \frac{u_{og}D_p\rho}{\mu} \frac{\mu}{\rho D_{AB}} \right)^{0.5}$$

$$Sh_p = c_1 \mathrm{Re}_p^{0.5} Sc_p^{0.5} \tag{9}$$

This approach suggests that mass transfer coefficients depend on bed particle size and not on the size of mass transferring particles.

Conclusions:

1. Mass transfer in tapered fluidized beds can be as good as the one expected in cylindrical fluidized beds.

2. Mass Transfer depends on the size of inert fluidized bed particles to a certain extent.

Acknowledgement

The authors thank Universiti Teknologi PETRONAS for the encouragement and financial support for carrying out this work.

Nomenclature:

A_{avg} Average cross sectional area for gas flow
A_{in} Area of cross section for flow in the inlet to the bed
A_{out} Area of cross section for flow at the top of the bed
D_{AB} Diffusion Coefficient
D_{NB} Diameter of Napthalene ball
D_p Diameter of inert particles in the fluidized bed
k_g Mass transfer coefficient
M_N Molecular weight of Naphthalene
P_s Vapor pressure of Naphthalene
Q_g Gas flow rate
R Gas Constant
Re_{NB} Reynold Number
Re_p Reynold Number
Sh_{NB} Sherwood Number
Sh_p Sherwood Number
Sc Schmidt Number
T_b Temperature of the bed
U_{og} Superficial gas velocity based on average area
W_{NB} Weight of naphthalene particle

Greek Letters

ΔW_{NB} Naphthalene particle Weight loss
Δt Time of mass transfer process
ε Bed voidage
μ Viscosity of gas
ρ Density of gas
T Time of exposure for mass transfer

References

LaNauze, R.D., Fundamentals of coal combustion, in Fluidization (2nd Edition) edited by Davidson, J.F, Clift R and Harrison, D., academic press (1985) 631-674

Das, Makhan Lal, Studies in gas-solid tapered fluidized beds, Ph.D. Thesis, IIT Delhi, India (1999)

Froessling, N, Gerland Beitr.Geophys, 52, (1938)170

Improving Fluidized-Bed Performance by Optimizing the Particle Size Distribution

J. Ruud Van Ommen, DelftChemTech, Delft University of Technology, Delft, Netherlands

R. Beetstra, DelftChemTech, Delft University of Technology, Delft, Netherlands

Naoko Ellis, Department of Chemical and Biological Engineering, University of British Columbia, Vancouver, BC, Canada

John Nijenhuis, DelftChemTech, Delft University of Technology, Delft, Netherlands

Manipulating the structure of multiphase reactors introduces extra degrees of freedom allowing decoupling of conflicting design objectives, such as high mass transfer versus low pressure drop (packed bed) and high gas flow rate versus small bubble size (fluidized bed). In this way, structuring facilitates reaching goals such as process intensification and more sustainable operation. At Delft University of Technology, much effort has been put over the past years in the research and development of structured reactors, such as structured fixed bed reactors, structured slurry bubble columns, and structured gas solid-fluidized beds. In this paper, we will focus on the latter category.

In a gas-solid fluidized bed, structuring can be imposed either (1) via the gas phase or (2) via the particles. Moreover, we can modify a) the geometry or b) the dynamic behaviour of each of the two phases. This gives in total four possible combinations. In this paper, we will discuss the effect of Approach (2).a): changing the geometry of the particles. Our goal is to design mixtures of particles with optimal fluidization properties. Using high-throughput experimentation, the relevant hydrodynamic data can be obtained in a limited period of time. This approach is demonstrated by assessing the influence of the particle size distribution on fluidized bed hydrodynamics of Geldart A powders. By manipulating the width of the particle size distribution of alumina powder the bubble diameter is reduced up to 40%. The addition of fines to a given particle size distribution also decreases the bubble diameter up to 40%, whereas the addition of coarse particles hardly influence the bubble size. At low gas velocities the bubble size was found to increase with fines addition. We will show that this approach can equip chemical engineers with practical tools to improve the operation of fluidized bed reactors.

See more of this Session: Gas/Solid Mixing and Heat/Mass Transfers in Fluidized Beds
See more of this Group/Topical: Particle Technology Forum

Preparation of Pharmaceutical Micro- and Nano-Composites Using Supercritical Fluid Extraction of Emulsions

Johannes Kluge, Institute of Process Engineering, ETH Zurich, Zurich, Switzerland

Francesco Fusaro, Institute of Process Engineering, ETH Zurich, Zurich, Switzerland

Nathalie Casas, Institute of Process Engineering, ETH Zurich, Zurich, Switzerland

Gerhard Muhrer, Chemical and Analytical Development, Novartis Pharmaceuticals, Basel, Switzerland

Marco Mazzotti, Institute of Process Engineering, ETH Zurich, Zurich, Switzerland

Particles of the biocompatible and biodegradable polymer poly(lactic-co-glycolic) acid, PLGA, constitute a widely studied pharmaceutical carrier system used to achieve e.g. controlled release or targeted delivery of active pharmaceutical ingredients. The manufacturing of the corresponding drug-polymer nano-composites represents an interesting challenge to drug formulation technology. The present study evaluates a novel technique for the production of such particles, namely supercritical fluid extraction of emulsions, SFEE.

For SFEE processing, drug and polymer are mixed in an organic solution that is dispersed in water, forming an oil-in-water emulsion stabilized by a suitable surfactant. In a second step, the emulsion is mixed with supercritical CO2, thereby extracting the organic solvent from the emulsion droplets, leading to the co-precipitation of drug and PLGA as composite particles. These form a stable and solvent-free suspension in the inert aqueous phase, from which they may be recovered upon depressurization.

The present study investigates the influence of operating conditions on the achieved particle size distributions. Further, the incorporation of various pharmaceuticals has been achieved by applying different strategies depending on the nature of the drug molecule. For instance, the encapsulation of water-soluble lysozyme may be achieved through double-emulsion and suspension-emulsion systems. On the other hand, hydrophobic compounds such as the anti-inflammatory drug ketoprofen (KET) are encapsulated in PLGA using a common organic solution.

For the latter system, also the product stability in aqueous suspension has been investigated. While composites of KET and PLGA are stable at low KET content, we have observed the recrystallization of KET in supersaturated co-formulations exceeding the solubility limit of KET in PLGA. By equilibration of PLGA particles with KET crystals, the solubility of KET in PLGA may be studied. The method has been used to compare the solubilities of racemic KET and enantiopure S-Ketoprofen, and to assess their dependency on temperature.

See more of this Session: Particle Formation in Supercritical Fluids for Food and Pharmaceuticals
See more of this Group/Topical: Particle Technology Forum

A New Granulation Method in Supercritical Fluid

Gary Liu, Particle Technology, DuPont, Wilmington, DE

Erik Gommeren, DuPont, Wilmington, DE

Oliver Gutsche, Stine Haskell Research Center, DuPont, Newark, DE

Many techniques are known for preparing granules from powdered materials such as wet granulation, solvent granulation and melt granulation. Wet granulation and solvent granulation require the addition of a liquid binder which aggregates the active materials, carriers and other ingredients into granules. After granulation, the liquid generally must be removed by a separate drying step. Melt granulation is similar to wet granulation, but uses a low melting point solid material as a binder. The solid binder in melt granulation is melted and acts as a liquid binder thereby aggregating the powdered active material and other ingredients into granules. Binder thereby, is permanently incorporated into the granules when the granules cool.

Each of these granulation techniques has certain limitations or disadvantages. Wet granulation requires a liquid to be added, and the liquid must subsequently be removed. Therefore, a drying step is needed which further complicates the manufacturing process and risks more contamination. Melt granulation on the other hand avoids the problems associated with adding liquids by incorporating a low melting solid binder. However, the low melting point binder must be heated to at least its softening point and melted during the granulation process, which will strictly limit the number of active materials which can be granulated.

This paper presents a new method for the granulation of fine particles in supercritical CO2 environment, which has the effect of suppressing melting point on organic crystals, as well as the effect of plasticizing certain polymers. Both jet-milled lactose and a crop protection active have been successfully agglomerated into free-flowing granules at temperature as low as 39ºC using a polymer binder: polyethylene glycol (PEG 8000), which has a melting point above 60ºC. Therefore the new granulation method can overcome some limitations that conventional granulation methods have, and being capable of handling especially thermally sensitive active materials used in pharmaceutical and food industries.

See more of this Session: Particle Formation in Supercritical Fluids for Food and Pharmaceuticals
See more of this Group/Topical: Particle Technology Forum

A Novel Technique for the Production of Drug Nanoparticles and Their Encapsulation Using Rapid Expansion of Water-in-Supercritical CO2 Microemulsion

Sunmook Lee, Institute of Industrial Science, The University of Tokyo, Tokyo, Japan

Yuichiro Otsuji, Institute of Industrial Science, The University of Tokyo, Tokyo, Japan

Masahiro Ikeda, Department of Mechanical and Biofunctional Systems, Institute of Industrial Science, The University of Tokyo, Tokyo153-8505, Japan

Atsushi Tsutsumi, Institute of Industrial Science, The University of Tokyo, Tokyo, Japan

A novel technique for the production of drug nanoparticles and their encapsulation was exploited using rapid expansion of water-in-supercritical CO2 microemulsion. The aqueous cores of reverse micellar nanodroplets filled with gadolinium-diethylenetrianminepenaacetic acid (Gd-DTPA), an anticancer drug used in neutron capture therapy, were formed and dispersed in the supercritical CO2. Paraffin, which forms a homogeneous phase with the supercritical CO2, functioned as a coating material of the core drug material Gd-DTPA. Rapid expansion of the supercritical microemulsion into the atmosphere was carried out for the simultaneous processes of the production and the encapsulation of Gd-DTPA with paraffin in nanosized dimension. The morphology of particle samples was examined with SEM and the coating of the particles were investigated with TEM and EDS analysis. The results showed that this technique is very promising for the production and the encapsulation of nanoparticles in nanoscale.

See more of this Session: Particle Formation in Supercritical Fluids for Food and Pharmaceuticals
See more of this Group/Topical: Particle Technology Forum

Solubility and Concentration Effects On Precipitation in Supercritical Antisolvent Processes

Rajeshwar B. Chinnawar, Chemical Engineering, Auburn University, Auburn University, AL

Daniel L. Obrzut, Chemical Engineering, Auburn University, Auburn University, AL

Christopher B. Roberts, Chemical Engineering, Auburn University, Auburn University, AL

Steve R. Duke, Chemical Engineering, Auburn University, Auburn University, AL

Studies were performed in a high magnification visualization system to determine the effects of solubility, concentration, and mass transfer on size distributions of particles produced using the supercritical antisolvent (SAS) precipitation process. The precipitation vessel includes clear windows for process visualization and filters for particle collection. Visualizations of the spray were obtained at various distances from the nozzle to measure the jet breakup lengths and droplet sizes. Solutions of polymethyl methacrylate in the solvents ethanol and acetone were processed via SAS using compressed carbon dioxide. Also, copolymers from the monomers methyl methacrylate and vinyl pyrrolidone were processed using SAS to observe the effects of their solubility in ethanol. The solubilities of the copolymers were influenced by the methyl methacrylate to vinyl pyrrolidone ratio. The process operating conditions (pressure, temperature, density) affects SAS spray characteristics and the resulting particles. Furthermore, studies have been conducted using planar laser induced fluorescence (PLIF) in SAS precipitation with a fluorescent molecule, poly[methylmethacrylate-co-(7-(4-trifluoromethyl)coumarin methacrylamide)], to obtain in situ concentration fields of the solute to characterize the mass transfer. The precipitated particles from these studies were analyzed with a scanning electron microscope and particle size distribution analysis. Relations between the process characteristics, the solvent-solute pairings, and the particle properties will be discussed.

See more of this Session: Particle Formation in Supercritical Fluids for Food and Pharmaceuticals
See more of this Group/Topical: Particle Technology Forum

The Relevance of Solvent Condensation During Rapid Expansion of Supercritical Solutions

Victor Stepanov, Munitions Engineering and Technology Center, US Army, RDECOM ARDEC, Picatinny, NJ

Lev N. Krasnoperov, Department of Chemistry and Environmental Science, New Jersey Institute of Technology, Newark, NJ

Abstract

The advantages of using supercritical fluids as solvents in chemical processing have been demonstrated for a wide spectrum of processes such as extraction, synthesis, and crystallization. These novel methods take advantage of the unique characteristics of supercritical fluids such as gas-like diffusivity and pressure dependent solvent strength. Such properties offer unique processing capabilities not feasible with conventional liquid solvents.

In the field of recrystallization, rapid expansion of supercritical solutions (RESS) has received much attention due to the unparalleled ability to generate nanoparticles of a wide range of materials with exceptional control of the particle size and purity, while processing at relatively mild conditions without detrimental effects on the final product. Since the solvents utilized (e.g. CO2) are typically gases at ambient temperature and pressure, another benefit is the ability to produce neat product free of residual solvents. Pharmaceutical compounds are a good example of a type of materials that can benefit from RESS as many such compounds tend to be thermally labile and elimination of toxic solvents from processing is greatly desired.

Understanding of the particle formation mechanism during RESS has significantly evolved over the past two decades. Mathematical models of fluid expansion and the resulting solute nucleation and particle growth have increased in complexity and account for more of the phenomena occurring within the expanding jet. Nevertheless, due to the complex nature of such expansion processes an accurate mathematical representation does not yet exist. Currently the predicted particle size is typically significantly smaller than that experimentally obtained.

To date, as a result of added complexity, models do not account for a significant aspect of RESS, in particular the condensation of solvent during expansion. As a result, processing conditions considered in such models lie outside of the relevant range for many processes of interest.

In the present paper the impact of CO2 condensation on the particle growth is examined. In this work the recrystallization of cyclotrimethylene trinitramine (RDX) from CO2 was investigated. The pre-expansion temperature and pressure were studied in the range from 50 to 90oC and around 100 to 350 bar respectively. The effect of the discharge pressure was studied in the range from atmospheric pressure to around 60 bar. Within the span of the experimental conditions both liquid and solid-phase solvent condensate was encountered. Based on empirical data as well calculations particle growth mechanisms are proposed.

See more of this Session: Particle Formation in Supercritical Fluids for Food and Pharmaceuticals
See more of this Group/Topical: Particle Technology Forum

Production of Hydrogel Nanoparticles by Inverse-Miniemulsion Polymerization and Supercritical CO2 Drying

Melinda G. Hemingway, Chemical Engineering, Auburn University, Auburn, AL

Ram B. Gupta, Chemical Engineering, Auburn University, Auburn, AL

Hydrogel nanoparticles can be successfully produced by polymerization in inverse miniemulsions, for use in a variety of applications including diagnosis, drug delivery, separation, and absorption. Unfortunately, conventional drying techniques result in particle agglomeration due to the sticky nature of the wet hydrogel particles. This work utilizes supercritical CO_2 drying to obtain free flowing dry hydrogel nanoparticles. Using inverse miniemulsion technique (cyclohexane continuous phase, water dispersed phase, and a non-ionic surfactant), nanoparticles (500 nm diameter) of polyacrylamide are produced. The polymerized miniemulsion is injected into supercritical CO_2 which results in rapid removal of cyclohexane, water, and surfactant. The morphology, particle size and size distribution of the nanoparticles are determined using dynamic light scattering and scanning electron microscopy. The effect of supercritical CO_2 flow on the drying efficiency is examined and compared with via freeze-drying and spray-drying techniques. Dry nanoparticles obtained from of supercritical CO_2 drying show much lower agglomeration.

See more of this Session: Particle Formation in Supercritical Fluids for Food and Pharmaceuticals
See more of this Group/Topical: Particle Technology Forum

Production of PLGA Particles by Using Supercritical Carbon Dioxide Anti-Solvent Method

Ruey Chi Hsu   S Hang Wang
Department of Chemical and Materials Engineering, Chang Gung University
Kwei-Shan, Tao-Yuan, Taiwan 333, R.O.C.

Abstract

Synthetic biodegradable poly(D,L-lactide-co-glycolide)(PLGA) which is derived from lactic acid (LA) and glycolic acid (GA), and their copolymers . They had been widely used for temporary therapeutic application such as drug delivery devices. The biodegradable material used for drug delivery devices which is specified on 5-10μm. In this study, we investigated the use of supercritical CO_2 as anti-solvent in PLGA/ ethyl acetate solution system to manufacture PLGA particle. In order to determine the pressure-volume expansion-temperature behavior of the ethyl acetate/supercritical CO_2, the modified Peng-Robison equation of state was used. A novel experimental apparatus, which is involving, view window and stirrer has been used to carry out the study. According the experimental results, a range form 5 to 10μm PLGA particle can be obtained by controlling the operation factors such as pressure, temperature, concentration, pressurized rate and agitation, etc. Furthermore, a recrystallized result of PLGA5050 was approved by XRD and DSC test.

Keywords : Supercritical CO_2, poly(D,L-lactide-co-glycolide), Gas Anti-Solvent, Ethyl acetate

1. Introduction

Gas Anti-Solvent(GAS) process utilizes the solubility of a gas or a supercritical fluid to dissolve and expand in organic solvent at moderate pressures and temperatures. Due to the dissolution of the compressed medium in the organic solvent, the expanded solvent has a lower solvating capacity for the solute. When the solution is expanded sufficiently, the organic solvent can no longer dissolve the solute, and then the precipitation will occur. The both of immiscibility of the solute and miscibility of the organic solvent with the compressed medium are the basic requirement for applying GAS.

Supercritical CO_2 is a clean and versatile solvent, It can be obtained a good miscibility of Supercritical CO_2 with the most organic solvent by turning the temperature and pressure. The use of supercritical fluids for particle formation was first proposed by Krukonis in 1984. Now, the technique has been successfully used to produce numerous compounds. According to polymeric substances have little to no solubility in the supercritical CO_2, thus CO_2 used in GAS is a universal process for precipitation polymeric particle.

The poly(D,L-lactide-co-glycolide)(PLGA) is a typical synthetic biodegradable aliphatic that derived from lactic acid (LA) and glycolic acid (GA), and their copolymers. They had been widely used for temporary therapeutic application such as tissue regeneration, wound closure and drug delivery devices etc. Control release has been a key point for the pharmaceutical development. The shell material and the uniform particle size play an important role in releasing process. PLGA which is used in a host of Food and Drug Administration (FDA) approved therapeutic devices, owing to its biodegradability and biocompatibility. In general, the PLGA used for drug delivery devices which are specified on 5-10μm. In this study, we investigated the use of supercritical CO_2 as anti-solvent in PLGA/ethyl acetate solution system to manufacture PLGA particle.

2. Experimental

2.1 Material

The degradation of PLGA copolymers are most depends on the copolymer composition, in where the decrease of molecular weight during degradation has been found to be accelerated with increasing glycolic acid content .In this work, PLGA copolymers with molar lactic/glycolic ratios of 50:50 were selected to investigate. The PLGA (PLGA50:50) was purchased from Sigma, USA. The molecular weight was Mw=400,000~75,000. According the manufacture method of PLGA by Gilding and Reed [1] which is the first synthetic method. Then, the modern process is based on this synthetic method. In this method the crude PLGA were extracted with ethyl acetate and Petroleum ether mixture. In another method ethyl acetate was used. So, in this study the PLGA is dissolved with ethyl acetate. The ethyl acetate (HPLC grade) was purchased from J. T. Baker and CO_2 (99.9%) was purchased from Yuan Ann, Taiwan.

2.2 Apparatus and Procedure

All the precipitation experiments were carried in batch mode. The first step was loaded the PLGA/ethyl acetate solution into the precipitation chamber. Then were rising temperature and filling up this vessel with CO_2 by various pressurized rate to desired operation conditions. The experimental apparatus is schematically shown in Fig.1. CO_2 from a cylinder with siphon attachment is passed through a cooling coil and compressed to the operating pressure by a high pressure diaphragm pump (LEWA, EK-1). The compressed fluid is heated by passing through a jacked type heat exchanger and then flow to a reservoir until the desired pressure was achieved. Then open the valve which between the reservoir and the precipitation chamber to control the pressurized rate for precipitation chamber. The precipitation chamber consists of three view windows to record the procedure of precipitation and one stirrer to enhance the mass transfer. The cleaning step was followed with PLGA particles were formed, ethyl acetate was removed that was dissolved in supercritical CO_2 by continuously purge with fresh CO_2. A wet flow meter(Ritter, Germany) was used to measure the overall supercritical CO_2 delivered. Furthermore, two stages depressurization was carried out in this study to ensure the complete removal of ethyl acetate from the vessel.

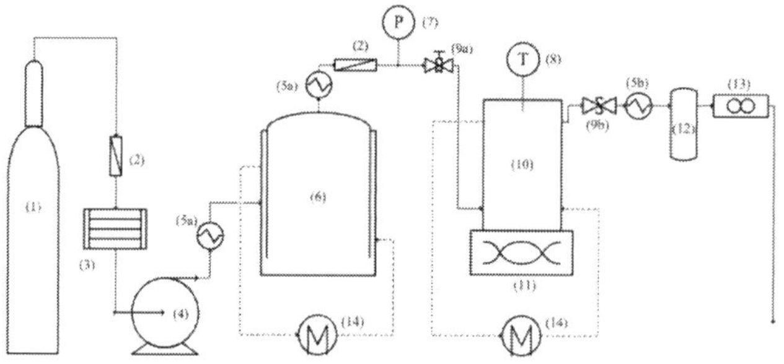

(1) CO_2 supply | (2) Filter | (3) Cooling uni
(4) High pressure pump | (5a、5b) Heating tape | (6) pressure reservior
(7) Pressure gauge | (8) Temperature gauge | (9a、9b) Flow control valve
(10) Precipitation vessel | (11) Magneto stirrer | (12) Separation vessel
(13) Wet flow meter | (14) Temperature controller

Fig.1 Flow diagram of GAS process

3. Results and Discussion

3.1 The volumetric expansion

The volumetric expansion of the solvent by the supercritical CO_2 is important for solute precipitation to occur. Vazquez et al. [2] have reported the data for phase behavior of binary CO_2 –ethyl acetate systems at 295K, 303K and 313K. According these data, the Peng-Robinson equation of state [3] was used to predict the volumetric expansions. The binary interaction parameter k_{ij} of 1.362 as suggested by Vazquez et al. [2] .

The relationship between for a given temperature(T) and specific volume(V), pressure(P) is as below

$$p = \frac{RT}{(v-b)} - \frac{a}{v(v+b)+b(v-b)} \tag{1}$$

Here, a and b are the characteristic parameters of the mixture given be following relation

$$a = \sum\sum y_i y_j a_{ij} \tag{2}$$

$$b = \sum y_i b_i \tag{3}$$

$$a_{ij} = (1-k_{ij})\sqrt{a_i a_j} \tag{4}$$

$$a_{ii} = \frac{0.45724 R^2 T_{ci}^2}{P_{ci}} \left[1 + (0.3764 + 1.54226\omega_i + 0.26992\omega_i^2)(1 - \sqrt{\frac{T}{T_{ci}}})\right]^2 \tag{5}$$

$$b_{ii} = \frac{0.0778 R T_{ci}}{P_{ci}} \tag{6}$$

Where T_{ci}, P_{ci} and ω_i are critical temperature, critical pressure, and accentric factor of component i, respectively.

The relative total volume expansion of the solvent was derivated from Kordikowski et al.[4] and expressed as ΔV/V×100% :

$$\frac{\Delta V}{V}=\frac{V_L(T,P,x_1)-V_2(T,P_o)}{V_2(T,P_o)}=\frac{1}{1-x_1}\cdot\frac{v_L(T,P,x_1)}{v_2(T,P_o)}-1 \qquad (7)$$

where V_L : total volume of the liquid phase

V_2 : total volume of the pure solvent at the same temperature and reference pressure P_0

v_L : molar volume of the liquid solution

v_2 : molar volume of the solvent, 2

x_1 : mole fraction of antisolvent, 1 in the solution

P_o: reference pressure, atmospheric pressure

The results of relative volume expansion at different equilibrium pressure for 295, 303, 313 and 323K isotherms, as shown in Fig.2. For all these isotherms, such as 303K isotherm, the relative volume expansion is extremely high above the critical pressure, due to the high solubility of the ethyl acetate in CO_2. The lower temperature has the higher expansion ratio under the same pressure. A higher relative volume expansion is suited for GAS process, as it gives rise to higher supersaturation, which lead particle formed easily.

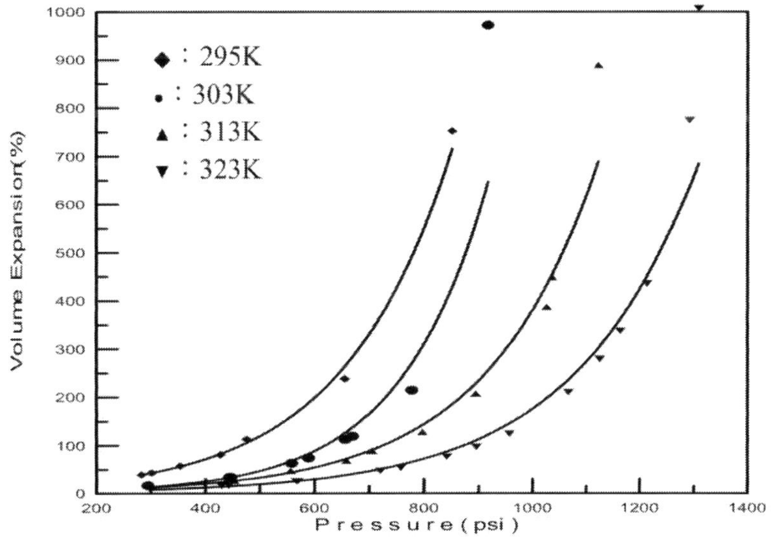

Fig.2 Relative volumetric expansion of ethyl acetate in supercritical CO_2 as a function of pressure

3.2 Effect of the CO_2 addition rate

The effect of CO_2 addition rate has direct influence on the supersaturation rate. The higher addition rate, will be had higher supersaturation rate. This effect has already been reported for GAS process by Gallagher et al.[5]. They suggested that the particle size will be smaller and monodisperse at high supersaturation rate. On the other, at low addition rate, a large particle and even film will be formed. Fig.3 presents PLGA particles that was formed at 1500 psi and CO_2 addition time 3, 20 and 60 sec, respectively. From the results, it can be found that the sphere particle will be formed at high CO_2 addition rate (short addition time, 3 sec). Furthermore, the low CO_2 addition rate (long addition time, 60 sec) appears a film. Thus, all the discussion in this work has been achieved at CO_2 addition time 3 sec.

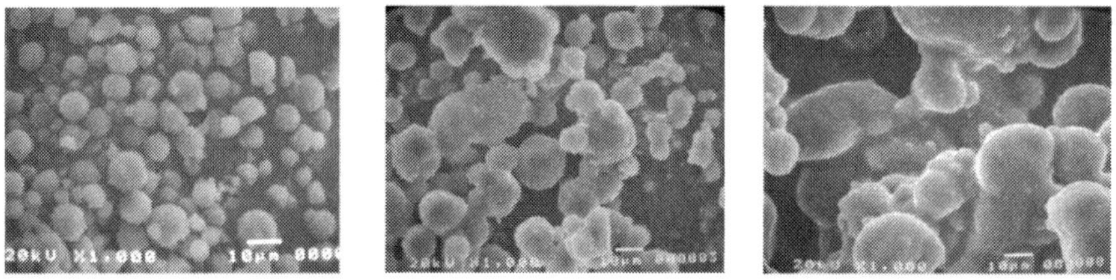

3 sec　　20 sec　　60 sec

Fig.3 Particle produced at 1500 psi and different addition time

3.3 Effect of agitation

The agitation has an influence on the mass transfer rate of CO_2 solution and the equilibrium between phases. Thus if there is a good mixing the time to produce homogeneous volumetric expansion will minimum [5]. In this work, the agitation was carried by powerful magneto-stirred supplier to drive the stirrer which in the precipitation chamber as shown in Fig.1. According this design, the stirring rate can not be measurement. Fig.4 shows the PLGA particles that were produced under agitation and no agitation. It can be obtained that agitation is suitable for the formation PLGA particle. Agitation benefits both of the dissolution of the CO_2 and enhances the PLGA dispersion in the expansion solution. These factors lead to the precipitation of PLGA particle as shown in Fig. 4(a). On the other hand, most of PLGA agglomerated under no agitation and leads to formation of a film as shown in Fig. 4(c). Thus, all the following discussion has been achieved under agitation condition.

(a) fast agitation　　(b) slow agitation　　(c) no agitation

Fig. 4 PLGA particle produced at different agitation condition

3.4 Effect of PLGA concentration

The higher PLGA concentration will lead the higher viscosity and surface tension. These factors have direct on the particle formation. Higher initial concentration implicates bigger agglomeration and leads the formation bigger particle or film. Fig. 5 shows the morphology of PLGA particle obtained from initial solutions of 0.025, 0.01, and 0.005 g/ml, respectively. For the initial concentration 0.025 g/ml the film is observed. For the concentration 0.1 wt% the particle is obtained. Finally, for the concentration 0.005 g/ml the more uniform particle is appeared. Although concentration 0.005 g/ml seem the best condition for particle formation, the more CO_2 is necessary to purge the ethyl acetate. According to above discussion, the optimum concentration is 0.005 g/ml.

(a) 0.025 g/ml (b) 0.01 g/ml (c) 0.005 g/ml

Fig.5 PLGA particle produced at various initial concentrations

3.5 Effect of temperature

It is important that to note that the properties of supercritical fluid depends strongly on the temperature and pressure. However, as previous discussed that the relative volume expansion is extremely high above the critical pressure. Since the volume expansion depends on the temperature stronger than that on the pressure above the critical pressure. Fig. 6 shows the PLGA particle obtained at 1500 psi and various temperatures of 293, 303 and 313K, respectively. It can be found that the particle size will be both of smaller and more uniform at lower temperature. It was resulted from the lower temperature has higher volumetric expansion under the same pressure. The higher volumetric expansion leads that the PLGA was dispersed uniformly and precipitated quickly, and then the small and uniform particle was produced.

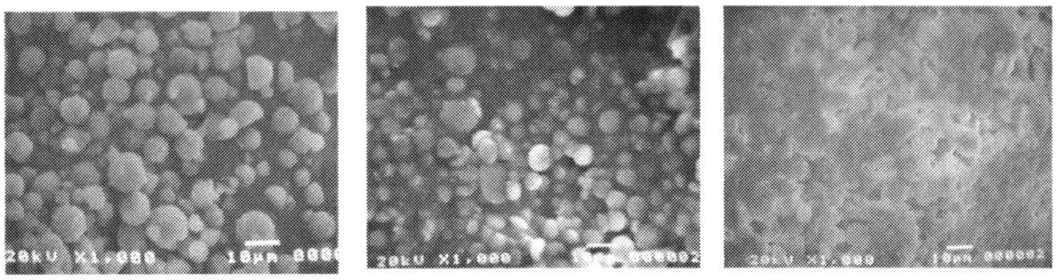

(a) 293K (b) 303K (c) 313K

Fig.6 PLGA particle produced under different temperature

3.6 Effect of pressure

Although the effect of pressure has not significant influence on volumetric expansion, it has a close relation with the density of supercritical fluid. The density effected especially for solubility. According to regular solution theory, two compound have the closer solubility parameter, will have better miscibility. Thus it was necessary to adjust the pressure, let the solubility parameter of CO_2 which was closed to this value. Fig.7 shows the PLGA particle obtained from pressure of 1000, 1500 and 2000 psi, respectively.

It was found that the middle pressure 1500 psi is suitable to produce PLGA particle, the low pressure 1000 psi the size of PLGA particle is bigger. The higher pressure 2000 psi the particle size does not decrease with the pressure increase. Thus 1500 psi is considered as optimum.

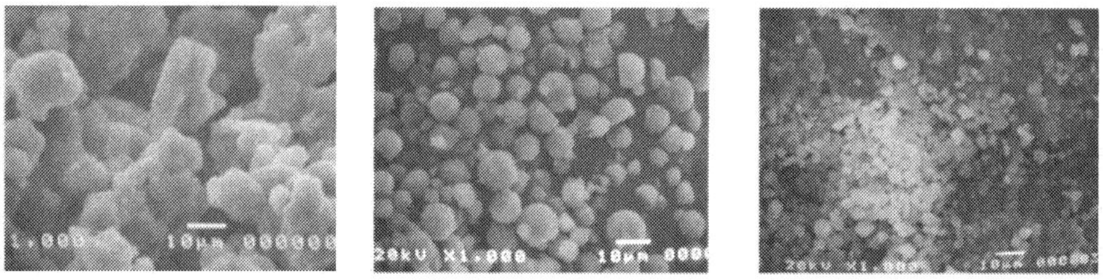

(a) 1000 psi (b) 1500 psi (c) 2000 psi

Fig.7 PLGA particle produced under different pressure

3.7 The particle size distribution

According to previous discussion, the optimum operation conditions were at pressure 1500 psi, temperature 293 K, concentration 0.005 g/ml, agitation, and the pressure elevated time 3 sec. Fig.8 show the size distribution of PLGA particle which was manufactured under above conditions. The particle size distribution was ranged from 5 to 10μm. The particle size distribution is suitable for the specification of used for drug delivery devices.

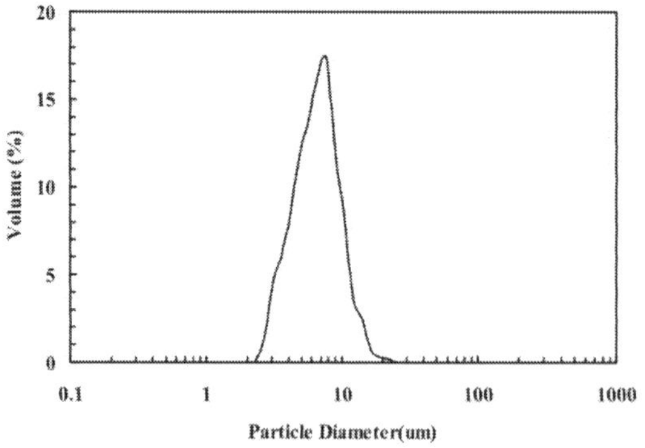

Fig.8 Particle size distribution at 1500 psi, 293 K , 0.005 g/ml, agitation and the pressure elevated time 3 sec.

3.8 Induced crystallization of PLGA5050

It is important to note that the crystallinity of biodegradable material for drug delivery devices which effect the degradation. Fig.9 and 10 shows the DSC and XRD for the change of crystallinity of PLGA. It can be found that PLGA had induced crystallinity during the SAS process. It is due to the PLGA used in this study is PLGA5050, which was 50:50 poly(D,L-lactide)/poly(glycolide). In the copolymer, poly(glycolide) is amorphous ,on the other hand poly(D,L-lactide) have the tendency to crystallize. Supercritical CO2 enhance this tendency of crystallization since it increases the free volume of the amorphous regions of the polymer. This allows the ordinarily rigid chains to rearrange into a more energetically favored state [7]. Moreover, supercritical CO2 plasticizes the amorphous regions of the poly(glycolide) and give the poly(D,L-lactide) chains increased mobility for chain extension to increase the crystallization opportunity of poly(D,L-lactide) in the PLGA5050.

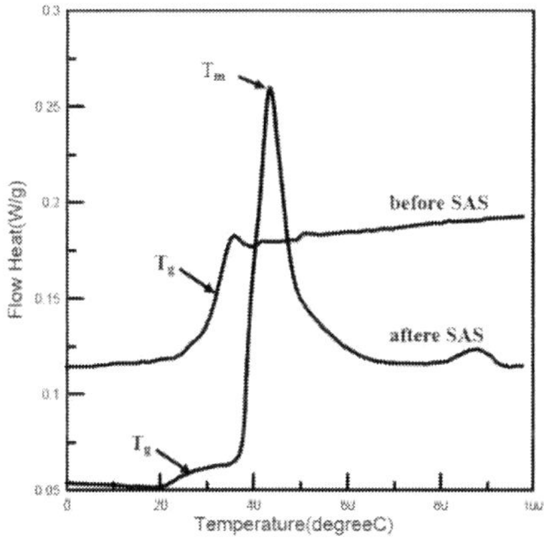

Fig.9 The comparison of DSC between before and after antisolvent process.

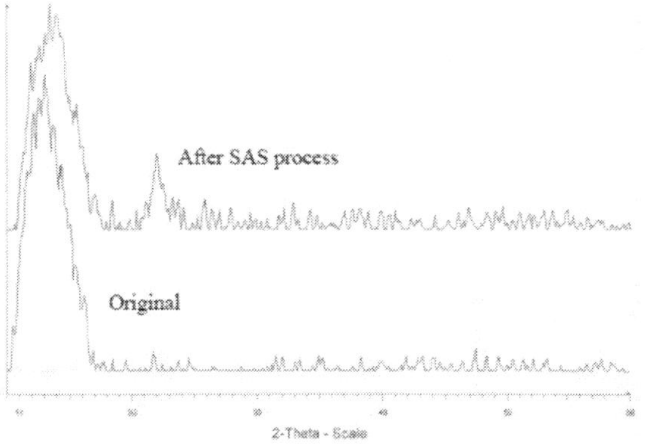

Fig.10 The comparison of XRD between before and after antisolvent process.

4. Conclusions

The 5-10μm PLGA particles can be successfully produced by supercritical CO_2 antisolvent precipitation process from PLGA/ ethyl acetate solution. The optimum operation conditions were at pressure 1500 psi, temperature 293K, concentration 0.005 g/ml, agitation, and the pressure elevated time 3 sec. The CO_2 addition rate is one very important factor in this process. Furthermore, a recrystallized result of PLGA5050 was approved by XRD and DSC test.

5. References

1. Gilding, D. K. and A. M. Reed (1979),"Biodegradable polymers for use in surgery-poly glycolic/poly lactic acid homo and copolymers", Polymer, 20, pp. 1459-1462.
2. Vazquez, M., D. Barbosa, O. P. Ferreira and J. Mendon(2000),"High pressure phase equilibrium data for the systems carbon dioxide/ethyl acetate and carbon dioxide/isomer acetate at 295K, 303K and 313K", Fluid Phase Equilibria,175, pp.19–33.
3. Chattopadhyay P. and , R. B. Gupta(1998), "Supercritical CO_2 based production of fullerene nanoparticles", Proceedings of the 5th Conference on Supercritical Fluid and Their Applications, pp.1-19.
4. Kordikowski , A. , A. P. Schenk , R. M. Van Nielen and C. J. Peters(1995) ,"Volume expansions and vapor-liquid equilibria of binary mixtures of a variety of polar solvents and certain near-critical solvents", Journal of Supercritical Fluids, 8, pp. 205-216.
5. Gallagher, P. M., V. J. Krukonis and G. D. Botsaris(1991), "Gas anti-solvent(GAS) recrystallization: Application to particle design", AIChE Symp. Ser., 284, pp. 96-103.
6. Cocero M. J. and S. Ferrero(2002), "Crystallization of β-carotene by a GAS process in batch effect of operating conditions", Journal of Supercritical Fluids, 22, pp. 237-245 .
7. Gross, S. M, G. W. Roberts and J. Desimone(2000),"Crystallization And Solid-State Polymerization of Poly(Bisphenol A Carbonate) Fabricated by Supercritical CO_2", Macromolecules, 33, pp. 40-45.

On Significance of Thermophysical Properties of Energetic Materials

Dabir S. Viswanath, Nuclear Science and Engineering Institute, University of Missouri, Columbia, MO

Veera Boddu, Environmental Processes Branch, US Army Engineer Research and Development Center, Champaign, IL

Few energetic materials were used in times of war and peace before the end of the cold war era. However since the end of the cold war era and the current terrorist activities, significant research is taking place in the synthesis, production, and use of such materials. Information on thermophysical properties of energetic materials is essential for:

1. Military needs to minimize the collateral damage to all sites other than the targeted sites,

2. Determining the fate of these chemicals in the environment,

3. Detecting hidden explosives - extremely accurate data on vapor pressure is required for development of sensors,

4. Proper disposal of these hazardous materials,

5. Preventing smuggling of these materials in their original form or as bombs, etc.

6. Testing predictive methods associated with the testing and fielding of new energetic materials to minimize the cost and time associated with an experimental program. For example, prediction of crystal density paves the way in predicting detonation velocities. Enthalpies of formation and sublimation are critical in assessing the potential energy release and performance in a gun or a warhead.

7. Developing models to predict accurately physicochemical properties. The models currently used to predict properties are far from satisfactory.

8. Studying syntheses pathways, reaction mechanisms and products, exhaust plume signature, etc.

9. To develop new energetic materials or improve the existing ones.

10. To improve the sensitivity/insensitivity of the existing energetic materials.

11. To develop energetic compounds for space applications. Specific impulse, ballistic and mechanical properties in addition to hazard and safety factors are important.

One of the important areas in the study of energetic materials is the detonation performance. Detonation performance is described via detonation velocity, pressure and energy, chemical structure and particle size. In assessing the detonation characteristics, the Chapman-Jouguet [C-J] pressure is perhaps the principal parameter. In addition as the detonation process involves a phase change from solid to gaseous/liquid phases which encompasses the thermal and diffusional processes, a number of transport and kinetic parameters are essential to describe the detonation process. This presentation outlines the need for theoretical and empirical methods of prediction of properties directly connected to performance of energetic materials.

See more of this Session: Thermophysical Properties
See more of this Group/Topical: Particle Technology Forum

Application of COSMO-RS Approximation to the Prediction of Physical Properties of Nitro-Compounds: Water Solubility

Leonid Gorb, SpecPro, Inc,, Vicksburg, MS

Frances Hill, US Army ERDC, Viksburg, MS

Yana Kholod, Chemistry, Jackson State University, Jackson, MS

Eugene Muratov, Department of Molecular Structure, A.V. Bogatsky Physical-Chemical Institute, Odessa, Ukraine

Anatoly G. Artemenko, Department of Molecular Structure, A.V. Bogatsky Physical-Chemical Institute, Odessa, Ukraine

Victor E. Kuz'min, Department of Molecular Structure, A.V. Bogatsky Physical-Chemical Institute, Odessa, Ukraine

A number of studies have shown that nitro compounds, as well as their metabolites of environmental transformation, by-products of synthesis or incomplete combustion are harmful for the biosphere due to their toxicity. Therefore, remediation of these contaminants is a point of prominent concern for environmental science. The environmental fate of chemicals, as well as the efficiency of in-solution treatment is determined largely by their solubility in water. In particular, high water solubility (SW) leads to expeditious distribution in water, i.e. chemicals that rapidly and completely dissolve in water are transported along with the general water flow. Low SW is generally associated with high affinity for adsorption to solids in water (soil particles or sediment). In addition, high water solubility is associated with absorption of pollutants by living organisms. Hence, solubility of hazardous chemicals is one of the most fundamental physical properties of ecological importance.

Water solubility values for twenty seven nitro compounds with experimentally measured values were computed using the conductor-like screening model for real solvent (COSMO-RS) based on the density functional theory and COSMO technique. We have found that the accuracy of the COSMO-RS approach for prediction of water solubility of liquid nitro compounds is impressively high (the errors are lower than 0.1 LU). However, for some solid nitro compounds, especially nitramines, there is sufficient disagreement between calculated and experimental values. In order to increase the accuracy of predictions the quantitative structure-property relationship (QSPR) part of the COSMO-RS approach has been modified. The solubility values calculated by the modified COSMO-RS method have shown much better agreement with the experimental values (the mean absolute errors are lower than 0.5 LU). Furthermore, this technique has been used for prediction of water solubility for an expanded set of twenty three nitro compounds including nitroaromatic, nitramines, nitroanisoles, nitrogen rich compounds and some their nitroso and amino derivatives with unknown experimental values. The solubility values predicted using the proposed computational technique could be useful for the determination of the environmental fate of military and industrial wastes and the development of remediation strategies for contaminated soils and waters. This predictive capability is especially important for unstable compounds and for compounds that have yet to be synthesized.

Finally, an application of COSMO-RS approximation to the prediction of other environmentally important physical properties is briefly discussed.

See more of this Session: Thermophysical Properties
See more of this Group/Topical: Particle Technology Forum

Ignition Mechanisms of Metal Powders in Electrostatic Discharge

Ervin Beloni, Chemical, Biological, and Pharmaceutical Engineering, New Jersey Institute of Technology, Newark, NJ

Edward L. Dreizin, Chemical, Biological, and Pharmaceutical Engineering, New Jersey Institute of Technology, Newark, NJ

Ignition of reactive powders in electrostatic discharge (ESD) is commonly used to assess the material reactivity and safety of its handling. Despite the widespread use of this test and despite availability of several respective commercial devices, the results of testing are often inconsistent between different facilities or even between different operators using the same setup. These discrepancies are largely due to poorly established criteria of what does or does not constitute powder ignition. The problem of defining ignition criteria stems from poor understanding of the mechanisms of powder initiation by electric sparks. Our recent work suggests that Joule heating of the powder by the spark current may be the primary effect of the spark discharge. The powder heated by the passing current can ignite either inside the holding cup or outside of it, being ejected by the shock wave produced by the spark. In both cases, ignition is driven by heterogeneous exothermic reactions either on the particle surfaces or on the reactive interfaces in composite materials. Ignition occurs when the kinetics such reactions and the resulting particle self-heating becomes sufficiently rapid due to the achieved particle temperature. For pure metal particles, it is important that the oxidizer is available to maintain the reaction. This paper is aimed to validate this proposed ignition mechanism and experimentally investigate ESD ignition of different metal powders. The ESD ignition experiments were performed with Al powders with different particle sizes as well as with powders of Ti and Fe. The spark energy is measured from the recorded current and voltage traces. The emission produced by the igniting or heated particles at different wavelengths is recorded using filtered photomultipliers. The streaks produced by the igniting particles are visualized using a digital camera. The ESD ignition experiments are supplemented by the ignition experiments using the same powders coated onto an electrically heated filament. In the filament heating experiments, the heating rates are varied systematically to assess the ignition reaction kinetics. The ignition kinetics is in turn used to interpret the results of the ESD ignition tests.

See more of this Session: Thermophysical Properties
See more of this Group/Topical: Particle Technology Forum

Vapor Pressures and Melting Points of Select Munitions Compounds

Siri Chakka (Speaker), Veera M. Boddu*, Stephen W. Maloney
Environmental Processes Branch
U.S. Army Engineer Research and Development Center
Construction Engineering Research Laboratory (ERDC-CERL)
P.O. Box 9005, Champaign, IL 61826-9005

Rebecca K. Toghiani
Department of Chemical Engineering
Mississippi State University, Starkville, MS 39762

Reddy Damavarapu
Energetics, Warheads & Technology Directorate; AMSRD-ARR-AEE-E
U.S. Army Armament Research, Development and Engineering Center (ARDEC)
Pica tinny Arsenal, NJ 07806-5000

ABSTRACT

Thermophysical and transport properties of munitions compounds are required for understanding their environmental distribution, and for development of potential environmental treatment processes. Informed decisions on development of waste treatment technologies can be made by close examination of physical properties of munitions compounds. Physicochemical properties such as melting point, boiling point and vapor pressure are important in understanding a munitions dispersion and fate within the environment. The physicochemical properties (both model predictions and experimental data) of 15 energetic materials were collected from literature. The energetic materials considered during this study are dinitroanisole (DNAN), *n*-methyl-*p*-nitroaniline (MNA), nitro-triazolene (NTO), triaminotrinitrobenzene (TATB), cyclotetramethylene-tetranitramine (HMX), cyclotrimethylenetrinitramine (RDX), trinitrotoluene (TNT), China Lake-20 (CL-20), Diamino-Dinitroethylene (DADE, FOX 7), 1,3,3-Trinitroazetidine (TNAZ), Pentaerythritol tetranitrate (PETN), 2,4-Dinitrophenol (DNP), 1-Methyl-2,4,5-trinitroimidazole (MTNI), Triacetone Triperoxide (TATP), and 2,4,6-trinitrophenyl-N-methylnitramine (TETRYL). Results of model predictions are compared with available experimental vapor pressure, meting point, and boiling point data, and the data gaps are identified.

Introduction

Development of insensitive munitions (IM) is a key part of modern explosives research. Informed decisions on development of waste treatment technologies can be made by close examination of physical properties of the munitions compounds. The overall goal is to examine physical properties of new munitions compounds to help develop more efficient insensitive munitions. The fate and distribution of an IM compound released into the environment are primarily controlled by: (1) the physicochemical and transport properties of the compound and (2) the prevailing environmental conditions at the points of discharge. Thus, accurate determination of physicochemical properties is critical in developing valid environmental models and impact assessments. However, the predictive/interpretive value of environmental studies is seriously compromised if the physicochemical data upon which they rely are of questionable or unknown quality.[1]

Physicochemical properties such as melting point (MP), boiling point (BP), and vapor pressure (VP) are important in understanding a munitions dispersion and fate within the environment. BP, MP, and VP are important in predicting and discovering the relationship between the munitions and temperature and their effects upon solubility. The BP of a munitions compound is the temperature at which the VP of the liquid equals the environmental pressure surrounding the liquid. The BP also indicates the volatility of a compound. VP is the pressure of a vapor in equilibrium with its non-vapor phases. The VP is important in evaluating the distribution of a compound into the environmental compartments of soil, air, and water and its life within these compartments.

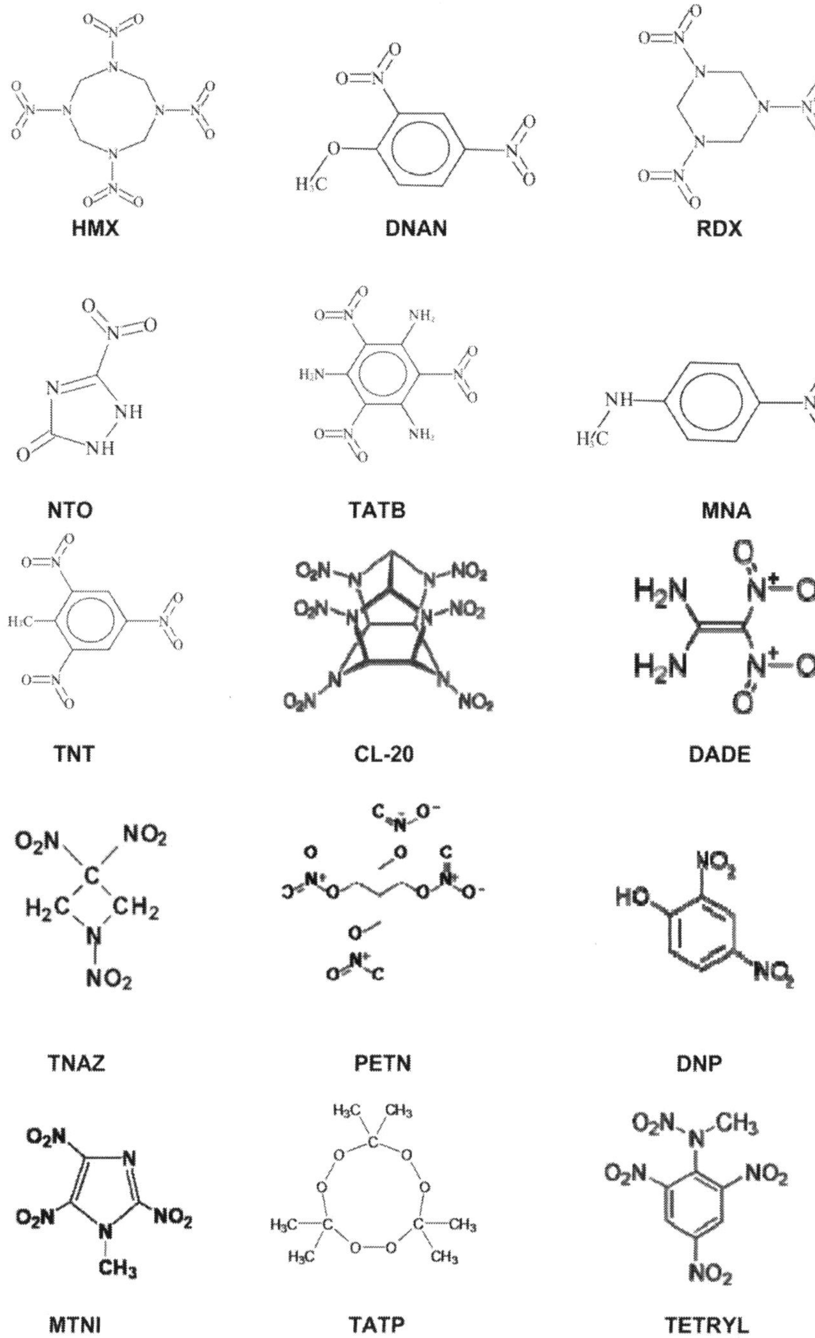

Figure 1. Structures of compounds of interest in this work

Model Predictions

EPI Suite as a means to predict physical and physicochemical properties is a relatively new means of studying organic materials.[4] It is developed by the EPA's Office of Pollution Prevention Toxics and Syracuse Research Corporation.

The MP value is estimated by both the Joback method[2] and by a correlation drawn between the MP and BP values.[3] The Joback method is a group contribution method which uses basic structural information of a chemical molecule such as basic functional groups and adds parameters to the groups to calculate thermophysical and transport properties as a function of the sum of group parameters. This method assumes that there are no interactions between the groups and therefore use only additive contributions and no contributions for interactions between groups. The two MP values derived by both methods are then compared to the chemical structure of the organic compound and the magnitude of the difference of both values and a suggested MP is estimated.

The BP is estimated by the Stein and Brown method.[4] The method uses the correlations equations (1)-(3), where "g_i" is the group increment value and "n_i" is the number of times the group occurs in the compound. The Stein and Brown method works on the basis of the group contribution method covering 85 different chemical groups. It is generally applicable to organic compounds with specific structural features and substituent groups. For a derivation set of 4,426 compounds, predicted boiling points had average absolute error of 15.5K (3.2%). For an independent test set of 6,584 compounds, the average absolute error was 20.4K (4.3%).[5] The resulting BP value is then corrected with either equation 2 or 3 based on the temperature.

$BP = 198.2 + \sum n_i g_i$ (1)
$BP(corr.) = BP - 94.84 + 0.5577\ BP - 0.0007705\ (BP)2$ [$BP<700^oK$] (2)
$BP(corr.) = BP + 282.7 - 0.5209\ BP$ [$BP>700^oK$] (3)

The VP value is estimated by the Antoine method[6], the modified Grain method, and the Mackay method.[2] The Antoine equation works best for liquids that boil below 200^oC and have vapor pressures above 10^{-2}kPa and 25^oC. The Antoine equation is said to be the most reliable three-parameter equation for representing vapor pressure as a function of temperature from the melting point to 85% of the critical temperature and provides an adequate fit for pressures in the range 1-100kPa.[5] L(x) is the isothermal retention index. A, B, and C are constants that depend on the substance and the stationary base used for Antoine's equation.

$$L(x) = A + \frac{B}{T_C + C}$$

The COSMOtherm (Toghiani et al.)[22] was used to predict vapor pressures for 7 compounds, as summarized in Table I. These predictions are much larger than those found in experimental reported data and the variance between the data points indicates that low volatility compounds are difficult to gather vapor pressure data from. These data are summarized in Figure 2 for the compound RDX.

Table I. Antoine Coefficients Generated Using COSMOtherm[22]

Compound	A_i	B_i	C_i
DNAN	18.01	7087.5	-43.94
DNP	17.72	7092.4	-60.25
MNA	18.19	6757.9	-61.43
MTNI	17.99	7080.5	-41.10
NTO	20.63	7334.04	-76.29
TATB	18.27	6756.2	-62.45
ε-CL20	18.74	9247.9	-38.79
RDX	18.56	7718.5	-42.65
HMX	20.24	11881.4	-35.96

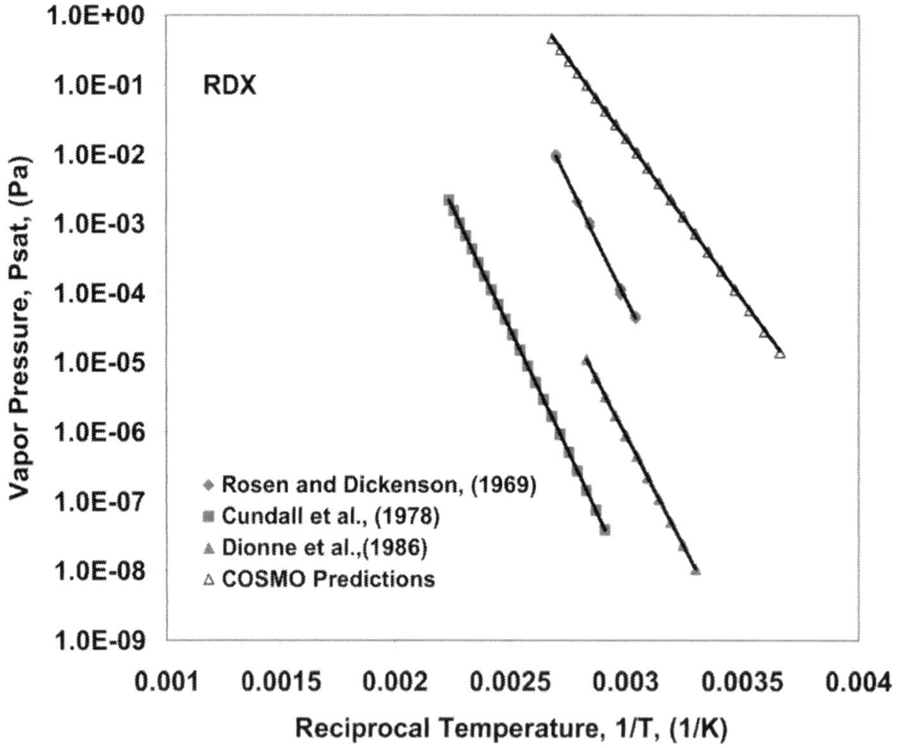

Figure 2. **Vapor Pressure of RDX – Comparison with Literature Data**

For higher boiling, less volatile substances, the Grain method gives more reliable estimates. The Grain method is also more reliable for estimating the vapor pressure of solids as it shows the vapor pressure lowering upon solidification of the supercooled liquid.[5] The Grain method uses an approximation to the Watson[5] expression for the temperature dependence of enthalpy, which takes care of the problem that is caused when the Antoine equation introduces unreasonable assumptions about the change in enthalpy with temperature.[5] The program then calculates a suggested VP based on which method is recommended for the chemical's class.

Experimental Values

Rosen and Dickenson[7] determined VP by the Langmuir method from measurements of the rates of sublimation at constant temperature.[7] They calucated VP values for RDX, HMX, and TATB from the observed weight loss of the heated sample in a vacuum using the expression below, where P is pressure in terms of torr, G is weight loss in grams/cm^2/sec., T is the absolute temperature, and M is molecular weight.

$$P = 17.14 \times G \times (T/M)^{1/2}$$

The BPs taken from Toghani et al.[22] paper for CL-20, RDX, HMX, DNAN, DNP, MNA, MTNI, NTO, and TATB were estimated by the Stein and Brown modification of the Joback method. The modification incorporates additional functional groups, including nitrogen groups in a ring.[8]

Results

Table II. Gathered BP, VP, and MP

	BP (°C)	VP (mmHg)	MP (°C)
DNAN	588[8] 319.62 (Stein&Brown) 206 @ 12mmHg (Exper.)	29936.32 @ 532.85 °C[8] .000145 (Modified Grain)	359.9[13] 96.56 (Weighted) 94.5 (Exper.)
MNA	527[8] 259.24 (Stein&Brown)	31286 @ 474.85 °C[8] .000839 (Modified Grain)	425.15[15] 60.96 (Weighted) 152 (Exper.)
NTO	568[8] 389.94 (Stein&Brown)	64521 @ 555.85°C[8] 5.83E-7 (Modified Grain)	539.35[17] 547.9[18] 161.19 (Weighted)
TATB	481.26 (Stein&Brown)	42391 @ 639.85 °C[8] 0.733E7 @ 129.3 °C[7] 0.746E7 @ 129.3 °C[7] 1.83E7 @ 136.2 °C[7] 1.93E7 @ 136.2 °C[7] 10.3E7 @ 150.0 °C[7] 9.42E7 @ 150.0 °C[7] 9.73E7 @ 150.0 °C[7] 32.2E7 @ 161.4 °C[7]	594-599[19] 203.85 (Weighted) 350 (Exp.)

		32.3E7 @ 161.4 °C[7] 45.8E7 @ 166.4 °C[7] 167E7 @ 177.3 °C[7] 1.58E-11 (Modified Grain) 3.00E-3 @ 175 C (Exper.)	
HMX	709[8] 436.41 (Stein&Brown)	39765 @ 640.85 °C[8] 0.0324E7 @ 97.6 °C[7] 0.164E7 @ 108.2 °C[7] 0.390E7 @ 115.6 °C[7] 0.385E7 @ 115.6 °C[7] 0.419E7 @ 115.6 °C[7] 2.83E7 @ 129.3 °C[7] 2.87E7 @ 129.3 °C[7] 2.41E-8 (Modified Grain) 3.3E-14 (Exper.)	553.15[12] 182.89 (Weighted)
RDX	627[8] 353.43 (Stein&Brown)	43516 @ 569.85°C[8] 3.5E7 @ 55.7 °C[7] 3.24E7 @ 55.7 °C[7] 3.42E7 @ 55.7 °C[7] 8.21E7 @ 62.6 °C[7] 7.14E7 @ 62.6 °C[7] 8.63E7 @ 62.6 °C[7] 69.3E7 @ 78.2 °C[7] 78.7E7 @ 78.2 °C[7] 155E7 @ 85.3 °C[7] 735E7 @ 97.7 °C[7] 667E7 @ 97.7 °C[7] 702E7 @ 97.7 °C[7] 1.34E-6 (Modified Grain) 4.1E-8 (Exper.)	478.5[11] 205.5 (Exp.)
TNT	364.14 (Stein&Brown)	1.72E-5 (Grain) 8.02E-6 (Exper.)	124.36 (Weighted) 80.1 (Exper.)
CL-20	862[8]	36688 @ 784.85°C[8]	513[10]
TATP	712[8] 168.45 (Stein&Brown)	$\ln P(atm) = 22.73 - 9695.5/T$ @ 12-97°C[21] $\ln P(atm) = 15.29 - 6978.8/T$ @ T> 97°C[21] $\log_{10} P(Pa) = 19.791 - 5780/T$ @ 12-60°C[20] 2.24 (Grain)	13.07 (Weighted)
DNP	575[8] 332.13 (Stein and Brown)	51994 @ 543.85 °C[8] 1.29E-5 (Modified Grain) 3.9E-4 @ 20°C (exper)	363-364[14] 118.46 (Weighted) 115.5 (Exper)

TNAZ	311.56 (Stein&Brown)	.000221 (Grain)	95.15 (Weighted) 101[9]
PETN	363.86 (Stein and Brown)	4.21E-6 (Grain) 5.45E-9	119.48 (Weighted) 140.5 (Exper)
MTNI	629[8] 402 (Stein & Brown)	41040 @ 571.85°C[8] 4.14E-7 (Grain)	355.15[16] 148.36 (Weighted)
TETRYL	432.11 (Stein & Brown)	1.17E-7 (Grain)	159.69 (Weighted) 131.5 (Exper)
DADE	287.51 (Stein & Brown)	.00104 (Grain)	83.37

Conclusions:

Vapor pressure, boiling and melting point data are important for predicting thermodynamic behavior of materials. Literature review shows that the vapor pressure, boiling and melting point data for explosive materials, in particular for the new insensitive munitions is very limited. The discrepancy between model predictions and available experimental data is rather significant. This brief survey reveals the need for experimental vapor pressure measurements and improved predictive models.

References

1. Kuhne, R.; Breitkopf, C.; Schurmann, G.; "Error propagation in fugacity level-III models in the case of uncertain physicochemical compound properties." Environ. Toxicol. Chem. 16 (1997) 2067-2069.
2. Reid, RC. "The Properties of Gases and Liquids." NY: McGraw Hill, 1987
3. Neely, W.B.; Blau, G.E.; "Environmental Exposure From Chemicals." CRC Press, 1985
4. "Estimation Software." Syrres.com. 1999. Syracuse Research Corporation. 15 February 2008. <http://www.syrres.com/esc/est_soft.htm>.
5. Boethling, R.S.; Mackay, D.; "Handbook of property estimation methods for chemicals: environmental and health sciences." CRC Press (2000) 33-34
6. Lyman, W. J., Handbook of Chemical Property Estimation Methods. Washington, DC: American Chemical Society, 1990
7. Rosen, J.M.; Dickinson, C.; "Vapor Pressures and Heats of Sublimation of Some High Melting Organic Explosives." J. Chem. Eng. Data, 14.1 (1969) 120-127
8. Toghiani, R. K.; Toghiani, H.; Maloney, S. W.; Boddu, V. M.; "Prediction of Physicochemical properties of energetic materials" Fluid Phase Equilibria. 264 (2008) 86-92
9. Sikder, N.; Sikder, A.K.; Bulakh, N.R.; Gandhe, B.R.; "1,3,3-Trinitroazetidine (TNAZ), a melt-cast explosive: synthesis, characterization and thermal behavior." J. Haz. Mat.113 (2004) 35-43.

10. Andelkovic-Lukic, M.; "New high explosive—polycyclic nitramine hexanitrohexaazaisowurtzitane (HNIW, CL-20)" Naucno-Tehnicki Pregled. 50.6 (2000) 60–64.
11. Hall, P.G.; "Thermal decomposition and phase transitions in solid nitramines." Trans. Faraday Soc. 67.3 (1971) 556–562
12. Maksimov, Y.Y.; Russ. J. Phys. Chem. 66.2 (1992) 280–281.
13. Beringer, F.M.; Brierley, A.; Drexler, M.; Gindler, E.M.; Lumpkin, C.C.; "Diaryliodonium salts II. The phenylation of organic and inorganic bases." J. Am. Chem. Soc. 75 (1953) 2708.
14. Katritzky, A.R.; Scriven, E.F.V.; Majumder, S.; Akhmedova, R.G.; Akhmedov, N.G.; Vakulenko, A.V.; "Direct nitration of five membered heterocycles." ARKIVOC 3 (2005) 179–191
15. MSDS for *N*-methyl-4-nitroaninline, http://ptcl.chem.ox.ac.uk/MSDS/ME/N-methyl-4-nitroaniline.html, accessed on 04/22/09.
16. Cho, J.R.; Kim, K.J.; Cho, S.G.; Kim, J.K.; "Synthesis and characterization of 1-methyl-2,4,5-trinitroimidazole (MTNI)." J. Heterocyclic Chem. 29.1(2002), 141–147.
17. Liu, Z.R.; Shao, Y.H.; Yin, C.M.; Kong, Y.H.; "Measurement of the eutectic composition and temperature of energetic materials., Part 1, The phase diagram of binary systems." Thermochim. Acta 250 (1995) 65–76.
18. Kim, K.J.; Kim, M.J.; Lee, J.M.; Kim, S.H.; Kim, H.S.; Park, B.S.; "Solubility, density, and metastable zone width of the 3-nitro-1,2,4-triazol-5-one + water system." J. Chem. Eng. Data. 43 (1998) 65–68.
19. Zeman, S.; "The thermoanalytical study of some amino derivatives of 1,3,5-trinitrobenzene." Thermochim. Acta. 216.1/2 (1993) 157–168.
20. Oxley, J.C; Smith, J.L; Shinde, K.; Morgan J.; "Determination of Vapor Density of Triacetone Triperoxide (TATP) Using a Gas Chromatography Headspace Technique." Propellants, Explosives, Pyrotechnics. 30.2 (2005) 127-130.
21. Egorshev, V.; Sinditiskii, V.; Smironov, S.; Glinkovsky, E.; Kuzmin, V.; "A Comparative Study on Cyclic Acetone Peroxides." Presentation in "New Trends in Research of Energetic Materials."Czech Republic, 2009.
22. Toghiani, R.K.; Toghiani, H.; "Prediction of Physciochemical Properties of Energetic Materials for Identification of Treatment Technologies for Generated Waste Streams." Draft Project Report submitted to ERDC-CERL. Mississippi State University. August 2009.

Mesoscale Modeling of Thermo-Mechanical Response: Constant-Energy Dissipative Particle Dynamics Simulations

John K. Brennan, Weapons and Materials Research Directorate, U.S. Army Research Laboratory, Aberdeen Proving Ground, MD

Martin Lisal, E. Hala Laboratory of Thermodynamics, Academy of Sciences of the Czech Republic, Prague, Czech Republic

Nanocomposites such as energetic materials (EMs) are inherently heterogeneous, containing several species types of disparate shapes and sizes. EMs are often comprised of micro-sized crystallites (or grains) held together by a small amount of polymer binder, and in some instances contain metal nanoparticles. Accordingly, mechanical stimulation (e.g., shock or shear) of these materials can incite responses over a wide range of spatial and temporal scales. For example, localized regions of elevated thermal energy, or hot spots, often occur near microscale defects in the material such as voids and grain boundaries; however the energy transfer processes are atomistically governed. While modeling and simulation of such phenomena can offer fundamental insight, it is computationally challenging. Computing power and resources continue to grow, but there still exists much behavior that cannot be studied using atomistic simulation techniques. Moreover, field-based mesoscale modeling techniques have fidelity limitations with respect to gaining a fundamental understanding.

To address these issues, we recently embarked on a study of the thermo-mechanical response of crystalline nanocomposites using a particle-based mesoscale modeling technique, namely, the constant-energy Dissipative Particle Dynamics method (DPDe) [1, 2]. A constant energy approach is required since materials subjected to mechanical stimuli will inevitably respond via energy exchange and transfer processes. In contrast to the standard Dissipative Particle Dynamics method [3, 4] that conserves only momentum, the DPDe method conserves both momentum and energy. The DPDe method assigns an internal energy to each particle, allowing particles to exchange both momentum and thermal energy. This particle internal energy is included as a separate equation of motion along with the equations of motion for the particle's position and momentum. The atomic degrees-of-freedom are explicitly resolved or coarse-grained into the particle internal energy, and included in the simulation in an averaged way.

A number of modeling challenges exist for nanocomposites, including the development of an accurate mesoscale model that can capture the following known thermo-mechanical responses: (i) phase transitions; (ii) structural rearrangements (plastic deformation, shear banding); and (iii) chemical decomposition reactions. The model parameters should be physics-based so that direct links to real material properties or higher resolution modeling are possible. The standard DPD model [5] likely cannot capture such features, fortunately though, promising alternatives exist [6-8]. Technical challenges also exist for such studies. Since while numerical integrator algorithms work quite satisfactorily for DPD [5] and DPDe [8] simulations under normal conditions, at extreme conditions such as high densities, much shorter integration timesteps are required to maintain numerical stability. In order to maintain a reasonable timestep that allows for the simulation of mesoscale events, higher quality integrators will be explored.

In this work, we will present the status of this ongoing project, including progress on nanocomposite model development, mesoscale simulation of phase transitions, and improved integrator algorithms. To date, progress has been encouraging, where we find the DPDe method to be a viable tool for simulating the thermo-mechanical response of crystalline nanocomposites to mechanical stimuli.

References

[1] Bonet Avalos, J., and A.D. Mackie, Europhysics Letters, 40, 141 (1997).

[2] Español, P., Europhysics Letters, 40, 631 (1997).

[3] Hoogerbrugge, P.J., and J.M.V.A. Koelman, Europhys. Lett., 19, 155 (1992).

[4] Koelman, J.M.V.A., and P.J. Hoogerbrugge, Europhys. Lett., 21, 363 (1993).

[5] Groot, R.D., and P.B. Warren, J. Chem. Phys., 107, 4423 (1997).

[6] Pagonabarraga, I., and D. Frenkel, Molec. Simul., 25, 167 (2000).

[7] Pagonabarraga, I., and D. Frenkel, J. Chem. Phys., 115, 5015 (2001).

[8] Groot, R.D., and S.D. Stoyanov, Phys. Rev. E, 78, 051403(1-9) (2008).

[9] Mackie, A.D., J. Bonet Avalos and V. Navas, Phys. Chem. Chem. Phys., 1, 2039 (1999).

See more of this Session: Thermophysical Properties
See more of this Group/Topical: Particle Technology Forum

Hydrodynamic Simulations of Energetic Materials at the Grain Scale

A. Dawson, L. Costa
Institute for Multi-Scale Reactive Modeling, Picatinny Arsenal, NJ 07806, USA

Abstract

To date, relatively little work has been done modeling energetic materials at the mesoscale. Any comprehensive work in the area of multi-scale modeling must include work at this scale. Modeling energetic materials as a continuum using a conventional programmed burn algorithm is insufficient to predict non-ideal effects in explosives. A more physically based method involves using a Monte Carlo based particle packing algorithm to generate an initial representation of an explosive's crystal morphology that is then transferred into a hydrodynamic code and run using any number of equations of state, strength and chemistry models. This method will lead to more physically based reactive flow models and improved predictive capabilities for the energetic materials community.

In this work, we present several computational simulations involving a strong shock wave passing through a granular representative volume element (RVE) of an insensitive high explosive. Several reactive burn models are explored and basic post-processing diagnostics are performed on the RVE to extract metrics of interest. Comparisons are made between purely continuum level calculations and calculations performed at the grain scale. Hydrocodes are shown to be an acceptable tool for the modeling of energetic formulations at the grain scale by capturing important statistical features present in mesoscale shock dynamics.

Keywords: Mesoscale; Grain Structure; Energetic Materials; CTH

1. Introduction

The use of multi-scale methods in computational modeling of detonation phenomena in heterogeneous energetic materials has received growing interest over the past decade [1-4]. The increased focus on insensitive explosive formulations has led to formulations that exhibit non-ideal initiation and wave propagation behavior. Coupled, dynamic effects at the atomistic, mesoscale, and continuum levels must be incorporated into modern reactive flow calculations in order to understand and predict these non-ideal effects. The atomistic and continuum descriptions of detonation behavior are relatively mature fields, however a large amount of work in the mathematical homogenization between scales is only now becoming tractable with the decreasing cost and increased availability of high performance computing resources.

Solid heterogeneous energetic materials are composites that may contain a wide variety of constituents with a range of thermal, mechanical, and chemical properties. These mixtures may include polycrystalline organic materials, metal particles, polymers with wide ranges of molecular weights, and other additives. Also the size of the solid granular constituents in a formulation is typically distributed over a range of values. The particle size distribution of a granular bed determines its packing properties and therefore its bulk density. The bulk density of an explosive has always been a property of interest from the continuum view as it affects the energy available per unit volume. In many explosive formulations, the bulk density can be increased to very near the material's theoretical maximum density (TMD) and cause the formulation to fail to initiate and propagate a reactive detonation wave. The continuum hypothesis fails to directly predict this. It may naively be expected that an optimal formulation choice would be at its highest density allowing for maximum volumetric energy density. Experiments have shown that the inclusion of a small percentage of voids at the sacrifice of bulk density increases the material's sensitivity and improves initiation behavior. This surprising behavior can be explained by considering the explosive microstructure closely.

The non-uniform particle size distribution creates stochastic microstructures that are averaged in the continuum view. The structural heterogeneities produce energy localizations that may lead to a variety of physical phenomena including: frictional effects, fracture, material jetting, viscous heating, interfacial heating and void collapse [5]. The nonlinear coupling of these phenomena with chemical kinetics produces deviations from classical ideal detonation theories. Fluctuations occur throughout the material's thermal, mechanical, and chemical fields chaotically. These fluctuations are responsible for most non-ideal effects. Take for example existence of a "failure distance" where the reactive detonation wave fails to propagate after a certain distance occurs due to these fluctuations decreasing in amplitude across spatial and temporal dimensions at the mesoscale. The energy density of a typical energetic material is enough to propagate a wave indefinitely in the continuum view. Results of so-called "programmed burn" calculations implemented into production hydrodynamic codes are examples of such an idealization.

Phenomena occurring at the mesoscale have characteristic sizes and times well above the atomistic and well below the continuum regimes. This work focuses on a slice of the mesoscale spectrum, considering granular micromechanics and reactive flow at "grain scale" dimensions (10-300 μm). The thermal, mechanical, and chemical fields have been shown to follow and evolve along statistical paths that can be remarkably and unambiguously described by a probability density estimate [2,5]. This intuitively suggests that modern stochastic calculus can provide a route to describing the statistics of these fields and their evolution across time. Owing to the youth of these efforts, the parameter space is still a canvas of infinite possible field variables. In this work we extend the work of others in the field, including Baer et al [3-5], and make comparisons between a select few of these parameters, some that have yet to be considered in the literature known to the authors at the time of this publication.

2. Computational Setup

The Eulerian shock physics code CTH [6] was used to simulate grain structures like those seen in Fig. 2.1 (a). CTH is a finite-difference/finite-volume code that solves analogs of the Lagrangian conservation equations in one step and then rezones to an Eulerian grid. CTH is well suited to handle problems involving strong shocks and large deformations in solids.

Packing grains into a three-dimensional domain in a physically realistic manner is non-trivial. Several methods were considered. The chosen method is based on Newton's laws of motion implemented in the particle packing software MacroPac [7]. MacroPac uses a Monte Carlo (MC) method combined with a user defined distribution to pack grains into a pre-defined space. For the runs presented below, approximately 1,000 individual grains were packed using a known particle distribution [8]. MacroPac's output was imported into CTH, and the void space was filled with an estane binder. The overall volume percentage of material packed in the domain was 25%. This was done to maintain the realistic size distribution without allowing many smaller grains into the problem that could not be properly resolved by CTH. The domain size used to define the representative volume element (RVE) was 0.1 x 0.1 x 0.5 cm, giving it a bulk density of 1.4569 g/cc. A section of the RVE is shown in Fig. 2.1 (b).

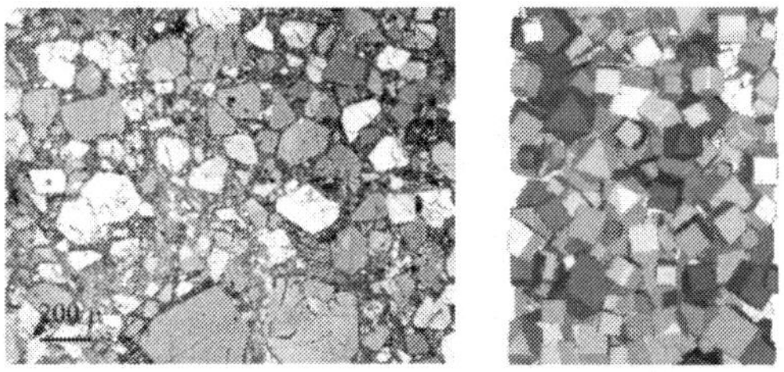

(a) (b)

Fig. 2.1 (a) Explosive grains in a binder and (b) computational representation of grains

A method of shock initiation was needed to observe both the mechanical and chemical effects of a shock and subsequent detonation wave passing through the RVE. To achieve a mechanical wave strong enough to 'ignite' the reactive burn model, a small detonator was initialized below the grains using a programmed burn algorithm. The detonator was given the Jones-Wilkins-Lee (JWL) Equation of State (EOS) for HMX. This detonation sent a strong shock wave through the RVE.

Several EOS were used for the energetic grains. CTH has several different reactive burn EOS implemented into the code, including History Variable Reactive Burn (HVRB), Ignition and Growth Reactive Burn (IGRB), Arhennius Reactive Burn (ARB) and Forest Fire Reactive Burn (FFRB). Of these, the first three have libraries with pre-defined parameters for different explosives. Because of the desire to compare HVRB, IGRB and ARB, parameters for pentaerythritol tetranitrate (PETN) were used, as PETN is the only material for which pre-defined parameters exist in all three models. The EOS for the binder was a user defined Mie-Gruneisen EOS known to properly characterize estane [9]. The strength models used for all materials were elastic-perfectly plastic.

The computational grid had a uniform cell size of 5 μm and was made up of 40 million cells. The average grain was roughly cubic in shape and was defined with 36 cells across its length. The smallest grains were defined with roughly eight cells across their length. The problems were run to 1.5 μs and used an average of 1,000 CPU hours each. Periodic boundary conditions were used in the directions lateral to the shock while outflow boundary conditions were used in the directions normal to the shock.

Fixed Eulerian tracers were distributed uniformly throughout all dimensions of the domain for post-processing the data produced by CTH. Tracers were placed at 25 μm intervals and were not placed on the domain boundaries. Each of the over 300,000 tracers tracked Pressure, Temperature, Density, Sound Speed, Reaction Extent, Kinetic and Internal Energy. By taking samples every 10 ns, 320 million tracer data points were produced. MatLab [10] was used to post-process the data.

3. Comparison of Existing Reactive Flow Models

Results from CTH's reactive flow models varied widely. Fig. 3.1 shows reaction extent in the RVE at 0.7 μs for three different reactive flow models.

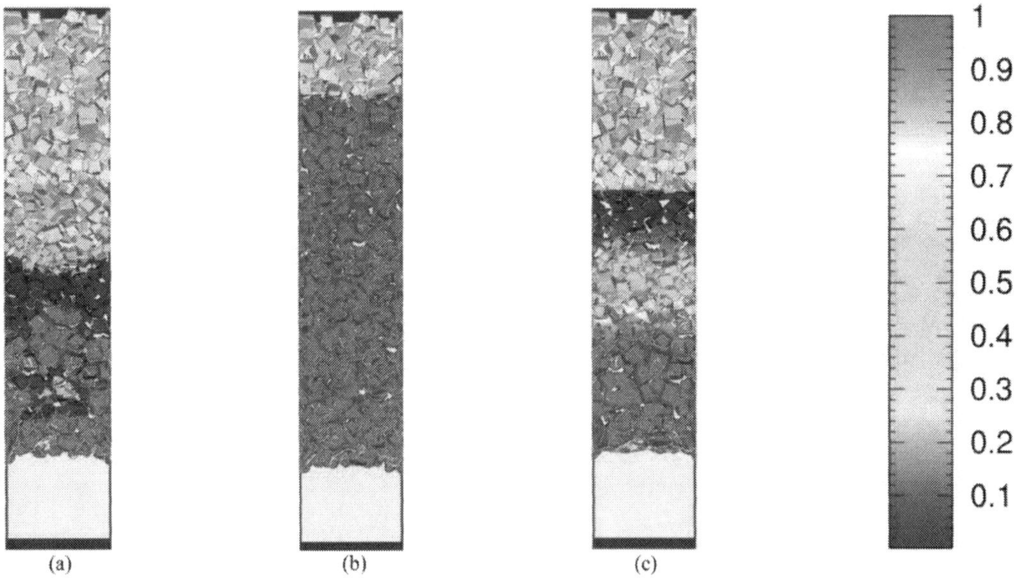

Fig. 3.1 Reaction extent at 0.7 μs in the (a) ARB, (b) HVRB and (c) IGRB models

Of the three reactive burn models, the Ignition and Growth model is the only one to show anything resembling a detonation wave. The Ignition and Growth model is a well respected reactive burn model that has been widely used for quite some time.

$$\frac{d\lambda}{dt}=(1-\lambda)^{S_0}G_0\left(\frac{\rho}{\rho_0}-1-a_0\right)^{y_0}+(1-\lambda)^{S_1}\lambda^{q_1}G_1P^{y_1}+(1-\lambda)^{S_2}\lambda^{q_2}G_2P^{y_2}$$

Equation 3.1 Ignition and Growth Reactive Burn model

Where λ is reaction extent, ρ is density, P is pressure and S, G, a, y, and q are adjustable parameters.

The HVRB model shows virtually no gradient in reaction extent and seems to be transitioning the grains to the full reacted product EOS as soon as the mechanical shock wave passes through them.

$$\lambda=1-\left(1-\frac{\phi^M}{X}\right)^X \qquad \phi=\frac{1}{\tau_0}\int_0^t\left(\frac{P-P_i}{P_R}\right)^Z dt$$

Equation 3.2 History Variable Reactive Burn model

Where τ is 10^{-6}, M, X and Z are adjustable parameters and P_i is a threshold pressure for initiation and P_r determines the pressure dependence of time [11].

The Arhennius model's PETN parameters in CTH are not well suited to handle strong shocks. Although its applicability in solid state kinetics is often criticized, we would like to stress that this model's failure in this current work is particularly due to calibration to a desensitized PETN. Arrhenius or Arrhenius-type models are well suited for reaction steps in solid kinetics. The interested reader is referred to Vyazovkin and Wight's review on solid state kinetics for a thorough explanation of the applicability of Arrhenius type equations for solid systems [12]. Experiments used to calibrate reactive flow models do not isolate elementary reactions steps. The addition of transient nucleation, diffusion, surface adsorption and desorption and solid state phase changes to kinetic rate laws need to be accounted for in a physically correct RVE if one were to use Arrhenius parameters determined for elementary steps.

$$\frac{d\lambda}{dt}=(1-\lambda)F\exp\left(\frac{-\Theta_0}{T}\right)$$

Equation 3.3 Arrhenius Reactive Burn model

Where F is a frequency function and Θ is a measure of the activation energy.

Based on the behavior observed in the three models, the decision was made to progress with the IGRB model in the short term. Use of this model will allow for exploration of statistical phenomena that exist in its well-defined reaction wave.

4. Impact of Granular Structure

In order to validate the use of granular structure in energetic materials in hydrocodes a similar problem was run as a continuous block of energetic material. The domain consisted of a solid block of PETN modeled using the IGRB model. This run showed a shock wave passing through the material followed by an accelerating reaction wave. As expected, the reaction wave 'caught up' to the shock front, thus creating a fully developed detonation wave. This process is seen in Fig. 4.1 (a) where Pressure is plotted along the z-axis (the axis normal to the shock front) for various times. Fig. 4.1 (b) shows the same plot for the RVE containing granular structure.

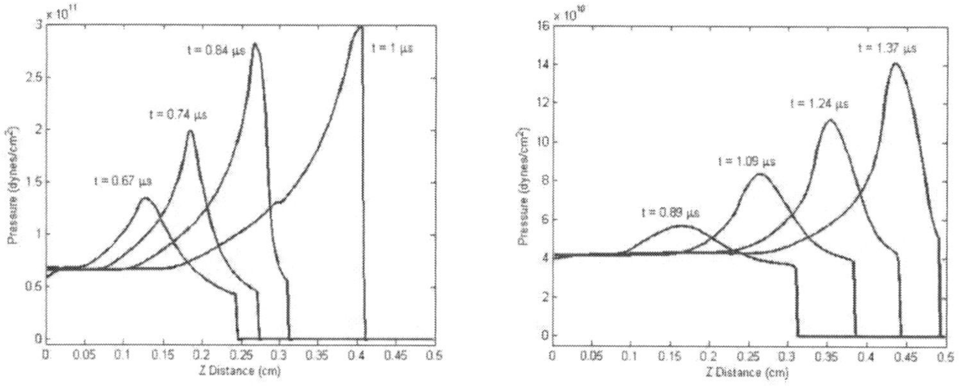

(a) (b)

Fig. 4.1 Pressure vs. Distance for (a) solid PETN and (b) a granular RVE

Fig. 4.2 (b) shows that the same process described above is captured by the granular structure. This demonstrates that hydrocodes are indeed able to simulate the propagation of mechanical and reactive waves through granular structure. The three most noticeable differences between the two plots are the pressure values, speed of the trailing reaction wave and the smoothness of the reaction wave in the granular RVE versus the sharpness of the reaction wave in the solid explosive. Peak pressure values in the solid block of PETN are approximately twice the value of those in the RVE. Also, the speed of the trailing reaction wave is significantly slower in the granular RVE. Both these differences are to be expected as the RVE has a significantly lower energetic material density than that of the solid block. The pressure profiles in Fig. 4.1 are averages of tracer data for each XY plane in the Z direction. It is apparent that the reaction wave in the RVE is less sharp than that of the solid. This smoothening is a function of the presence of the grains and the explicit presence of the binder being averaged over the XY directions. In reality, sharp, turbulent-like features exist locally nucleating and coalescing throughout the XY plane as the shock progresses.

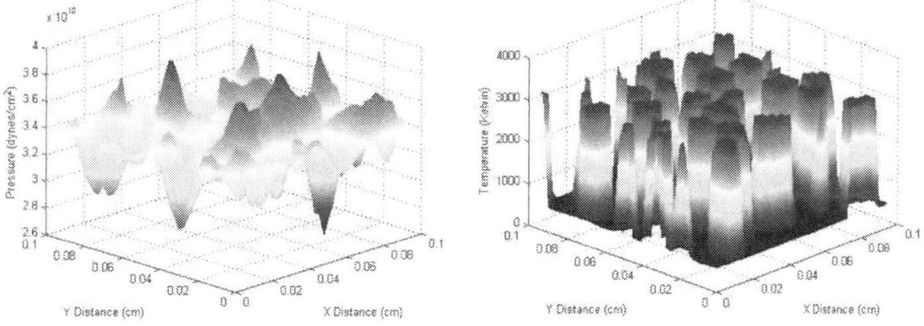

(a) (b)

Fig. 4.2 Plots in the X-Y plane at fixed Z values for (a) Pressure at 0.59 μs and (b) Temperature at 1.39 μs

The pressure plot in Fig. 4.2 (a) is taken immediately after the mechanical shock passes through an unreacted X-Y plane of the RVE. Clear fluctuations in the pressure field are shown that arise from multiple shock interactions, shock impedance mismatches, and material deformation. All of these are mechanical consequences of the stochastic microstructure. The surface plot in Fig. 4.2 (b) shows high temperature regions occurring within the PETN grains and low temperature regions in the surrounding binder material as the initial shock moves through the material. The temperature field at this time step helps to illustrate the grain locations in a representative XY slice. Upon inspection of Figs. 4.2a and 4.2b, it is apparent that the pressure fluctuations are not directly associated with a material but are dependent on a complex set of local and history-dependent phenomena. According to the IGRB pre-defined reaction thresholds, reaction is initiated in the PETN, raising the temperature as material is converted to follow the reacted products' EOS. This work's calculations neglected thermal conduction as a mechanism of heat transfer in the RVE. We do expect thermal conduction to play a role in the time evolution of our state variables and will be considered in future work.

Temperature fields at the mesoscale follow predictable statistics [2,5]. This is paramount to developing a stochastic reactive flow model. Turbulence models have long used probability density function (PDF) methods to capture subscale eddy formation and other turbulent dynamics with great success [13]. Direct experimental confirmation of predictable statistical behavior in subsonic turbulent flow fields has long been available. Experiments that capture mesoscale dynamics in detonating condensed phase energetic materials are not yet mature. We therefore are limited to traditional experimental data and making sound judgments over a matrix of numerical experiments. A framework that homogenizes the grain scale with the continuum by interrogating arbitrary PDFs of appropriate state variables allows for continual improvement of physics at scales below the RVE. Consider the development of a proper model constructed from an atomistic simulation of crystal dislocations moving in a single grain, compounding, and leading to a microscopic initiation event. Once these statistics are known, this model can then be homogenized up to influence the shape of the grain scale PDF in a computational cell of a macroscopic simulation.

Our studies predict statistical distributions that fit reasonably well using Gaussian mixture distributions. The MatLab kernel smoothing density estimate function, ksdensity, was used to estimate the PDF of 2D (XY) slices of tracer data restricted to a positive support with no censoring. The Gaussian mixture distribution function also included in MatLab, gmdistribution, uses the Expectation Maximization algorithm to optimize a Gaussian mixture distribution of an arbitrary number of Gaussians [10]. Fig. 4.3 (a) shows the ksdensity estimate for a pressure PDF of an XY plane and the gmdistribution fit. Fig. 4.3 (b) shows the time evolution of a PDF fit and leads to some important features. The PDF clearly undergoes a characteristic drift toward higher pressures and backward diffusion as the reactive wave progresses through the microstructure. This drift and diffusion can be captured by contemporary methods in stochastic calculus once appropriate assumptions are made. Those assumptions, including arguments for or against independent and stationary statistics will first need to be rigorously validated.

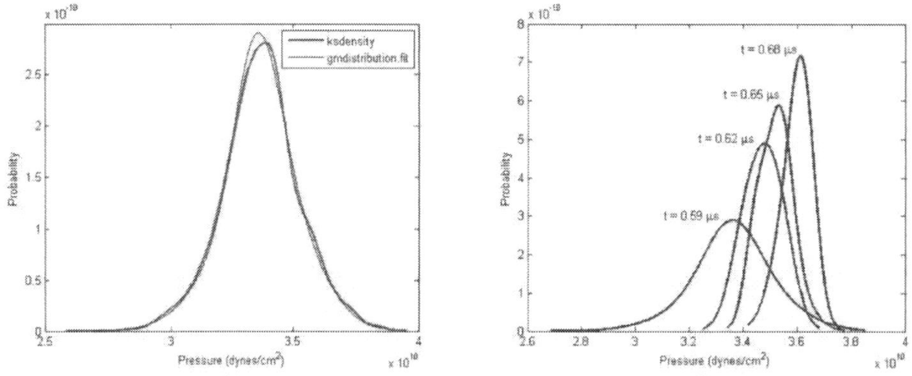

(a) (b)

Fig. 4.3 (a) Kernel smoothing density estimate and (b) Gaussian mixture distribution functions for 2D slice of the RVE

5. Conclusions

Several reactive burn models were evaluated for use in grain scale modeling. In the short-term, the IGRB model was chosen with pre-defined parameters for PETN. However, future work will focus on the Arrhenius type models appropriately modified for solids. Comparisons were made between solid block and granular computations. Hydrocodes demonstrated the ability to model both mechanical shock and reaction waves through granular structures. These computations showed stochastic fluctuations in state variables not seen in continuum representations. Finally, statistical representations in the form of PDFs were extracted from the data in an effort to show the direction of future work.

As part of the Institute for Multi-Scale Reactive Modeling the authors will work toward developing a stochastic reactive flow model based on statistics derived from these types of numerical experiments at the grain scale. Short term goals include choosing improved material models for pure constituents of the RVE and closing in on an optimal parameter space of stochastic state variables. Studies of improved viscoelastic models for the polymeric binder and crystal strength models that are parameterized through ab initio methods are forthcoming. Longer term goals include the homogenization of atomistic and microscale dynamics into models describing phenomena occurring at levels below the grain scale.

Acknowledgements

The authors are grateful to Mel Baer for invaluable assistance in this work. They would also like to thank Greg Fast for providing a great deal of value in a short amount of time. Finally, thanks are due to the Research and Development Branches of the Energetics, Warheads, and Environmental Technology Directorate of the U.S. Army's Armament Research, Development and Engineering Center. Computational resources were provided by the US Army High Performance Computing and Modernization Office (HPCMO) at the Army Research Laboratory.

References

[1] Baer, M. R., "Modeling Heterogeneous Energetic Materials at the Mesoscale", Thermochimica Acta, (384), pp. 351-367, 2002.

[2] Howe, P. M., Benson, D. C., "Exploitation of Some Micromechanical Concepts to Develop an Engineering Model of Shock Initiation", *Twelfth International Detonation Symposium*, San Diego, California, 2002.

[3] Swift, D. C., et al., "Mesoscale Modeling of Reaction in HMX Based Explosives", *Twelfth International Detonation Symposium*, San Diego, California, 2002.

[4] Reaugh, J. E., "Grain-scale Dynamics in Explosives", Lawrence Livermore National Laboratories, UCRL-ID-150388, September 2002.

[5] Baer, M. R., "Mesoscale Modeling of Shocks in Heterogeneous Reactive Materials", In: Yasuyuki Horie (ed.), Shock Wave Science and Technology Reference Library, Volume II, pp. 321-356, New York: Springer Berlin Heidelberg, 2007.

[6] McGlaun, J. M., Thompson, S. L. and Elrick, M. G., "CTH: A Three-Dimensional Shock Wave Physics Code", *International Journal of Impact Engineering*, 10, pp. 351-360, 1990.

[7] MacroPac, Intelligensys, Ltd., UK.

[8] Skidmore, C. B., et al., "The Evolution of Microstructural Changes in Pressed HMX Explosives", *Eleventh International Detonation Symposium*, Snowmass, Colorado, (ONR 33300-5), pp. 556-564, 1998.

[9] M. R. Baer, personal communication.

[10] MatLab, The MathWorks, USA.

[11] E. S. Hertel, Jr., and G. I. Kerley, "CTH Reference Manual: The Equation of State Package", Sandia National Laboratories report SAND98-0947, 1998.

[12] Vyazovkin, S., Wight, C. A., "Kinetics in Solids", Annu. Rev. Phys. Chem., (48), pp. 125-149, 1997

[13] Pope, S. B., Turbulent Flows, Cambridge: Cambridge University Press, 2000.

Thermophysical Property Predictions of Energetic Materials From Atomistic Molecular Dynamics Simulations

Nandhini Sokkalingam, Department of Chemical Engineering, Detroit, MI

Jeffery J. Potoff, Chemical Enginering, Wayne State University, Detroit, MI

The enormous volume of damage inflicted upon both personnel and equipment by unintended detonation of munitions has pushed the defense community towards new age energetic materials. Munitions employed in the warfield today are designed to meet the standards of Insensitive Munitions (IM) for the benefit and safety of the soldier. IM compounds exhibit low shock sensitivity and high thermal stability over traditional explosive compounds, such as TNT. In addition to the ability to resist unintended detonation, the environmental performance new energetic materials is also of great concern. The objective of this research is to determine how some energetic materials behave in the environment. The fate of any compound in soil, water or the atmosphere can be determined by predicting the partition or distribution coefficient. Two key partition coefficients used by the scientific community to assess a compound's impact in the environment are octanol-water partition coefficients and Henry's law constants (air-water partition coefficient). A wide variety of experimental techniques exist to measure the partition coefficient but atomistic simulation is attractive as a predictive tool for studying the environmental impact of these compounds due to three primary reasons: the hazardous nature of the explosives, long experimental time scales involved in testing of the materials and the costs associated with it. Furthermore, compounds can be evaluated for environmental hazards via simulation before synthesis, potentially leading to significant cost savings.

In this work force fields (molecular models) have been developed for six energetic compounds, 2,4-dinitroanisole (DNAN), N-methyl-p-nitroaniline (MNA), dinitropyrazole (DNP), nitrotriazolone (NTO), methyl-trinitroimidazole (MTNI) and TATB (trinitrotriaminobenzene) categorized by the army as potential insensitive munitions compounds. These force fields are used to predict pharmacokinetic properties such as octanol-water partition coefficients and Henry's law constants along with other thermophysical properties such as vapor-liquid equilibria, vapor pressure, critical parameters and normal boiling points. The validation of the developed force fields is accomplished with the scarce experimental data available. The octanol-water partition coefficients and Henry's law constant predicted for DNAN and MNA are in good agreement with the experiment. For DNP, NTO, MTNI and TATB, materials for which limited experimental data are available, the predicted partition coefficients agree well with values calculated with group contribution methods.

See more of this Session: Thermophysical Properties
See more of this Group/Topical: Particle Technology Forum

Liquid Flame Spray Deposited Nanocoatings On Painted Metal

Mikko Aromaa, Aerosol Physics Laboratory, Tampere University of Technology, Tampere, Finland

Miia Luhtalampi, Tampere University of Technology, Tampere, Finland

Mika Valden, Tampere University of Technology, Tampere, Finland

Jyrki M. Mäkelä, Aerosol Physics Laboratory, Tampere University of Technology, Tampere, Finland

Objects often have some surface finishing, e.g. paint or lacquer. There is a need to alter the surface properties of these kinds of surfaces as well. Here, we introduce the Liquid Flame Spray (LFS) method to alter the contact angle of surface treated metal. There will be great additional value for painted surfaces, as well. For example roof tops and cars need not to be cleaned that often if there is a hydrophobic coating on top for the running water will clean the surface. Previously, we have applied TiO_2 coatings on metal and glass in order to have photocatalytic surfaces [e.g. Keskinen et al. 2006]. Now, we painted metal sheets using ordinary paint and lacquered some of those as well. TiO_2 coating is deposited on the surface straight from the LFS. The painted surface is quite porous, which increases the yield. However, painted surfaces cannot take up as much heat as clean metal surfaces. Therefore, the processing temperature needs to be controlled more carefully. TiO_2 particles are produced from Titanium(IV) isopropoxide dissolved in isopropanol. The solution is sprayed into a hydrogen-oxygen flame and the particles form via a gas phase synthesis inside the flame. The particles are then deposited on the substrate to form a coating by sweeping the surface through the flame. First tests show an improvement in contact angle from a reference of 110° to 150°.

Keskinen H., Mäkelä J. M., Aromaa M., Keskinen J., Areva S., Cilâine V. T., Rosenholm J.B., Pore V., Ritala M., Leskelä M., Raulio M.,Salkinoja-Salonen M. S., Levänen E. and Mäntylä T. (2006) Deposition of Titania and Titania-Silver Nanoparticles by Liquid Flame Spray and Their Application as a Photocatalyst, Catalysis Letters, vol. 111, pp.127-132

See more of this Session: Gas Phase Synthesis of Nanoparticles
See more of this Group/Topical: Particle Technology Forum

Synthesis and Covalent Surface Functionalization of Non-Oxidic Iron Core-Shell Nanomagnets for Large Scale Magnetic Separations

Inge K. Herrmann, Department of Chemistry and Applied Biosciences, ETH Zurich, 8093 Zurich, Switzerland

Robert N. Grass, Department of Chemistry and Applied Biosciences, ETH Zurich, 8093 Zurich, Switzerland

Wendelin J. Stark, Department of Chemistry and Applied Biosciences, ETH Zurich, 8093 Zurich, Switzerland

Successful use of magnetic nanoparticles for target isolation strongly relies on rapid response of the nanomagnets to magnetic field gradients (high saturation magnetization) and additionally requires stability under a wide range of different conditions. Nanomagnets with optimal magnetic properties for above applications can be obtained by the use of metals (Ms, bulk ≤ 220 emu/g). Unfortunately, naked metal nanoparticles are chemically highly active (typically pyrophoric) and readily oxidize upon contact with air or water, generally resulting in a complete loss of magnetism. Up to now, this has promoted the widespread use of moderately magnetic oxide nanoparticles, such as magnetite (Ms, bulk ≤ 92 emu/g). In this work, we show the production of iron-based nanomagnets by a flame spray synthesis process under reducing atmosphere [1-3]. The reducing flame spray synthesis setup allows the continuous production of carbon coated iron carbide particles with a mean diameter of 30 nm at a production rate of > 10 g/h [3]. The carbon-encapsulated iron carbide nanoparticles have an exceptionally high saturation magnetization of 140 emu/g and are highly air stable (up to 200°C). Additionally, the carbon surface of these iron carbide nanomagnets could be covalently functionalized with various linkers by following the procedures initially described by Grass et al.[1]. The high saturation magnetization, thermal and chemical stability, low raw material costs, and the efficient synthesis process make the herein described carbon encapsulated iron carbide nanoparticles an attractive low cost alternative for currently applied nanomagnets. Figure: Reducing Flame Spray Synthesis setup (left) allows a continuous production of highly magnetic core/shell nanomagnets (center). The nanomagnets can be completely recovered from suspension within a few minutes (right). [1] Grass R. N., Athanassiou E.K., Stark W.J., (2007), Angew. Chem. Int. Ed., 46, 4909-12. [2] Athanassiou E.K., Grass R.N., Osterwalder N., Stark W.J., (2007) , Chem. Mater., 19 (20), 4847-53. [3] Herrmann I. K., Grass R.N., Mazunin D., Stark W.J., under review.

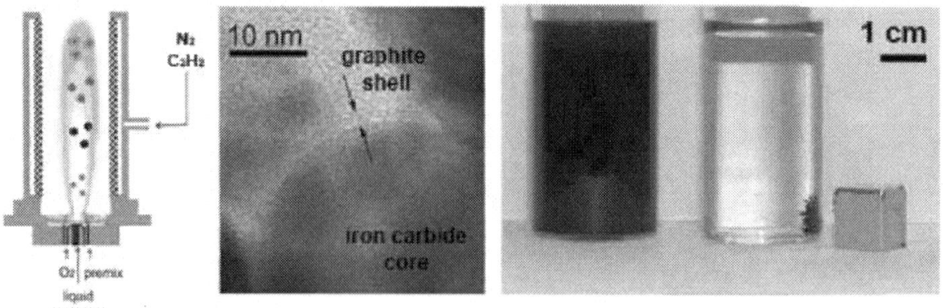

See more of this Session: Gas Phase Synthesis of Nanoparticles
See more of this Group/Topical: Particle Technology Forum

Impact of Fractal-Like Morphology On Surface Oxidation of Nanoparticles Synthesized Via Aerosol Route: A Kinetic Monte Carlo Study

Dibyendu Mukherjee, Department of Chemical and Biomolecular Engineering, Material Research and Innovation Laboratory(MRAIL), University of Tennessee, Knoxville, TN

Matthew Wang, Department of Chemical and Biomolecular Engineering, Material Research and Innovation Laboratory(MRAIL), University of Tennessee, Knoxville, TN

Bamin Khomami, Department of Chemical and Biomolecular Engineering, Material Research and Innovation Laboratory (MRAIL), University of Tennessee, Knoxville, TN

Nanoparticle manufacturing via aerosol routes has become increasingly conducive for the fabrication of nano-structured materials and thin films with controlled porosity and specific surface coatings. This either calls for production of spherical nanoparticles for low surface area compact structures (electronic device fabrications, sensors, etc.) or fractal-like aggregates for high surface area reactive structures (nanocatalysts, energetic materials, H2 storage, etc.). Yet, the complex inter-play between various chemical/physical pathways such as collision, coalescence and surface reactivity leading to the evolution of these structures is not clearly understood. Here, we present a kinetic Monte Carlo (KMC) based model that simulates the growth and evolution of fractal-like nanoparticles by particle-particle collision, subsequent non-isothermal coalescence and surface reaction, i.e., oxidation. The interesting aspect of the present KMC model lies in its ability to capture with high fidelity the interactions between competing mechanisms like the exothermic nature of particle coalescence events coupled with the inter-particle coagulation events and enhanced surface oxidation at nano-scale, all in the framework of complex particle morphology.

Specifically, the objective of the present study being focused on its ability to capture the influence of most energetically active processes namely, surface energy driven coalescence and oxidation on the particle evolution, we use surface fractal dimension, Ds varying between 2 (perfect spheres) to 3 (complete fractals) to define the morphological complexity of our nanoparticle systems. Results from our simulation indicate that at intermediate and later stages of nanoparticle growth, wherein the role of competing events such as collision, exothermic coalescence and surface oxidation becomes significant, the particle Ds value increasingly deviates from the spherical assumption, thereby increasing the surface area. This in turn strongly enhances surface area driven oxidation. In turn, the heat of reaction triggers particle temperature rise that quickly couples non-linearly to the coalescence driven energy release. This mechanism enhances particle coalescence, therefore, drastically driving the resulting oxidation. This self reinforcing process is eventually quenched by the oxide shell formation and heat loss to surroundings. The aforementioned non-linear coupling between coalescence and oxidation on complex particle surface morphologies in the framework of particle collisions based on a kernel modified by the fractal-like nano-aggregate structures, makes the physics of the present study highly intriguing and challenging. To demonstrate the fidelity of our new approach, we present results from the simulation of Al nanoparticles grown in air at different temperatures and pressures. Our results indicate that particle evolution and extent of oxidation differ significantly in the case of fractal-like nanoparticle structures from those with the spherical assumptions. The results are also consistent with experimental studies of Al nanoparticle oxidation.

See more of this Session: Gas Phase Synthesis of Nanoparticles
See more of this Group/Topical: Particle Technology Forum

Reducing Reaction Mechanism for Synthesis of TiO2 Nanoparticles

Maulik Mehta, Chemical and Biological Engineering, Iowa State University, Ames, IA

R. O. Fox, Department of Chemical & Biological Engineering, Iowa State University, Ames, IA

Venkat Raman, Aerospace and Engineering Mechanics, University of Texas at Austin, Austin, TX

Using computational fluid dynamics to model nanoparticle synthesis in chemical reactors has been proven to be an important tool for understanding and implementing coupling between transport processes and chemical reactions[1]. Accurate modeling of synthesis in chemical reactors involves a detailed reaction mechanism. In developing a chemical mechanism we propose a list of species and take into account all feasible reactions. This leads to a large and highly detailed reaction mechanism involving species and reactions that might not have a significant effect on the reaction path of the species of our interest. For example West et al.[2] have proposed a reaction mechanism for TiO2 nanoparticle production from TiCl4 that involves 25 species and 51 reactions. Computational fluid dynamics techniques are used to couple the reaction mechanism with the transport processes but due to the large number of species and reactions involved, coupling this chemistry with detailed flow solvers is computationally very expensive. Thus, for practical simulations reducing these highly detailed chemical kinetics to moderate size is very important.

Reducing the number of species and consequently the reactions is non-trivial. We have to make sure that taking out species from the mechanism does not lead to inaccurate estimation of the desired species in comparison to the detailed set. Finding out how sensitive our species of interest are with respect to the reaction rate constants will tell us which reactions are important for their formation. Thus, the reaction set can be reduced by taking into account only reactions that are highly sensitive to the formation of the desired species.

The objective of this work is to develop a sensitivity analysis tool that given a reaction mechanism set, complete with reaction rate constants, can give us the sensitivity of the species with respect to specified reactions. Thus, reducing the number of species that the flow solvers has to track to make simulating nanoparticle synthesis more feasible and computationally tractable. We carry out sensitivity analysis on the detailed reaction mechanism given by West et al.[3] for the combustion synthesis of TiO2 from TiCl4 that includes 30 species and 66 reactions.

References:

1. Rodney O. Fox, "CFD models for analysis and design of chemical reactors," Advances in Chemical Engineering, vol. 31, Ed. G. B. Marin, Elsevier, pp. 231-305 (2006).

2. R. H. West, M. S. Celnik, O. R. Inderwildi, M. Kraft, G. J. O. Beran, and W. H. Green, "Toward a Comprehensive Model of the Synthesis of TiO2 Particles from TiCl4," Ind. Eng. Chem. Res., 46(19):6147-6156 (2007).

3. R. H. West, R. A. Shirley, M. Kraft, C. F. Goldsmith, W. H. Green, "A Detailed Kinetic model for Combustion Synthesis of Titania from TiCl4 ," Preprint. http://como.cheng.cam.ac.uk/index.php?Page=Preprints&No=55

See more of this Session: Gas Phase Synthesis of Nanoparticles
See more of this Group/Topical: Particle Technology Forum

Design of Gas-Phase Coating of Nanoparticles

Beat Buesser, Particle Technology Laboratory, Institute of Process Engineering, Department of Mechanical and Process Engineering, ETH Zurich, Zurich, Switzerland

Sotiris E. Pratsinis, Particle Technology Laboratory, Institute of Process Engineering, Department of Mechanical and Process Engineering, ETH Zurich, Zurich, Switzerland

Nanoparticles are coated to make them compatible to their host matrices. With such core-shell particles, desirable properties of core (e.g. dielectric, scattering or opacifying performance) particles are conserved while modifying their surface by the shell material. For example, rutile TiO_2 particles are made by the "chloride" or "sulfate" process and are coated with silica, alumina and other oxides by wet impregnation processes to prevent the photocatalytic reaction of pigmentary TiO_2 with the organic host matrix in paints (Clark, 1975).

Aerosol coating is attractive as it contributes to gas-phase synthesis of functionalized nanoparticles in one-step facilitating their economic manufacture. Here aerosol coating of TiO_2 core particles with SiO_2 shells is investigated by a combination of particle dynamic models and computational fluid dynamics, accounting for SiO_2 monomer generation and growth by coagulation and sintering. The evolution of the SiO_2 coating particle population is simulated distinguishing monomers and particles with the bimodal model of Jeong and Choi (2003). At industrial scale manufacture of nanoparticles, such aerosols can encounter high concentration dynamics resulting in much faster growth kinetics and even gelation (Buesser *et al.*, 2009).

Figure 1 shows the modeled pathways of the coating process. The uncoated TiO_2 core particles, N_c, (white) enter the reactor at the bottom where they have been produced by flame spray pyrolysis (Mädler *et al.*, 2002). Coating starts by injecting vapor of coating shell precursor, C, HMDSO (hexamethyldisiloxane) through a torus ring after the TiO_2 particles have reached their optimal primary particle size that no longer changes by sintering (Teleki *et al.*, 2009). That way TiO_2 primary particles and aggregates have attained a self-preserving distribution that can be represented by a single size facilitating the simulations. Coating monomers (blue), N_{f1}, are generated by gas-phase reaction of HMDSO and form smooth coating shells, V_s, on the core particles by coagulation. Further downstream, coating monomers grow by coagulation and sintering to coating particles, either aggregates or agglomerates, N_{f2}, which create rough coating films, V_r, by coagulation with core particles. Rough coating shells sinter to smooth ones depending on temperature and primary particle size (Teleki *et al.*, 2005).

The aerosol coating model is combined with computational fluid dynamics to account for the non-premixed gas-phase reaction of the coating shell precursor and the flow and temperature fields defining the important mixing of the core and coating particle aerosol inside of the coating reactor (Teleki *et al.*, 2009). The influence of TiO_2 particle size and concentration, process temperature and cooling rate, degree of mixing and HMDSO precursor concentration on SiO_2 coating thickness and texture and process efficiency are investigated. The simulations are compared to experimental data of nanometer thin SiO_2-coated flame-made rutile TiO_2 particles (Teleki *et al.* 2008).

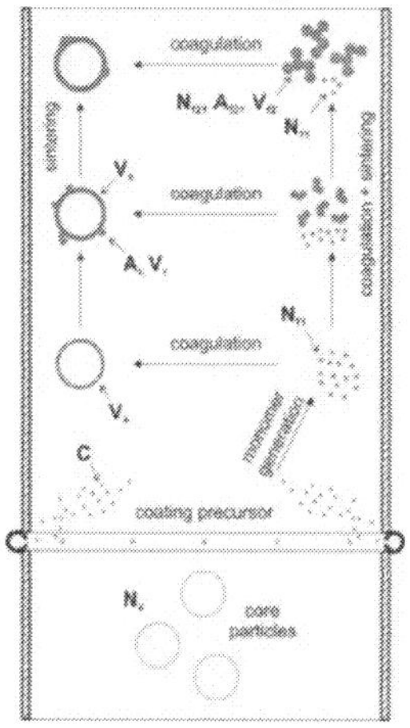

Figure 1 Schematic of the modeled pathways of aerosol coating processes.

Financial support from the Swiss National Science Foundation (SNF) grant # 200021-119946/1 is gratefully acknowledged.

Buesser, B., Heine, M.C., Pratsinis, S.E (2009), "Coagulation of highly concentrated aerosols" *J. Aerosol Sci.*,

40, 89-100

Clark, H. B., Ed. (1975). Titanium dioxide pigments. Treatise on Coatings, 3, Pigments Part I. New York, Marcel Dekker

Jeong, J. I. & Choi, M. (2003). A simple bimodal model for the evolution of non-spherical particles undergoing nucleation, coagulation and coalescence. *J. Aerosol Sci.* **34**: 965-976

Mädler, L., Stark, W. J. & Pratsinis, S. E. (2002). Rapid synthesis of stable ZnO quantum dots. *J. Appl. Phys.* **92**: 6537-6540

Teleki, A., Pratsinis, S. E., Wegner, K., Jossen R. & Krumeich, F. (2005), "Flame-coating of titania particles with silica" *J. Mater. Res.,* **20**, 1336-1347

Teleki, A., Heine, M. C., Krumeich, F., Akhtar, M. K. & Pratsinis, S. E. (2008). *In-situ* coating of flame-made TiO_2 particles by nanothin SiO_2 films. *Langmuir* **24**: 12553-12558

Teleki, A., Buesser, B., Heine, M. C., Krumeich, F., Akhtar, M. K. & Pratsinis, S. E. (2009), "Role of gas-aerosol mixing during in situ coating of flame-made titania particles" *Ind. Eng. Chem. Res.*, **48**, 85-92

See more of this Session: Gas Phase Synthesis of Nanoparticles
See more of this Group/Topical: Particle Technology Forum

Universal Correlation and Mechanism for the Antibacterial Activity of Ag Nanoparticles

Sotiris E. Pratsinis, Department of Mechanical & Process Engineering, ETH Zurich, Zurich, Switzerland

Georgios A. Sotiriou, Department of Mechanical & Process Engineering, ETH Zurich, Zurich, Switzerland

Uncoated and SiO_2-coated Ag/SiO_2 nanoparticles were made by flame spray pyrolysis (FSP) of a Ag nitrate and hexamethyldisiloxane (HMDSO) containing solution. The product particles were characterized by STEM/TEM, EDX spectroscopy, X-ray diffraction, N_2 absorption (BET) and O_2 chemisorption. The Ag clusters exhibited a bimodal distribution as determined by microscopy and XRD. The conductivity of aqueous solutions containing the uncoated Ag/SiO_2 nanoparticles was measured as well as their antibacterial activity against the Gram negative bacterium *Escherichia coli*. Low Ag-content particles had more exposed Ag surface area. The antibacterial performance of Ag/SiO_2 nanoparticles was determined solely by the exposed Ag surface area in the solution over a wide range of particle concentrations and Ag-content. Additionally, by hermetically coating the Ag nanoparticles by a SiO_2 layer, inhibition of the antibacterial activity can be established. Finally, Ag ions rather than Ag nanoparticles determine the antibacterial performance of the Ag/SiO_2 nanoparticles, as it was shown by monitoring bacterial growths in solutions containing similar Ag^+ ions concentrations, in the presence and absence of Ag nanoparticles.

See more of this Session: Gas Phase Synthesis of Nanoparticles
See more of this Group/Topical: Particle Technology Forum

Particioning of Elements and Phases in Flame-Synthesized Complex Metal Oxide Particles

Bing Guo, Department of Mechanical Engineering, Texas A&M University, College Station, TX

Partitioning of elements and phases in flame-synthesized complex metal oxide particles has not been fully studied. In this study, the partitioning of elements and phases was investigated from a thermodynamic standpoint.

The results suggest that

See more of this Session: Gas Phase Synthesis of Nanoparticles
See more of this Group/Topical: Particle Technology Forum

Development of High Density Circulating Fluidized Beds for Steam Gasification of Coal or Biomass

Guoqing Guan, Department of Mechanical and Biofunctional Systems, Institute of Industrial Science, The University of Tokyo, Tokyo153-8505, Japan

Masahiro Ikeda, Department of Mechanical and Biofunctional Systems, Institute of Industrial Science, The University of Tokyo, Tokyo153-8505, Japan

Chihiro Fushimi, Department of Mechanical and Biofunctional Systems, Institute of Industrial Science, The University of Tokyo, Tokyo153-8505, Japan

Atsushi Tsutsumi, Department of Mechanical and Biofunctional Systems, Institute of Industrial Science, The University of Tokyo, Tokyo153-8505, Japan

Toshiyuki Suda, IHI Corporation, Yokohama 235-8501, Japan

Satoru Matsuda, Clean Gas Group, Energy Technology Research Institute, National Institute of Advanced Industrial Science and Technology, Ibaraki 305-8569, Japan

Yoshizo Suzuki, Clean Gas Group, Energy Technology Research Institute, National Institute of Advanced Industrial Science and Technology, Ibaraki 305-8569, Japan

In order to increase the power generation efficiency, it is possible to use an advanced-integrated gasification combined cycle (A-IGCC) or advanced-integrated gasification fuel cell (A-IGFC) system making use of exergy recuperation technology for recycling of waste heat from gas turbine or fuel cells for steam gasification of coal and biomass. In this system, using a circulating fluidized bed (CFB) gasifier with high solid mass flux and large solid holdup to take the place of traditional fluidized bed gasifiers such as spouted bed gasifier is expected to more effectively utilize the heat energy produced by the combustion of the tar in the riser or from the waste steam for the pyrolysis and gasification of coal and biomass. In this study, a novel circulating fluidized bed (CFB) system, which composed of a downer (as a pyrolyzer), a bubbling fluidized bed (as a gasifier), and a riser (as a combustor), was set up and the fundamental characteristics were investigated. The objective is to find the optimum operation conditions and structure to realize a particle circulating with a high density and a high flux in the CFB system.

See more of this Session: Fluidization and Fluid-Particle Systems for Gasification and Biomass Utilization
See more of this Group/Topical: Particle Technology Forum

Carbon Negative Biomass Chemical Looping (BCL) Process - Reactor Design and Life Cycle Analysis

Fanxing Li, Chemical and Biomolecular Engineering, The Ohio State University, Columbus, OH

Hyung Rae Kim, Chemical and Biomolecular Engineering, The Ohio State University, Columbus, OH

Liang Zeng, Chemical and Biomolecular Engineering, The Ohio State University, Columbus, OH

Luis Gilberto Velazquez Vargas, The Babcock and Wilcox Company, Barberton, OH

Fei Wang, Chemical and Biomolecular Engineering, The Ohio State University, Columbus, OH

L.-S. Fan, Chemical and Biomolecular Engineering, The Ohio State University, Columbus, OH

The Biomass Chemical Looping (BCL) process utilizes iron oxide as the oxygen carrier to produce hydrogen and electricity from cellulosic biomass, lignin-rich non-fermentable residues, and agricultural residues with 100% CO_2 capture. Preliminary experimental results indicate that the BCL process concept is feasible. In this paper, the design of the BCL reactors is proposed. Studies on the hydrodynamic behaviors of the BCL reactors using the proposed design are performed with a cold flow model. An ASPEN Plus® model is developed to evaluate the performance of the BCL process. Life cycle analysis (LCA) on the BCL process is also conducted. Process analyses show that close to 70% process efficiency (HHV basis) can potentially be achieved for hydrogen and electricity co-production from cellulosic biomass using the BCL process. Moreover, the process is carbon negative with a net CO_2 emission of less than -1.3 tons per ton of biomass converted. The high energy conversion efficiency and carbon negative nature of the BCL process results from the process intensification, energy management optimization, and in-situ CO_2 separation.

See more of this Session: Fluidization and Fluid-Particle Systems for Gasification and Biomass Utilization
See more of this Group/Topical: Particle Technology Forum

Computational Simulation of Fluidization Dynamics Inside a Biomass Gasifier

Paper 169852

Peter Blaser, CPFD Software, LLC, Albuquerque, NM
Ravi Chandran, ThermoChem Recovery International, Inc., Baltimore, MD

Abstract

Scale-up of biomass gasification technology, from experiments to pilot plants to commercial units, is an involved, time-consuming and expensive process. With advances in modeling and high speed computing, it becomes possible to dovetail experimental and simulation work to enhance the speed as well as reduce the overall cost of technology commercialization. This paper presents a computational model for simulating the complex, three-dimensional, transient, multiphase gas-solids behavior inside a commercial biomass gasification unit using the computational particle fluid dynamics numerical method. Results include complex gas-solids flow phenomena including bed expansion, bubble behavior and solids circulation profiles.

Introduction

Advances in biomass production and conversion promise a biofuture with the hallmark of sustainability, energy security and climate stability. There exist biochemical and thermochemical platforms to convert biomass into valuable products. At present, the suite of technologies in the biochemical platform can convert a portion of select biomass into a limited number of products. In contrast, gasification – a technology in the thermochemical platform – can convert all, or almost all, of any carbonaceous biomass into a wide range of downstream value added products. Gasification is an enabling technology that generates a syngas comprising H_2 and CO from biomass, engenders a polygeneration optionality and offers the potential to transform regional economies into sustainable communities. The benefits include the displacement of fossil fuels, production of renewable biofuels and biochemicals, generation of green electricity, reduced environmental impact and improved energy efficiency.

Biomass gasification typically is accomplished under high temperatures in a fluidized bed. Fluidized beds are used for their large heat capacities resulting in nearly isothermal conditions when well mixed. Important elements in a biomass gasifier include overall unit geometry, fluidization media, biomass injection, fluid-particle separation systems such as cyclones, heating components, particle properties such as density and particle size distribution (PSD) and bed fluidization quality and uniformity. These elements do not work in isolation, but rather as components of a whole.

Scale-up, from experiments to pilot plants to commercial units, is an involved, time-consuming and expensive process. With advances in modeling and high speed computing, it becomes possible to dovetail experimental and simulation work to enhance the speed as well as reduce the overall cost of technology commercialization. Computational models help understand, predict, design and optimize full-scale, commercial biomass gasifier unit operations. This paper presents a computational model of the fluidization dynamics inside a commercial black liquor steam reformer.

Computational model

The overall gasifier geometry is shown in Figure 1. Fluidization steam enters through a gas sparger at the bottom of the vessel. Liquid biomass, or black liquor, is injected from discrete nozzles above a gas sparger. Three bundles of pulse heater tubes add the necessary heat to the system. Gases and solids exit into one of four internal cyclones. Solids are returned to the fluidized bed via the cyclone dip legs.

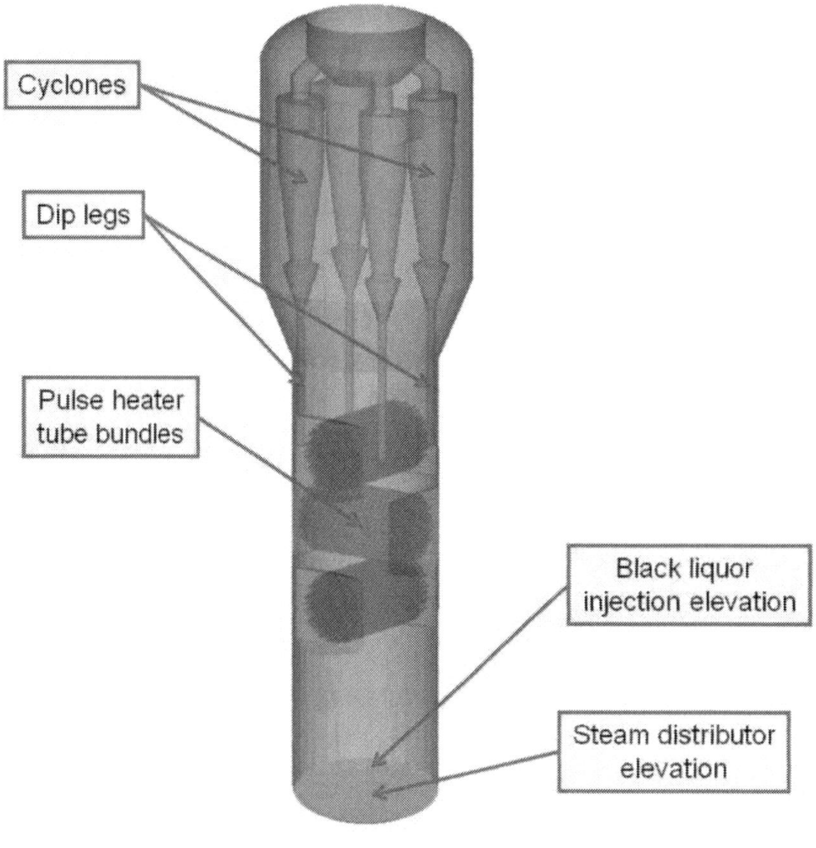

Figure 1: Geometry

The computational particle fluid dynamics (CPFD[1]) numerical method[2] was used to solve for the three-dimensional, transient, multiphase, gas-solids flow inside the steam reformer. The gas flow is computed in an Eulerian frame of reference while the particles are treated as discrete. Both phases are tightly coupled in a bi-directional manner. Details of the numerical method can be found in the literature[3,4].

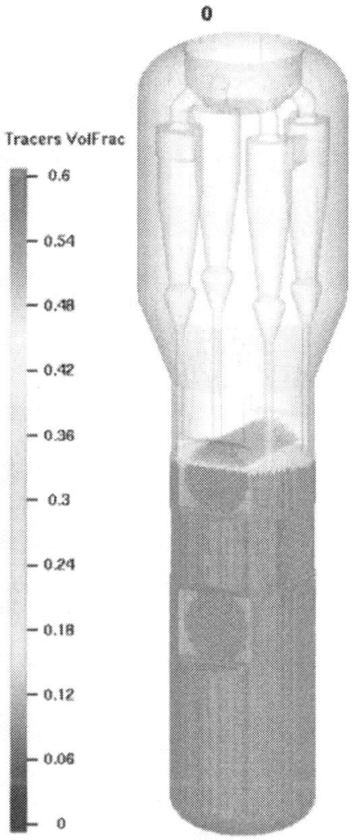

Figure 2: Particle initialization

A 235,000 cell grid was cut from a 437,000 cell Cartesian mesh. Additionally, the motions of approximately 2.2 million computational particles were computed. A grid generator methodology appropriate to multiphase numerical method was used; CPFD grid walls act as both boundaries of fluid control volumes and obstacles for particle bounce models.

Steam was injected through a distributor while black liquor, or liquid biomass, was injected through various injection nozzles. Not all liquor injectors were utilized, mimicking the unit operating conditions. The commercial unit operator has a choice of active black liquor injection nozzles and flow rates through each. The gas exited the domain at the cyclone inlet horns. Additionally, a minimal amount of gas was returned to the bed with the particles at the cyclone dip legs. The bed material is sodium carbonate with a density of 2275 kg/m^3 (142 lb/ft^3) and a defined particle size distribution. The initial bed loading is shown in Figure 2.

The calculation was compressible and isothermal. While the thermal behavior is extremely important to the performance of the actual unit, this work is focused on the ability of the model to predict the dynamics of gas-solids fluidization. Additionally, it was assumed that the liquid biomass gasified immediately upon entering the fluidized bed. Ongoing efforts include expanding this work to include thermal behavior, gasification and gas-solids chemistry.

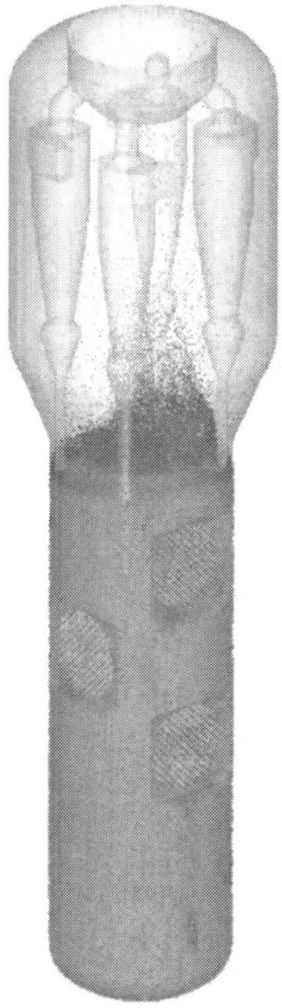

Figure 3: Steady-state bed height

Results

A three-dimensional, transient, multiphase calculation of the isothermal gas-solid flows in the steam reformer was performed for one minute of real time. Figure 3 shows the expanded bed height at under quasi-steady operating conditions. A definite bed level is observed to be higher than the static bed height due to the presence of bubbles in the system. It is desirable that the expanded bed height be above the top heater tube bundle, for optimal heat transfer, yet far below the cyclone inlets, to avoid excessive entrainment into the cyclones. Figure 3 also shows the presence of bed particles above the expanded bed height; these particles are ejected out of the bed by the bursting of bubbles at the bed surface. Fine particles are entrained with the gas flows entering the cyclones.

Many simulations were carried out to understand fluidization dynamics and a snapshot of the results are presented herein.

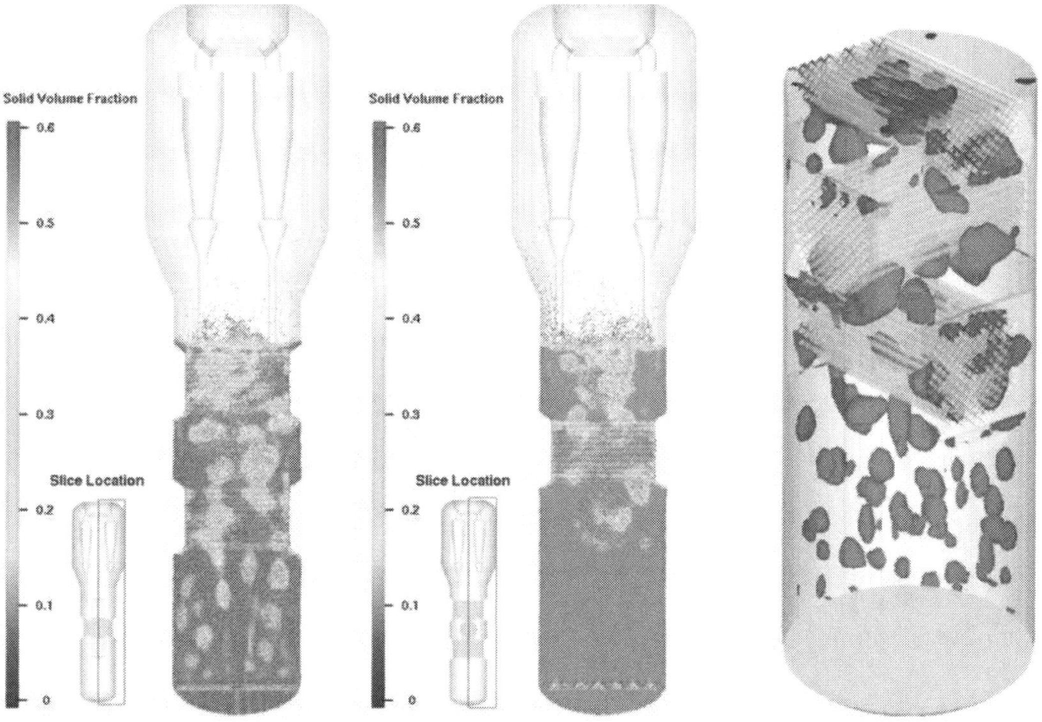

Figure 4: Particle volume fraction inside the steam reformer

Figure 4 shows three views of the particle volume fraction inside the steam reformer. The images on the left and center of Figure 4 show that a different bubbling pattern can be observed at different planes inside the unit. The differences in bubbling are due to the geometric details of the steam sparger and black liquor injectors. The image on the right of Figure 4 shows all the locations where the solid volume fraction is less than 0.3. The bubbling appears quite uniform through the cross-section, however bubbles tend to grow as they rise up the bed due to gas compressibility and expansion effects.

Figure 5 shows how the bubbling behavior inside the steam reformer can change as a function of the input conditions. The images on the left and right of Figure 5 represent different steam and black liquor injection rates. This shows that the multiphase bed dynamics are a function of the inlet conditions. The two states observed in Figure 5 show a transition from smaller bubbles to fewer, larger gas streams.

An understanding of the bed dynamics is vital when performing scale-up operations from a pilot plant to commercial designs. Figure 6 shows how the bed circulation changes as a function of a change in steam injection, black liquor flow and particle size distribution. The circulation patterns have an impact on mixing and thermal behavior of the fluidized bed. Extrapolating flow patterns from smaller-scale studies to commercial scales would be poor engineering practice.

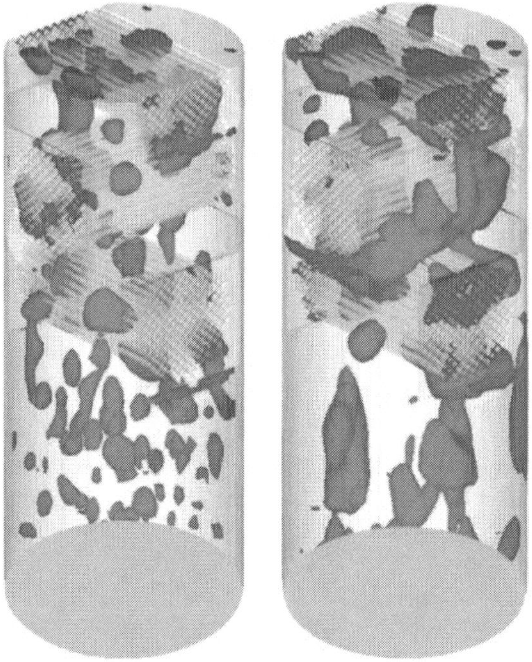

Figure 5: Sensitivity of bubble dynamics to process conditions

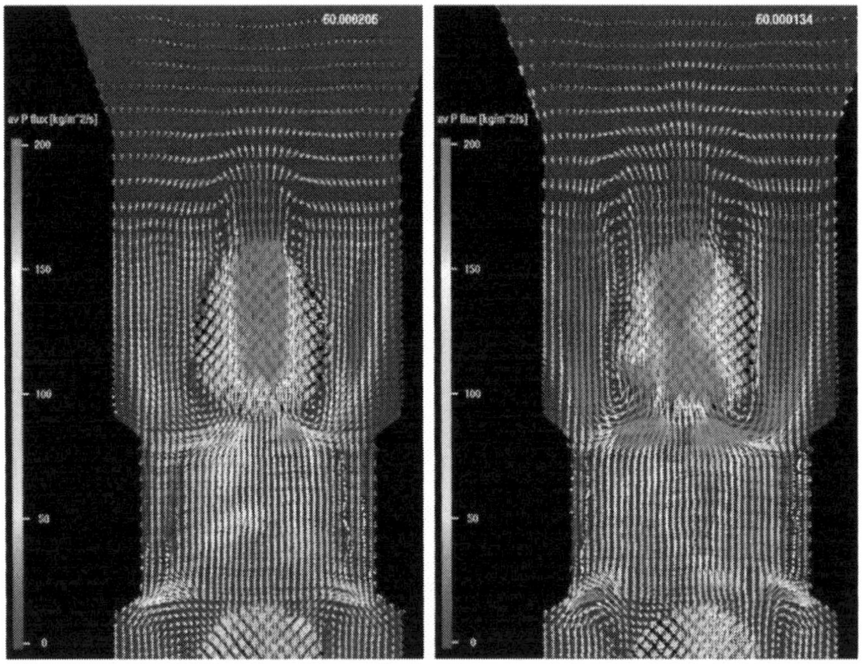

Figure 6: Sensitivity of bed circulation to process conditions

Discussion

These results have presented a computational model for the three-dimensional, transient, isothermal flow inside a commercial biomass gasifier unit. Understanding the gas-solids flow is vital for engineers designing full-scale units. The multiphase flow is influenced by geometry, gas injection locations, gas flow rates, biomass injection location, biomass flow rates, operating pressure, as well as particle density, mean particle size and particle size distribution. A computational model is an excellent tool for understanding the interplay of these factors on unit performance.

However, actual gasifiers utilize thermal and chemical processes. The use of computational models for prediction of chemical processes inside gasifiers is reported elsewhere[4]. Ongoing efforts are focused on gasification quantification inside a biomass gasification pilot plant. The pilot plant scale is excellent for validation of the computational model against known data such as thermal profiles and syngas production.

Conclusions

The production of green fuels, chemicals and power is enabled by biomass gasification technology. This thermochemical process can convert all, or almost all, of any carbonaceous biomass into a wide range of downstream value added products, enabling sustainable systems and carbon-neutral emissions. The proper scaling of commercial units, is vital for rapid and successful commercialization. This paper has presented a computational model for simulating the complex, three-dimensional, transient, multiphase gas-solids behavior inside a commercial biomass gasification unit. Such a computational tool enables decisions regarding design and selection of process parameters for reliable and confident commercial scale-up.

References

---

[1] CPFD is a registered trademark of CPFD Software, LLC.

[2] The computational particle fluid dynamic (CPFD) numerical method was developed by D. Snider at CPFD Software, LLC.

[3] Snider, D. M., 2001, "An Incompressible Three-Dimensional Multiphase Particle-in-Cell Model for Dense Particle Flows". Journal of Computational Physics **170,** 523–549.

[4] Snider, D., and Clark, S, "Three-dimensional prediction of a chemically reacting syngas gasifier with the CPFD Eulerian-Lagrangian method." AIChE proceedings, New York, NY, 2009.

Biomass Mixing and Segregation in a Fluidized Bed

Theodore J. Heindel, Department of Mechanical Engineering, Iowa State University, Ames, IA

Norman Keller, Department of Mechanical Engineering, Iowa State University, Ames, IA

Fluidized bed technology is useful for combustion, pyrolysis, and gasification of solid fuels such as biomass, which, as an alternative fuel, is important to industry because biomass is a potential alternative to petroleum-based fuels in the United States. There is usually a notable difference in the fluidization behavior between the solid fuel particle and the fluidized bed media (e.g., refractory sand) due to contrasting size, shape, and particle density; these differences can lead to poor solid fuel distribution and diminished performance. The hydrodynamics in a fluidized bed drive gas-solid contact, and thus, have a significant influence on fluidized bed performance. Although fluidized bed hydrodynamics are key parameters in their operation, they are still poorly understood, particularly when the solid fuel component like biomass is significantly different from the fluidized bed media.

This study provides a summary of the visual observations when biomass is mixed with inert fluidized bed material in a 9.5 cm diameter cold-flow fluidized bed. The biomass is composed of ground corncob in three different size ranges (200-300 micron, 500-600 micron, and 800-1000 micron), while the inert bed material is 500-600 micron glass beads. The ground corncob/glass bead mixtures are fluidized at several flow velocities and particle mixing and segregation is observed and documented.

See more of this Session: Fluidization and Fluid-Particle Systems for Gasification and Biomass Utilization
See more of this Group/Topical: Particle Technology Forum

Imposition of the Pressure-Dependent Wall Slip Condition in Mathematical Modeling of the Continuous Processing of Energetic Materials: Twin Screw Extrusion and Die Flows

AICHE, Nashville, Thursday, November 12, 2009: 12:30 PM
Lincoln C (Gaylord Opryland Hotel)

Dilhan Kalyon, <dkalyon@stevens.edu> Stevens Institute of Technology, Hoboken, NJ
Moinuddin Malik, HfMI, Stevens Institute of Technology, Hoboken, NJ
Michael Fair, US Army, ARDEC, Dover, NJ
David Fair, Picatinny Arsenal, Dover, NJ

A number of studies have indicated that various types of flow instabilities exhibited by polymeric melts and concentrated suspensions of polymeric suspensions are affected by the total normal stress at the wall (which for concentrated suspensions can be assumed to be equivalent to pressure) and the pressure dependence of the wall slip velocity [Tang and Kalyon (2008 a and b). The development of flow instabilities during the flow of concentrated suspensions of rigid particles incorporated into polymeric binders in simple capillary dies (cylindrical reservoir connected to a converging entry region into a straight circular tube) can involve time-dependent changes in the concentration of particles or changes in the shape of the extrudates emerging from tube flow. Time-dependent changes in the concentration of particles generally occur on the basis of time-periodic oscillations in extrusion pressure, upon the formation and break-up of mats of solids at the capillary entrance and the filtration of the binder phase [Yaras et al. (1994); Rough et al. (2000)].

The time-dependent changes in the concentration of rigid particles of extrudates of concentrated suspensions become especially significant with decreasing binder shear viscosity (filtration based flow instabilities are especially prevalent with Newtonian binders), increasing particle size, increasing capillary convergence ratio (reservoir diameter over the tube diameter) and decreasing flow rate [Yilmazer et al. (1989)]. On the other hand, with increasing shear viscosity of the binder phase, concentrated suspensions exhibit time and apparent shear rate-dependent changes in the shape of extrudates emerging from tube flow [Birinci and Kalyon (2006)] akin to flow instabilities associated with various types of extrudate shape distortions observed in the tube flow of polymeric melts [Benbow and Lamb (1963); Kalika and Denn (1987); Denn (2001); Robert et al. (2004); Hatzikiriakos and Migler (2005)].

The effects of the presence and the concentration of rigid particles on the development of shape distortions of the extrudates of suspensions are pronounced and contradictory. For an elastomer that does not exhibit flow instabilities over a wide range of shear rates in capillary or rectangular slit flows, the incorporation of particles induced extrudate shape distortions over the same apparent shear rate range [Tang and Kalyon (2004a)]. On the other hand, for other polymer melts that exhibit instabilities upon flow through cylindrical dies, the incorporation of particles reduces the range of apparent shear rates over which flow instabilities are observed at relatively low particle concentrations [Rosenbaum et al. (2000); Birinci and Kalyon (2006)] and eliminates them altogether at relatively high particle concentrations [Birinci and Kalyon (2006)].

The slip-based theoretical treatment of the development of flow instabilities of polymer melts in simple channels has generally relied on the considerations of the compressibility and various empirical non-monotonic wall slip velocity versus wall shear stress expressions [Hatzikiriakos and Dealy (1992a); Den Doelder et al. (1998); Georgiou (2003)]. For polymeric liquids the pressure dependence of the wall slip velocity can be significant [Vinogradov and Ivanova (1967); Hatzikiriakos and Dealy (1992b); Person and Denn (1997); Tang and Kalyon (2008)] with the slip velocity decreasing with increasing pressure. Since the flow curves characterized over relatively high apparent shear rates are subject to wall slip, and since such slip is a function of wall shear stress, as well as pressure, new methodologies, including the use of inverse problem solutions and squeeze flow, are necessary for the characterization of the parameters of pressure-dependent wall slip and shear viscosity material functions [Tang and Kalyon (2008)].

With suspensions the primary cause of wall slip is the apparent slip mechanism associated with the formation of a pure binder or binder-rich apparent slip layer at the wall [Reiner (1960)]. The apparent slip layer (Vand layer, i.e., a zone which is free of particles and thus consists solely of the binder) has a thickness, δ, at the wall. Since the thickness of the slip layer, δ, would be significantly smaller than the channel gap, the formation of the slip layer gives the appearance of wall slip; hence the "apparent slip" at the wall [Reiner (1960); Cohen and Metzner (1985); Yilmazer and Kalyon (1989); Kok et al. (2002); Tabuteau et al. (2004); Kalyon (2005)]. This mechanism is depicted in an exaggerated manner in Figure 1, which shows the apparent slip layer formed next to the wall of the tubular die. Such apparent wall slip may or may not be significant in comparison to the mean velocity in the channel, i.e., constituting the weak versus strong wall slip of the suspension. The rigidity of the particles can alter the dynamics of the formation of the apparent slip layer [Meeker et al. (2004); Adams et al. (2004)].

Unlike the earlier analyses in the treatment of apparent slip, here the binder itself is also considered to be subject to a slip condition with the wall slip velocity of the polymeric binder, u_{sb}, following a hyperbolic tangent-type dependence on the wall shear stress, τ_w [Tang and Kalyon (2004a); Tang and Kalyon (2008)]:

$$u_{sb} = \beta_b \tau_w^{s_1} \left(0.5 + 0.5 \tanh\left(\alpha(\tau_w - \tau_c)\right)\right) \quad (1)$$

where β_b and s_1 are the slip coefficient (referred to as Navier's slip coefficient for $s_1=1$) and slip exponent of the polymeric binder, respectively, and α is a positive constant (typically 1-20) describing the sharpness of the weak-to-strong slip transition in the slip velocity of the polymeric binder at the critical wall shear stress, τ_c.

As noted by Hatzikiriakos and Dealy (1992b), for viscoelastic polymer melts the slip velocity should be a function of the pressure and the first and second normal stress differences at the wall. For polymeric suspensions with rigid particles the development of normal stress differences would also be affected by the formation of anisotropic particle clusters [Nott and Brady (1994); Zarraga et al. (2000)]. For the materials considered here, i.e., highly filled suspensions of low aspect ratio particles, the normal stress effects are relatively small and only the pressure effect is considered. The slip coefficient for the binder, β_b, is assumed to vary inversely with pressure, akin to the slip of compressible gases during flow through simple conduits [Knudsen (1950)]:

$$\beta_b = \beta_0 \left(\frac{p_a}{p} \right)^{\kappa}, \tag{2}$$

where p is pressure at any location in the die, p_a is the atmospheric pressure, β_0 is the slip coefficient of the binder at atmospheric pressure. The exponent κ becomes equal to one for Knudsen flow but needs to be determined experimentally for a polymeric melt [Tang and Kalyon (2008)] and its suspensions with rigid particles with κ dependent on the concentration of particles, i.e., $\kappa(\phi)$. The available experimental data for polymer melts [Vinogradov and Ivanova (1967); Hatzikiriakos and Dealy (1992b)] indicate that the slip coefficient decreases with increasing pressure (thus κ is positive). The principal underlying mechanisms for pressure dependence of wall slip are considered to be the entrainment of air into the binder phase of the suspension and the residence time dependence of the establishment of the apparent slip condition as discussed later.

It is assumed that the Ostwald-de Waele or "power law" behavior represents the behavior of the shear viscosity of the binder phase in tube flow (where $u(r)$ is the axial z-velocity),

$\tau_{rz} = -m_b \left|\frac{du}{dr}\right|^{n_b - 1} \left(\frac{du}{dr}\right)$, with m_b and n_b the consistency index and the power law index parameters of the Ostwald-de-Waele "power-law" equation for the pure binder, respectively. The slip velocity, u_s, for the apparent slip mechanism [Kalyon, (2005)] subject to the slip of the binder at the tube wall (Figure 1) consists of the contributions of the slip of the binder, u_{sb}, and the contribution of the apparent slip mechanism, u_{sa}:

$$u_s = u_{sa} + u_{sb} = \frac{\tau_w^{s_b}}{m_b^{s_b}} \frac{R}{(s_b + 1)} \left[1 - \left(1 - \frac{\delta}{R}\right)^{s_b + 1}\right] + u_{sb} \qquad (3)$$

Considering that the reciprocal power-law index of the binder, i.e., $s_b = 1/n_b$, is positive and assuming an integer value - the use of Binomial Theorem provides $(1 - \delta/R)^{s_b + 1} \cong 1 - (s_b + 1)\delta / R$, the slip velocity at the interface between the apparent slip layer and the bulk of the suspension, u_s, i.e., at $r = R - \delta$, becomes:

$$u_s = u_{sa} + u_{sb} = \frac{\tau_w^{s_b}}{m_b^{s_b}} \delta + u_{sb} = \beta_a \tau_w^{s_b} + u_{sb} \qquad (4)$$

where β_a is the Navier's slip coefficient resulting from the apparent slip mechanism, i.e., $\beta_a = \delta / m_b^{s_b}$ [Kalyon (2005)]. A correlation for the apparent slip layer thickness, δ, and the Navier's slip coefficient $\beta_a = (D_p / m_b^{sb})(1 - \phi / \phi_m)$ can be used [Kalyon (2005)], where D_p is the particle diameter, ϕ is the volume percent of rigid particles, and ϕ_m is the maximum packing fraction of solids. This correlation for the mean value of the slip layer thickness (over the length of the die) applies for concentrated suspensions with rigid particles exhibiting low aspect ratios at $\phi < \phi_m$ and suggests that the flow of the concentrated suspension approaches plug flow as $\phi \rightarrow \phi_m$. This is seen below from the simplified analysis of the apparent wall slip of a Newtonian suspension with viscosity, μ_s and with a Newtonian binder with viscosity μ_b and on the basis of a no-slip condition at the wall for the Newtonian binder, which constitutes the apparent slip layer. The total flow rate, Q, versus the wall shear stress, τ_w, relationship for the apparent slip flow of a Newtonian suspension in a tube with radius, R, becomes:

$$Q = Q_s + \frac{\pi R^3 \tau_w}{4\mu_s} = \pi R^2 u_s + \frac{\pi R^3 \tau_w}{4\mu_s} = \pi R^2 \frac{\delta}{\mu_b} \tau_w + \frac{\pi R^3 \tau_w}{4\mu_s} \qquad (5)$$

where $Q_S = \pi R^2 \frac{\delta}{\mu_b} \tau_w$. The ratio of the flow rate due to slip Q_S over the total flow rate, Q, *i.e.*, Q_S/Q, indicates the relative importance of wall slip at any wall shear stress, τ_w and is given by:

$$\frac{Q_s}{Q} = \frac{\frac{\delta}{\mu_b}}{\left(\frac{\delta}{\mu_b} + \frac{R}{4\mu_s}\right)} = \frac{\frac{\delta}{R}}{\left(\frac{\delta}{R} + \frac{1}{4\eta_r}\right)} \qquad (6)$$

Upon replacing the relative viscosity of the suspension, $\eta_r = \mu_s / \mu_b$ with $\eta_r = (1 - \frac{\phi}{\phi_m})^{-2.0}$ [Krieger and Dougherty (1959)] and the apparent slip layer thickness, δ, with $\frac{\delta}{D_p} = 1 - \frac{\phi}{\phi_m}$ [Kalyon (2005)], Q_S/Q becomes:

$$\frac{Q_s}{Q} = \frac{1}{\left(1 + \frac{R}{4D_p}\left(1 - \frac{\phi}{\phi_m}\right)\right)}. \qquad (7)$$

Therefore, the application of the apparent slip mechanism to a Newtonian suspension suggests that as $\phi \rightarrow \phi_m$, $Q_s/Q \rightarrow 1$, i.e., the flow approaches plug flow with increasing concentration of the rigid particles and that apparent slip has a negligible effect for relatively small concentration of rigid particles, i.e., $\phi \rightarrow 0$, $Q_s/Q \rightarrow 0$ since $R >> D_p$. Such plug flow behavior is indeed observed during the Poiseuille flow of Newtonian suspensions with increasing ϕ, irrespective of the flow rate [Karnis and Mason (1967)].

Thus, with the contributions of the slip of the binder and the apparent slip mechanism the apparent wall slip velocity, u_s, of the suspension becomes:

$$u_s = \beta_0\left(\frac{p_a}{p}\right)^{\kappa(\phi)} \tau_w^{s_1}\left(0.5 + 0.5\tanh\left(\alpha(\tau_w - \tau_c)\right)\right) + \frac{D_p}{m_b^{s_b}}\left(1 - \frac{\phi}{\phi_m}\right)\tau_w^{s_b} \qquad (8)$$

The rheological characterization of the polymeric binder provides β_0, s_1, s_b, m_b and τ_c. Upon the characterization of the shear viscosity and the wall slip behavior of the binder of the suspension and the physical properties of the rigid phase (the harmonic mean particle diameter, D_p, volume fraction, ϕ, and maximum packing fraction of the particles, ϕ_m) the only unknown in Equation (8) that describes the wall slip behavior of the suspensions with the same binder is, $\kappa(\phi)$. If the slip of the binder is neglected, the apparent slip layer can be considered to be a function of pressure, i.e., the apparent slip layer thickness can be taken as:

$$\delta = \delta_a \left(\frac{P_a}{P + P_a}\right)^{\kappa} \quad (9)$$

where, κ is slip exponent, δ_a is the nominal slip layer thickness at the ambient pressure P_a and $P = P(z)$ is the bulk pressure at a section of the flow channel transverse to the main flow z-direction. The governing equations for mass and momentum conservation considering incompressible suspensions, with simplifications pertinent to the twin screw extrusion and die flow combinations are:

Continuity:

$$\frac{\partial u}{\partial x} + \frac{\partial v}{\partial y} + \frac{\partial w}{\partial z} = 0 \quad (10)$$

Momentum:

$$\frac{\partial}{\partial x}\left(2\eta\frac{\partial u}{\partial x}\right) + \frac{\partial}{\partial y}\left\{\eta\left(\frac{\partial u}{\partial y} + \frac{\partial v}{\partial x}\right)\right\} + \frac{\partial}{\partial z}\left\{\eta\left(\frac{\partial u}{\partial z} + \frac{\partial w}{\partial x}\right)\right\} - \frac{\partial p}{\partial x} = 0 \quad \text{(u, v, w; x, y, z)} \quad (11)$$

together with the conservation of energy, solved as a set of coupled differential equations in conjunction with the 3-D finite element method using Penalty-Galerkin formulation [Malik and Kalyon (2005) and Kalyon and Malik (2007)] of the governing equations. Some additional features come in to the computational algorithm due to matching of velocity components and shear stress at the nodes shared at the interfaces of the core flow and slip layer elements.

Some typical calculation results and comparisons with experimental results are shown in Figures 2-3 for two simple dies. In the first a circular die is used and in the second a thin rectangular slit die is used (both are instrumented for temperature and pressure). The effects of apparent wall slip and the role played by the pressure dependence are shown (Figures 2 and 3). Clearly, the comparisons with the experimental findings indicate that the apparent slip mechanism plays a significant role and the pressure dependence of the apparent slip layer mechanism renders the simulations more realistic.

REFERENCES

Adams, S., W. J. Frith and J. R. Stokes, "Influence of particle modulus on the rheological properties of agar microgel suspensions", J. Rheol. **48**, 6, 1195-1213 (2004).

Allende, M., and D. Kalyon, "Assessment of particle-migration effects in pressure-driven viscometric flows", J. Rheol. **44**, 79-90 (2000).

Aral, B. and D. M. Kalyon, "Effects of temperature and surface roughness on time-dependent wall slip in steady torsional flow of concentrated suspensions," J. Rheol. **38**, 957-972 (1994).

Aral, B. and D. M. Kalyon, "Rheology and Extrudability of Very Concentrated Suspensions: Effects of Vacuum Imposition," Plast. and Rubber Comp. Proc. and Appl. **24**, 201-210 (1995).

Benbow, J. J. and P. Lamb, "New aspects of melt fracture," SPE Trans. **3**, 7-17 (1963).

Bernstein, B., E. Kearsley and L. Zapas, "A study of stress relaxation with finite strain", Trans. Soc. Rheol. **7**, 391-410 (1963).

Birinci, E. and D. Kalyon, "Development of extrudate distortions in poly(dimethyl siloxane) and its suspensions with rigid particles," J. Rheol. **50**, 3, 313-326 (2006).

Brandrup, J., E. Immergut and E. Grulke, "Polymer Handbook", Fourth Edition, John Wiley and Sons, NY (1999).

Cohen, Y. and A. B. Metzner, "Apparent slip flow of polymer solutions, " J. Rheol. **29**, 67-102 (1985).

Denn, M. M. "Extrusion instabilities and wall slip," Annual. Rev. Fluid Mech. **33**, 265-287 (2001).

Den Doelder, C. J., Koopmans, R. J. and J. Molenaar, "Quantitative modeling of HDPE spurt experiments using wall slip and generalized Newtonian flow," J. Non-Newtonian Fluid Mech. **79**, 503-514 (1998).

Friend, J. P. and R. J. Hunter, "Plastic flow behavior of coagulated suspensions treated as a repeptization phenomenon," J. Colloid Int. Sci. **37**, 3, 548-556 (1971).

Gadala-Maria, F. and A. Acrivos, "Shear-induced structure in a concentrated suspension of solid spheres," J. Rheol. **24**, 799-814 (1980).

Gevgilili, H. and D. M. Kalyon, "Step strain flow: wall slip effects and other error sources," J. Rheol. **45**, 1-9 (2001).

Gevgilili, H., "Development of the wall slip condition and its ramifications for polymers and other complex fluids", PhD Thesis, Stevens Institute of Technology, Hoboken, NJ (2003).

Georgiou, G. C., "The time-dependent, compressible Poiseuille and extrusion-swell flows of a Carreau fluid with slip at wall," J. Non-Newt. Fluid Mech. **109**, 93-114 (2003).

Hatzikiriakos, S. G. and J. M. Dealy, "Role of slip and fracture in the oscillating flow of HDPE in a capillary," J. Rheol. **36**, 845-884 (1992a).

Hatzikiriakos, S. G. and J. M. Dealy, "Wall slip of molten high density polyethylenes. II. Capillary rheometer studies," J. Rheol. **36**, 4, 703-741 (1992b).

Hatzikiriakos, S. G. and K. B. Migler, Editors, "Polymer processing instabilities: Control and understanding", Marcel Dekker, New York (2005).

Jana, S., B. Kapoor and A. Acrivos, "Apparent wall-slip velocity coefficients in concentrated suspensions of noncolloidal particles", J. Rheol. 39, 1123-1132 (1995).

Kalika, D. S. and M. M. Denn, "Wall slip and extrudate distortion in linear low density polyethylene," J. Rheol. **31**, 815-834 (1987).

Kalyon, D. M., D. Yu and J. Yu, "Melt rheology of two engineering thermoplastics: Poly(ether imide) and Poly(2,6-Dimethyl-1,4-phenylene ether)," J. Rheol. **32,** 789-811 (1988).

Kalyon, D. M., R. Yazici, C. Jacob, B. Aral and S. W. Sinton, "Effects of Air Entrainment on the Rheology of Concentrated Suspensions during Continuous Processing," Polym. Eng. Sci. **31** 1386-1392 (1991).

Kalyon, D. M., P. Yaras, B. Aral, and U. Yilmazer, "Rheological behavior of a concentrated suspension: A solid rocket fuel simulant," J. Rheol. **37**, 35-53 (1993).

Kalyon, D., H. Gokturk, P. Yaras and B. Aral, "Motion Analysis of Development of Wall Slip During Die Flow of Concentrated Suspensions," Society of Plastics Engineers ANTEC Technical Papers, **41**, 1130-1134 (1995).

Kalyon, D. and H. Gevgilili, "Wall slip and extrudate distortion of three polymer melts. J. Rheol. **47**, 3, 683-699 (2003).

Kalyon, D., "Apparent slip and viscoplasticity of concentrated suspensions", J. Rheol. **49**, 621-640 (2005).

Kalyon D. M. and M. Malik, "An integrated approach for numerical analysis of coupled flow and heat transfer in co-rotating twin screw extruders", International Polymer Processing, 22, 293-302 (2007).

Kalyon, D. and H. S. Tang, "Inverse problem solution of squeeze flow for parameters of generalized Newtonian fluid and wall slip", Journal of Non-Newtonian Fluid Mech. **143**, 133-140 (2007).

Karnis, A. and S. G. Mason, "The flow of suspensions through tubes VI. Meniscus effects", J. Colloid and Int. Sci. **23**, 120-133 (1967).

Knudsen, M. "The kinetic theory of gases", Methuen, London, 3rd Edition (1950).

Kok Hartman, P. J. A., S. Kazarian, C.J. Lawrence and B. J. Briscoe, "Near-wall particle depletion in a flowing colloidal suspension", J. Rheol. **46**, 2, 481-493 (2002).

Krieger, I.M. and T. J. Dougherty, "A Mechanism for Non-Newtonian Flow in Suspensions of Rigid Spheres," Trans. Soc. Rheol. **3**, 137-152 (1959).

Laun, H. M., "Description of the nonlinear shear behavior of a low density polyethylene melt by means of an experimentally determined strain dependent memory function," Rheol. Acta **17,** 1-15 (1978).

Leighton, D., and A. Acrivos, "The shear-induced migration of particles in concentrated suspensions", J. Fluid Mech. **275**, 155-199 (1987).

Lu, G., D. Kalyon, I. Yilgör and E. Yilgör, "Rheology and processing of $BaSO_4$ filled medical-grade thermoplastic polyurethane", Polym. Eng. Sci. **44**, 10, 1941-1948 (2004).

M. Malik and D. Kalyon, "Three-dimensional Finite Element Simulation of Processing of Generalized Newtonian Fluids in Counter-rotating and Tangential Twin Screw Extruder and Die Combination", Int. Polym. Processing, 20, 398- 409 (2005).

Meeker, S. P., R. T. Bonnecaze and M. Cloitre, "Slip and flow in pastes of soft particles: Direct observation and rheology", J. Rheol. **48**, 1295-1320 (2004).

Mewis, J. "Rheology of concentrated dispersions", Adv. Colloid Int. Sci. **6**, 173-200 (1976).

Nott, P. R. and J. F. Brady, "Pressure-driven flow of suspensions: Simulations and theory," J. Fluid Mech. **275**, 157-199 (1994).

Osaki, K., "Proceedings of the Seventh International Congress on Rheology," C. Klason and J. Kubat, Eds, Tages Anzeiger, Götenburg, 104-109 (1976).

Person, T. J. and M. M. Denn, "The effect of die materials and pressure-dependent slip on the extrusion of linear low-density polyethylene", J. Rheol. **41**, 2, 249-265 (1997).

Phillips, R. J., R. C. Armstrong, R. A. Brown, A. L. Graham, and J. R. Abbott, "A constitutive equation for concentrated suspensions that accounts for shear-induced particle migration," Phys. Fluids A **4**, 30-40 (1992).

Reiner, M., "Deformation, Strain and Flow", H. K. Lewis and Co. Ltd., London (1960).

Richtmeyer, R. D. and K. W. Morton, "Difference methods for initial-value problems", 2nd Ed., Interscience Publishers, New York (1967).

Robert, L., Y. Demay and B. Vergnes, "Stick-slip flow of high density polyethylene in a transparent slit die investigated by laser Doppler velocimetry," Rheol. Acta **43**, 1435-1528 (2004).

Rosenbaum, E.E., S. K. Randa, S. G Hatzikiriakos, C. W. Stewart, D. L. Henry, M. Buckmaster, "Boron nitride as a processing aid for the extrusion of polyolefins and fluoropolymers", Polym. Eng. Sci. **40,** 179-190 (2000).

Rough, S. L., J. Bridgwater and D. I. Wilson, "Effects of liquid phase migration on extrusion of microcrystalline cellulose pastes", Int. J. Pharmaceutics **204**, 117-126 (2000)

Dealy, J. private communication (2006).

Tabuteau, H., J. Baudez, F. Bertrand, and P. Coussot, "Mechanical characteristics and origin of wall slip in pasty biosolids", Rheol. Acta **43**, 2, 168-174 (2004).

Tang, H. S. and D. M. Kalyon, "Time-dependent development of flow instabilities of non-Newtonian melts and suspensions with rigid particles", Proceedings of AICHE Annual Meeting, Austin, Texas (2004a).

Tang, H. S. and D. M. Kalyon, "Estimation of the parameters of Herschel-Bulkley fluid under wall slip using a combination of capillary and squeeze flow viscometers," Rheol. Acta **43**, 80-88 (2004b).

Tang, H. S. and D. M. Kalyon, "Unsteady circular tube flow of compressible polymeric liquids subject to pressure dependent wall slip", J. Rheol. **52**, (2) March/April (2008).

Tyrrell, J.W. G., and P. Attard, "Images of nanobubbles on hydrophobic surfaces and their interactions", Phys. Rev. Letters 87, 1761041-4 (2001).

Vinogradov, G. V. and L.I. Ivanova, "Viscous properties of polymer melts and elastomers exemplified by ethylene-propylene copolymer", Rheol. Acta **6**, 209-222 (1967).

Wagner, M. H., "Analysis of the time-dependent non-linear stress-growth data for shear and elongational flow of a low-density branched polyethylene melt", Rheologica Acta **15,** 136-142 (1976).

Yaras, P., D. M. Kalyon, and U. Yilmazer, "Flow Instabilities in Capillary Flow of Concentrated Suspensions," Rheol. Acta **33,** 48-59 (1994).

Yilmazer, U. and D. M Kalyon, "Slip effects in capillary and parallel disk torsional flows of highly filled suspensions," J. Rheol. **33**, 1197-1212 (1989).

Yilmazer, U., C. Gogos and D. M. Kalyon, "Mat Formation and Unstable Flows of Highly Filled Suspensions in Capillaries and Continuous Processors," Polymer Composites **10,** 4, 242-248 (1989).

Zarraga, I. E., D. A. Hill and D. T. Leighton, Jr., "The characterization of the total stress of concentrated suspensions of noncolloidal spheres in Newtonian fluids", J. Rheol. 44, 185-220 (2000).

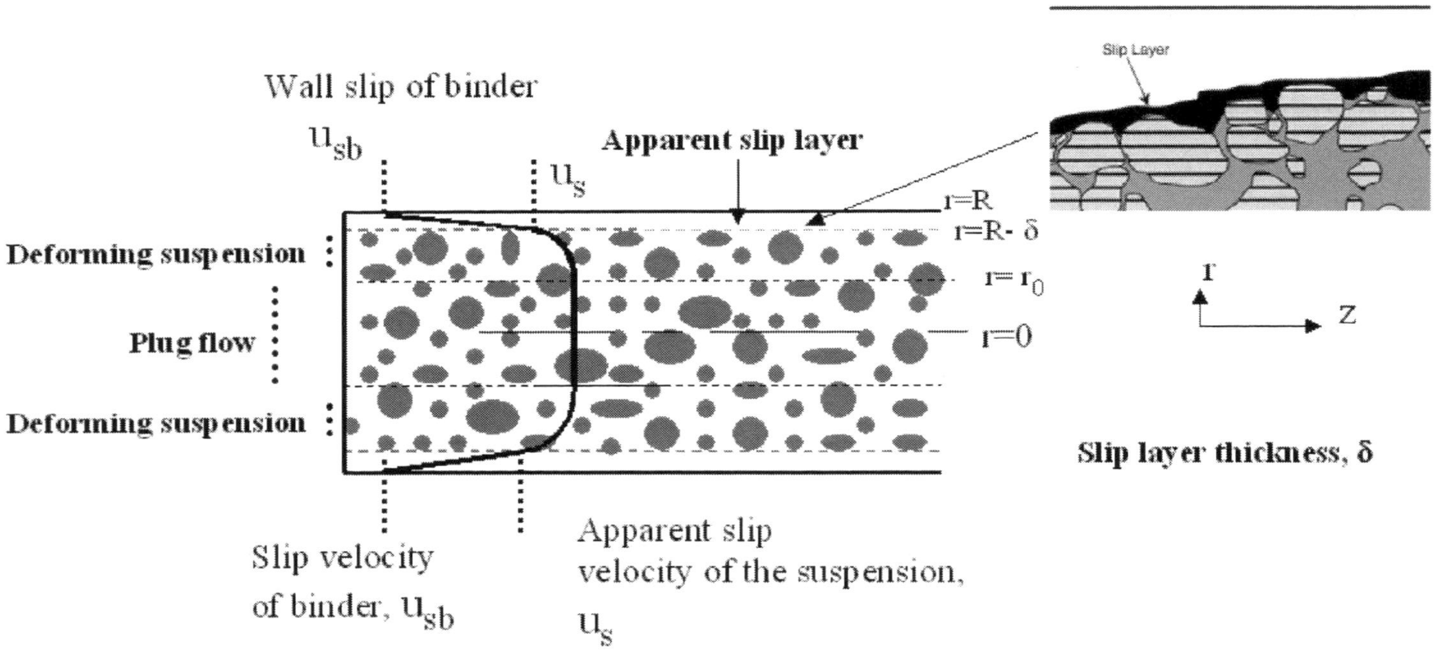

Figure 1: Apparent slip mechanism in a tubular die [Tang and Kalyon, J. Rheology (2008b)].

Fig.1

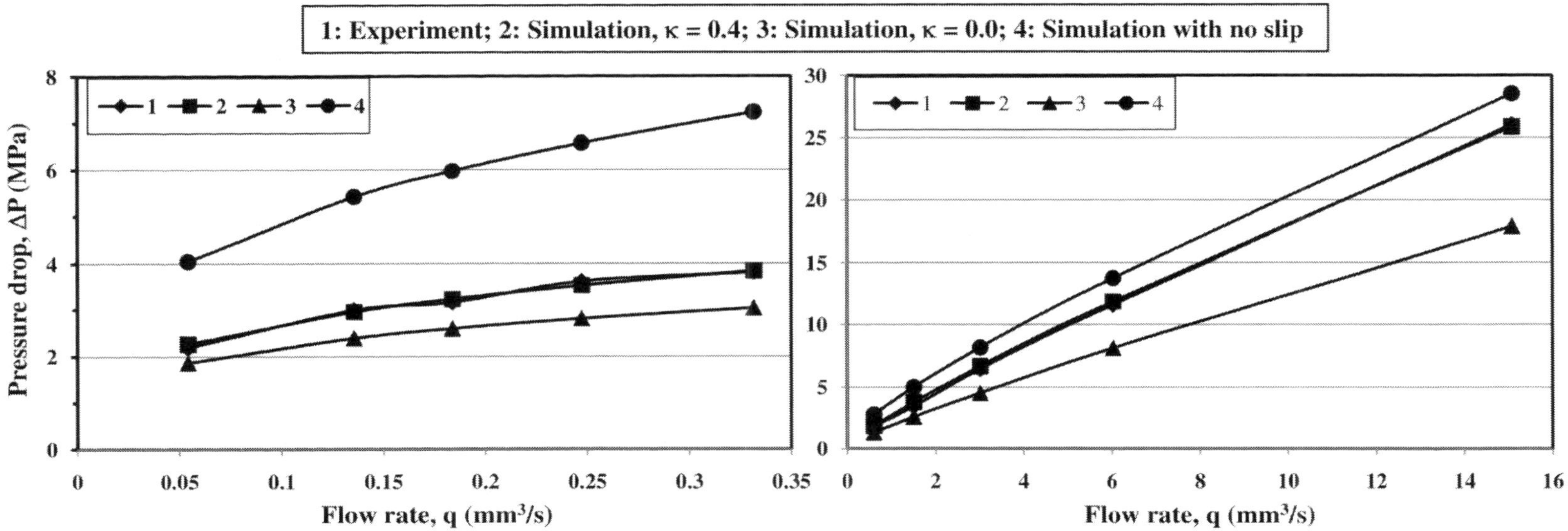

Suspension: m_0 = 3500 Pa.s^n, n = 0.336, τ_y = 1000 Pa
Binder: m_0 = 730 Pa.s^n, n = 0.265
Cartridge diameter = 0.375 inch
Capillary diameter = 0.0591 inch, Capillary L/D = 40
Nominal thickness of apparent slip layer, δ_a = 1.5 μm

Suspension: m_0 = 10424 Pa.s^n, n = 0.84, τ_y = 4300 Pa
Binder: m_0 = 26.2 Pa.s^n, n = 1.0
Cartridge diameter = 0.375 inch
Capillary diameter = 0.0675 inch, Capillary L/D = 35
Nominal thickness of apparent slip layer, δ_a = 0.629 μm

Fig. 2 Experiment vs Numerical Simulation: Suspension flow through converging capillary dies

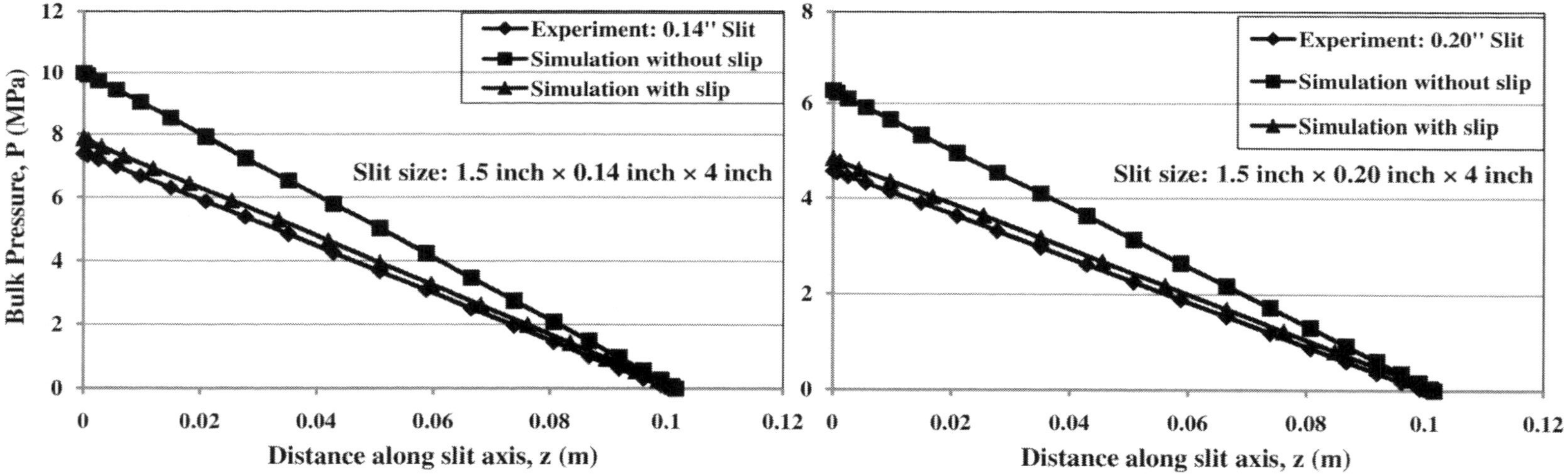

Suspension: m_0 = 61205 Pa.s^n, n = 0.21, τ_y = 20208 Pa; Binder: m_0 = 1490 Pa.s^n, n = 0.48
Nominal thickness of slip layer, δ_a = 1 μm; Inlet and slit wall temperature = 361 K; Flow rate = 25 lb/hr

Fig. 3 Experiment vs Numerical Simulation: Suspension flow through rectangular slits

Fig.3

Discrete Element Method Modeling of Preparation of Reactive Nanomaterials by Mechanical Milling

Priya R. Santhanam, Chemical, Biological, and Pharmaceutical Engineering, New Jersey Institute of Technology, Newark, NJ

Alexandre Ermoline, Chemical, Pharmaceutical, and Biological Engineering, New Jersey Institute of Technology, Newark, NJ

Mirko Schoenitz, Dept. of Chemical, Biological, and Pharmaceutical Engineering, New Jersey Institute of Technology, Newark, NJ

Edward L. Dreizin, Chemical, Biological, and Pharmaceutical Engineering, New Jersey Institute of Technology, Newark, NJ

Reactive nanocomposite powders with different compositions were recently prepared by arrested reactive milling, a mechanical milling technique. Shaker and planetary mills were used for laboratory scale production. A discrete element method (DEM) model was used to characterize the material evolution in the mill by tracking a milling progress function, defined by the energy transferred from the milling media to the material under processing. It was observed that the milling tools interact differently in shaker and planetary mills and contribute to the refinement of the milled material in different ways. Specifically, head-on collisions between milling balls controlled the rate of powder refinement in the shaker mill, while rolling motion of the balls caused the powder refinement in the planetary mill. Both shaker and planetary mills are widely used in laboratory experiments; however, attritor mills are better suited for scaled-up production of new materials and thus an accurate description of the respective milling process is needed. In this work, a DEM model is developed and used to assess the powder processing in an attritor mill. Previously, milling effectiveness of attritor and shaker mills were compared to each other based on both laboratory experiments and simplified models. However, the results of such comparisons were inconsistent with experimental observations related to mechanical alloying. In this effort, the commercial EDEM software by DEM Solutions Inc. was used. A model 01HD attritor mill by Union Process was described in the code. Predictions were made for the mill operated with milling media comprising steel balls of different diameters. The influences of milling parameters, including number and dimensions of the milling balls, rate of impeller rotation, and the impeller's vertical position on the energy transfer from the milling media to the powder were investigated. In addition, the energy transfer in the attritor mill was directly compared to that predicted by the same DEM code for shaker and planetary mills. Results and their implications for design of larger scale production facilities for preparation of reactive nanocomposite materials will be discussed.

See more of this Session: Processing and Safety
See more of this Group/Topical: Particle Technology Forum

Processing and Handling of Reactive and Structural Reactive Materials

Paul Redner, Energetics and Warheads Division, US Army RDECOM-ARDEC, Picatinny, NJ

Deepak Kapoor, Materials Division, US Army RDECOM-ARDEC, Picatinny, NJ

Chris Haines, US Army RDECOM-ARDEC, Picatinny, NJ

Darold Martin, US Army RDECOM-ARDEC, Picatinny, NJ

Joseph Paras, US Army RDECOM-ARDEC, Picatinny, NJ

Ryan Carpenter, US Army RDECOM-ARDEC, Picatinny, NJ

Brian Travers, Energetics and Warheads Division, US Army RDECOM-ARDEC, Picatinny, NJ

James Pham, Energetics and Warheads Division, US Army RDECOM-ARDEC, Picatinny, NJ

ARDEC is exploring the use of reactive materials, and structural reactive materials, in various applications such as the replacement of warhead components, particle impact mitigation sleeves (PIMS liners), reactive fragments, and reactive casings.

In order for these materials to be developed and transitioned effectively, the proper processing, characterization, and handling techniques and precautions need to be established. Effectively identifying all of these issues will ensure that we will see heavier usage of these materials in the near future to enhance the performance of munitions and to improve muntions' response to outside stimuli such as bullet impact (BI), fragment impact (FI) and cookoff. Achieving Insensitive Munitions (IM) compliance is a key goal while not reducing the overall effectiveness of the warhead.

This paper will discuss some of the materials and techniques ARDEC has explored in order to build the necessary capability to handle these types of materials, as well as the processing techniques used to make materials and components for specific end-item applications.

See more of this Session: Processing and Safety
See more of this Group/Topical: Particle Technology Forum

"Mathematical Models of the Coating of Energetic Grains Upon Diffusion of Deterrents and Experimental Validation"

Halil Gevgilili, Dilhan Kalyon, Semra Senturk

Highly Filled Materials Institute, Stevens Institute of Technology, Hoboken, NJ

Donald Chiu, Sam Moy, ARDEC,

US Army, Dover, NJ

Summary: The coating of energetic grains with deterrents and the determination of the coating thickness (the concentration distribution of the deterrent in the grain) are challenges which currently require lengthy trial and error procedures for correct implementation. Here series of analytical models were developed to describe the diffusion of a deterrent into grains which exhibit differing geometries. Two experimental methods involving fluorescence and thermo gravimetric analysis were also developed to facilitate the determination of the diffusion coefficient of the deterrent into the energetic grain using an inverse problem solution methodology on one hand and the validation of the mathematical models and the determined diffusion coefficients on the other. The theoretical results compared favorably with the experimental findings.

Introduction: For a successful coating operation various kinds of deterrents should be applied to surfaces of extruded energetic grains (ram extruded and twin screw extruded) to modify the performance of the energetic grains. There are a number of challenges including controlling the depth in the grain that the deterrent will diffuse as a function of time and operating conditions and what happens as a function of time as the energetic grain ages. Furthermore, the determination of the diffusion coefficient of the deterrent in the grain is not straightforward and requires special methodologies.

Mathematical Model: The penetration and diffusion of the deterrent liquid into the energetic grain on a pan coated with the deterrent liquid was considered for three distinct grain geometries: (a) cylindrical grain, (b) spherical grain. Furthermore, the determination of the diffusion coefficient was based on experiments involving specimens shaped as rectangular slabs of the energetic material and thus the diffusion of the deterrent in the rectangular slab needed to be considered also.

If we consider here just the rectangular slab case:

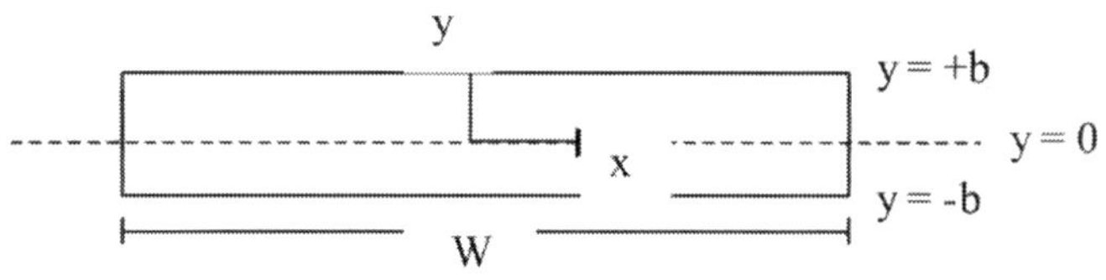

The diffusion of the coating material/solvent into the energetic grain then will proceed according to Fickian diffusion with typical diffusion eqVation given below: Consider that the slab is symmetric around the x-z pane and consider W>>2b, $\frac{dC_\alpha}{dx} << \frac{dC_\alpha}{dy}$

Equation of Continuity for Deterrent species α.

C_α is the concentration of w, deterrent in the particle.

$\frac{\partial C_\alpha}{\partial t} + \vec{\nabla} \cdot C_\alpha \vec{V}^* = -\vec{\nabla} \cdot \vec{J}_\alpha^* + R_\alpha$ R_α= Rate of reaction of α per unit volume

J_α^* = Flux of α

$V^* = \frac{\sum C_\alpha \vec{V}_\alpha}{C}$ V^*= Molar Average Velocity

$C = \sum_{\alpha=1}^{\infty} C_\alpha$ C= Molar Density of Solution

Consider a binary system with the deterrent diffusing into energetic grain with both walls wet with deterrent and without flow, the Fick's second law of diffusion becomes:

$\frac{\partial C_\alpha}{\partial t} = D_{\alpha p} \nabla^2 C_\alpha + R_\alpha$ $D_{\alpha p}$= Binary diffusion coefficient of α deterrent in propellant, p

Assumption: R_α; rate of reaction of α in energetic grain is zero, i.e., the deterrent diffuses into the energetic grain without reaction. R_α=0

$$\frac{\partial C_\alpha}{\partial t} = D_{\alpha p} \nabla^2 C_\alpha$$

where, t is time, $D_{\alpha p}$ is diffusion coefficient and C_{α} is concentration of deterrent with the initial condition: $C_{\alpha}=C_{\alpha 0}=0$ for $t \leq 0$ and at t=0 the wall comes into contact with the deterrent with concentration $C_{\alpha b}$.

Results and Discussion: Figure 1 shows the prediction of the effect of diffusion coefficient on concentration profile for slab thickness (2b) of 0.001m, 500 seconds after initiation of diffusion process. In 500 seconds, the concentration within the sample approaches close to a uniform profile if the diffusion coefficient is larger than $2*10^{-9}$ m^2/s. For smaller diffusion coefficients, it takes significantly longer durations of time to achieve a uniform concentration within the grain. Similarly, Figure 2 shows the effect of diffusion coefficient on the variation in the concentration as a function of time at the midpoint (y/b=0.5) of the slab (thickness =2b= 0.001m). If the diffusion coefficient is less than $5*10^{-10}$ m^2/s, little material diffuses up to the y/b=0.5 point within 500 seconds. Hence such a system will require much longer mixing time in the coating apparatus.

Figure 3 shows the theoretical propagation of the deterrent in the grain as a function of time for three different diffusion coefficients on 3-dimensional plots for better visualization of the entire process in time and spatial coordinates. If the diffusion coefficient is ~$5*10^{-9}$ m^2/s, the concentration of the deterrent approaches a uniform profile within 200 seconds while for $2*10^{-9}$ m^2/s, the concentration profile is still in evolution and for $1*10^{-9}$ m^2/s, little penetration is achieved within 200 seconds.

In order to determine the diffusion coefficient two step experimental procedure was developed: (1) A thermo-gravimetric analysis, TGA, based method for the determination of the diffusion coefficient of a deterrent diffusing in a stimulant grain (2) a fluorescence based method for the validation of the source code and the diffusion coefficient determined using TGA.

Experimental

Determination of the diffusion coefficient:

The demonstration of the methodologies are carried out using a realistic deterrent of dibutyl pythalate (which is also typically used as a plasticizer, and it is selected based on US Patent 6345577– Energetic Deterrent Coating for Gun Propellant) and a typical simulant formulation.

Thermogravimetric Analysis: Rectangular Samples with uniform porosity and thickness were immersed into the deterrent for different time intervals. Upon removal, excess amount of deterrent was removed from the surface of the coated inert slab. A small section (10-20 mg) is cut from the center of the slab. The sample was subjected to thermogravimetric analysis with the following thermal procedure: a) Heat to 210°C with 15°C/min ramp rate, b) Keep isothermal for 30 minutes, c) Heat to 580°C with 50°C/min ramp rate to purge the residues, d) Keep isothermal for 5 minutes. Repeat the procedure three times for each immersion time to get reproducibility. Fit the data to the analytical model of the diffusion into a rectangular slab to determine the diffusion coefficient

As anticipated the concentration of the deterrent in the stimulant slab increases with time of exposure of the surface of the slab that is in contact with the deterrent liquid. The concentration data were then converted into a relative concentration, C/C_o (Figure 5). Here C_o represents the concentration of the deterrent found at the surface of the grain. An inverse problem solution was developed for the fore mentioned diffusion problem facilitating the estimation of the diffusion coefficient. The diffusion coefficient of the deterrent in the stimulant slab was determined to be 5.57E-8 m^2/s. The agreement of the experimental data and the prediction of C/C_o using the analytical model for the slab and the diffusion coefficient obtained with the inverse problem solution is shown in Figure 6 and the agreement is favorable.

Fluorescence analysis: For the validation of the diffusion coefficient and the analytical model for the diffusion of the deterrent into a rectangular slab a set of fluorescence experiments were carried out. In these experiments a fluorescent tracer, pyrene was used as the fluorescent die and was dissolved in the deterrent with the concentration of 5 mg/mL. Simulant samples were dipped into the solution for different time intervals, approximately 1 mm slices were cut and analyzed under UV microscope (Figure 7). Coated sections appear as bright blue color upon UV illumination. UV fluorescence intensity was quantified as a function of pixel location and was subsequently converted to relative concentration versus radial location using Image J software developed by NIH (Figures 8 and 9). As indicated in Figure 10, at least three samples were tested for each immersion time for fluorescence intensity, and relative concentration averages are plotted as a function of location. The symmetric nature of the problem allows the combination of the measurements on two sides of the slab (Figure 11).

The summary of the relative concentration profiles, C/C_o, versus distance for all immersion times as determined by uv fluorescence is shown in Figure 12. The data are provided as 95 % confidence intervals determined according to Student's t-distribution. The comparison of the relative concentration profiles, C/C_o, versus distance for all immersion times as determined by UV fluorescence with the predictions of the analytical model for rectangular slab using the diffusion coefficient determined using thermogravimetric analysis (TGA). The agreement is acceptable and suggests that the methodologies developed can be utilized as routine tools for the determination of the diffusion coefficient of a deterrent in a energetic grain. They also validate our source codes and analytical models with one experimental system.

References:

1. B. Bird, W. Stewart and E. Lightfoot, "Transport Phenomena", 2nd Edition, Wiley, Hoboken, NJ (2002).
2. H.S. Carslaw and J.C. Jaeger, "Conduction of Heat in Solids", Oxford University Press, London (1959).
3. R. J. Cramer, S. Peters, R. Simmons, S. Mitchell, "Energetic Deterrent Coating for Gun Propellant" US Patent # 6345577 (2002).

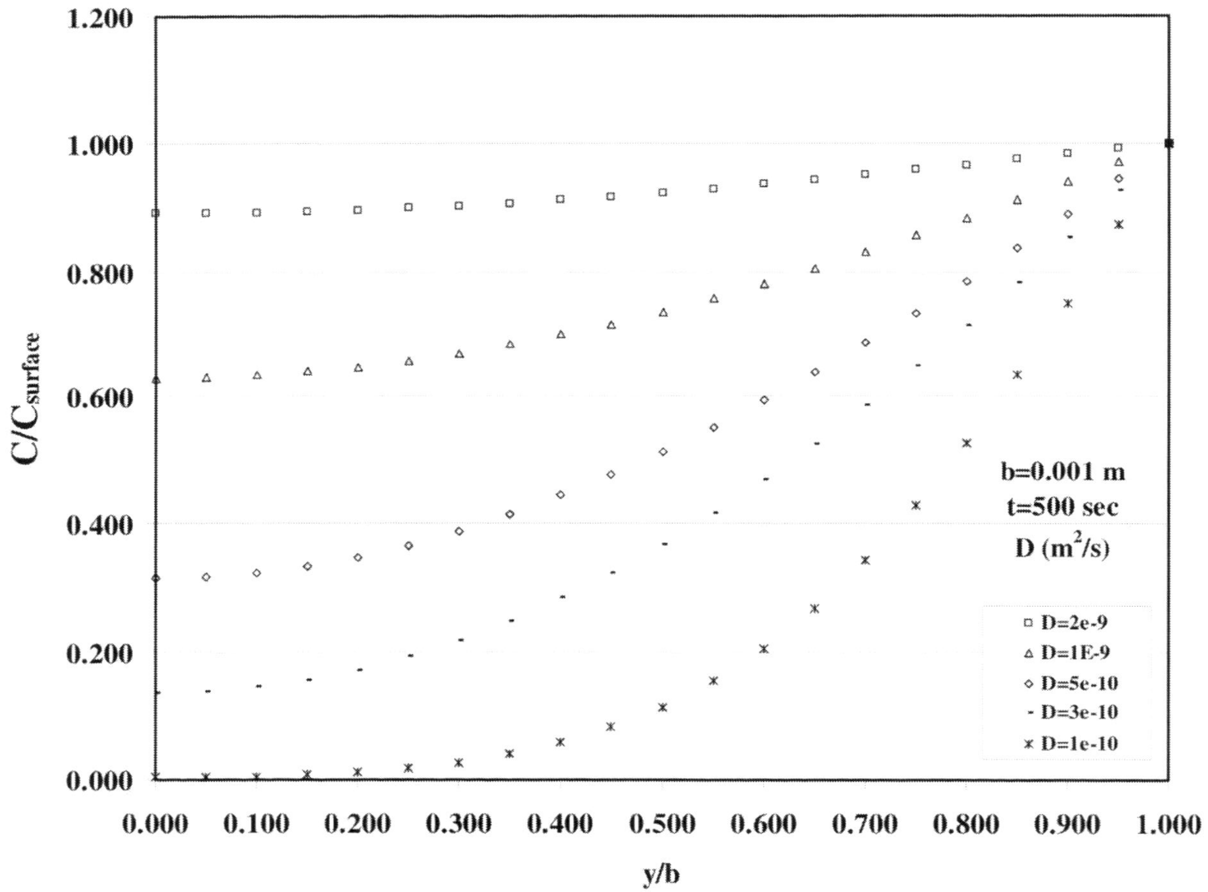

Figure 1: Effect of diffusion coefficient on the concentration profile for slab thickness (2b) of 0.001m, 500 seconds after initiation of diffusion process.

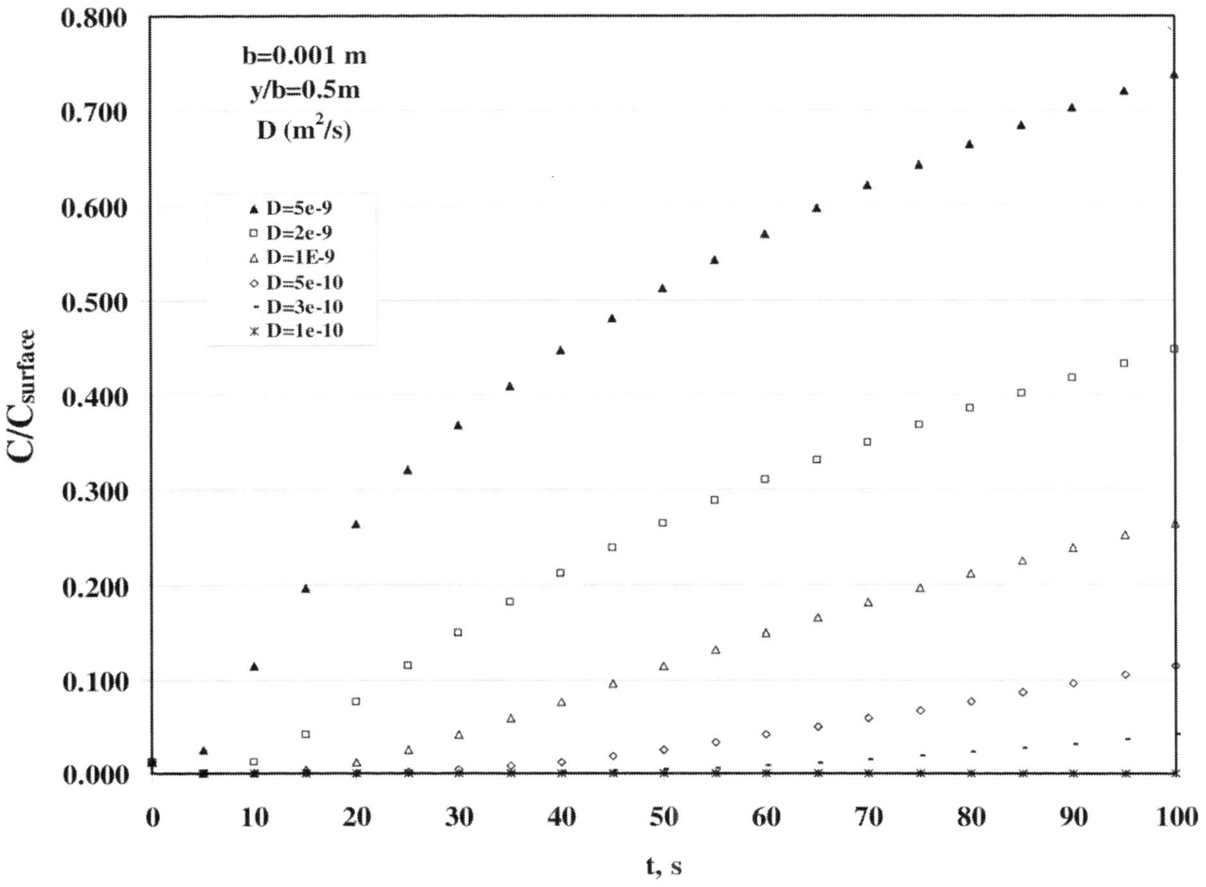

Figure 2: Effect of diffusion coefficient on the variation in the concentration as a function of time at the midpoint of the slab (thickness =2b= 0.001m).

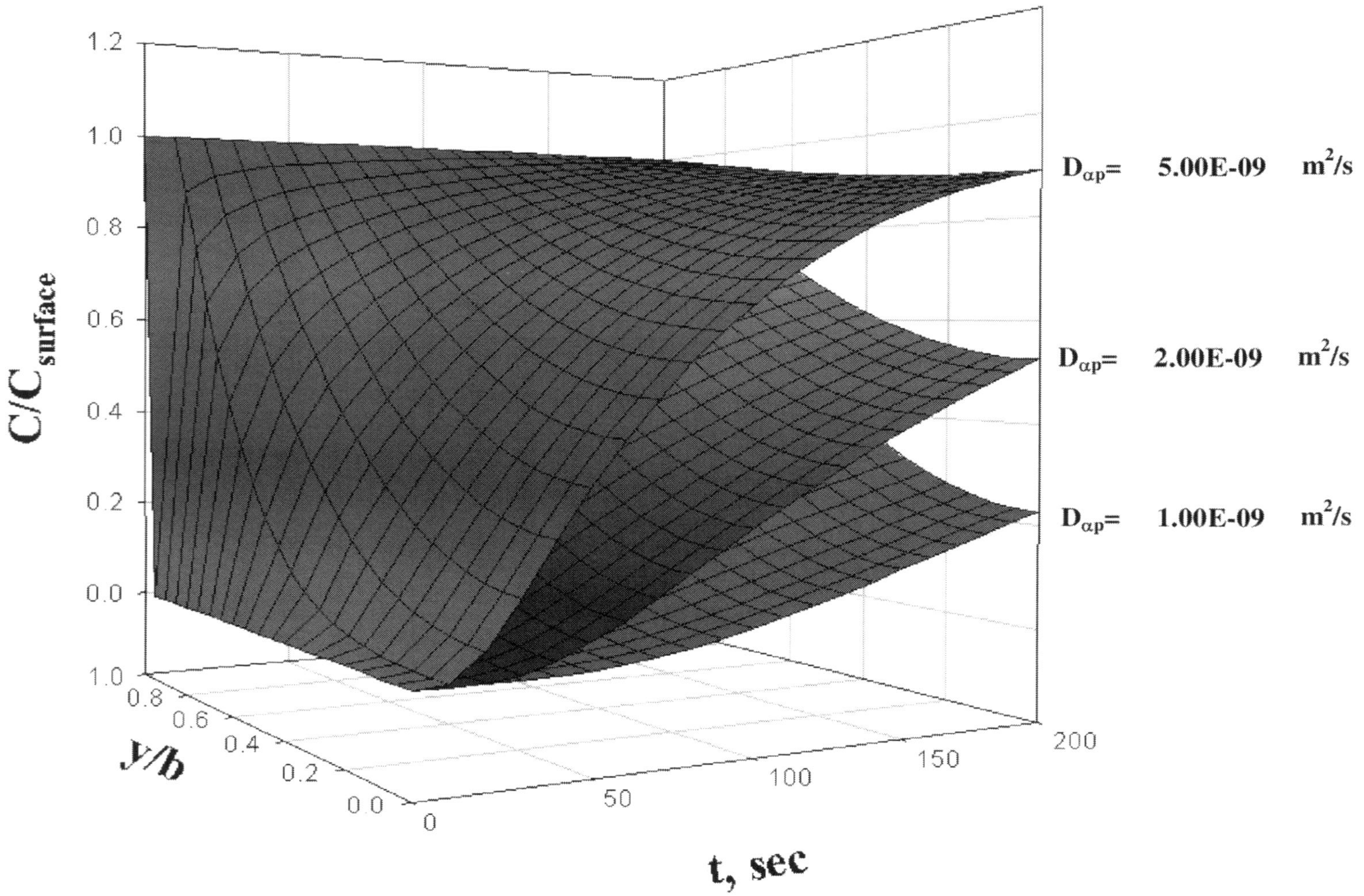

Figure 3: Variation of Concentration as a function of time and location for various diffusion coefficients

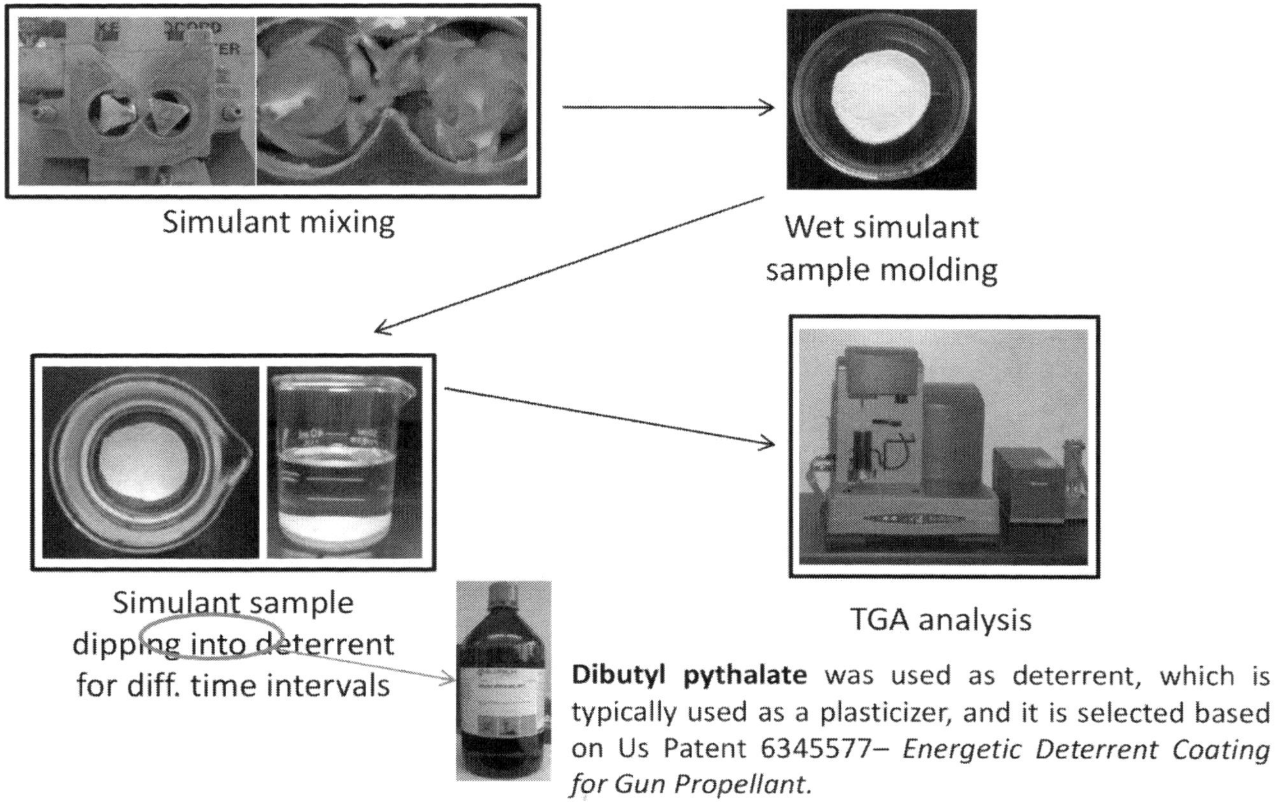

Figure 4: Experimental set-up

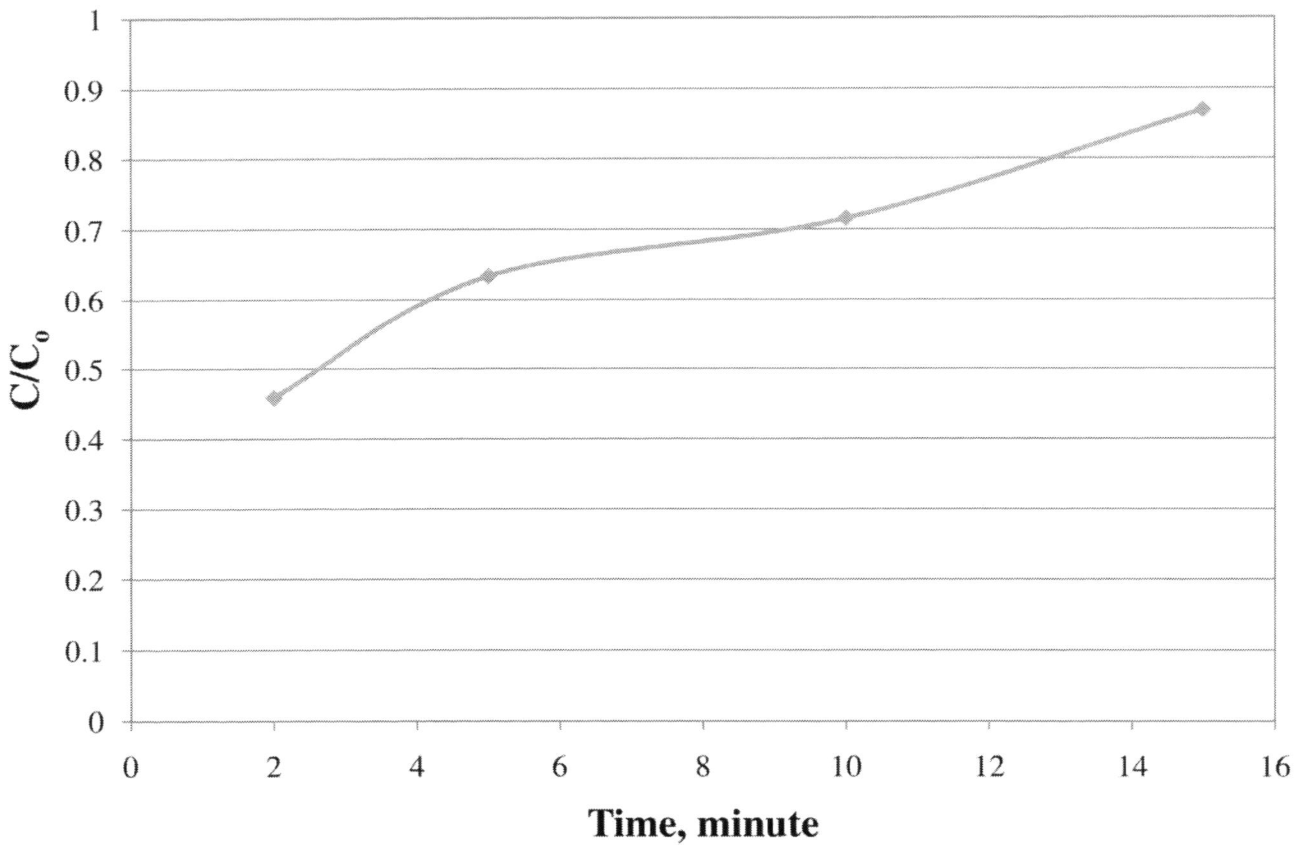

Figure 5: Average percent weight loss converted into relative concentration as a function of immersion time.

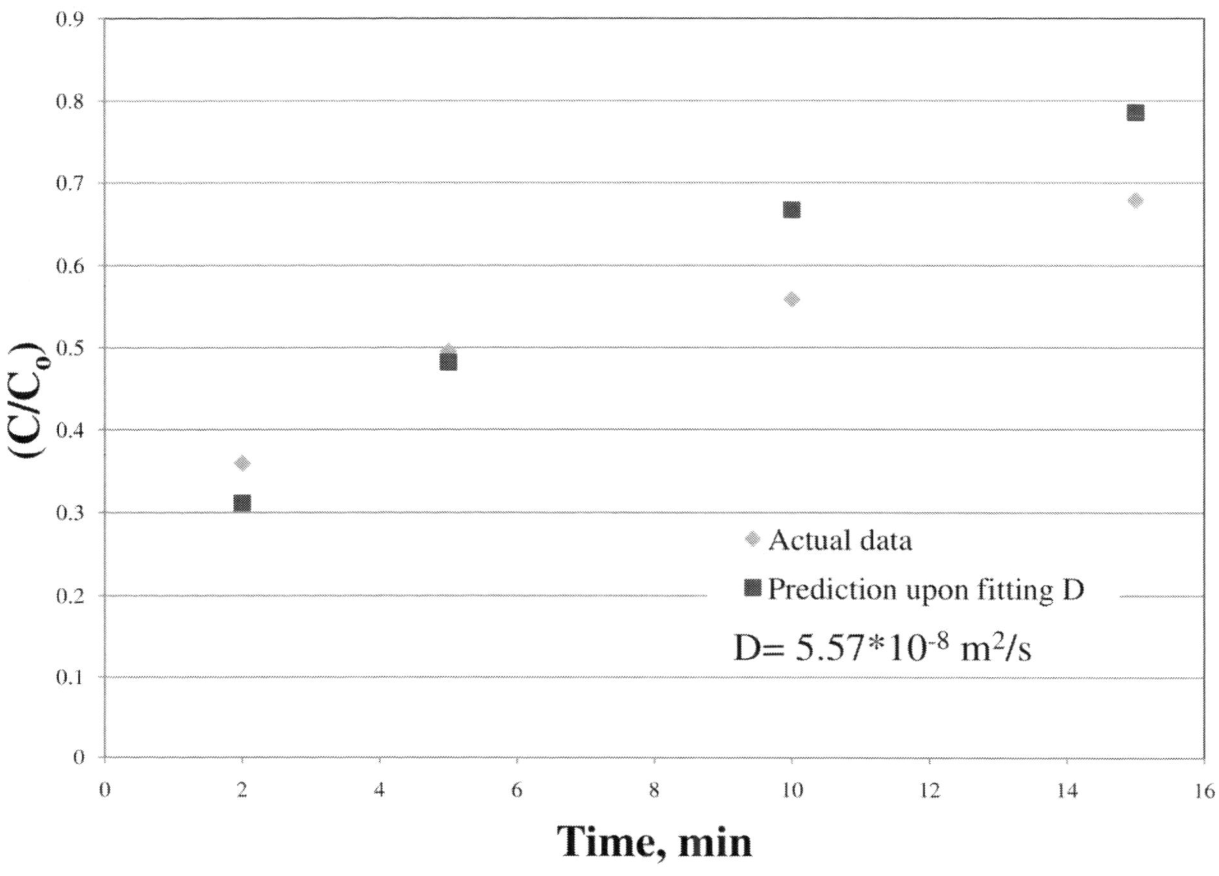

Figure 6: Relative concentration data obtained by TGA are fitted to the rectangular slab model to determine the diffusion coefficient

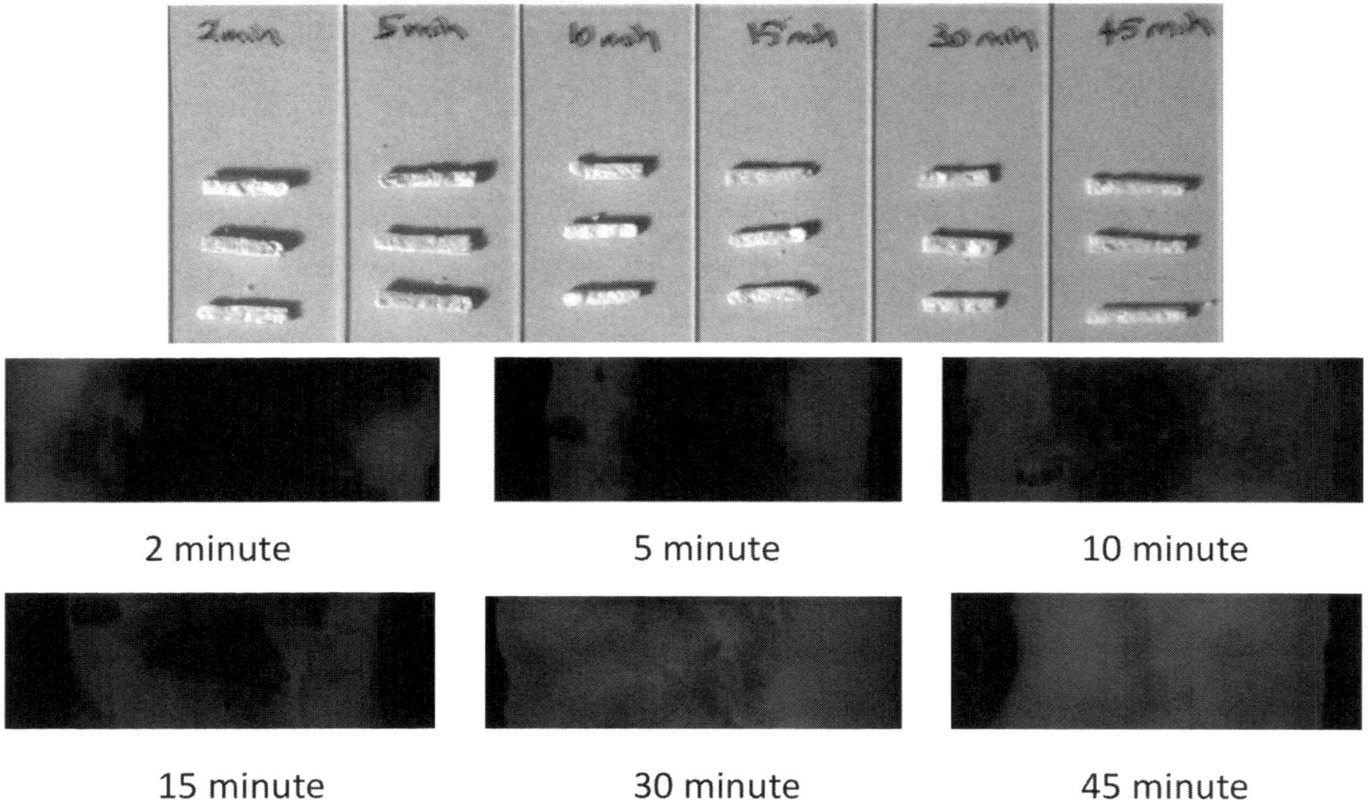

Figure 7: Fluorescence intensity determination

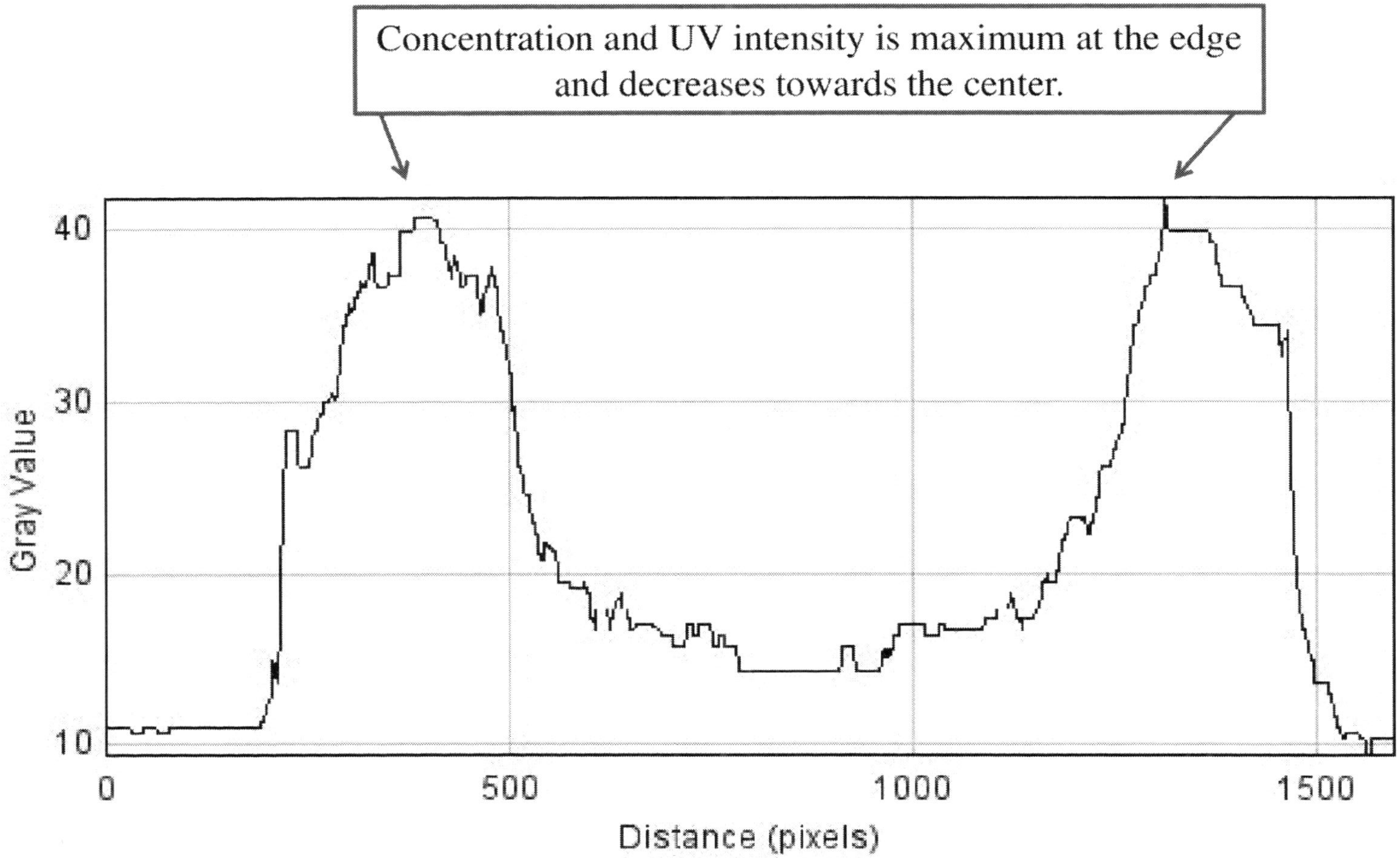

Figure 8: Typical UV fluorescence data at 2 minutes.

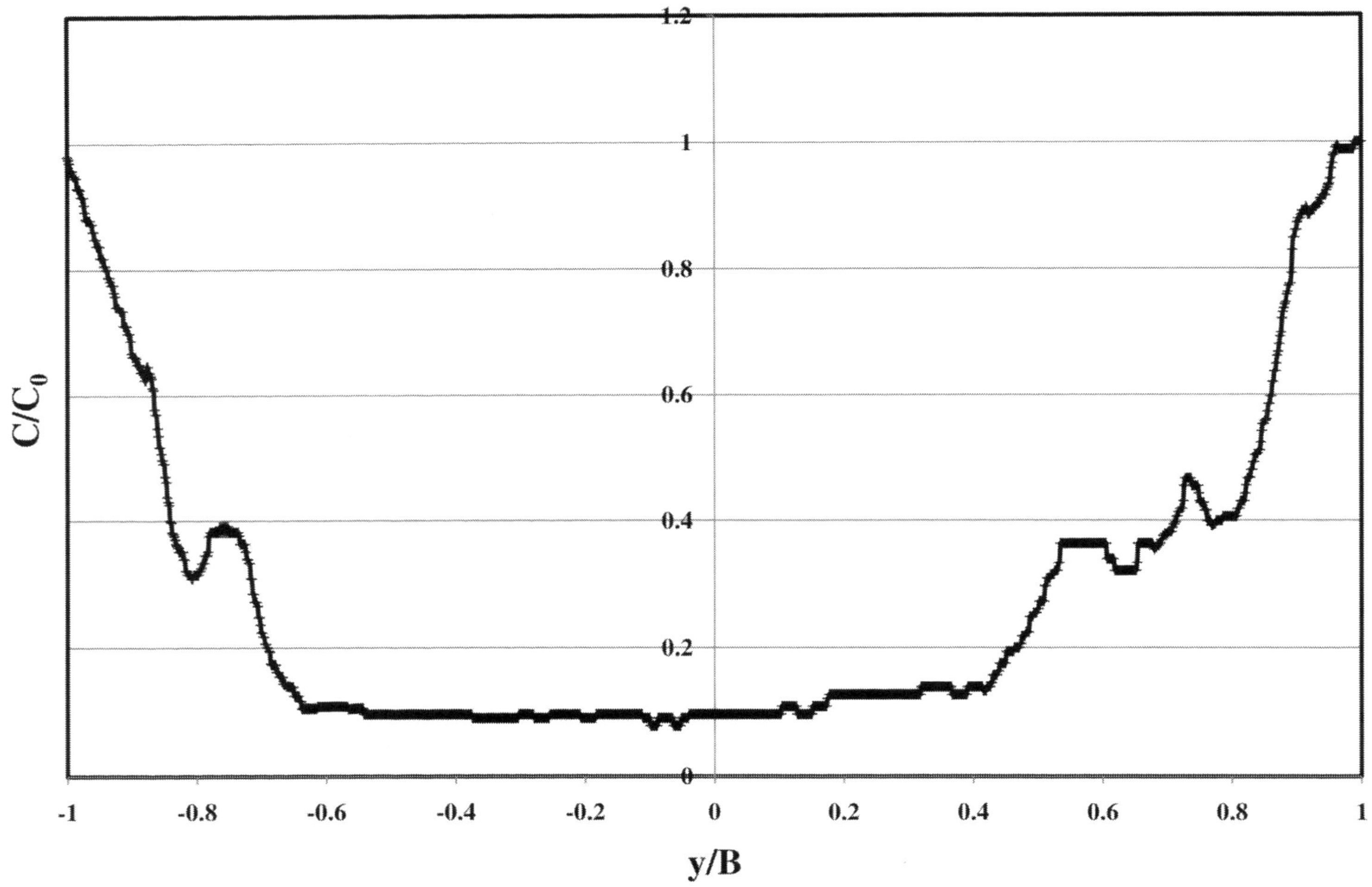

Figure 9: UV intensity converted into relative concentration at 2 min.

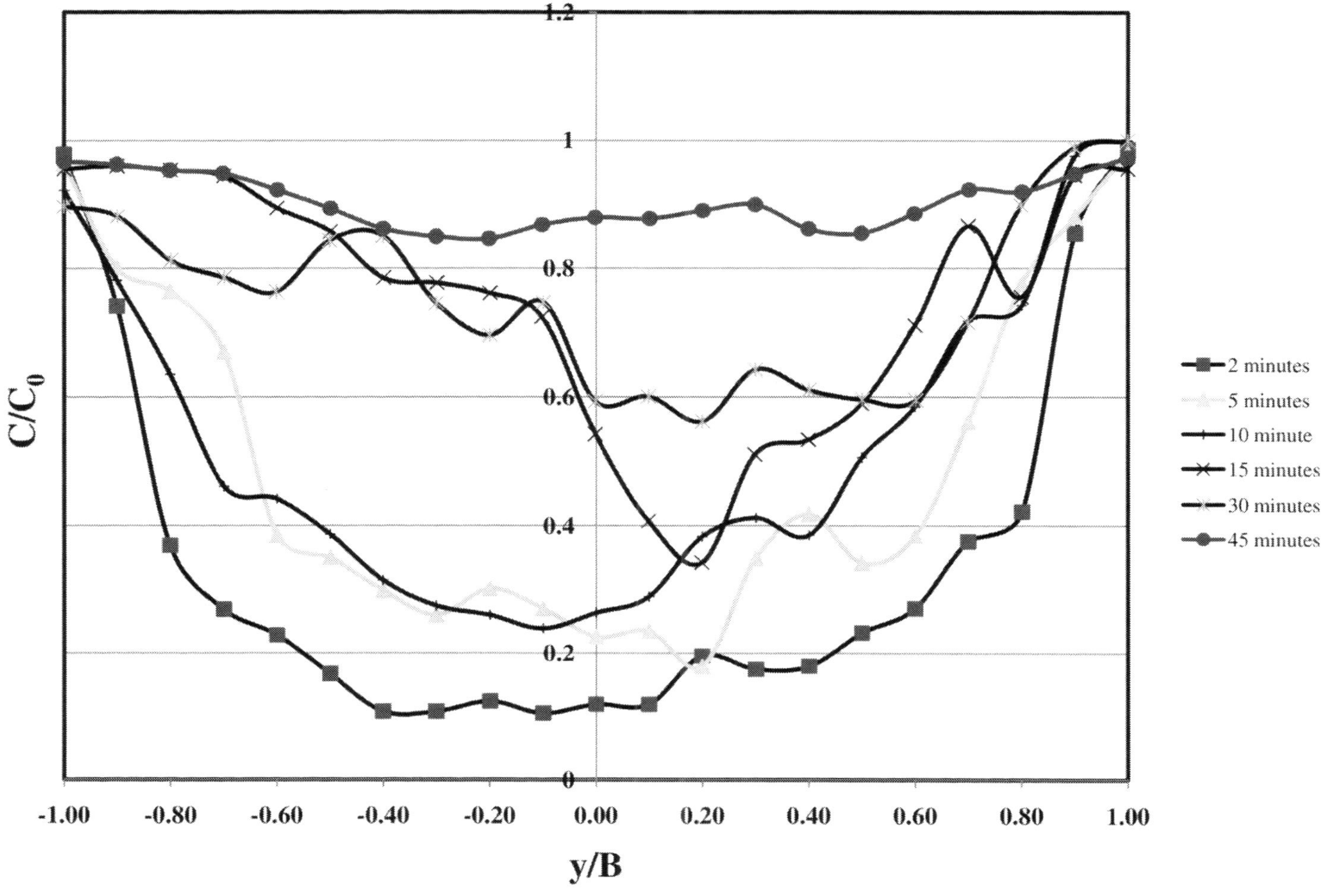

Figure 10: Relative concentration averages as a function of axial location.

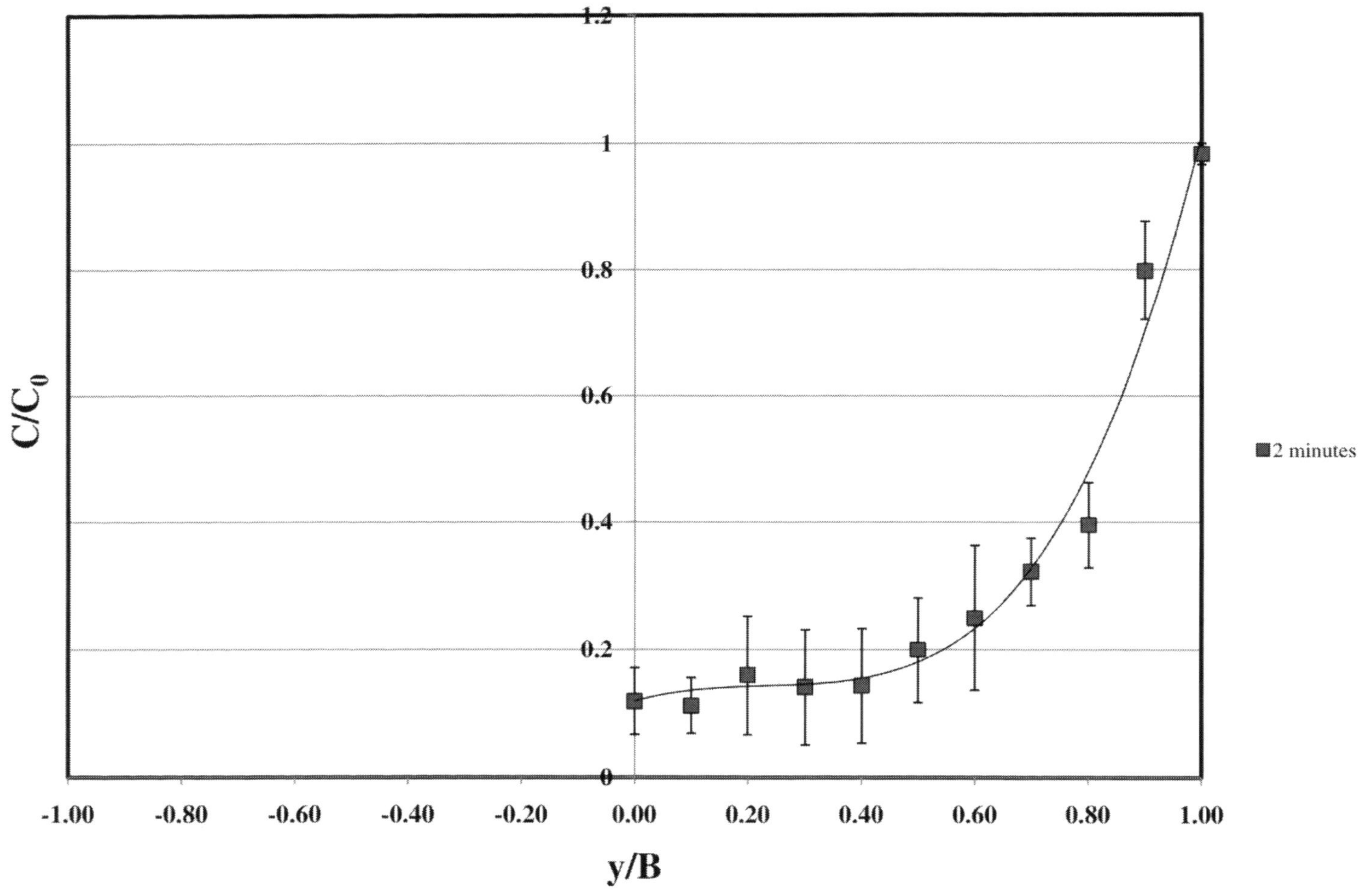

Figure 11: Experimental concentration distribution

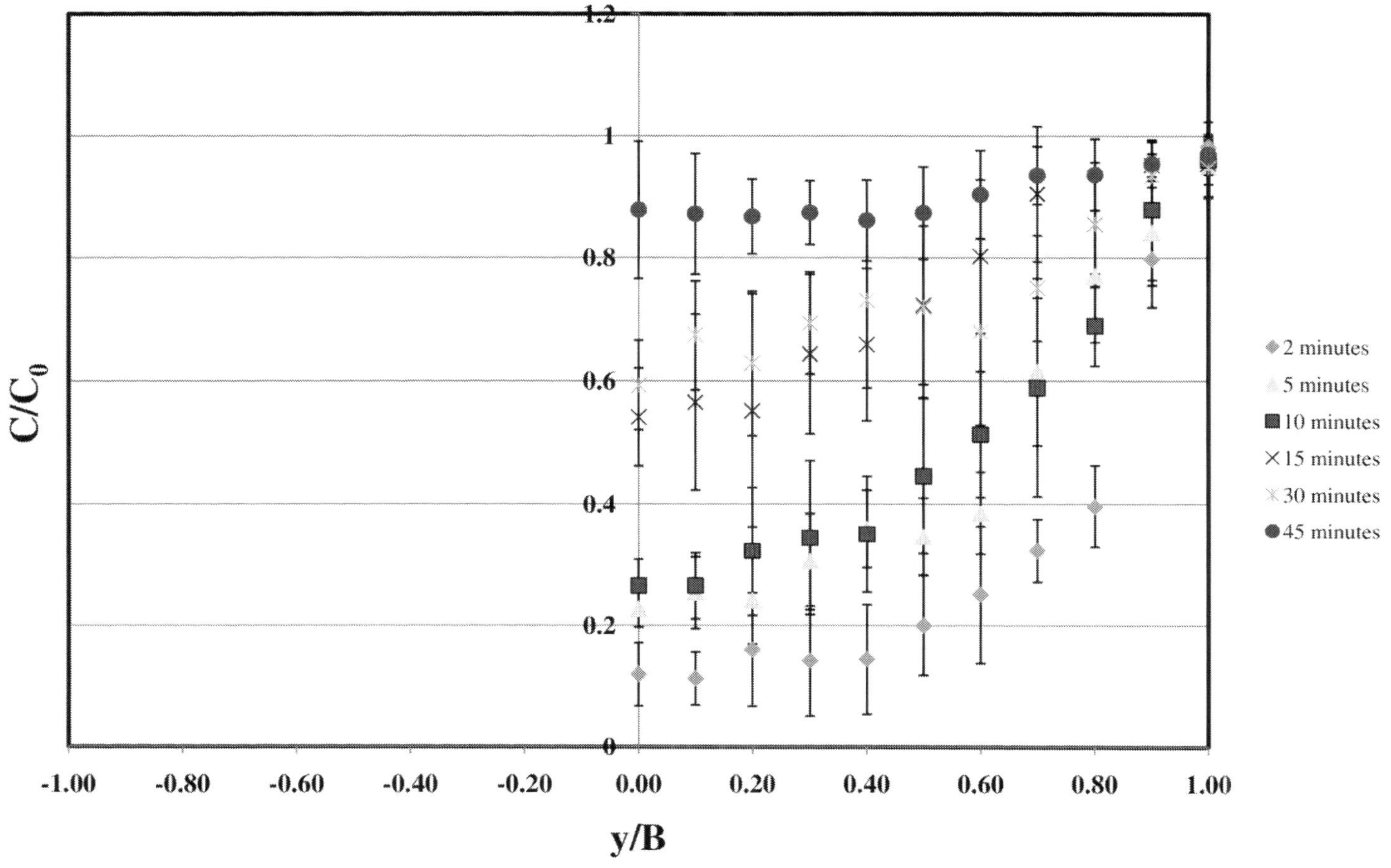

Figure 12: Summary of the relative concentration profiles for all immersion times as determined by UV fluorescence.

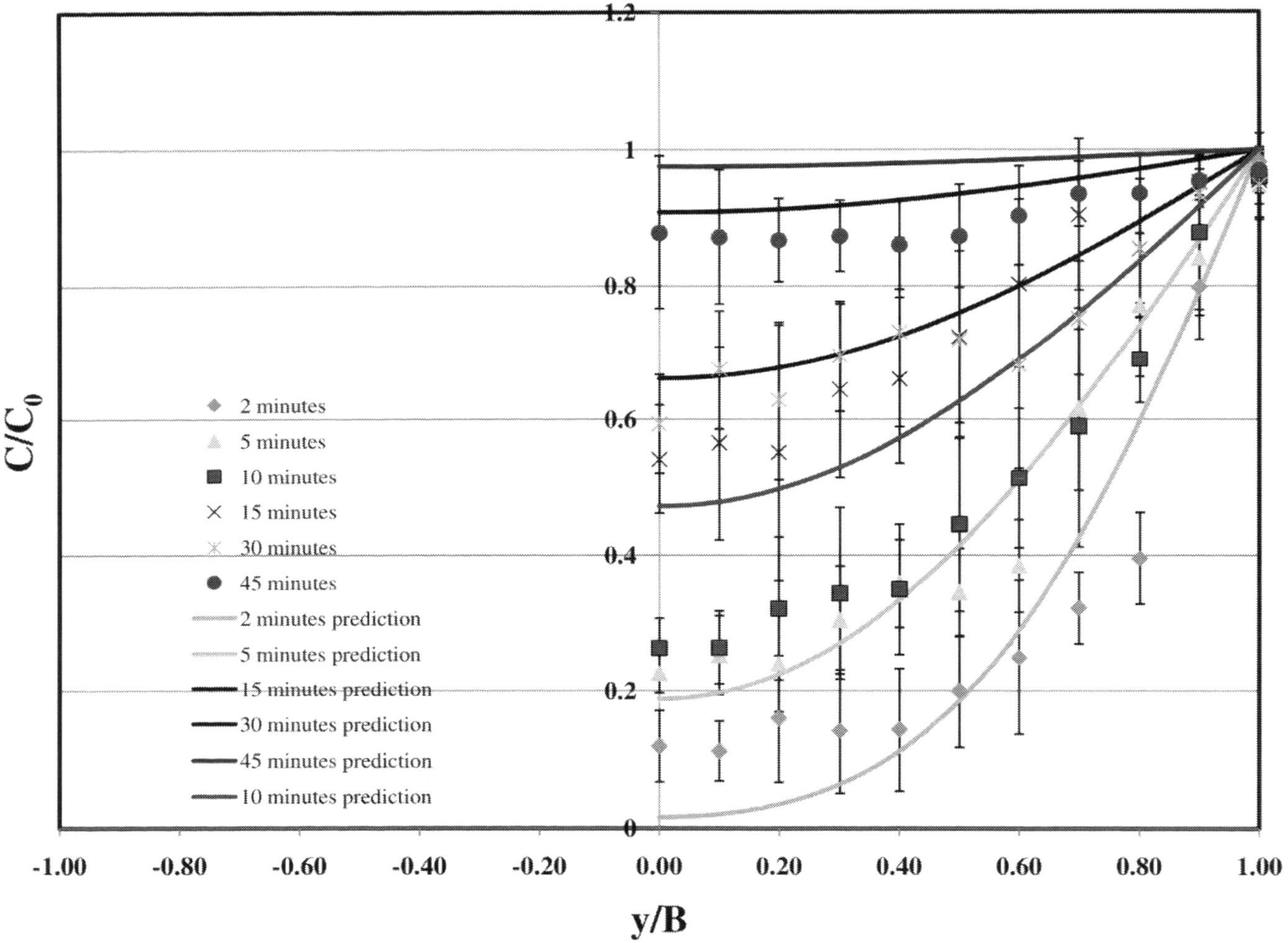

Figure 13: Summary of the relative concentration profiles for all immersion times as determined by UV fluorescence together with prediction of the rectangular slab model using the diffusion coefficient based on thermogravimetric analysis (TGA).

Theoretical Analyses of Co-Extrusion Process for Co-Layered Energetic Extrudates and Experimental Investigations of Co-Extrusion

Suzanne Prickett, Richard Muscato, Christopher Gonzalez,

Code 910 M, Naval Surface Warfare Center, Indian Head, MD

and

Dilhan Kalyon, Halil Gevgilili, Moinuddin Malik

Stevens Institute of Technology, Hoboken, NJ

Summary: Analytical and numerical solutions of the mathematical models of the co-extrusion process involving the rectilinear flow of multiple viscoplastic fluids subject to wall slip are to be presented along with experimental data that are collected using an industrial scale and well-equipped fully-intermeshing co-rotating twin screw extruder and a co-extrusion die.

Introduction: Multilayered coextrusion involves the bringing together of two or more fluids [1] to generate layered products with structural distributions and ultimate properties that depend on the properties of the individual fluids [2]. The fluids could be any one of many structured materials including polymeric melts, suspensions with polymeric binders and rigid particles, foams, nanocomposites, or gels. Typically multiple extruders (single or twin screw extruders) or gear pumps are used to pump the fluids into either an adopter section followed by a coextrusion die or directly into a coextrusion die [2, 3]. The basic steady state design problem deals with the determination of the pressure drop in the coextrusion die for targeted thicknesses of the different layers and the production rate. Coextrusion is followed by various processes including thermoforming and extrusion blow molding, to fabricate a wide range of multilayered products with each layer providing a separate function, i.e., acting as a barrier to gases, providing mechanical strength, enhanced appearance, cost reduction and printability. If the adhesion between two layers is not acceptable an adhesive tie layer is used as a third layer in between the two layers.

Various structured fluids including concentrated suspensions, gels, and nanocomposites exhibit viscoplastic behavior (i.e., exhibit a yield stress) and wall slip during flow. The viscoplasticity and wall slip of such structured fluids render the design of hardware and process optimization for their coextrusion difficult. The design and optimization tasks for coextrusion require a realistic understanding of the steady-state behavior of viscoplastic fluids in coextrusion flows. In the past a number of studies have developed analytical models of the coextrusion flow of two or more immiscible, incompressible, Newtonian or Ostwald-de Waele (power-law) type generalized Newtonian fluids under steady, laminar and fully-developed isothermal conditions in simple dies of constant slit or circular cross-sections subject to the no slip condition at the wall [4-6]. Various numerical methods were also used for the analysis of the coextrusion flow especially in conjunction with nonisothermal flow conditions [7-11]. Experimental studies have revealed that the stability of the interface in multilayered flows is a significant problem [12-13].

In the earlier studies of multilayered flows the boundary condition at the wall has generally been taken to be the conventional boundary condition of fluid dynamics, i.e., the no-slip condition versus the wall slip condition [14]. However, it is well known that various structured fluids including polymer melts [15-26], gels [27-31] and concentrated polymeric suspensions and pastes [32-42] exhibit a slip condition at the wall that can be quantified [14, 33-42]. Furthermore, the stability of the simple shear flow is intimately related to the development of the wall slip boundary condition along the length of the flow channel [43-44].

Here the steady and fully-developed coextrusion flow of two viscoplastic fluids subject to wall slip under isothermal and creeping flow conditions is considered. The model is specifically applied to the solution of the coextrusion problem in a thin slit die formed by two parallel walls, which are a distance h apart (Figure 1). The gap, h << the slit width, W, i.e., the width, W, is taken to be at least 10 to 20 times greater than the gap, h, to allow the flow to be considered one-dimensional. Simplifications of the model also provide the solutions for the fully-developed and isothermal coextrusion flows of any combination of Hershel-Bulkley, Bingham, power-law and Newtonian fluids with or without wall slip at one or both walls. The model can be adapted to other geometries. The methodologies can also be extended to the coextrusion of viscoplastic fluids subject to wall slip with the number of layers exceeding two. Furthermore, the coextrusion flow can also be solved using numerical simulation in conjunction with 3-D FEM. Finally, experimental studies of co-extrusion provide guidance on the validity of the assumptions that are carried out and allow the validations of the mathematical models of the process.

References:

1. B. Bird, R. Armstrong, O. Hassager, Dynamics of Polymeric Liquids, 1, 2nd Edition, Wiley, New York (1987).

2. Z. Tadmor and C. Gogos, Principles of Polymer Processing, 2nd Edition, Wiley-Interscience, Hoboken, NJ (2006).

3. D. G. Baird and D. I. Collias, Editors, Polymer Processing – Principles and Design, John Wiley and Sons, New York (1998).

4. C. D. Han and R. Shetty, Polym. Eng. Sci., 16, 697 (1976).

5. C. D. Han and R. Shetty, Polym. Eng. Sci., 18 187 (1978).

6. T. C. Yu and C. D. Han, J. Appl. Polym. Sci., 17, 1203 (1973).

7. M. E. Nordberg and H. H. Winter, Polym. Eng. Sci., 28, 444 (1988).

8. M Perez and J. R. Collier, Polym. Eng. Sci., 29, 1010 (1989).

9. S. Puissant, B. Vergnes, Y. Demay and J. F. Agassant, Polym. Eng. Sci., 32, 213 (1992).

10. A. Torres, A.N. Hrymak, J. Vlachopoulos, J. Dooley and B. T. Hilton, Rheologica Acta, 32, 513 (1993).

11. W. A. Gifford, Polym. Eng. Sci., 37, 315 (1997).

12. N. Minagawa and J. L. White, Polym. Eng. Sci., 15, 825 (1975).

13. C. D. Han and R. Shetty, Polym. Eng. Sci., 18, 180 (1978).

14. M. Mooney, J. Rheol., 2, 210 (1931).

15. J. J. Benbow and P. Lamb, SPE Trans., 3, 7 (1963).

16. W. R. Schowalter, J. Non-Newton. Fluid Mech., 29, 25 (1988).

17. M. R. Kamal, S. K. Goyal and E. Chu, AIChE J., 34, 94 (1988).

18. S. G. Hatzikiriakos and J. M. Dealy, J. Rheol., 36, 703 (1992).

19. Y. Chen, D. M. Kalyon and E. Bayramli, J. Appl. Polym. Sci., 50, 1169 (1993).

20. T. J. Person and M. M. Denn, J. Rheol., 41, 249 (1997).

21. M. Reimers and J. Dealy, J. Rheol., 42, 527 (1998) .

22. M. Denn, Annu. Rev. Fluid Mech., 33, 265 (2001).

23. H. Gevgilili and D.M. Kalyon, J. Rheol., 45, 467 (2001).

24. D. M. Kalyon and H. Gevgilili, J. Rheol., 47, 683 (2003).

25. D. M. Kalyon, H. Gevgilili and A. Shah, Int. Poly. Proc., 19, 129 (2004).

26. H. Park, S. Taek Lim, F. Smillo, J. M. Dealy, and C. G. Robertson, J. Rheol., 52, 1201 (2008).

27. T. Q. Jiang, A. C. Young, and A. B. Metzner, Rheol Acta, 25, 397 (1986).

28. A. Yoshimura and R. K. Prud'homme, J. Rheol., 32, 53 (1988).

29. H. J. Walls, S. B. Caines, A. M. Sanchez, and S. A. Khan, J. Rheol., 47, 847 (2003).

30. S. Adams, W. J. Frith, and J. R. Stokes, J. Rheol., 48, 1195 (2004).

31. J. Seth, M. Cloitre and R. T. Bonnecaze, J. Rheol., 52, 1241 (2008).

32. M. R. Kamal and T. Mutel, J. Polym. Eng., 5, 293 (1985).

33. U. Yilmazer and D. M. Kalyon, J. Rheol., 33, 1197 (1989).

34. D. M. Kalyon, P. Yaras, B. Aral and U. Yilmazer, " J. Rheol., 37, 35 (1993).

35. B. Aral and D. M. Kalyon, J. Rheol., 38, 957 (1994).

36. P. J. A. Hartman Kok, S. G. Kazarian, C. J. Lawrence, and B. J. Briscoe, J. Rheol., 46, 481 (2002).

37. R. Luckner and R. T. Bonnecaze, J. Rheo., 43, 735 (1999).

38. V. Bertola, F. Bertrand, H. Tabuteau, D. Bonn, P. Coussot, J. Rheol., 47, 1211 (2003).

39. H. Tang and D. M. Kalyon, Rheol. Acta, 43, 80 (2004).

40. D. Kalyon, J. Rheol., 49, 621 (2005).

41. C. Feger, J. D. Gelorme, M. McGlashan-Powell and D. M. Kalyon, IBM J. Research and Dev., 49, 699 (2005).

42. D. M. Kalyon and H. Tang, J. Non-Newtonian Fluid Mech., 143, 133 (2007).

43. H. S. Tang and D. M. Kalyon, J. Rheol., 52, 507 (2008).

44. H. Tang and D. M. Kalyon, J. Rheol., 52, 1069 (2008).

45. W. H. Herschel, R. Bulkley, Am. Soc. Test. Proc., 26, 621 (1926).

46. A. Lawal, D. M. Kalyon and U. Yilmazer, Chem. Eng. Comm., 122, 127 (1993).

47. A. Lawal and D. M. Kalyon, Polym. Eng. Sci., 34, 1471 (1994).

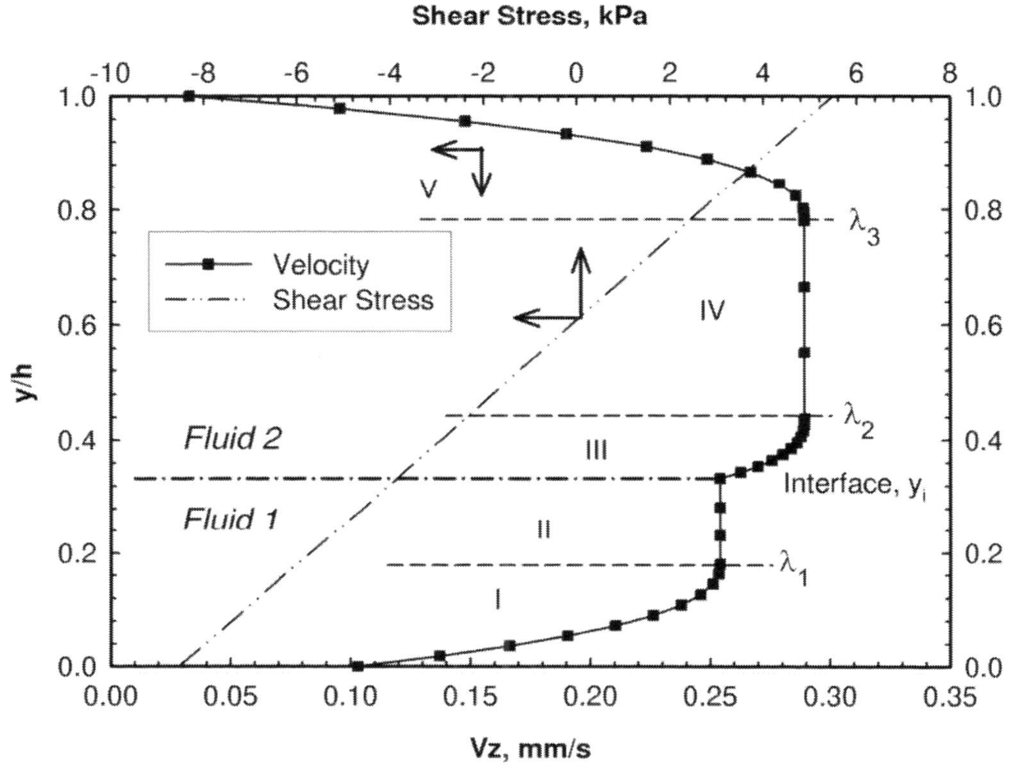

Figure 1: Co-extrusion of two concentrated suspensions in rectangular slit die subject to wall slip with differing wall slip coefficients.

Scale up of 2,6-Diamino-3,5-Dinitropyrazine-1-Oxide (LLM-105) and Its Precusors

Christopher M. Radack, Indian Head Div., Naval Surface Warfare Center, Indian Head, MD

Jerry S. Salan, Chemical Research and Development, Pfizer, Inc., Groton, CT

Kimberly Hanson, Naval Surface Warfare Center Indian Head Division, Indian Head, MD

The Naval Surface Warfare Center at Indian Head (IHDIV, NSWC) has recently produced multi-kilogram quantities of 2,6-diamino-3,5-dinitropyrazine-1-oxide (LLM-105) at the 100-gallon scale. First synthesized by Dr. Phil Pagoria at Lawrence Livermore National Laboratory, LLM-105 is being investigated as an energetic ingredient for military applications because of its unique combination of high-density, excellent performance, thermal stability, and insensitivity. Specifically LLM-105 is being investigated as an alternative to triaminotrinitrobenzene (TATB) in several DOD and DOE military applications.

The LLM-105 reaction sequence had been developed, resulting from a three year process development and scale-up endeavor. The current three-step reaction sequence involves the nitration of 2,6-dimethoxypyrazine (DMP) to form 2,6-dimethoxy-3, 5-dinitropyrazine (DMDP), which undergoes a subsequent amination to form 2,6-diamino-3,5-dinitropyrazine (ANPZ). The third and final step involves the oxidation of ANPZ to yield the product LLM-105. IHDIV, NSWC has established a robust scale-up methodology that can be implemented in further scale-up efforts and future production campaigns. The challenges encountered during the LLM-105 development and scale-up, paired with the scientific approach used to overcome each engineering challenge, are presented.

See more of this Session: Processing and Safety
See more of this Group/Topical: Particle Technology Forum

Comparison of the Mechanism of Toxicity of Binary Metal Oxide Nanoparticles Based On Dissolution and Oxidative Stress Properties

Lutz Mädler, Department of Production Engineering, University of Bremen IWT Foundation Institute of Materials Science, Bremen, Germany

Suman Pokhrel, University of Bremen IWT Foundation Institute of Materials Science, Bremen, Germany

Tian Xia, UCLA, Los Angeles

Monty Liong, Department of Chemistry and Biochemistry, UCLA, Los Angeles

B. Gilbert, Earth Science Division, Lawrence Berkeley National Laboratory, Berkeley, CA

J.I. Zink, Department of Chemistry and Biochemistry, UCLA, Los Angeles, CA

Andre E. Nel, Division of NanoMedicine, UCLA, Los Angeles

Nanomaterials (NM) are potential candidates with promising physicochemical properties and strong affinity towards biological substrates and processes. For this reason, several binary (TiO2, ZnO, and CeO2) were synthesized by Flame Spray Pyrolysis (FSP) technique to investigate the biological activities such as cellular uptake, subcellular localization, and toxic effects exhibited by these NMs based on oxidative stress and cytotoxicity. ZnO induced toxicity at the level of ≥ 50µg/ml concentration leading to the generation of reactive oxygen species (ROS), oxidant injury, excitation of inflammation and eventually cell death. Using ICP-MS and fluorescent-labelled ZnO, we found that NM dissolution could take part in culture medium and also in the endosomes. Fluorescent-labelled CeO2 nanoparticles were taken up into positive endosomal compartments free from toxicity and inflammation suppressing ROS production inducing cellular resistance to an exogenous source of oxidative stress. Fluorescent-labelled TiO2 was processed by the same uptake pathways as CeO2 but did not response towards any adverse or protective effects. These results demonstrate that metal oxide nanoparticles induce a range of biological responses that vary from cytotoxic to cytoprotective and can be properly understood by using a tiered test strategy developed for oxidative stress which could also successfully be extended towards cellular toxicity determination.

See more of this Session: Health and Environmental Effect of Nanoparticles
See more of this Group/Topical: Particle Technology Forum

Assessment of Toxicity of Metal Oxide Nanoparticles to Microbial Species

Bing Wu, Energy, Environmental and Chemical Engineering, Washington University, St Louis, MO

Yin Wang, Energy, Environmental and Chemical Engineering, Washington University, St Louis, MO

Manoranjan Sahu, Energy, Environmental and Chemical Engineering, Washington University, St Louis, MO

Rick Huang, Energy, Environmental and Chemical Engineering, Washington University, St Louis, MO

Xueyang Feng, Energy, Environmental and Chemical Engineering, Washington University, St Louis, MO

Pratim Biswas, Energy, Environmental and Chemical Engineering, Washington University, St Louis, MO

Yinjie Tang, Energy, Environmental and Chemical Engineering, Washington University, St Louis, MO

Nanomaterials have unique chemical and physical properties compared to the larger counterparts because of their high specific surface area and quantum size effects. To safely apply nanotechnology in biology, medicine, food and cosmetic industries, it is necessary to understand nanoparticles' toxicity and develop remediation methods to safely remove nanoparticles. This study focuses on the toxicity of metallic nanoparticles. Specifically, aerosol reactors were used to carefully synthesize nanomaterials with different metal compositions: TiO2 nanoparticles (NPs), Cu-doped TiO2 NPs, and ZnO NPs (all in 20-40nm size). The well characterized nanomaterials were then used to evaluate the toxicity by studying their effects on several microbial species: E.coli, a member of Enterobacteria that is commonly found in the lower intestine; Mycobacterium smegmatis, a model pathogenic strain used for the study of Tuberculosis; Shewanella oneidensis MR-1, a gram-negative environmental bacterium found in sediments; Cyanothece sp. ATCC 51142, a photosynthetic bacterium capable of hydrogen production; Saccharomyces Cerevisiae, a type of yeast often used for ethanol fermentation.

We have systematically examined microbial responses to nanomaterials by monitoring cell growth rates, morphological changes, key enzyme functions, gene expression, and photosynthesis activities (for Cyanothece sp.). Some of the results are:

1. TiO2 NPs did not affect the growth of any tested microbial species, while Cu-doped TiO2 NPs and ZnO NPs significantly reduced M. smegmatis growth (particles concentration = 0.02 g/L). Such inhibition was mainly mediated by released Cu2+ or Zn2+ from the NPs and addition of EDTA alleviated the NPs inhibitory effect on cellular growth.

2. TEM images indicated that nano-sized Cu-doped TiO2 and ZnO particles could not penetrate inside the cell and physically damage the cellular structure since they tended to aggregate in the medium solution, and thus the "nano" associated toxicity was reduced.

3. The tested microbial species responded to toxic NPs differently. M. smegmatis could not produce significant amount of extracellular polymeric substances (EPS, i.e. extracellular proteins and polysaccharides), so M. smegmatis was very sensitive to toxic NPs and may be used as a model strain for toxicity testing. On the other hand, S. oneidensis MR-1 and E.coli produced more EPS under the NP stress. The extracellular protein production improved the resistance of both bacteria to the toxic NPs.

4. S. oneidensis MR-1 reduced ionic metals released from NPs, thus it can be considered for metal oxide NP bioremediation.

5. Although metal oxide NPs did not inhibit cellular growth or protein functions due to particle aggregation in aqueous culture, NPs' toxicity can be clearly revealed if NPs were electrosprayed onto a bacterial biofilm

so that nano-sized NPs could be uniformly deposited on the cell surface. For example, electrospraying 20nm ZnO NPs dramatically reduced cellular viability compared with the control cases of electrospraying $ZnCl_2$ or buffer solutions. On the other hand, electrospraying large-size ZnO particles (480 nm) caused much less toxicity to cells. This observation implies that the exposure schemes of NPs to biological system are also important considerations for NPs toxicity.

See more of this Session: Health and Environmental Effect of Nanoparticles
See more of this Group/Topical: Particle Technology Forum

Good Dispersion and Intense Interaction Enhance Antibacterial Activity of Single-Walled Carbon Nanotubes

Yuan Chen, School of Chemical and Biomedical Engineering, Nanyang Technological University, Singapore, Singapore

Shaobin Liu, School of Chemical and Biomedical Engineering, Nanyang Technological University, Singapore, Singapore

Li Wei, School of Chemical and Biomedical Engineering, Nanyang Technological University, Singapore, Singapore

The toxicity and antibacterial activity of single walled carbon nanotubes (SWCNT) is related to their purity and aggregation. We previously developed a novel catalyst, cobalt-incorporated MCM-41 (Co-MCM-41) with controllable subnanometer scale metal clusters. It produces SWCNTs with a very narrow diameter distribution containing ca. 20 different (n,m) nanotubes. More recently, a centrifugation based purification protocol was developed to purify SWCNTs from Co-MCM-41 catalyst, by which SWCNT purity index based on absorption spectra of sodium cholate (SC) dispersed tubes is 0.268, approaching the previously estimated value (~0.325) for analytically pure SWCNTs. By tuning the centrifugation speeds, SWCNT samples containing different amounts of metallic resides (ranging between 19.28 at.% and 0.17 at.%) can be easily obtained. Furthermore, absorbed surfactant molecules such as SC can be eliminated by heating in air after removal of metallic residues. This allows us to re-dispersed "clean" SWCNTs in biocompatible surfactants.

In this study, we compared the antibacterial activity of purified SWCNT samples dispersed in saline solution 0.9 wt% NaCl, 0.1 wt% Tween + 0.9 wt% NaCl, and 1 wt% SC on gram-negative bacterium (E. coli K12) and gram-positive bacterium (B.subtilis). Our results showed that good dispersion in non-toxic surfactants can greatly improve SWCNT antibacterial activity. Moreover, the antibacterial activity is correlated with interaction intensity between SWCNT and bacteria. Investigation of bacteria membrane stuffiness by atomic force microscope suggests that bacteria with more flexible membranes are more vulnerable to SWCNT piercing which results in membrane damage leading to higher cell death rate. We also show that the antimicrobial properties of SWCNTs are independent of cobalt residue concentration on SWCNT samples.

Reference:

Chen, Y.; Ciuparu, D.; Lim, S.; Yang, Y.; Haller, G. L.; Pfefferle, L., Synthesis of Uniform Diameter Single-Wall Carbon Nanotubes in Co-MCM-41: Effects of the Catalyst Prereduction and Nanotube Growth Temperatures. J. Catal. 2004, 225, (2), 453-465.

Chen, Y.; Ciuparu, D.; Lim, S.; Yang, Y.; Haller, G. L.; Pfefferle, L., Synthesis of Uniform Diameter Single Wall Carbon Nanotubes in Co-MCM-41: Effects of CO Pressure and Reaction Time. J. Catal. 2004, 226, (2), 351-362.

Wei, L.; Wang, B.; Wang, Q.; Li, L.-J.; Yang, Y.; Chen, Y., Effect of Centrifugation on the Purity of Single-Walled Carbon Nanotubes from MCM-41 Containing Cobalt. Journal of Physical Chemistry C 2008, 112, (45), 17567-17575.

See more of this Session: Health and Environmental Effect of Nanoparticles
See more of this Group/Topical: Particle Technology Forum

Bioavailable Metal Speciation and Surface Redox Capacity in Toxicity of Nanoparticles

Bing Guo, Department of Mechanical Engineering, Texas A&M University, College Station, TX

Metal-mediated oxidative stress is an important mechanism of nanoparticle toxicity. Metal mediated oxidative stress is highly dependent on bioavailable metal speciation. However, currently the bioavailable metal speciation is not adequately characterized in toxicology of nanoparticles. In addition, the effect of surface redox capacity of elemental carbon nanometerials on bioavailable metal speciation is yet to be understood.

The objective of this study was to test the hypothesis that bioavailable iron speciation is a key factor of iron-mediated oxidative stress, and that carbon black nanoparticles can affect the bioavailable iron speciation through surface redox capacity. Cultured A549 human lung epithelial cells were exposed to (1) carbon black nanoparticles alone, (2) $Fe2O3$ nanoparticles alone, and (3) both types of particles simultaneously. Protein oxidation, lipid peroxidation, and cellular cell viability were measured after 25 hours of exposure. The reductive capacity of carbon black nanoparticles towards $Fe3+$ ions was measured separately in a cell-free assay.

Cells co-exposed to both carbon black and $Fe2O3$ particles showed significant oxidative stress. In the cell-free assay, significant reduction of $Fe3+$ ions by soot nanoparticle was found. On the other hand, surface oxidized carbon black particles did not show significant redox capacity, and co-exposure to $Fe2O3$ and oxidized carbon black particles did not generate significant oxidative stress. The results suggest that bioavailable metal speciation is an important parameter for assessing toxicity of nanoparticles, and that surface redox capacity should be characterized because it may play an important role in nanoparticle toxicity.

See more of this Session: Health and Environmental Effect of Nanoparticles
See more of this Group/Topical: Particle Technology Forum

Nanoparticle Transfer and Deposition Using An Integrated CFD Model of the Respiratory System

Aleck Alexopoulos, Chemical Process Engineering Research Institute, Thessaloniki, Greece

Paraskevi Karakosta, Department of Chemical Engineering, Aristotle University of Thessaloniki & Chemical Process Engineering Research Institure, Thessaloniki, Greece

Costas Kiparissides, Department of Chemical Engineering, Aristotle University of Thessaloniki & Center for Research & Technology Hellas, Thessaloniki, Greece

The transfer and deposition of nanoparticles in the respiratory system is of major interest for the development of targeted drug delivery formulations but also due to the increasing concerns over the potential toxicity of natural and engineered nanoparticles. Experimental and theoretical work have focused on different regions of the respiratory system, e.g., oropharyngeal, nasal, pulmonary, alveolar, where many aspects of the nanoparticle, NP, penetration are fairly well understood. However, several issues remain to be elucidated including the deposition of non-spherical NPs and nanofibers, dispersion and aggregation (e.g., of carbon nanotubes), changes in particle size and shape as well as aggregation state and finally the fate of deposited particles which is of major concern. Deposited particles can undergo disaggregation, can release beneficial drugs or harmful toxins, and, if sufficiently small, can penetrate into the bloodstream.

To further improve our understanding of NP penetration and deposition an integrated CFD model of the entire respiratory system, describing the flow of a dispersion of more particles has been developed through the nasal cavity and lungs. The flow through the pulmonary system is described down to the alveolar sac and individual alveoli level. The deposition of nanoparticles and nanofibers is approximated by a Lagrangian/Eulerian tracking scheme. Local deposition models of NPs are developed to provide a better insight into the deposition dynamics and the sticking efficiency of the NPs. These local deposition models utilize CFD solutions of the global geometry to supply inlet conditions for the local deposition problem geometry.

The nasal cavity consists of two nasal air paths converging to a single pathway at the back of the cavity which is then directed downwards to the trachea. The two nasal air paths are highly curved and convoluted in shape providing a total surface area of about 150 cm^2. The nasal walls are covered by a mucous layer which moves to the posterior clearing deposited particles. The flow and deposition of particles in the nasal cavity has recently been investigated by several groups (Liu et al., 2007; Shi et al., 2008; Wen et al., 2008). In this work, the nasal cavity geometry is reconstructed based on a series of medical images. A number of different computational grids are then constructed (from 0.3 10^6 polyhedral to 1.6 10^6 tetrahedral cells). Turbulent flow is described with a transitional k-ω model. The nasal deposition of large particles (i.e., 1-10μm) was found to be higher than the deposition of smaller particles. The particle deposition efficiency is described in terms of the impaction parameter, QD^2. Inlet particles are assumed to be either single- or multiple- sized. Comprehensive information on the axial and size distribution of deposited particles can be obtained. It was found that larger particles were mostly deposited in the anterior region of the nasal cavity while smaller particles were deposited less and more evenly throughout the nasal cavity.

The pulmonary system consists of a multitude of nonsymmetrical branches of progressively smaller diameter. There are a total of 23-24 branch generations leading to about 107 simple branches in the entire pulmonary system (Finlay, 2004). This limits the number of branch generations that can be completely simulated to around 5-6 (Longest and Vinchurkar, 2007; van Ertbruggen et al., 2005; Zhang et al., 2002). However, if a single pathline down to the alveolar sacs is considered, a successive simulation approach can be followed (Nowak et al., 2003). In the present work, a model of the pulmonary tract is developed based on 7 consecutive "blocks" of the pulmonary system. Each block consists of 4 generations of symmetrical branches with one side rotated 90 degrees relative to the other and is discretized into 2 10^5 tetrahedral cells. The inlet flow and particle motion conditions of each block are obtained from the outlet conditions of the previous block and the inlet conditions of the first block are obtained from the outlet conditions of the nasal cavity. Particle depositions in the pulmonary tract again favored the larger particles (i.e., 100nm-1μm), and only the smallest of particles (i.e., <100nm) could reach the lower respiratory tract and alveolar sacs at significant concentrations. Toxin (or drug) release from deposited particles and penetration through the underlying mucosal layer is described using a dual diffusion-layer model which accounted for the mucosal layer motion as well as the deposition location.

The alveolar sac model was integrated with the pulmonary model. Specifically, the outflow from the last block of the pulmonary model is taken as the inflow to an alveolar sac. Although each alveolar sac contains a number (i.e., 10-50) of alveoli, the description of flow to the alveolar sac and to the individual alveoli is simplified by considering a model of a single alveoli. The single alveoli model is comprised of a spherical volume which changes in size during inhalation due to inflow from the alveoli mouth. Nanoparticles also enter the alveoli and their deposition is controlled by Brownian motion and inertial forces.

The proposed integrated respiratory model describes the flow, penetration, and deposition of nanoparticles in the respiratory system accounting for the influence of the nasal cavity and the 23-24 generations of pulmonary branches including the alveolar sacs. Particle deposition is driven by Brownian motion for the smaller particles and by inertial forces for the larger particles. Sticking efficiencies for non-spherical NPs and nanofibers can be estimated from local deposition models. The present integrated model can approximately describe the transfer and deposition of nanoparticles and nanofibers throughout the respiratory system.

References

Finlay, W.H. "Lung Deposition Simulation" in A.J. Hickey (ed.) Pharmaceutical Inhalation Aerosol Technology, Marcel Dekker, NY 2004.

Liu Y., E.A. Matida, J. Gu, and M.R. Johnson, 2007, "Numerical simulation of aerosol deposition in a 3-D human nasal cavity using RANS, RANS/EIM, and LES", Aerosol Sci., 38, 683-700.

Longest, P. and Vinchurkar, S., 2007, "Validating CFD predictions of respiratory aerosol deposition: Effects of upstream transition and turbulence", J. of Biomechanics, 40(20, 305-316

Nowak N., P.P. Kakade, and A.V. Annapragada, 2003, "Computational Fluid Dynamics Simulation of Airflow and Aerosol Deposition in Human Lungs", Annals of Biomed. Engineering, 31, 374-390.

Shi, H.W., C. Kleinstreuer and Z. Zhang, 2008, "Dilute suspension flow with nanoparticle deposition in a representative nasal airway model", Physics of Fluids 20, 013301.

van Ertbruggen, C., C. Hirsch, and M. Paiva, 2005, "Anatomically based three-dimensional model of airways to simulate flow and particle transport using computational fluid dynamics", J. Appl. Physiol. 98, 970-980.

Wen J., K. Inthavong, J. Tu and S. Wang, 2008, "Numerical simulations for detailed airflow dynamics in a human nasal cavity", Respiratory Physiology & Neurobiology.

Zhang Z., C. Kleinstreuer, and S.K. Chong, 2002, "Aerosol deposition efficiencies and upstream release positions for different inhalation modes in an upper bronchial airway model" Aerosol Sci. & Technol., 36, 828-844.

See more of this Session: Health and Environmental Effect of Nanoparticles
See more of this Group/Topical: Particle Technology Forum

Design, Development, and Scale-up of A Reactive Crystallization for Particle Size Control: A Double Dip Pipe (DDP) Salt Formation Process

Fang Wang, Chemical Development, Gilead Sciences, Foster City, CA

Ernest Carra, Chemical Development, Gilead Sciences, Foster City, CA

Michael Hawley, Material Science, Pfizer, Groton, CT

Alison Acciacca, Material Science, Pfizer, Groton, CT

A Double Dip Pipe (DDP) crystallization process for an active pharmaceutical ingredient (API) has been designed and demonstrated successfully in the lab. This new process design is based on the learning obtained from the previous semi-batch process and Impinging Jet crystallization process. Solubility data for both the salt and the freebase have enabled us to choose the appropriate solvent for this process. The DDP crystallization process together with a switch of solvent has offered significant improvement in the particle size and the breadth of distribution at the lab scale. This improvement has allowed us to crystallize this API directly within the particle size target without milling at a 20 g scale. With the assistance of computational fluid dynamic (CFD) modeling, this process was also successfully scaled up in the pilot plant with some modification to the configuration at a 30 kg scale while meeting all specifications and targets with a higher than expected yield.

See more of this Session: Particle Formation and Crystallization Processes From Liquids, Slurries and Emulsions
See more of this Group/Topical: Particle Technology Forum

Tailoring the Size Distribution of Nanoparticles Via An Emulsion-Assisted Precipitation Process: A Model-Based Analysis

Michael Fricke, PCP, MPI Magdeburg, Magdeburg, Germany

Andreas Voigt, Process Systems Engineering, University of Magdeburg, Magdeburg, Germany

Kai Sundmacher, Max Planck Institute for Dynamics of Complex Technical Systems, Magdeburg, Germany

Several techniques for the synthesis of nanoparticles on the laboratory-scale have been developed in recent years [1]. With respect to industrial-scale processes it is of major importance to identify a suitable production route for the formation of narrowly distributed nano-scale particles. The precipitation of crystals induced by chemical reaction in conventional stirred tank reactors is a widely used industrial process. Due to the ionic character of the participating reactants a very fast and irreversible chemical reaction takes place. The production of sparingly soluble product is followed by subsequent nucleation and growth of particles. Additionally, agglomeration of single crystals in highly concentrated reactors influences the mean particle size. Furthermore, differences in time constants between chemical reaction and micro mixing cause inhomogeneities in crystallization kinetics. The combination of both effects results in broad size distributions of micrometer particles. In order to produce narrowly distributed particle on the nanometer scale emulsion-assisted precipitation techniques overcome the drawbacks of bulk precipitation processes. In particular the application of precipitation reactions in water pools of water-in-oil(w/o)-emulsions with a mean droplet size on the submicrometer scale is of interest for industrial applications [2].

This contribution provides a population balance model-based analysis of an emulsion-assisted precipitation process. We consider a case where the precipitation reaction is induced by mass transfer of one of the reactants from the continuous phase of the emulsion across the liquid-liquid interface into the droplets, while the other reactant is located in the aqueous dispersed droplet phase. The actual precipitation reaction takes place in the droplets. This technique can be considered as a semi-batch process. In a first process step a well-prepared emulsion is filled into the reactor, whereas one of the reactants is solved in the aqueous core of the droplets. In a second step the other reactant is added as a pure substance into the reactor by a dosing pump.

A mathematical model of this process consisting of mass balances in the continuous and dispersed phase and a population balance in the droplet phase has been derived. Assuming a spherical shape and symmetric boundary condition at the droplet interface the description of spatial effects of mass and population dynamics on the radius coordinate in a spherical coordinate system is sufficient. Consequently, a 1+1 dimensional population balance equation for the particle formation was deployed, whereas aggregation and breakage of particles are disregarded. As shown experimentally by Sathaye et al. [3] nucleation at the liquid-liquid interface should not be neglected in emulsion assisted precipitation processes. Therefore, two nucleation mechanisms, namely homogeneous nucleation in the volume of the droplet and heterogeneous nucleation at the droplet interface are both incorporated in the population balance. Hence, the interplay of different nucleation mechanisms and particle growth on the one hand, and the influence of process parameters such as the droplet size, concentration levels and dosing rate on the particle properties on the other hand are investigated by computer simulations.

It is shown that the predominant nucleation mechanism determines the width and the modality of the resulting particle size distribution. At fixed nucleation and growth kinetics, the particle size and the polydispersity of the distribution can be varied by adjusting the process parameters. We show for example that for the production of small primary particles with a narrow size distribution a reduction in size of the emulsion droplets is favorable.

A number of model parameters need more experimental input because a detailed modeling of interfacial phenomena is necessary to assess nucleation and mass transfer at the liquid-liquid interface. In order to prove interfacial transport and nucleation mechanisms, applicable experiments have to be developed. The derived model will be validated by experiments using a suitable reaction system.

[1] Niemann, B.; Rauscher, F.;Adityawarman, D.; Voigt, A.; Sundmacher, K., Chem. Eng. Process., 2006, 45, 917-935.

[2] Sager, W.; Eicke, H.-J.; Sun, W., Colloid Surface A, 1993, 79, 199-216.

[3] Sathaye, S.D.; Patil, K.R:; Paranjape, D.V.; Mitra, A.; Awate, S.V.; Mandale, A.B., Langmuir, 2000, 16, 3487-3490.

See more of this Session: Particle Formation and Crystallization Processes From Liquids, Slurries and Emulsions
See more of this Group/Topical: Particle Technology Forum

Impact of Seed Particle Size Distribution Uncertainty On the Supersaturation Controlled Batch Cooling Crystallization Time: Experimental and Model Based Evaluation

Levente L. Simon, Institute for Chemical and Bioengineering, Safety and Environmental Technology Group, ETH Zurich, Zurich, Switzerland

Yannick Elias, Institute of Bio and Chemical Engineering, ETH Zurich, Zurich, Switzerland

Zoltan K. Nagy, Chemical Engineering Department, Loughborough University, Loughborough, England

Konrad Hungerbuehler, Institute for Chemical and Bioengineering, ETH Zurich, 8093 Zurich, Switzerland

This project investigates the sources of uncertainty for supersaturation controlled crystallization batch time. Since variation of the seed particle size distribution (PSD) is propagated in the final batch time (assuming constant yield operation), the first part of the contribution presents the systematic evaluation of sieving for seed preparation using the QICPIC from Sympatec and Morphologi G3 from Malvern, units which are based on digital imaging. The second part of the contribution quantifies the impact of seed PSD on the crystallization time considering supersaturation controlled batch cooling operation. This is carried out using a population balance model solved using the method of characteristics which assumes size-dependent crystal growth and constant supersaturation, thus an analytical solution is obtained [1]. The main findings are that the PSD of the powder collected from the sieves is shifted towards large sizes, and does not correspond to the nominal sieve specifications; furthermore, the batch-to-batch variability of sieving is low and often bimodal distributions are obtained due to fines. For a fixed sieving recipe, the crystal shape influences the powder PSD obtained on sieves. Related to the crystallization operation, it is concluded that, for size dependent crystal growth, fines increase the crystallization time with as much as 40%; however if these grow above 50-100 microns filtration problems are avoided. Moreover, the exact PSD of the powder obtained by sieving should be determined by an additional sizing technique to provide high quality initial conditions for on-line, population balance model based PSD estimation applications.

[1] Aamir E, Nagy ZK, Rielly CD. Combined Quadrature Method of Moments and Method of Characteristics Approach for Efficient Solution of Population Balance Models for Dynamic Modeling and Crystal Size Distribution Control of Crystallization Processes, INDUSTRIAL & ENGINEERING CHEMISTRY RESEARCH, 48,18,8575-8584,2009

See more of this Session: Particle Formation and Crystallization Processes From Liquids, Slurries and Emulsions
See more of this Group/Topical: Particle Technology Forum

Workflow for the Design of Seeding Application in Batch Crystallization

Kui S. Kwok, Department of Chemical and Biomolecular Engineering, Hong Kong University of Science and Technology, Hong Kong, China

R. Lakerveld, Process & Energy, Delft University of Technology, Delft, Netherlands

Ka M. Ng, Department of Chemical and Biomolecular Engineering, Hong Kong University of Science and Technology, Hong Kong, China

Herman J.M. Kramer, Process & Energy, Delft University of Technology, Delft, Netherlands

Peter J. Jansens, Process & Energy, Delft University of Technology, Delft, Netherlands

Alex H. C. Chan, Department of Chemical and Biomolecular Engineering, Hong Kong University of Science and Technology, Hong Kong, China

Christianto Wibowo, ClearWaterBay Technology, Pomona, CA

Abstract

A workflow procedure that guides the decision making for seeding application in batch crystallization is presented to facilitate the production of crystals of desired quality, which include mean crystal size, crystal size distribution, crystalline state and impurity level. Such a workflow is embedded in the framework related to the design of a complete crystallization-based process. As a basis for decision-making in the workflow, the effect of seed production method on seed quality and how such seed quality affects the crystallization operation and the product quality are first summarized and presented in the form of a causal table, which is made according to experimental results and model-based simulations from literature. The developed workflow consists of three major steps. The first step involves the definition of process objectives and gathering of relevant thermodynamic and kinetic information of the system. In the second part, preliminary experiments are performed for seed preparation and for determining the safe operating region of the seeding application. A model equation is also used to calculate the optimum seed mass for a given seed size, based on which seeded batch crystallization are then conducted to produce crystals. The quality of such crystals is evaluated in the last step with the target values. If the objectives are not met, the seed quality is changed repeatedly with the aid of causal table and model equation such that product crystals of desired quality are finally obtained. Two selected binary systems, succinic acid-water and paracetamol-water, are used to illustrate the workflow.

See more of this Session: Particle Formation and Crystallization Processes From Liquids, Slurries and Emulsions
See more of this Group/Topical: Particle Technology Forum

Processing techniques for improved batch cooling crystallization of RDX

Ruslan Mudryy & Victor Stepanov

U.S. Army, Armaments Research, Development and Engineering Center,

Picatinny Arsenal, NJ 07806-5000, USA

INTRODUCTION

Batch cooling crystallization is widely used for the production of energetic materials. In the crystallization process, the ability to control the supersaturation level is of high importance as it can have a significant impact on the product crystal characteristics. In the present study, *in situ* FTIR spectroscopy[1] was employed for the on line measurement of solution concentration during cooling crystallization. Kinetic curves were measured based on the isothermal desupersaturation of solutions following controlled seeding. Based on the isothermal desupersaturation kinetics obtained at multiple temperatures, crystallization kinetic parameters were determined. Kinetic data is critical in optimizing cooling profiles, where, typically, the supersaturation level is maintained at a desired level during the entire cooling process. In this work, cyclotrimethylene trinitramine (RDX) growth kinetics were calculated using the diffusion layer model described by Chernov[2].

Maintaining a desired supersaturation level is one means of preventing undesired nucleation events during cooling which can ultimately result in poor product homogeneity. It is also desired to prevent nucleation in the first place by seeding a supersaturated solution prior to the onset of spontaneous nucleation. Seeding is a simple and effective means to control the final product size, size distribution, and may also impact the morphology of the product. Another means of controlling the initial precipitation is via ultrasonic agitation of supersaturated solutions [3]. It is well known that in the presence of ultrasound spontaneous nucleation can occur at very low supersaturation levels, therefore, nucleation can be triggered as desired within the broad metastable zone. Furthermore, much better product size uniformity as well as control of the crystal habit may be also achieved via ultrasonic agitation.

Cooling crystallization takes advantage of the temperature dependence of solute's solubility. The solubility generally decreases with decreasing temperature forming a supersaturated solution. A supersaturated solution is not at equilibrium and in order to relieve the supersaturation and move towards equilibrium spontaneous nucleation occurs. Supersaturation is the driving force for nucleation and crystal growth during cooling crystallization process. Although necessary for crystallization, excessive supersaturation resulting from uncontrolled cooling can lead to undesired nucleation, which can result in poor quality product with wide size distribution.

It is known from the literature [4] that RDX crystallization from cyclohexanone is influenced by impurities in the system which can lead to large the formation of inclusions within the product crystals as well as adversely affect the crystal habit. An impurity that is particularly problematic in this respect is TAX which is a common byproduct from the Bachmann synthesis process. It was also shown[4] that presence of impurities in the solution caused nucleation at low levels of supersaturation. Consequently, the presence of such an impurity can lead to multiple nucleation events, resulting in poor crystal size distribution. Furthermore, the presence of TAX was linked to the formation of crystals with an undesired flat, rectangular morphology. One way to eliminate the effect of TAX is to saturate the cyclohexanone solvent with water (~3w%). Water is believed to bind with TAX and considerably decreases the effects of this impurity. In the current work crystallization with and without water was studied.

EXPERIMENTAL

Experimental Set-up

An automated one liter reactor, Mettle Toledo RC-1, was used for the batch cooling crystallization experiments. The apparatus consists of a one-liter jacketed Pyrex vessel, automated stirring control and temperature with 0.1 °C precision. An in situ attenuated total reflectance IR (ATR-IR), Mettler Toledo ReactIR -4000 instrument equipped with a diamond probe (DiComp™) was used for monitoring of the solute concentration. The ATR-IR generated spectra were collected in the range 650-4000 cm^{-1}. An polarizing optical microscope, Nicon Eclipse E600POL, with digital image acquisition was employed for analyzing the crystal size and morphology. Ultrasonic agitation was achieved using a probe sonicator Cole Parmer Ultrasonic Processor with 20 kHz frequency and probe a 1.25 cm diameter.

Materials

BAE Holston grade RDX synthesized via the Bachman route was used as the precursor in the recrystallization experiments. RDX contained around 10 wt. % of HMX. Reagent grade cyclohexanone (Aldrich) and cyclohexanone saturated with deionized water (~3 wt. %) were used as the process solvent.

Impact of Cooling Rate

A series of experiments was conducted to study the effect of cooling rate on the nucleation process. 110 grams of RDX and 570 g of cyclohexanone were loaded into 1 liter reactor and heated to 85°C to ensure that all solids were dissolved. This experiment was performed without water. Next, the temperature of the solution was brought down to the estimated saturation value, 71°C. A linear temperature cooling profile was programmed on the RC-1 system beginning at the saturated temperature and ending at 25°C. The linear cooling at 0.1 K/min, 0.2 K/min and 0.4 K/min rates were studied. Stirring speed remained constant at 150 rpm during the entire crystallization process. The Solution concentration was monitored by in situ ATR-IR measurements. At the onset of nucleation, the RDX maximum absorbance peak (1270 cm^{-1}) begins to decrease. The temperatures corresponding to the onset of nucleation at the various cooling rates are shown in Figure 1.

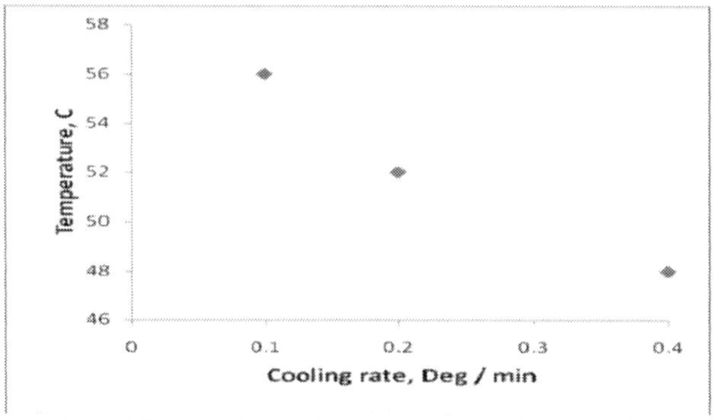

Figure 1. Temperature at the onset of nucleation as a function of cooling rate.

It can be seen that, as expected, precipitated RDX crystals become detectable at lower temperature for higher cooling rates. This is in agreement with the conventional observation of the existence of an induction time, a time from the moment the solution becomes supersaturated to the detection of crystals, during which nucleation and growth of crystals to detectable size takes place. As the cooling rate increases, assuming a characteristic induction time to be nearly constant, a lower temperature can be reached within this time period.

The effect of the cooling rate on the crystal size is shown in Figure 2. It can be seen that crystallization at the lower cooling rate resulted in larger crystals. This result is consistent with the classical nucleation theory, according to which nucleation occurring at a higher level of supersaturation should result in a larger number of nuclei and therefore smaller crystal size of the product. The results in Figure 1support this interpretation as the lower temperature at the onset of nucleation corresponding to the higher cooling rate should result in nucleation at a higher level of supersaturation. No significant impact on the morphology of the crystals was observed.

Figure 2. RDX crystals produced by cooling crystallization with linear cooling profiles of 0.2 K/min (left) and 0.4 K/min cooling rate.

Influence of Ultrasound

A series of cooling crystallization experiments were performed using power ultrasound to initiate nucleation in supersaturated solutions. An experiment was performed where a solution of RDX in cyclohexanone was supersaturated by linear cooling at 0.4 K/min rate. Ultrasound was applied for ten seconds at the temperature seven degrees below the saturation temperature. No precipitate was visible prior to ultrasound application. The influence of ultrasound power at 10 and 20 W levels on the crystallization was investigated and compared to that under silent conditions (0 W).

It was observed that with ultrasound nucleation can be induced at supersaturation levels lower than required for spontaneous precipitation under agitation with a stirrer. Furthermore, the crystal habit of the product crystals showed substantial change compared to crystals grown under silent conditions as can be seen in Figure 3.

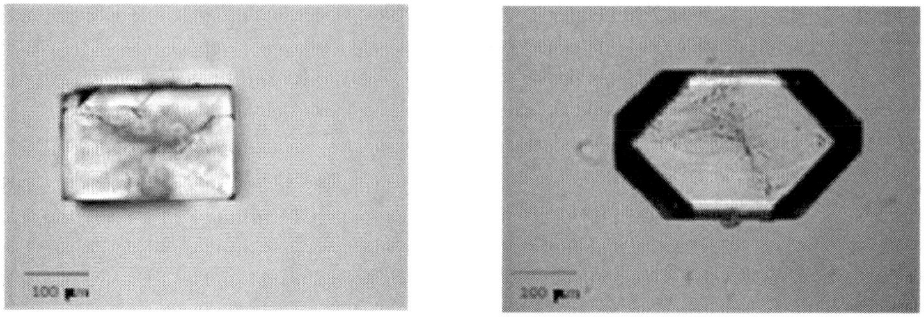

Figure 3. RDX crystallized without (left) and with ultrasound (10 W).

The intensity of the ultrasound during nucleation was found to impact nucleation. At the lower power level, 10 W, larger crystals were formed compared to ultrasound at 20 W as is shown in Figure 4. It can be concluded that at the higher power level the nucleation rate is higher, resulting in a larger number of nuclei and consequently larger number of smaller product crystals.

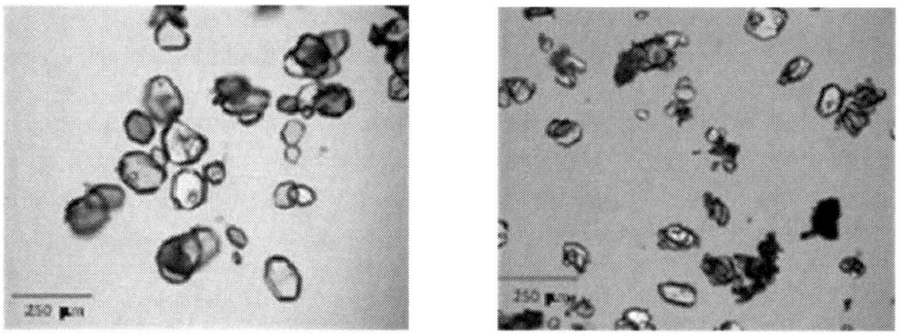

Figure 4. RDX crystallized with 10 W (left) and 20 W (right) ultrasound induced nucleation.

Crystallization with External Seeding

External seeding was employed to eliminate spontaneous nucleation of RDX during recrystallization. Typically a known amount of seeds with a mean crystal size of around 3 μm were dispersed in a small volume of water using a bath sonicator and added to the crystallizer at a temperature around 5 degrees below the saturation temperature. The solution was linearly cooled to near 25 °C prior to sampling of the product. Crystals obtained without and with seeding are shown in Figure 5. It can be seen that a more uniform crystal size distribution is obtained with seeding.

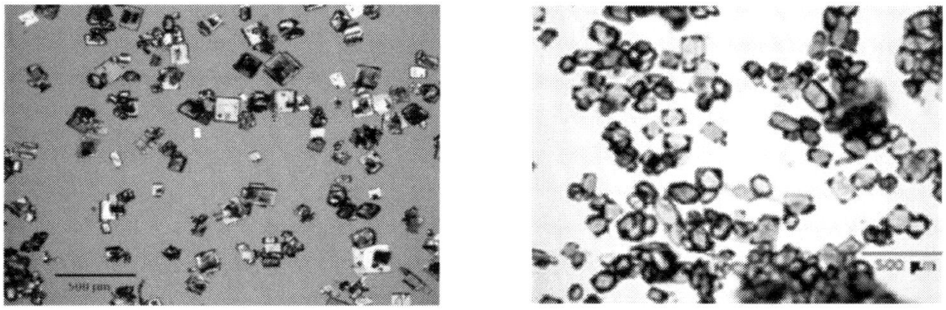

Figure 5. RDX crystals grown without (left) and with seeding (right).

The product crystal size from a batch crystallizer can be controlled by the mass of added seeds. Assuming that nucleation is eliminated, amount of seeds needed to get a final product with desired size was calculated from the followed the mass-balance equation:

$$M_s = \frac{Y \cdot L_s^3}{(L_p^3 - L_s^3)} \tag{1}$$

M_s - mass of seeds, Y -crystal yield, L_s -size of seeds, L_p -product crystal size.

Crystallization of RDX with 5 and 50 mg of loaded seeds was performed. The calculated mean crystal size based on the experimental conditions used is around 50 and 25 μm for 5 and 50 mg of loaded seeds respectively. The corresponding crystal sizes obtained experimentally are in good agreement with the expected sizes as can be seen in Figure 6. This is a good indication that nucleation was suppressed.

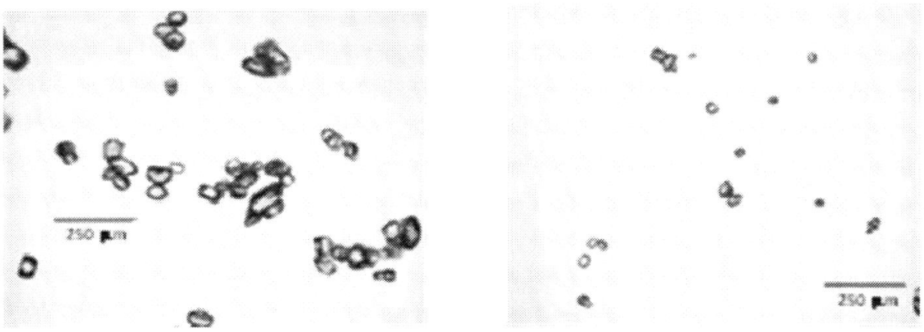

Figure 6. RDX recrystallized with 5 mg (left) and 50 mg (right) of seed loading.

Crystal Growth Kinetics

There are a number of experimental methods that can be used to obtain kinetics. We have used ReactIR instrument described above for determination of crystal growth rate. Measurements of growth kinetics were performed by collecting isothermal desupersaturation curves of seeded supersaturated solutions of RDX in cyclohexanone. RDX seed crystals of known total mass and size (3μm) were added to the solution. Growth of the seed crystals caused a decline in the concentration which was measured by the ReactIR probe as function of time.

Isothermal desupersaturation curves were measured at two temperatures, 40 °C and 55°C. The obtained concentration profiles are shown in Figure 7.

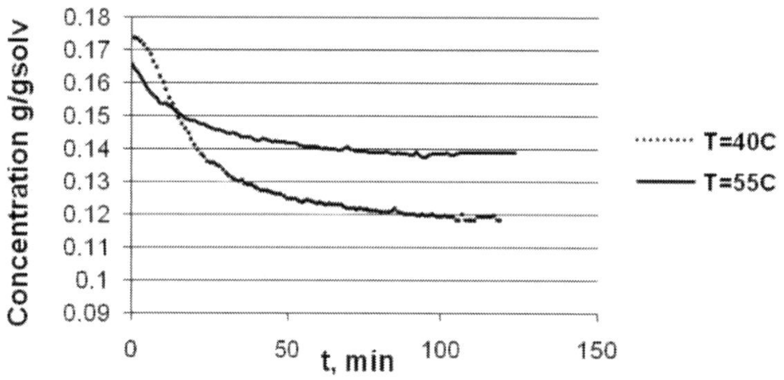

Figure 7. Isothermal desupersaturation curves measured at 40 °C and 55°C.

The supersaturation level at the solution temperature corresponding to moment of seeding was calculated according to the following expression:

$$S = \frac{c - c^*}{c^*} \cdot 100\% \tag{2}$$

where c^* is the solute concentration at saturation and c is the actual concentration.

A number of models are available to estimate the growth kinetics from the desupersaturation data [5,6]. In the present study, crystal growth kinetics was estimated from the diffusion layer model described by Chernov [2]. The basis of the diffusion layer model is that solute diffuses through the boundary layer and is then incorporated into the crystal. The overall crystal growth rate, R_g may be expressed in terms of mass flux [5].

$$R_g = \frac{1}{A_T} \cdot \frac{dm}{dt} \tag{3}$$

where $\frac{dm}{dt}$ is the mass rate of deposition and A_T is the total crystal surface area available for growth. Assuming that negligible nucleation occurs after the seeds are added, the change in solution concentration dc at any instant is proportional to the mass deposition dm on existing crystals, i.e.:

$$dm = -M \cdot dc$$

where M is the mass of the solvent. Taking this into account, and that $dc = d(c - c^*) = d\Delta c$, the mass flux equation can be rewritten as :

$$R_g = -\frac{1}{A}\frac{d\Delta c}{dt} \quad \text{where} \quad A = \frac{A_T}{M} \tag{4}$$

Where A_T is the total crystal surface area per unit mass of solvent. The value of dc/dt is obtained from desupersaturation curve and from the total crystal surface area from which the crystal growth rate, R_g, can be determined. It is known from the literature [2, 5] that the empirical expression for the relationship between supersaturation and crystal growth has a form:

$$R_g = k_g \cdot \Delta c^g \tag{5}$$

The mass growth rate coefficient, k_g, and the growth order, g, can be determined by linearizing the empirical growth equation as follows:

$$Ln(R_g) = Ln(k_g) + \mathrm{g} \cdot Ln(\Delta c) \tag{6}$$

With calculated kinetic parameters g and k_g it is possible to calculate the crystal growth rate for different concentrations or supersaturation levels. We have employed the following approach to estimate the total crystal surface area. This approach assumes that total surface area is evaluated as variable area as function of time and according to [6]:

$$A(t) = A_0 \cdot \left[\frac{W + (\Delta c_0 - \Delta c(t)) \cdot M}{W}\right]^{\frac{2}{3}} \tag{7}$$

where Δc_0 is the initial supersaturation, $\Delta c(t)$ is the time dependent level of supersaturation, W is the mass of seeds added and A_0 is the calculated initial seed surface area. The mass growth rate coefficient k_g is temperature dependent, and its temperature dependence conforms to the Arrhenius equation:

$$k_g = \tilde{A} \cdot \exp\left(-\frac{E_g}{RT}\right) \tag{8}$$

where $\tilde{A}$ is assumed to be constant, E_g is the activation energy and R is gas constant.
Experimental data at a minimum of two temperatures are necessary for the determination of $\tilde{A}$ and E_g parameters. Following the procedure described above and the isothermal desupersaturation data shown in Figure 7 the crystallization kinetics of RDX in cyclohexanone were calculated. The experimental conditions are described in Table 1. T* is temperature at which solution is saturated, c is initial concentration and S is the supersaturation at the moment of seeding.

Table 1. Experimental conditions.

Experiment 1	Experiment 2
T=55 °C	T= 40 °C
T*=62°C	T*= 51°C
S=15%	S=25%
c=0.161 grams/grams solv	c=0.128 grams/grams solv

The growth rate of a crystal is often described in terms of a linear growth rate G and has dimensions of length per unit time:

$$G = \frac{dL}{dt} \tag{9}$$

The linear growth rate G is directly related to the mass growth rate R_g : according to $G = \frac{\beta}{3\alpha \cdot \rho} \cdot R_g$,

where $\rho = 1.80 \frac{g}{cm^3}$ is density of RDX, α is the volume shape factor and β is the area shape factor.

It has been assumed in this study that seeds have a spherical shape and $\frac{\beta}{\alpha} = 6$ for spheres. The linear growth rate G was calculated from the isothermal desupersaturation curves and is shown as a function of supersaturation in Figure 8.

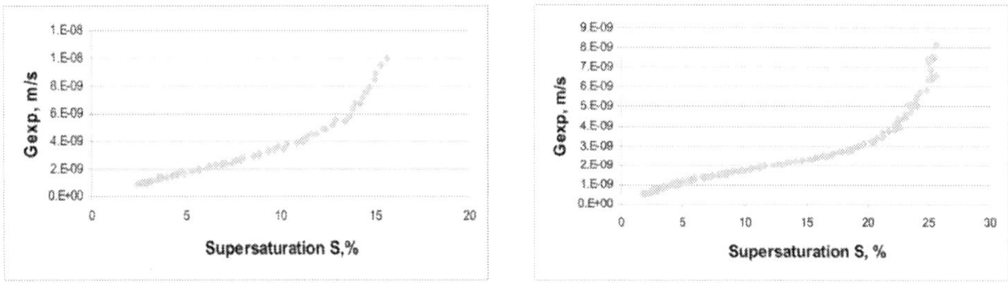

Figure 8. Calculated linear growth rate as a function of supersaturation based on the 55 °C (left) and 40 °C desupersaturation curves.

The parameters g and k_g were then calculated according to equation 6. The parameter values were obtained from the linear fits of experimental data shown in Figure 9.

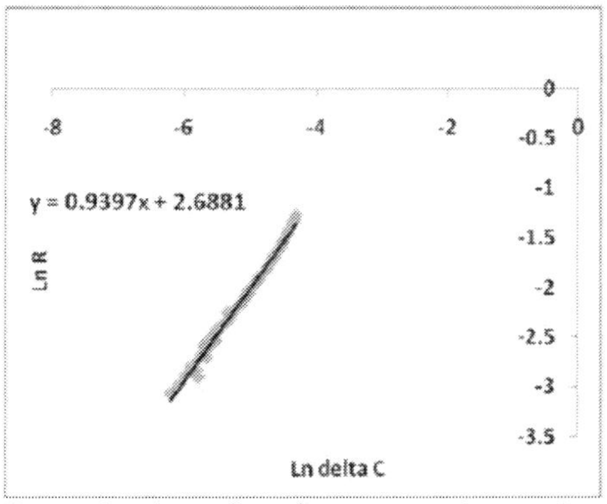

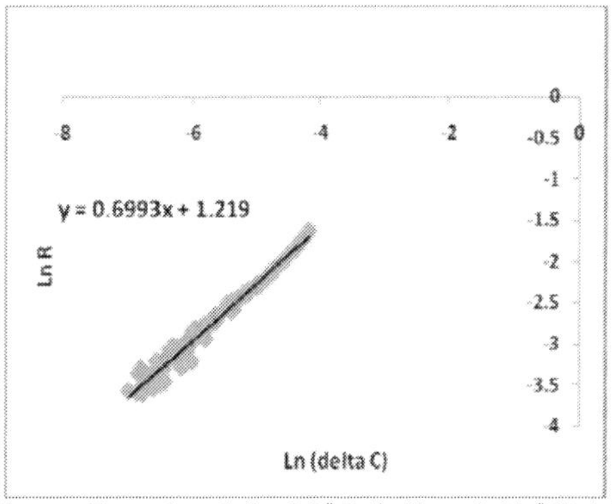

Figure 9. Linear fits of mass growth rate plotted according to Equation 6 for 55 °C (left) and 40 °C(right) desupersaturation data.

The calculated values are tabulated in Table 2.

Table 2 Calculated values of the growth rate coefficient k_g and growth order g determined graphically from the 40 °C and 55 °C desupersaturation data.

T=55 °C	**T=40 °C**
g =0.94	g =0.7
k_g=0.245 $\frac{grams}{m^2 \cdot s \cdot \left({grams}/{grams_{solv}} \right)^g}$	k_g=0.056 $\frac{grams}{m^2 \cdot s \cdot \left({grams}/{grams_{solv}} \right)^g}$

From the growth rate coefficient k_g obtained at two different temperatures (55 °C and 40 °C) and using the Arrhenius equation (8) we can estimate the activation energy according to:

$$E = \frac{R \cdot T_1 \cdot T_2}{T_2 - T_1} \cdot Ln \frac{k_2}{k1} \tag{10}$$

The calculated value of the activation energy of RDX crystallization in cyclohexanone is: $E = 91.4 \frac{kJ}{mole}$. This is in good agreement with results obtained by others for this system [4].

Using the obtained kinetic parameters, the growth rate coefficient k_g can be calculated for a wide range of temperatures. Once growth rate coefficient k_g has been determined for desired temperature, crystal growth kinetics can be estimated using empirical equation 5.

Conclusions

The goal of this work was to improve crystal quality of RDX produced by batch cooling crystallization. Several techniques were demonstrated to impact the product characteristics of RDX such as the crystal habit, crystal size, and size uniformity.

Techniques such as ultrasonic power and seeding were shown to be effective at controlling the supersaturation. Ultrasound was shown to induce nucleation at supersaturation levels below that of silent systems. Ultrasound power was demonstrated to be an effective control parameter which can influence the nucleation rate in a supersaturated solution and therefore control the crystal size of the product. It was also shown that with seeding spontaneous nucleation can be practically eliminated and consequently the product uniformity can be greatly improved.

In order to take full advantage of these techniques the crystal growth kinetics are required. In cooling crystallization, growth kinetics can be used in determining optimal cooling profiles such that secondary nucleation events are prevented. RDX crystal growth kinetic parameters in cyclohexanone were determined from experimentally obtained isothermal desupersaturation data incorporated into the diffusion layer model. Such kinetic parameters may be used in generating optimal cooling profiles.

References

[1] G. Fevotte, New perspectives for the on-line monitoring of pharmaceutical processes using in situ infrared spectroscopy, *International Journal of Pharmaceutics* (2002), vol. 241, pp. 263-278.

[2] A. Myerson, Handbook of Industrial crystallization (2nd edition),Butterworth Heinemann,2002, p. 56

[3] N. Amara, B. Ratsimba, A. Wilhelm, H.Delmas, Crystallization of potash alum: effect of power ultrasound, *Ultrasonics Sonochemistry* (2001), vol. 8,issue 3, pp.265-270

[4] J. ter Horst, Molecular Modeling and Crystallization, Univ. Press Science Publishers, 2000, p.41-57

[5] A.Jones, J.Mullin , Programmed cooling crystallization of potassium sulphate solutions, *Chemical Engineering Science* (1974), vol.29, pp.105-118

[6] J. Garside, L. Gibilaro, N. Tavare, Evaluation of crystal growth kinetics from a desupersaturation curve using initial derivatives, *Chemical Engineering Science* (1982), vol. 37, № 11, pp. 1625-1628

Population Balance and Online Analytics for Batch Crystallization of Adipic Acid

Dominik Fuchs, Department of Biochemical and Chemical Engineering, Technische Universtät Dortmund, Dortmund, Germany

Feelly Ruether, Department of Biochemical and Chemical Engineering, Technische Universtät Dortmund, Dortmund, Germany

Gabriele Sadowski, Department of Biochemical and Chemical Engineering, Technische Universtät Dortmund, Dortmund, Germany

Batch crystallization of organic components is a gentle and effective separation process. Due to consecutive treatment as filtration, formulation etc., control of retrieved particle size distribution is of great interest. Because online and in-situ analytics are indispensable to gain knowledge about crystal birth and growth behaviour, in the presented work concentration data as well as information about evolution of the solid phase are retrieved during batch experiments by means of the online analytical techniques ATR-FTIR-spectroscopy and FBRM particle measurement. Besides these batch techniques, independent growth measurements for crystallization of adipic acid were also performed in a single crystal growth cell. Growth rates for every crystal face were obtained by means of optical measurements. The obtained face-dependent growth parameters are taken into account by the presented population balance, allowing the modelling of concentration run during batch crystallization as well as size and shape evolution during batch experiments. Supersaturation during crystallization is calculated in terms of activities by use of the PC-SAFT equation of state. While calculated concentration runs are directly comparable to concentration data retrieve by ATR-FTIR-spectroscopy, a comparison of calculated and measured size distribution efforts further considerations. The FBRM-particle probe provides a statistical chordlength distribution. Calculated particle distributions were processed in an analogue statistical way, with knowledge of size and shape of the entire crystal population, to obtain a comparable statistical chordlength distribution. A comparison of measured (solid line) and calculated concentration (dashed line) run is shown in fig.1, analogue fig.2 shows a comparison of measured (solid line) and calculated (dashed line) chordlength distribution for batch crystallization of adipic acid from solution with 1-propanol.

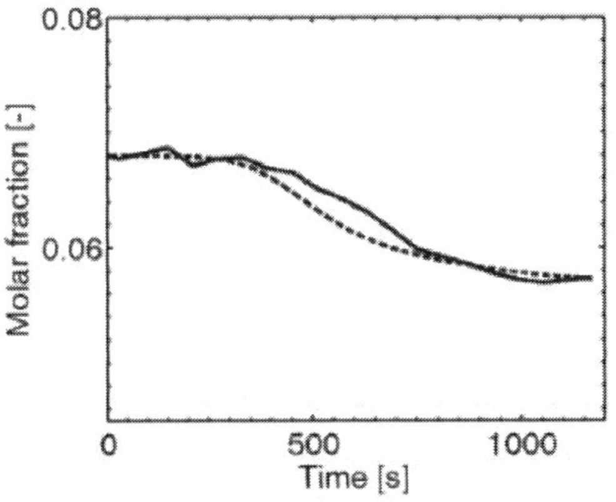

Fig. 1:Concentration run during batch experiment

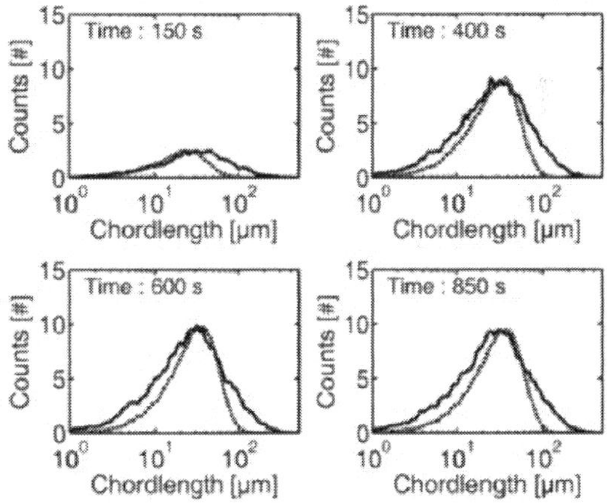

Fig. 2:Chordlength distribution during batch experiment

See more of this Session: Particle Formation and Crystallization Processes From Liquids, Slurries and Emulsions
See more of this Group/Topical: Particle Technology Forum

SYNTHESIS OF NON-RADIOACTIVE SLURRIES TO SIMULATE THE PROCESSING BEHAVIOR OF PARTICLES IN RADIOACTIVE WASTE SLURRIES – 626g

David C. Koopman, Savannah River National Laboratory, Aiken, SC
Daniel P. Lambert, Savannah River National Laboratory, Aiken, SC
Russell E. Eibling, Savannah River National Laboratory, Aiken, SC
J. David Newell, Savannah River National Laboratory, Aiken, SC
Michael E. Stone, Savannah River National Laboratory, Aiken, SC

ABSTRACT

Process development using non-radioactive analogs to high-level radioactive waste slurries is an established cost effective alternative to working with actual samples of the real waste. Current simulated waste slurries, however, do not capture all of the physical behavior of real waste. New methods of preparing simulants are under investigation along with mechanisms for altering certain properties of finished simulants. These methods have achieved several notable successes recently in the areas of rheology and foaminess. Particle size is also being manipulated more effectively than in the past, though not independently of the rheological properties. The interaction between rheology and foaminess has exhibited counter-intuitive behavior with more viscous slurries being less foamy even though drainage of liquid from the foam lamellae should be inhibited by higher viscosities.

INTRODUCTION

Legacy radioactive high-level waste (HLW) generated at the Savannah River Site (SRS) during Cold War production of enriched uranium and plutonium is currently being processed into a stable borosilicate glass waste form for long term storage. A majority of the legacy waste is stored as a mixture of insoluble hydroxides and hydrous oxides in large cylindrical storage tanks at SRS. Over 90% of the waste solids are non-radioactive chemical byproducts derived from fuel rod targets and purification chemistry. Uranium and plutonium are two of the major radionuclides by mass in the balance of insoluble material. The 3.5-4.9 thousand cubic meter (900,000-1,300,000 gallon) carbon steel waste storage tanks also contain 5-7M sodium solutions rich in hydroxide, nitrate, and nitrite anions that are contaminated with soluble radioactive isotopes of cesium and strontium.

The majority of the laboratory and pilot scale process development work being performed at the Savannah River National Laboratory (SRNL) in support of nuclear waste processing is done with non-radioactive simulated waste slurries, or "simulants". These simulants match the bulk chemical properties of the non-radioactive components in the insoluble and aqueous phases of the actual radioactive wastes. Following simulant testing, limited proof of concept testing is done in special facilities using radioactive samples from the actual waste tanks. Cost is one of the major factors driving this approach. There is roughly a ten-fold cost advantage to performing the preliminary testing with the non-radioactive analogs to the real waste. The success of this strategy is dictated in part by how well the processing properties of the non-radioactive simulant slurries match those of the actual waste slurries.

Actual waste slurries were produced at SRS from about 1955-1990. Many have sat relatively undisturbed in the storage tanks since they were originally precipitated by adding caustic to the acidic waste streams. The freshly precipitated slurry particles are not in thermodynamic equilibrium. Atomic scale processes, such as Ostwald ripening, act to change the chemical and physical properties of the particles over time. For example, aluminum precipitates primarily as Gibbsite, $Al(OH)_3$, but a significant fraction converts to Boehmite, AlOOH, over time. Similar changes are likely occurring with the amorphous hydrous ferric oxide precipitate, $Fe_2O_3 \cdot xH2O$. The value of x appears to decline with increasing storage time. The percent crystallinity of the precipitated solids is low when examined by X-ray diffraction, and this is particularly true of the ferric oxide. The waste solids are very difficult to quantify with respect to speciation, crystal forms, etc. due to the presence of over fifty elements. The insoluble solids are believed to include double salts as well as simpler compounds involving single cations. The waste is, however, relatively free of any organic species (there is a small amount of oxalate).

TECHNICAL

Several different approaches have been taken historically toward preparing simulated waste slurries. All of the approaches used in the past dozen years involve some precipitation of the species using similar chemistry to that which formed the radioactive waste solids in the SRS tank farm. All of the approaches add certain chemical species as commercially available insoluble solid compounds. The number of species introduced in this manner, however, has varied widely. All of the simulant preparation approaches make the simulated aqueous phase by adding the appropriate ratios of various sodium salts. The simulant preparation sequence generally starts with an acidic pH and ends up with a caustic pH (typically in the 10-12 range).

Presently, all recipes first precipitate MnO_2 from $Mn(NO_3)_2$ and $KMnO_4$ by combining two aqueous solutions. Reagent grade MnO_2 did not behave prototypically during chemical pre-treatment steps. Following MnO_2 precipitation, hydrous ferric oxide and nickel hydroxide are simultaneously precipitated in the presence of the MnO_2 solids from a solution of their nitrates. Nitrates of many other elements may or may not be present and precipitated at this point including Al, Ba, Ca, Cu, Cr, Gd, La, Mg, Pb, Sn, Zn, and Zr. Alternatively, these species may be added as mixtures of oxides, hydroxides, sulfates, oxalates, carbonates, etc. as indicated by thermodynamic modeling and the bulk chemical distribution of the elements.

Following precipitation of the nitrate salts using sodium hydroxide and sodium carbonate, excess nitrate is removed by repeated additions of inhibited water (0.001 M NaOH, 0.001 M $NaNO_2$) and allowing the insoluble solids to settle, then decanting the aqueous supernate. Typically 3-5 wash cycles are performed. Certain minor species, such as silica and TiO_2, are added as finely-milled insoluble solids based on data indicating that the proper form or speciation would not be obtained from a nitrate precipitation. Finally, soluble sodium salts are added to adjust the composition of the aqueous phase to match that of the HLW slurry.

Significant differences in physical properties of the simulated wastes are observed depending on the preparation route selected. Similar differences are observed with respect to the actual HLW waste slurry being modeled. The rheological properties of the slurries tend to be more viscous when more species are precipitated together. Batch precipitations tend to produce more viscous slurries than continuous precipitations. Batch precipitations occur over a wide range of pH from less than three to about ten. Continuous precipitations occur at the target pH of the pre-washed final slurry (typically about ten). Continuous precipitations tend to produce a narrower range of particle sizes than batch precipitations, Figure 1:

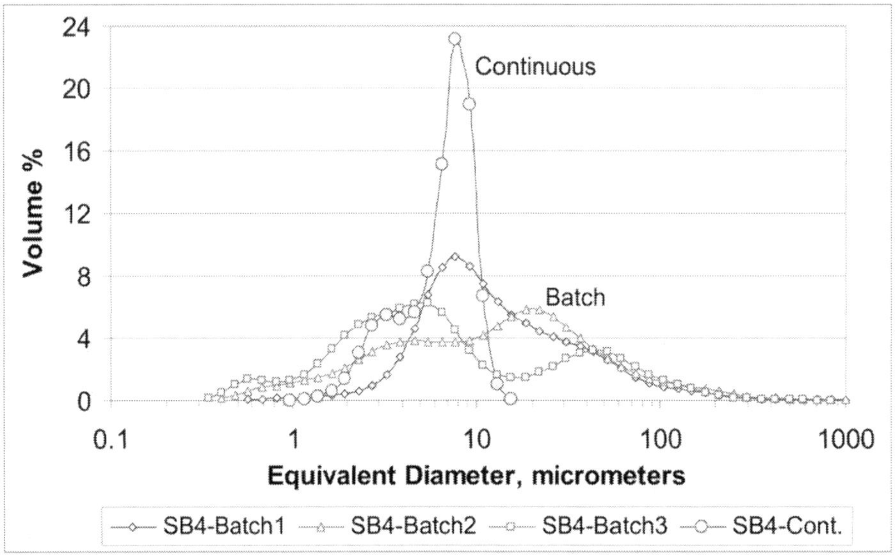

Figure 1. Impact of continuous precipitation geometry on the particle size distribution of the final simulated slurry solids compared to batch precipitation.

Note that while the sodium hydroxide driven precipitations in Figure 1 were either batch or continuous, the subsequent carbonate strike, washing, and trimming steps were performed using equivalent batch processing methods.

The particle size distribution of the MnO_2 in the first precipitation step depends significantly on the presence or absence of the nitrates of the cationic species being formed in the second-step precipitation, Figure 2:

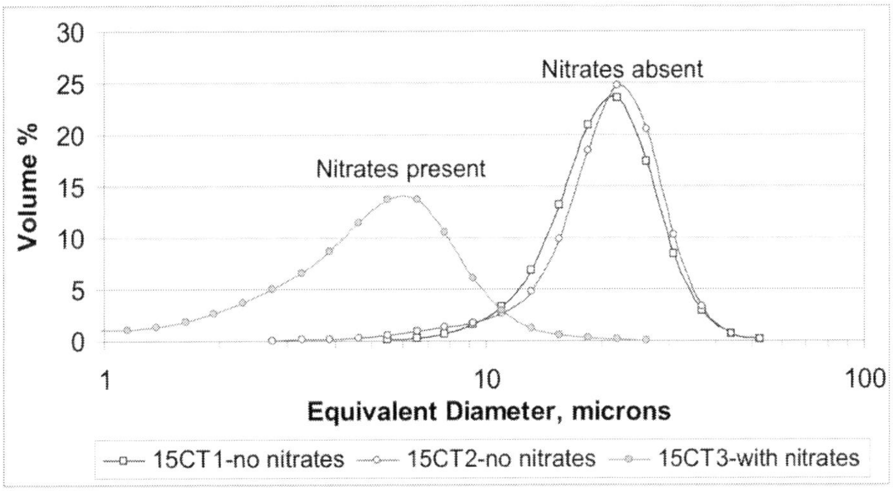

Figure 2. Impact of the presence of additional metal nitrates on the particle size of batch precipitated MnO_2.

Precipitation of iron, aluminum, and other waste hydroxides from the nitrate solution using NaOH proceeded relatively unaffected by the particle size distribution of the MnO_2 particles formed in the first stage precipitation. The conclusion was that the MnO_2 particles were not acting as generic seed crystals for the other precipitated solids. This conclusion was qualitatively supported by visual observations of centrifuged simulant solids which showed a distinct region of darker, denser solids enriched in MnO_2 that segregated from the majority of the other solids near the bottom of the centrifuge tubes.

The sequence of steps involved in producing the final simulant (two precipitations, a carbonate strike, batch washing, and final trimming) typically lasted about three weeks and exhibited a gradually evolving particle size distribution, Figure 3:

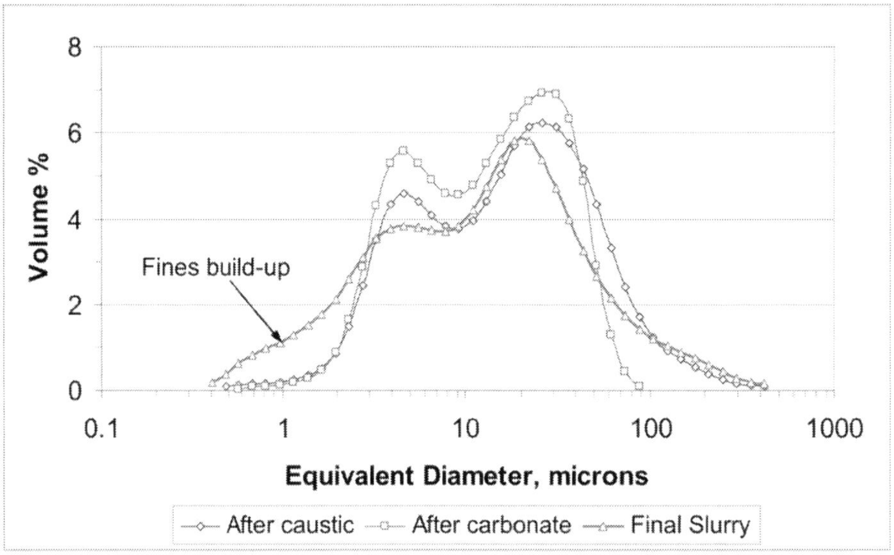

Figure 3. Evolution of particle size distribution during post-precipitation processing.

One trend of particular concern was the gradual accumulation of fines as the number of processing steps increased. There was statistical correlation between increasing fines content (fraction of particles less than 1-2 microns in diameter) and undesirable increases in the apparent viscosity of the final simulants.

Equipment scale was of particular importance in controlling the rheological properties of batch-produced simulants. Although every effort was made to ensure comparably high degrees of mixing intensity at every scale, the final simulant slurry products became progressively more viscous with increasing vessel (batch) size. The Bingham plastic model was fit to flow curve data obtained using a rotational Haake rheometer operating in the Searle mode (concentric cylinder geometry with rotating inner bob with fixed outer cup). Variations in the plastic viscosity (consistency) parameter are relatively minor compared to the differences in the yield stress. Yield stress values obtained from this model are given in Figure 4 covering then range 2-120 dm^3 (L) of precipitation batch size.

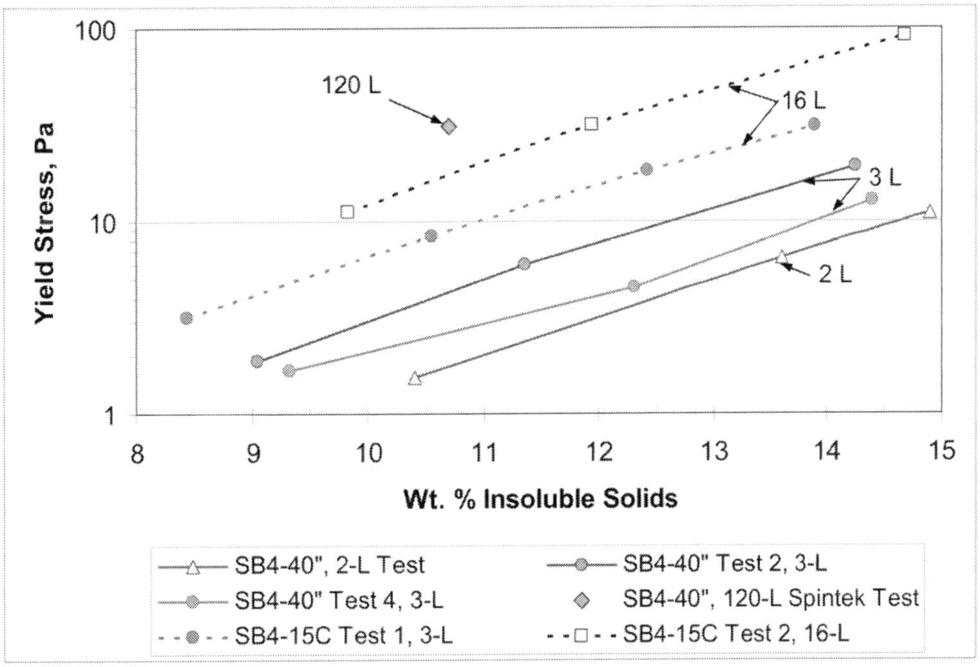

Figure 4. Impact of scale on the Bingham plastic yield stress of batch-produced simulants of high-level waste slurries.

Typically, starting yield stress values in the subsequent pilot plant equipment are desired to be in the 1-10 Pa range. The slurry products prepared at 16 L and 120 L scale were sufficiently difficult to process in the lab-scale pilot plant equipment at the desired wt.% insoluble solids that they had to be blended off with lower viscosity simulants prior to use or put directly into one of the laboratory residue streams.

Switching to a continuously fed stirred precipitator was comparable to about an order of magnitude reduction in vessel size in the batch precipitations. An example of the impact of this change on rheology is shown in Figure 5. Sixteen liters of simulant were prepared by the two different methods with order of magnitude differences in the Bingham plastic yield stress.

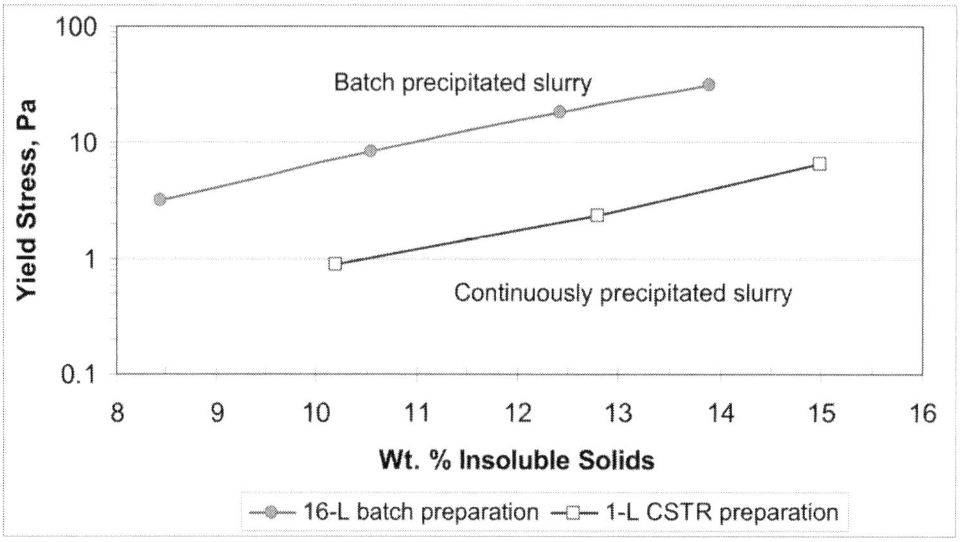

Figure 5. Impact of continuous versus batch precipitation of 16 liters of simulated HLW slurry on the dominant rheological property, Bingham plastic yield stress.

In some simulant preparations, a larger fraction of the insoluble solids were added as reagent chemicals instead of as nitrate salts, for example 50% or more of the insoluble solids. In these simulants, the particle size distributions of the commercial insoluble solids had a significant impact on the rheological properties of the simulants. These simulants were often insufficiently viscous to perform studies on the ability of the slurry to suspend glass frit particles which are added prior to vitrification. Thermal treatments of the freshly precipitated slurries to simulate aging in the waste tanks have only produced minor changes in the physical properties.

High shear mixers have been used to increase the apparent viscosity of simulants that are within a factor of two to three of the target rheological behavior. The high shear mixers seem to increase the Bingham plastic yield stress with increasing time for a while (30-60 minutes) after which the rate of change in yield stress with respect to time becomes nearly zero. It has now been found that ultrasonication can produce nearly a ten-fold increase in the Bingham plastic yield stress in just a few minutes, and that the high shear mixers are not able to achieve the same increases.

The radioactive waste simulants being produced today mimic the chemical processing behavior of real waste essentially as well as can be quantified and with appropriate allowances for the absence of the radioactive species. Differences still remain in physical behavior, however. Currently, a program to improve physical properties of simulants is underway. One goal is to investigate how to produce particles that are foamier under boiling conditions. Real HLW slurries tend to foam when boiled under caustic conditions, while simulants do not (both seem to foam comparably following acidification during chemical pretreatment, but prior to that the two are different). Research at the Illinois Institute of Technology (Bindal et al.) has established that biphyllic particles are stabilizing slurry foams. Simulants form such particles during the acidification phase of waste processing (acidified simulants are foamy during gas generation and boiling), but apparently do not have enough biphyllic particles as made. Studies on foaminess have been conducted that show that it is possible to alter this characteristic to some extent. Additional studies using the continuous precipitator operated over a designed partial factorial matrix of conditions are underway to determine whether or not certain properties can be adjusted by varying the operational mode of the equipment.

REFERENCES

Koopman, D., D. Lambert, and M. Stone, "Successes and Emerging Issues in Simulating the Processing Behavior of Nuclear Waste Blends", *Proceedings of the 2009 AIChE Annual Meeting*, New York, NY: AIChE. 2009.

Bindal, S., A. Nikolov, D. Wasan, D. Lambert, and D. Koopman, "Foaming in Simulated Radioactive Waste", *Environmental Science and Technology*, **35**, 3941-3947, 2001.

Green Synthesis Improvements of 3,3' Diaminoazoxy Furazan (DAAF)

Kimberly Hanson, Ordnance Department, Indian Head Division Naval Surface Warfare Center, Indian Head, MD

Shilpa P. Lad, Ordnance Department, Indian Head Division Naval Surface Warfare Center, Indian Head, MD

Elizabeth Francois, DE-1 Engineering, Los Alamos National Laboraties, Los Alamos, NM

David Chavez, DE-1 Engineering, Los Alamos National Laboraties, Los Alamos, NM

High explosive syntheses are often not able to benefit from a greener synthesis due to constraints on yield and impurity profiles. However, Los Alamos National Laboratories (LANL), was able to replace the original synthesis of 3,3'-Diaminoazoxyfurazan (DAAF) with an environmentally friendly synthesis while improving the processing time and maintaining purity profiles and yield without producing a hazardous waste stream. The original route historically required the oxidation of diaminofurazan (DAF) with hydrogen peroxide and concentrated sulfuric acid. The resulting product, DAAF, was impure, containing energetic intermediates which adversely affect the safety margins of this explosive. Therefore, recrystallization was necessary. The aforementioned route utilized to manufacture 5 kilograms (kg) of DAAF generated 40 gallons of hazardous acidic waste (pH=0) containing hydrogen peroxide, sulfuric acid and high explosive (<1%). The recrystallization step produces another 40 gallons of hazardous waste containing DMSO, water and high explosive (<1%). Additionally, the hazardous waste proved difficult to store as solar heat caused the remaining peroxide to disproportionate and create gas in the sealed drum, thus creating a pressurization hazard. Replacing the hazardous reagents with more benign, environmentally friendly reagents proved beneficial to the environment, shows an increase in yield, and improves the purity profile of the resulting DAAF by eliminating the energetic intermediates.

See more of this Session: Environment & Lifecycle
See more of this Group/Topical: Particle Technology Forum

Degradation and Transformation Studies of 2,4 -Dinitroanisole (DNAN) On Nanoparticles of Magnesium and Zinc Oxide Nanomaterials

Veera Boddu, Environmental Processes Branch, US Army Engineer Research and Development Center, Champaign, IL

Anthony J. Bednar, Environmental Laboraory, US Army Engineer Research and Development Center, Vicksburg, MS

Rober A. Kirgan, Environmental Laboratory, US Army Engineer Research and Development Center, Champaign, IL

Dabir S. Viswanath, Nuclear Science and Engineering Institute, University of Missouri, Columbia, MO

Mo Qasim, US Army ERDC, Vicksburg, MS

Understanding the degradation and transformation mechanisms and pathways of the new 'Insensitive Munitions' (IM) will foster development of efficient processes for mitigation of Army waste streams prior to introduction of these compounds into service. Residuals from the IM formulations may soon appear in waste streams from industrial facilities and in environmental matrices. It is timely and appropriate to explore novel nanomaterials for their surface functionality and reactivity for degradation of IM compounds and thereby facilitating development of innovative treatment technologies. This presentation focuses on results of studies on synthesis of nanomaterials of magnesium and zinc oxides and degradation studies of one of the insensitive munitions compound 2,4-Dinitroanisole (DNAN). Preliminary results indicate that the primary ring cleavage occurs through hydroxylation of DNAN. Denitration and amination of the DNAN are also observed.

See more of this Session: Environment & Lifecycle
See more of this Group/Topical: Particle Technology Forum

Abstract #167253

Analysis of Di- and Tri- Nitroaromatic Analogues of TNT: Rates and Reactivities of Transformation Pathways

Mohammad (Mo) Qasim[1], US Army ERDC, Vicksburg, MS

Leonid Gorb3; Minori Uchimiya 1;Veera Boddu2; Sandra Brasfield1; Jerzy Leszczynski1,4; Matthew Middleton1,5; Brett Moore6; Lyssa Taylor6 1USAE ERDC, 3909 Halls Ferry Road, Vicksburg, MS 39183; 2ERDC-CERL, Champaign, IL; 3SpecPro, Huntsville, AL

This research compared 2,4,6-trinitrotoluene (TNT) to 1,3,5-trinitrobenzene (TNB); 2,4,6-trinitrophenol (picric acid or TNP); 1,3-dinitrobenzene (DNB); 2,4- and 2,6 dinitrotoluene (DNT) and 2,4-dinitroanisole (DNAN). Theoretical calculation plus selective spectroscopic verification provided a valid approach to the premise that *molecular structure determines transformation pathways* that can be theoretically predicted and spectroscopically verified. This combination of computational chemistry methods and spectroscopic tools, utilized to evaluate rates and mechanisms of these aromatic compounds and their transformation intermediates as relating to structures and possible environmental impacts, demonstrated this combination's useful and proactive role for studying competing pathways, including those of compounds with toxic potential.

TNT and the selected derivatives were reacted via: i) alkaline hydrolysis with sodium hydroxide (NaOH); ii) free radical initiators through photolysis, using monochromatic irradiation at respective absorption maxima, and iii) Fenton reactions in complex competing transformation mechanisms.

UV/Vis was used to measure concentrations and follow courses of chemical reactions through changes in shape and wavelength. Concentration measurements of parent, intermediate and final compounds were observed through changes in height of their spectra where maximum absorption occurs. Changes in spectral shape (appearance/disappearance of shoulders; site of wavelengths of maximum absorption) indicated the appearance and disappearance of intermediate and final transformations. A Stopped Flow (SF) Spectrophotometer was used to measure rate constants and to follow rates of fast-forming intermediates, indicating steps of the reactions through changes in slope. A monochromatic irradiation system designed for performing photochemical reactions was constructed and used to carry out photochemically induced free radical reactions at wavelengths of maximum absorption

UV/Vis spectra were taken and major peaks of parent compounds identified prior to addition of NaOH. Upon addition of varied concentrations of NaOH, both transformation of parent compound peaks and formation of new transition product peaks were observed. These peaks were further evaluated via SF spectroscopy to obtain reaction kinetics. UV/Vis data were supported by corresponding results with SF spectra. UV/Vis and SF data were then compared to the CC data for all possible transition states, intermediates, and final products to discern which compounds comprise chemically and energetically feasible transformation products.

Only alkaline hydrolysis produced consistent and effective transformation intermediate and final products in this research. However, irradiation of the product generated by alkaline hydrolysis at 450 nm (wavelength of maximum absorption) caused complete disappearance of the spectra.

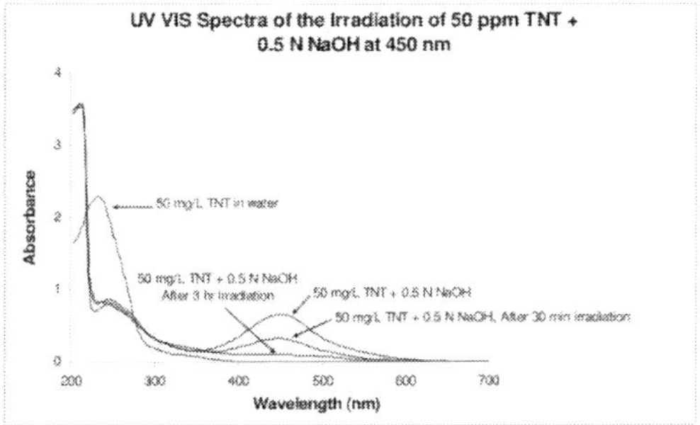

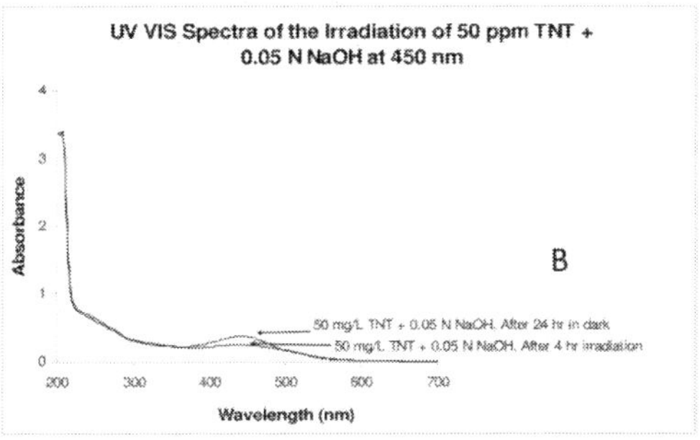

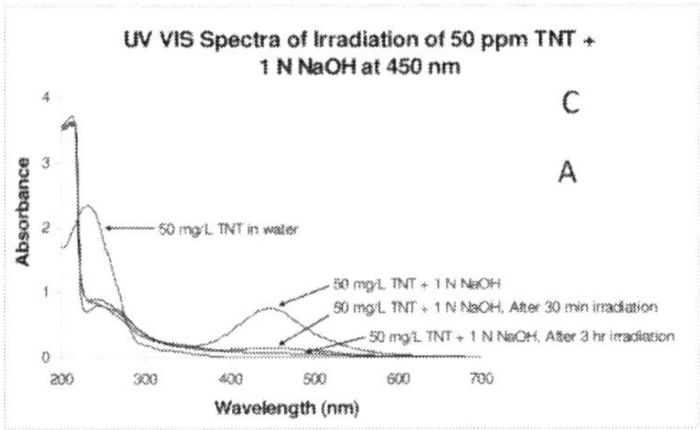

UV Vis Spectra of TNT after Treatments with Sodium Hydroxide and Irradiation at 450 nm. *UV Vis spectra of 50 ppm TNT treated with (A) 0.05N NaOH; (B) 0.5N NaOH; (C) 1N NaOH followed by irradiation at 450 nm were collected at intervals over a 24 hour time period. Irradiation speeded the transformation at 450 nm where product via alkaline hydrolysis appeared and disappeared.*

Neither Fenton Reagent nor photolysis was effective in transforming TNT at its wavelength of maximum absorption (232 nm).

Quantum and classical mechanical computational (MOPAC) and DFT methods, performed at the B3LYP (Beck 3 Parameter, Lee, Yang, Parr), were used to characterize selected properties: heat of formation; HOMO and LUMO energies; charge density of the g system; and dipole moment prior to our experimental approach, in which two experimental transformation methods compared the effectiveness of alkaline hydrolysis with photolysis.

Comparisons of the following groups were based on molecular structure and number/position of substituents: i) TNT and TNB; ii) DNAN, DNB, DNT and DNP; iii) TNT and DNT; iv) DNB and TNB. Significant trends were revealed among these groups corresponding to substituent nature, position, and number of amino and hydroxyl groups . HOMO/ LUMO energies reveal aspects of molecular structure and can thereby indicate aspects of potential toxicity and recalcitrance. Our CC and experimental data revealed that type, number and position of substituents determine reaction rates and reactivities and, therefore, potential environmental impacts.

Although additional compounds are shown, the table below shows position and number of substituents in amine and hydroxyl groups, representing trends in TNT, TNB, TNP, DNB and both 2,4- and 2-6 DNT.

Energies and Other Properties of TNT and Other Compounds

Energies and Some Properties of TNT and Related Compounds								
Compound		MM2 Total Steric Energy (kcal/mole)	Orbital Level	B3LYP HOMO / LUMO Energies (eV)	B3LYP Heat of Formation (kcal/mole)	AM1 Minimal Energy HOMO / LUMO Energies (eV)	AM1 Minimal Energy Heat of Formation (kcal/mole)	AM1 Minimal Energy Dipole Charge (debye)
2,4,6-Trinitrotoluene	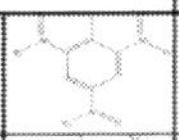	6.4610	N=43 N=42	LUMO -0.13410 HOMO -0.31784	17.112	LUMO -2.43 HOMO -11.72	42.259	0.941
2-Amino-4,6-dinitrotoluene	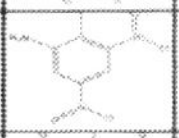	-0.8787	N=38 N=37	LUMO 0.10816 HOMO -0.25269	11.106	LUMO 1.63 HOMO -9.53	28.579	7.191
4-Amino-2,6-dinitrotoluene	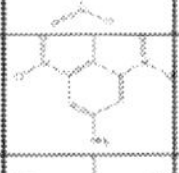	-0.2180	N=38 N=37	LUMO -0.10493 HOMO -0.24772	14.878	LUMO -1.65 HOMO -9.45	31.067	5.459
2,6-Diamino-4-nitrotoluene	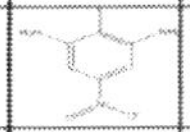	-4.9658	N=33 N=32	LUMO -0.08143 HOMO -0.21899	10.370	LUMO -0.66 HOMO -8.58	18.968	7.116
2,4-Diamino-6-nitrotoluene	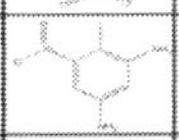	-4.8493	N=33 N=32	LUMO -0.07746 HOMO -0.21289	13.676	LUMO -0.62 HOMO -8.41	20.823	6.766
2,4,6-Triaminotoluene	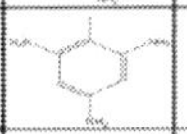	-9.0304	N=28 N=27	LUMO 0.02846 HOMO -0.18403	17.562	LUMO 1.31 HOMO -7.66	15.490	0.094
2,4,6-Trinitrophenol	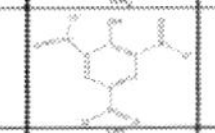	3.9152	N=43 N=42	LUMO -0.14893 HOMO -0.30900	-23.797	LUMO -2.53 HOMO -11.41	0.939	1.53
2-Amino-4,6-dinitrophenol	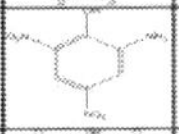	2.5323	N=38 N=37	LUMO -0.12048 HOMO -0.24461	-33.552	LUMO -1.73 HOMO -9.38	-12.485	6.017
4-Amino-2,6-dinitrophenol	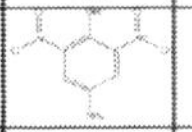	-5.6038	N=38 N=37	LUMO -0.12335 HOMO -0.24039	-23.599	LUMO -1.87 HOMO -9.31	-8.378	6.142
2-Hydroxy-4,6-dinitrotoluene	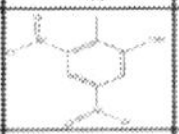	-3.1551	N=38 N=37	LUMO -0.11512 HOMO -0.27606	-31.936	LUMO -1.95 HOMO -10.47	-13.646	5.397
4-Hydroxy-2,6-dinitrotoluene	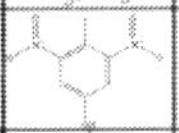	-8.1751	N=38 N=37	LUMO -0.11212 HOMO -0.27246	-27.052	LUMO -1.86 HOMO -10.37	-11.786	3.539
2,6-Dihydroxy-6-nitrotoluene	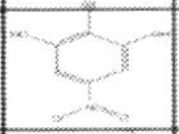	-4.9939	N=33 N=32	LUMO -0.09423 HOMO -0.24687	-78.082	LUMO -1.20 HOMO -9.69	-66.280	7.223
2,4-Dihydroxy-6-nitrotoluene	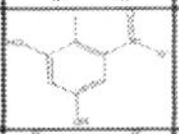	-2.1985	N=33 N=32	LUMO -0.08960 HOMO -0.24033	-73.676	LUMO -1.19 HOMO -9.56	-65.168	4.836
1,3,5-Trinitrobenzene	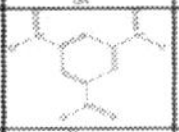	3.2040	N=40 N=39	LUMO -0.14149 HOMO -0.33665	19.386	LUMO -2.68 HOMO -12.75	44.662	0.002
5-Amino-1,3-dinitrobenzene	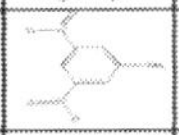	-6.3134	N=35 N=34	LUMO -0.11424 HOMO -0.25560	15.581	LUMO -1.71 HOMO -9.67	32.294	6.748
5-Amino-1-hydroxy-3-nitrobenzene	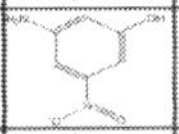	-5.7349	N=30 N=29	LUMO -0.09051 HOMO -0.23030	-26.639	LUMO -0.93 HOMO -9.04	-18.748	7.26
2,4-Dinitrotoluene	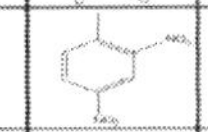	-2.2109	N=35 N=34	LUMO -0.11546 HOMO -0.30488	10.118	LUMO -1.85 HOMO -11.03	28.291	5.655
2,6-Dinitrotoluene	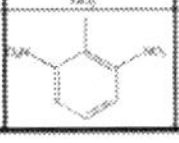	2.3992	N=35 N=34	LUMO -0.11119 HOMO -0.29770	15.027	LUMO -1.82 HOMO -10.95	31.201	3.508

Data in the Table suggest that TNB has a slightly higher heat of formation and lower steric energy (MM2) than TNT due to inductive, steric, and electrostatic effects of the methyl group of TNT. The same trend for heat of formation was obtained through DFT calculations where replacing the methyl group with an

OH^- considerably lowered the heat of formation. For example, addition of OH^- to form picric acid (TNP) drives the heat of formation drastically lower than in either TNT or TNB. Compared to TNT and TNP, TNB has the highest heat of formation but the lowest steric, as well as the lowest HOMO/LUMO energies, as expected. The heat of formation decreased with higher concentrations of OH^-. Also, heat of formation is consistent with HOMO/LUMO ratios [1].

In our research, AM1 and DFT B3LYP consistently showed that substituting an amine for the TNT nitro group results in the following: i) HOMO energy values are always lower when the amine is substituted on the ortho C_2 rather than on the para C_4 position with respect to the methyl group, because amines add electrons to the aromatic ring, thereby activating them. This lowering of energy to the methyl group ortho position is due to both inductive and steric effects of the methyl group. (The resonance effect is the same for both ortho and para positions.) ii) For both mono- and di- substituted amino and hydroxyl groups, HOMO and formation energies are always lowest when the substitution is on the ortho rather than the para position. For example, 2-amino is lower than 4-amino and 2-hydroxy is always lower than 4-hydroxy. iii) The same trend pertains to diamino and dihydroxy substitutions on TNT. That is, 2,6-diamino and 2,6-dihydroxy are always lower than 2,4-diamino and 2,4-dihydroxy substitutions. iv) The opposite effect is observed for the nitro groups due to their electron-withdrawing nature. For example, 2,6-dinitrotoluene has higher steric energy, heat of formation, and HOMO energy than 2,4-dinitrotoluene [1].

HOMO/LUMO energies increased while the heat of formation decreased with subsequent nitro group replacement by amino or hydroxyl groups. The greatest lowering of the heat of formation occurred with greatest hydroxyl group substitution. HOMO energy is lower when amines or hydroxyl groups are substituted at the ortho (2,6-) rather than the para (4-) position. This is true for both mono- and di-substitutions. For example, 2-amino and 2-hydroxyl positions are always of lower HOMO energy than 4-amino-dinitrotoluene and 4-hydroxyl-dinitrotoluene. Also 2,6-diamino-nitrotoluene and 2,6-hydroxyl-nitrotoluene are always of lower HOMO/LUMO energies than 2,4-dinitro-nitrotoluene and 2,4-dihyroxyl-nitrotoluene. On the other hand, due to the electron-withdrawing nitro groups, the trend is reversed in the dinitrotoluenes with the 2,4-dinitrotoluene always having lower HOMO/LUMO and formation energies than 2,6-dinitrotoluene [1].

In summary, effective TNT transformation intermediates and final products were produced only by alkaline hydrolysis. However, after alkaline hydrolysis was carried out, monochromatic irradiation at 450 nm enhanced TNT transformation. Alkaline hydrolysis of TNT, TNB and TNP form in the same absorption visible region (430 to 460 nm) suggesting a common intermediate with additional hydroxylation substituting for nitro groups and consequent loss of aromaticity—thus resulting in degradation of the parent compound [1]. Theoretical prediction and analysis coupled with UV Vis and SF spectroscopic corroboration provided a valid approach, illustrating that modes of transformation are dependent upon structural characteristics. Thus, understanding structural variations and changes reduces time and cost of predicting, identifying and analyzing most likely competing transformation pathways in the environment.

(Information on DNAN transformations as well as more detailed information on selected TNT analogues will be given in the presentation. Stopped Flow comparative data and trends will also be presented although they are not included in this abstract.)

Reference:

Qasim, Mohammad, Moore, B., Taylor, L. Honea, P. Gorb, L. Leszczynski, J., "Structural Characteristics and Reactivity Relationships of Nitroaromatic and Nitramine Explosives – A Review of Our Computational Chemistry and Spectroscopic Research," *International Journal of Molecular Sciences*, ISSN 1422-0067, 2007, 8, 1234-1264.

Abstract ID#: 167253

Password: 740274

Title: Analysis of Di- and Tri- Nitroaromatic Analogues of TNT: Rates and Reactivities of Transformation Pathways

Session Selection: 03E03 Thermophysical Properties

Preferred Presentation Format: Oral

Lead Presenter's Email: mohammad.m.qasim@usace.army.mil

Comments to Organizers: Mohammad (Mo) Qasim (Speaker)1; Sandra Brasfield1; Veera Boddu2; Leonid Gorb3; Jerzy Leszczynski1,4; Matthew Middleton1,5; Brett Moore6; Lyssa Taylor6 1USAE ERDC, 3909 Halls Ferry Road, Vicksburg, MS 39183; 2ERDC-CERL, Champaign, IL; 3SpecPro, Hunts

Related to existing technology or advanced technology associated with the overall use and supply of energy or global climate change research? Yes

First author

Presenting Author

Mo Qasim

Principal Investigator

US Army ERDC

Vicksburg, MS 39180

Phone Number: 601-634-3422

Email: Mohammad.M.Qasim@usace.army.mil

Safe Optimization and Scale-up of Novel Energetic Materials

Shilpa P. Lad, Chemical Development Branch, Indian Head Division Naval Surface Warfare Center, Indian Head, MD

Jerry S. Salan, Chemical Research and Development, Pfizer, Inc., Groton, CT

New energetic ingredients are consistently arising in the energetics community, some with potential to make a positive impact in energetic formulations and their respective systems. As formulation requirements for testing and evaluation increase, so do the required quantities of the novel energetic ingredients being investigated. Process development is inherent with the scale-up of energetic materials. During scale-up processing parameters are evaluated for optimization. The challenge with scale-up of energetic materials is providing a process that is both safe and economical. Comprehensive knowledge of the energetic process is necessary for effective process development. The use of automated lab reactors , reaction calorimetry, and in situ analytical techniques accelerates the chemical engineer's process understanding, thus enabling safe optimization and scale-up of novel energetic ingredients.

See more of this Session: Environment & Lifecycle
See more of this Group/Topical: Particle Technology Forum

Crystallization Processes for Energetic Materials

Andrew G. Pearsall, Chemical Development, Naval Surface Warfare Center, Indian Head, MD

Tod S. Ricks, Chemical Development, Naval Surface Warfare Center, Indian Head, MD

Shilpa P. Lad, Chemical Development, Naval Surface Warfare Center, Indian Head, MD

Control over crystallization operations of energetic materials plays a key role in obtaining ingredients with desirable crystal characteristics for formulations. At Indian Head Division, Naval Surface Warfare Center, with the aid of sophisticated laboratory equipment, greater control and understanding of crystallization processes is being obtained. This equipment is being used across several projects aimed at quickly and safely fielding new energetic materials or modified forms of more mature explosives. The different technologies allow for in-situ particle size analysis of crystals in solution, real time images of these crystals, and infrared reflectance spectroscopy and subsequent concentration measurements of materials in solution.

See more of this Session: Environment & Lifecycle
See more of this Group/Topical: Particle Technology Forum

Experimental Analysis of Fluidization of Nano Particles

Talin Avanesian, Chemical and Biological Engineering, Illinois Institute of Technology, Chicago, IL

An experimental study was conducted to investigate the structure and behavior of nano particle agglomerates in the fluidized bed systems. A series of fluidization experiments were performed in a glass column using Aerosil® R974 and Aerosil® 200 nano particles. It was observed that silica nano particles fluidize smoothly through formation of agglomerates. No bubbles or dead regions were observed during fluidization. The minimum fluidization velocity was found to be around 0.27 cm/s for Aerosil® R974 nano particles. Very high bed expansion (up to 8 times the initial bed height) was observed in the fluidization of Aerosil® R974 hydrophobic nano particles. During the fluidization experiment, samples of nano particle agglomerates were taken from the freeboard of the fluidized bed by an adhesive tape. The agglomerate sizes were measured using an Olympus SZx12 optical microscope. The measured agglomerates sizes were in the range of 1-450 ìm. It was found that the agglomerate size depends on the gas flow rate, fluidization time and height of sampling. The agglomerate sizes decreased by increasing gas flow rate. A decrease in agglomerate size was observed with fluidization time until an equilibrium agglomerate size was achieved. The samples were taken from the higher regions of the bed freeboard had smaller agglomerate size.

The terminal velocities of the agglomerates were calculated for the measured agglomerate sizes with the assumption of a sphericity of 0.7 for the agglomerates. The results showed that 92% of agglomerates have terminal velocities below the superficial gas velocity and can be elutriated from the fluidized bed, indicating that the measured agglomerate sizes are reasonable estimates of the nano particle agglomerates in the fluidized bed.

See more of this Session: Fluidization and Handling of Submicron and Nano Particles
See more of this Group/Topical: Particle Technology Forum

Enhanced Nanofluidization by Alternating Electric Fields

Daniel Lepek, Department of Chemical Engineering, The Cooper Union, New York, NY

Jose Manuel Valverde, University of Seville, Seville, Spain

Robert Pfeffer, Chemical Engineering, Arizona State University, Tempe, AZ

Rajesh Dave, Otto H. York Department of Chemical, Biological and Pharmaceutical Engineering, New Jersey Institute of Technology, Newark, NJ

Experimental results are shown on a proposed technique to enhance the fluidization behavior of nanoparticle fluidized beds. This technique consists of the application of an alternating electric field to the nanofluidized bed. Nanoparticle agglomerates are naturally charged by contact and tribo-charming mechanisms and are therefore agitated by the action of the externally applied field. This agitation contributes to the increased drag on the agglomerates, and thus enhances gas-solid mixing. Three different electric field configurations have been tested: co-flow field, cross-flow field, and variable field configurations. According to our observations, the best results are obtained for the variable field configuration. In this arrangement, the electric field strength is higher at the bottom of the fluidized bed. The larger agglomerates, which usually sink to the bottom during fluidization due to stratification, become the most agitated. In addition, the weaker electric field at the top of the fluidized bed causes the smaller agglomerates near the free surface to be weakly excited. Consequently, fluidization is greatly enhanced while excessive elutriation is avoided. It is demonstrated that this technique is even suitable to achieve expanded fluidization states of unsieved nanopwder samples even though the fluidization state returns to be heterogeneous after the electric field is turned off.

See more of this Session: Fluidization and Handling of Submicron and Nano Particles
See more of this Group/Topical: Particle Technology Forum

Atomic Layer Deposition of Functional Coatings On Nanoparticles Using a Micro-Jet Assisted Fluidized Bed Reactor

David M. King, Department of Chemical Engineering, University of Colorado, Boulder, CO

J. Ruud Van Ommen, Delft ChemTech, Delft University of Technology, Delft, Netherlands

Robert Pfeffer, Chemical Engineering, Arizona State University, Tempe, AZ

Alan W. Weimer, Department of Chemical Engineering, University of Colorado, Boulder, CO

The gas-phase fluidization behavior of nanoparticles has been rapidly developing in recent years. Atomic layer deposition (ALD) is a gas-phase reactive process by which nanoscale functional layers can be chemically bonded to the surfaces of fine particles. ALD is typically performed in stainless steel fluidized bed reactors at temperatures between 100 and 500°C, and under vacuum conditions (~1 Torr). It is well known that shear forces must be imparted to the nanoparticles in order to overcome attractive interparticle forces and break apart large nanoagglomerates which form due to van der Waals and other attractive forces.. Previously, mechanical agitation was delivered to the ALD fluidized bed reactor using a vibrating tray or a stirring unit. A high-pressure micro-nozzle within the bed allows for the delivery of large shear forces and enables the use of glass reactors to monitor the fluidization behavior of powders during ALD processes. Nozzle diameter, pressure and relative flow rate were studied at a variety of conditions to optimize nanoparticle fluidization behavior in the presence of reactive precursors. Since metal oxides are typically deposited using water as an oxygenation source, an in-depth study of the temperature-dependent water removal rate from a jet-fluidized bed reactor is presented. The temperature-dependent implications are discussed in the context of metal-containing precursors that may undergo thermal decomposition at elevated temperatures. The transient fluidization behavior of nanoparticles in the presence of surface reactions is presented for the first time.

See more of this Session: Fluidization and Handling of Submicron and Nano Particles
See more of this Group/Topical: Particle Technology Forum

Effect of the Micro-Packing Status On the Bulk Behaviour of Loose Agglomerated Nano-Powders

Guangsheng Gu, Dept. of Chemical Engineering, Tsinghua University (Current affiliation: Rice Univ), Beijing, China

Yao Wang, Dept. of Chemical Engineering, Tsinghua University, Beijing, China

Fei Wei, Dept. of Chemical Engineering, Tsinghua University, Beijing, China

Nano-agglomerated Fluidized Bed (NAFB) has been extensively studied in recent years, in which nano-powder is the key factor. Nano-powders are classified as belonging to Geldart-C group. Some of them show good agglomerate flow or fluidization behaviour, while many others are difficult to flow. The Geldart classification, which separates powders by the density and average diameter of the primary particles, is not general enough to explain the diversity. The effect of the micro-packing status (MPS) and inter-agglomerate forces (van der Waals force) are analyzed in this work. The results show that both the fluidization behaviour and the maximum angle of stability of nano-powders are related to the MPS, i.e., whether nano-powders behave like Geldart-C group or not is dependent on MPS. This is helpful to understand the bulk behaviour of nano powders more correctly, and could be considered as an extension of the Geldart classification.

See more of this Session: Fluidization and Handling of Submicron and Nano Particles
See more of this Group/Topical: Particle Technology Forum

Nanostructured and Nanophased Materials for Army Weapons Systems

Paul Redner, Energetics and Warheads Division, US Army RDECOM-ARDEC, Picatinny, NJ

Deepak Kapoor, Materials Division, US Army RDECOM-ARDEC, Picatinny, NJ

Chris Haines, US Army RDECOM-ARDEC, Picatinny, NJ

Darold Martin, US Army RDECOM-ARDEC, Picatinny, NJ

Joseph Paras, US Army RDECOM-ARDEC, Picatinny, NJ

Ryan Carpenter, US Army RDECOM-ARDEC, Picatinny, NJ

The potential application of nanotechnology and advanced materials in both DoD weapons systems and commercial products is great. ARDEC is exploring a wide variety of materials and technologies to solve current technology gaps. Addressing these gaps will enhance lethality and survivability of the Warfigher. ARDEC is combining internal capabilities with those of key industry and academic partners to address these gaps.

This talk will describe some of the recent research and successes under a variety of efforts. Nanocomposite superthermites, the replacement of single-point failure materials, the use of nanoporous silicon as an alternative to nano-aluminum and near-term and mid-term goals to develop new materials for the Energetics, Reactive Materials, Armor/Anti-armor and Power & Energy product areas will be discussed.

See more of this Session: Nano-Energetic Materials
See more of this Group/Topical: Particle Technology Forum

Aluminum-Metal Reactive Nanocomposite Materials

Yasmine Aly, Dept. of Chemical, Biological, and Pharmaceutical Engineering, New Jersey Institute of Technology, Newark, NJ

Mirko Schoenitz, Dept. of Chemical, Biological, and Pharmaceutical Engineering, New Jersey Institute of Technology, Newark, NJ

Edward L. Dreizin, Chemical, Biological, and Pharmaceutical Engineering, New Jersey Institute of Technology, Newark, NJ

Multiple Al-based nanocomposite reactive materials have been explored recently, including thermite compositions, coated powders, and multilayered foils. In most cases, a highly exothermic reaction is expected to occur between aluminum and one of the additional components resulting in an accelerated ignition. However, adding solid oxidizers inevitably reduces the overall energy density of the metal fuel additives. Aluminum combustion was also shown to be enhanced when it is alloyed or mechanically alloyed with titanium and magnesium. Exothermic formation of an Al-Ti intermetallic phases and selective oxidation of Mg were proposed to increase the overall reactivity of the respective alloys. A new type of Al-based reactive nanocomposite materials is explored in this project. In these materials, aluminum is mixed on the nanoscale, but not alloyed, with a metal, rather than oxide. The materials prepared and characterized are micron-sized powders in which each particle has aluminum matrix and nano-sized inclusions of the metal additive. The bulk compositions are aluminum-rich to take advantage of the high aluminum combustion enthalpy. Such materials have higher energy density than nanocomposite thermites with similar morphology and are expected to be more reactive than alloyed powders with the same bulk compositions. It is expected that the reactivity of such composite materials will be improved compared to both pure aluminum and to its homogeneous alloys because of several reasons. Early selective oxidation of the added metal can offer an energy boost during the heating of respective particles. An exothermic intermetallic reaction can occur; this reaction would be particularly important for such systems as Al-Ni and others with relatively high reaction enthalpy. The kinetics of aluminum oxidation, controlled by the growth kinetics for various aluminum oxide polymorphs are also expected to change as a result of the combined oxidation processes of different metals and because of formation of ternary oxides, less protective, than pure alumina. Finally, an increase in the surface area of the reactive interface, including multiple grain boundaries in the prepared nanocomposite materials is expected to increase the rate of their heterogeneous oxidation. Different mechanisms are expected to improve reactivity of Al-metal nanocomposite materials for different metal additives. In this project, nanocomposite materials in the systems of Al-Zn, Al-Fe, and Al-Ni were prepared by mechanical milling of aluminum powders mixed with powders of the respective metal additives. For each material, structure and composition were characterized using electron microscopy and x-ray diffraction. Reactivity of the prepared powders was studied using thermal analysis and heated filament ignition experiments. It was observed that oxidation kinetics of the prepared aluminum-metal reactive nanocomposite materials differ substantially from the kinetics of aluminum oxidation. It was also observed that reduced ignition temperatures are achieved for different Al-metal nanocomposite materials. Detailed studies of the reaction processes are carried out to establish mechanisms of accelerated oxidation kinetics and reduced ignition temperatures for different materials.

See more of this Session: Nano-Energetic Materials
See more of this Group/Topical: Particle Technology Forum

Mechanisms of Multistep Reactions in Nanocomposite Metal-Oxide Systems

Alexandre Ermoline, Chemical, Pharmaceutical, and Biological Engineering, New Jersey Institute of Technology, Newark, NJ

Mirko Schoenitz, Dept. of Chemical, Biological, and Pharmaceutical Engineering, New Jersey Institute of Technology, Newark, NJ

Edward L. Dreizin, Chemical, Biological, and Pharmaceutical Engineering, New Jersey Institute of Technology, Newark, NJ

Nanocomposite thermites received substantial attention due to their increased specific reactive interface and, therefore, higher reactivity compared to regular thermite powders. Such materials with reactive components mixed on the nanometer scale are expected to increase performance characteristics of the current metalized energetic formulations in pyrotechnics, explosives and propellants. A great deal of experimental research has been conducted on preparation of different materials systems and characterizing their combustion performance. In particular, many compositions (e.g., Al-MoO3, Al-CuO, Al-Bi2O3 and others) were produced using arrested reaction milling (ARM). The structure, morphology, and reactivity of these materials were characterized using electron microscopy, X-ray diffraction analysis, Differential Scanning Calorimetry and Thermo-Gravimetric Analysis (DSC/TGA), as well as customized ignition and combustion tests. Data on ignition of nanocomposite thermites at various heating rates and in different environments, and measurements describing their combustion dynamics have been collected. The experiments showed that these materials are useful as rapidly igniting, high energy density additives to various energetic compositions, generating hot spots to promote the reaction of the entire formulation. Thus, detailed description of their ignition kinetics is of significant interest. However, current theoretical models of reaction kinetics in thermites are quite limited. In most cases, the model is reduced to a single Arrhenius-like term. Models including several independent kinetic terms have also been proposed, but are unlikely to accurately describe the multistep reactions over a wide range of reaction rates, occurring when a material is being thermally activated. An alternative approach consists of a mechanistic description of various processes occurring in the thermally activated materials, and accounting for the effect of these processes on the material temperature history leading to ignition. Recent research on ignition of aluminum, a common component of many nanocomposite thermites, showed that the rates of diffusion through different Al2O3 polymorphs and rates of polymorphic phase transitions control aluminum ignition kinetics. In nanocomposite thermites these processes in the alumina, which is the reaction product, are also expected to be important. The situation is further complicated by structural and compositional changes occurring in the oxidizer, the metal oxide component of the thermite. These changes affect primarily the concentration of oxygen ions available to diffuse through the aluminum oxide to react with aluminum. Recently, a modeling approach incorporating many of the above mentioned factors was discussed for the Al-MoO3 system. The model considered reduction of MoO3 to form various MoOx oxides (x<3) and diffusion of oxygen ions to aluminum through different, evolving layers of aluminum oxide polymorphs. Similar processes, i.e., diffusion through various alumina polymorphs and step-wise reduction of the oxidizer, such as CuO, Bi2O3, Fe2O3, and others, control reaction rates in respective Al-metal oxide systems. The current work develops the modeling effort describing these two types of processes as defining the rates of heterogeneous thermite reactions. Kinetic parameters, available from the literature as well as from recent ignition and DSC/TGA measurements, are analyzed and selected for use in the model. Two types of nanocomposite structures are explored: planar alternate layers of aluminum and molybdenum oxides, and spherical inclusions of molybdenum oxide domains in aluminum matrix. Effect of size distribution of the planar layers and spherical domains on the ignition kinetics is considered. The results are compared with DSC/TGA measurements performed at different heating rates.

See more of this Session: Nano-Energetic Materials
See more of this Group/Topical: Particle Technology Forum

Mechanism for Rapid Mechanical Activation of Gasless High-Density Energetic Mixtures

Jeremiah D.E. White, Department of Chemical and Biomolecular Engineering, University of Notre Dame du Lac, Notre Dame, IN

Robert V. Reeves, School of Mechanical Engineering, School of Aeronautics & Astronautics, Purdue University, West Lafayette, IN

Steven F. Son, School of Mechanical Engineering, School of Aeronautics & Astronautics, Purdue University, West Lafayette, IN

Alexander S. Mukasyan, Department of Chemical and Biomolecular Engineering, University of Notre Dame, Notre Dame, IN

The ignition mechanism and ability to control the initiation and combustion processes in advanced heterogeneous high-density energetic mixtures (HDEMs) is of particular interest because of a variety of potential applications, including chemical energy storage and micro-scale energetic systems, materials synthesis, as well as propellants in rocket, air-breathing and other energetic formulations. In this work, HDEMs of metal-based compositions are in specific focus. The ignition (or so-called thermal explosion, TE) characteristics of such systems have been extensively investigated during the last ten years both in inert and reactive atmospheres. In gasless heterogeneous systems, the ignition temperature is typically related to specific features of the system phase diagram. For example, the eutectic melting point (or m.p. of the least refractory reagent), at which a significant increase in reactant contact area occurs, typically determines the system ignition temperature. This conclusion presents challenges to control the ignition characteristics for such gasless HDEMs.

The influence of rapid high-energy ball milling on the ignition characteristics of a gasless heterogeneous reactive system, i.e. Ni-Al, has been investigated. It was shown that 5-15 minutes of such mechanical treatment leads to a significant decrease of ignition temperature, which appears to be well below the eutectic in the considered binary system. Thus the thermal explosion process is defined by solid-state reactions. Several hypotheses have been suggested to explain this effect, i.e. mechanical activation (MA) of chemical reaction. One is related to a substantial increase of contact area between reactive particles and fresh inter-phase boundaries formed in inert atmosphere ball milling. A second considers a substantial decrease in effective activation energy of interaction between the reactants. A third takes into account the formation of non-equilibrium solid solutions. Finally, theories suggest that defects and strain imparted on the crystalline structure are the driving force.

In this work a set of advanced analytical tools, such as High-Speed Thermal Imaging, Time-Resolved X-ray Diffraction (TRXRD), and High-Temperature Eletcro-Thermal Analysis was used in order to address the question about the contribution of the aforementioned hypotheses on the apparent mechanism of MA in such systems.

See more of this Session: Nano-Energetic Materials
See more of this Group/Topical: Particle Technology Forum

The Effect of Particle Morphology on the Combustion Properties of MIC Materials

Zac Doorenbos[a], Jan Puszynski[a], and Deepak Kapoor[b]
[a]Chemical and Biological Engineering Department, Rapid City, SD 57701
[b]Armament Research, Development and Engineering Center, Picatinny Arsenal, NJ 07806

Abstract

Lamellar and spherical aluminum nanoparticles with a specific surface area of 26 m^2/g and a reactive aluminum content of approximately 76% were mixed with metal oxides, Bi_2O_3 and Fe_2O_3, to form Metastable Intermolecular Composites (MIC). The lamellar aluminum was generated using a wet attrition process developed at the South Dakota School of Mines and Technology in conjunction with Innovative Materials and Processes. This study investigates the effects of fuel particle morphology on combustion velocity, pressure generation, recoil force and electrostatic discharge on MIC materials. The use of lamellar aluminum has shown an increase in the ESD ignition threshold over spherical aluminum in testing of loose powder MIC materials. Results from combustion experiments of MIC materials using 80 nm and lamellar aluminum nanoparticles showed comparable results.

Introduction

In recent years, research in the area of Metastable Intermolecular Composite (MIC) materials has focused on the use of nano-scale reactants to modify the combustion characteristics. By decreasing the particle size of the fuel and/or oxidizer from micron to nano scale the combustion velocity can be increase from a few meters to tens or hundreds of meters per second depending on the type of MIC material and its components [1]. Most MIC materials use nanometer size aluminum powder with the most common being 80 nanometer spherical aluminum particles. The aluminum is mixed with oxides such as molybdenum trioxide, bismuth trioxide, copper oxide, or iron oxide. The cost of 80 nm spherical aluminum powder is very high and its market availability is limited. The physical properties of 80 nm aluminum include a specific surface area in the range of 22-24 m^2/g and a reactive aluminum content of approximately 76-80%. The increase in the combustion velocity of MIC materials is due to the increase in interfacial surface area that is provided by the nanoparticles. Recent publications have shown a comparison of various MIC materials and the effects of fuel and oxidizer particle size along with the addition of binders on the combustion properties [2-3].

MIC materials such as aluminum-bismuth trioxide, aluminum-copper oxide, and aluminum-molybdenum trioxide are very sensitive to electrostatic discharge (ESD) when in a loose powder form. A study on the investigation of ignition characteristics of MIC materials was conducted by Bulian et al.[4]. They showed that the use of flake aluminum instead of spherical aluminum increased the ESD ignition energy threshold from 0.125 μJ to approximately 100 μJ [4] for loose powder.

Dreizin et al. from New Jersey Institute of Technology (NJIT) has been performing arrested reactive milling of MIC materials [5-6] for the formation of lamellar structures. In this case though both the fuel and oxidizer were flake like in structure. Arrested reactive milling uses ball milling to increase the surface area of both the fuel (Al) and oxidizer (MoO_3) [5-6]. These researchers were able to generate fully dense Al-MoO_3 MIC materials using micron size reactants that have combustion properties similar to those of MIC materials using nanoscale reactants. An alternative route for the formation of aluminum with similar specific surface area and reactive aluminum content to 80 nm spherical aluminum has been investigated at the South Dakota School of Mines and Technology in conjunction with Innovative Materials and Processes (IMP). This nano-size aluminum with a lamellar structure has been formed using attritor. The comparison of combustion characteristics such as, recoil force, pressure generation and combustion velocities for MIC materials derived using either spherical nano- or flake- aluminum mixed with Bi_2O_3 or Fe_2O_3 will be discussed in this paper.

Experimental

Milling of Aluminum Flakes

Milling of the aluminum precursor to generate flakes was conducted using a laboratory scale attrition mill (Model 01HDDM purchase from Union Process) in a 110 or 750 cc milling container. Several different precursors were investigated, including micron size aluminum powder, but the best material turned out to be uncoated aluminum flakes with the average size of 11 micron purchased from Alfa Aesar. The aluminum flakes were milled using 0.4 mm zirconia grinding media in anhydrous IPA for several hours. The specific surface area of aluminum flakes can be tuned by adjusting the following milling parameters: total milling time, grinding media size, grinding media to aluminum weight ratio in the milling container and the shaft speed of the attrition mill. Anhydrous IPA must be used to ensure that the reactive aluminum content stays above 75%.

Mixing of MIC Materials

The following reactants and solvents were used in the preparation of Al/Bi_2O_3 and Al/Fe_2O_3 mixtures: a) 80 nm aluminum (NovaCentrix), b) flake aluminum (IMP), c) 20 - 40 nm Fe_2O_3 (Alfa Aesar), d) Bi_2O_3 (Alfa Aesar 3-4m^2/g), e) isopropyl alcohol (IPA) and f) acetone

All of the MIC materials were mixed using a combination of ultrasonic and vortex mixing with acetone or IPA as the solvent. The mixing procedure is as follows: 1) weigh out desired amount of aluminum and add to vial, 2) add 1-2 ml of solvent (dependent on amount of aluminum), 3) sonicate for 10 minutes using a sonic bath, 4) add desired amount of oxidizer (Fe_2O_3 or Bi_2O_3) and additional solvent, 5) sonicate for 10 minutes, 6) mix for 10 minutes using vortex mixer at a speed of 3000, 7) pour mixture into grounded stainless steel pan for drying.

The optimum ratio of fuel to oxidizer for each of the MIC material was previously determined [1, 3]. For the Al/Fe_2O_3 mixture the optimum ratio of Al to Fe_2O_3 is stoichiometric whereas for the Al/Bi_2O_3 system the optimum mixture is between 14-15 wt% aluminum. When calculating the amount of aluminum, the oxide layer that is present on the surface of the aluminum must be taken into account. This correction increases the aluminum concentration from 25 wt% to 31 wt% based on 76% reactive aluminum.

Confined Combustion Velocity Measurement

The combustion velocity of the Al/Fe_2O_3 material was measured by packing an acrylic tube (1/8" ID, 1.5" Length) with approximately 150 mg of Al/Fe_2O_3 for a packing density of approximately 12%. The packed tube was then inserted into the acrylic block shown in Figure 1 so that the end opposite to ignition can be closed off. By closing off the end opposite ignition un-reacted material cannot be ejected from the tube due to a buildup in pressure during combustion. The combustion of the MIC material was recorded using a high speed camera (Fastcam Ultima APX) recording at 30,000 fps. This setup cannot be used to measure the combustion velocity of Al-Bi_2O_3 due to the high speed and large amount of pressure that is generated. Our devices is not able to handle the pressure that are generated by Al-Bi_2O_3 and at the recording speed needed to measure the combustion velocity of a few frames can be recorded which does not allow for accurate measurement.

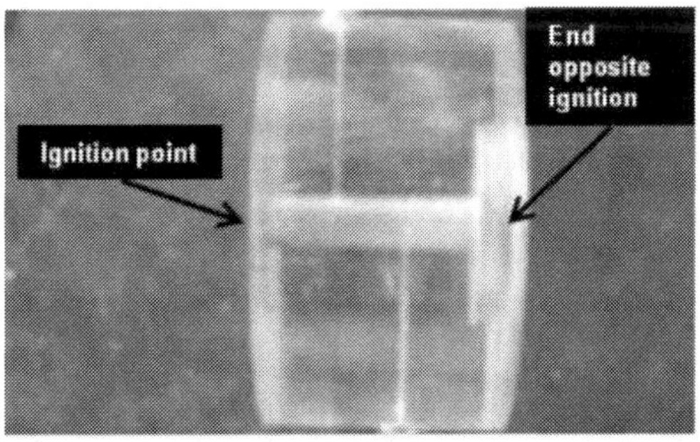

Figure 1: *Photograph of acrylic block use to hold 1/8" tube packed with MIC material for combustion velocity tests.*

Measurement of pressure generation

In certain applications, a generation of pressure during MIC combustion is important. For comparison of the flake and spherical aluminum Bi_2O_3 the pressure generation was measured by ignition of 50 mg of MIC material in a 2.77 cc stainless steel chamber. A schematic of the setup is shown in Figure 2. Pressure generation and recoil force was not measured for the Al-Fe_2O_3 mixtures since this reacting system does not generate any appreciable gas quantity. The pressure was measured using a piezoelectric pressure transducer with a range of 0 – 5000 psi. The signal was recorded using an oscilloscope.

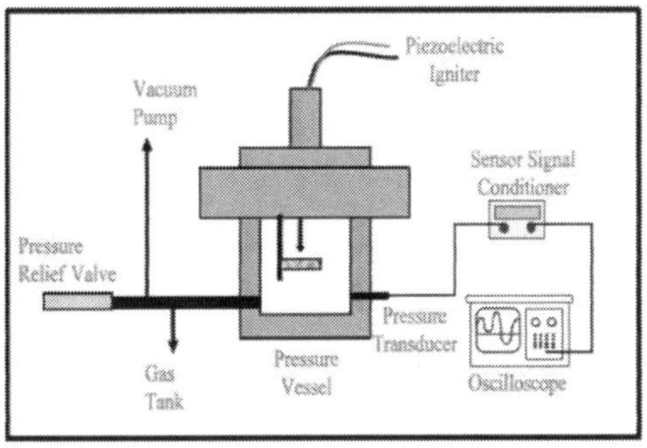

Figure 2: *Schematic of a constant volume pressure cell.*

Measurement of recoil force

Recoil force measurements were conducted on flake Al-Bi_2O_3 and 80 nm Al-Bi_2O_3 MIC mixtures for comparison of generated specific impulses. The recoil force measurement setup is shown in Figure 3 and uses a force sensor with a range of 0 to 22,000 Newton purchased from Kistler (Model 9212 High Impedance Load Cell).

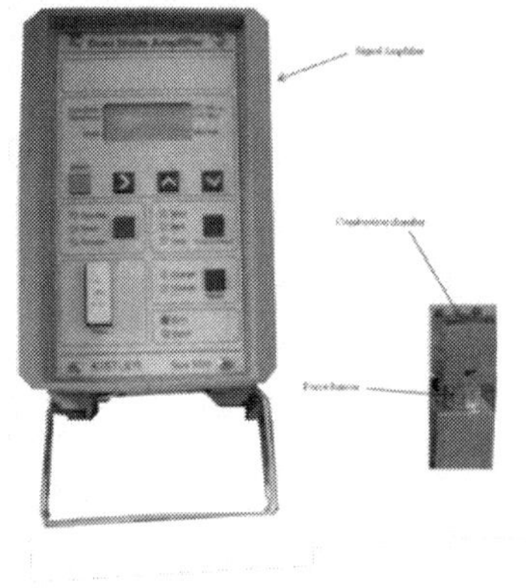

Figure 3: *Recoil force measurement setup.*

The MIC material (5 – 50 mg) was loaded into the combustion chamber (¼" diameter by ¼" deep) as a loose powder and ignited using a NiCr ignition wire. The signal amplifier was attached to an oscilloscope where the signal was recorded.

Electrostatic Discharge (ESD) Testing

MIC materials can be exposed to electrostatic discharge during processing and handling for certain applications. ESD testing of the materials was done in a powder form. Testing has shown that the consolidation or addition of coatings can reduced the ESD sensitivity of the MIC materials. Testing was conducted to determine what is the effect particle morphology has on the ESD sensitivity of MIC materials using the setup shown in Figure 4 purchased from Electrotech Systems, Inc.

Figure 4: *ESD discharge gun model 930D-FTS.*

This ESD discharge gun has interchangeable capacitors ranging from 0.1 to 20 nano-Farads (nF) with a range of 0 to 25 kV depending on the capacitor in the gun. The duration time of the discharge can be increased by the addition of a 1.5 or 5.0 kΩ resistance. The powder samples were tested twenty times by discharging electricity at each energy level. If no ignition occurred, the energy was increased and the sample was retested. This was repeated until ignition occurred. Once ignition was detected, a second identical experiment was conducted to verify the ignition energy threshold. The ignition energy was calculated from $E=CV^2/2$, where E is the energy in Joules, C is the capacitance in Farads, and V is the voltage in Volts.

Results and Discussion

Flake Aluminum

The generation of flake aluminum with a specific surface area of at least 26 m^2/g was accomplished using an attrition mill. The milling parameters that affect the surface area of the end product were discussed in the experimental section with the main parameter being the milling time. Experiments have shown that as the size of the grinding media is increased the milling time to reach the desired surface area increases. Figure 5 shows the effect of milling time on the specific surface area and, the reactive aluminum content of the aluminum flakes.

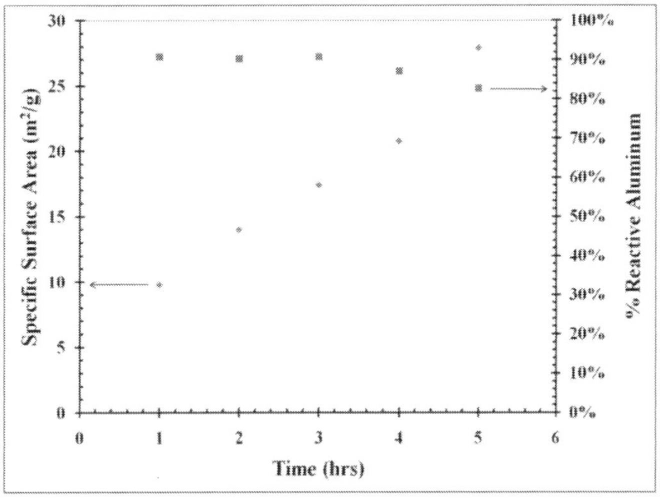

Figure 5: *Specific surface area and percent reactive aluminum versus milling time.*

The aluminum that was generated using this milling method had a specific surface area of 26.3 m^2/g as determined by BET analysis. The reactive aluminum content was approximately 76% as determined by a volumetric method. SEM images of the flake aluminum after milling are shown in Figure 6.

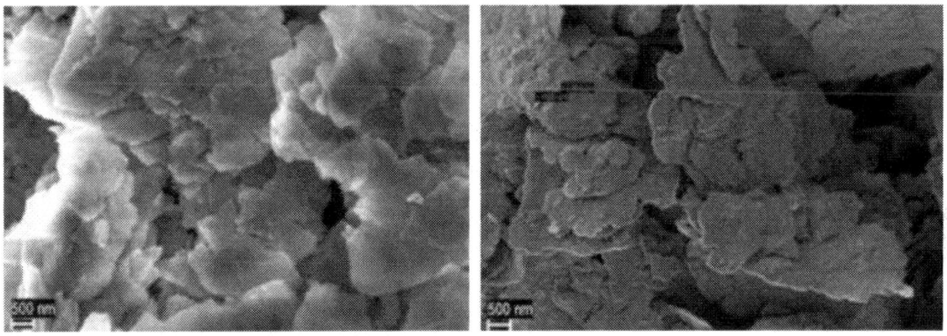

Figure 6: *SEM image of milled aluminum exhibiting flake like morphology.*

Combustion velocity

Due to the high combustion front velocities in the Al-Bi_2O_3 system, which may range from 600 to 800 m/s depending on any additional coatings of the aluminum[3], this system was not tested due to recording camera limitations. The combustion velocity of Al-Fe_2O_3 was measured in a tube with one end close as described before, using the existing high speed camera. The confined combustion velocity of the 80 nm Al mixed with 20-40 nm Fe_2O_3 was determined to be between 38 – 40 m/s. The flake aluminum iron oxide mixture showed a combustion velocity of approximately 35 m/s.

Pressure generation and recoil force

Pressure generation curves for flake and 80 nm Al mixed with Bi_2O_3 are shown in Figure 7. Even though the flake Al-Bi_2O_3 mixtures shows a higher maximum pressure generation (average of 6.43 psi/mg MIC) then the 80 nm Al-Bi_2O_3 (average of 5.05 psi/mg MIC) for the same amount of material, the 80 nm Al-Bi_2O_3 system shows a higher initial pressure gradient after ignition. This faster response could indicate that the reaction between the 80 nm Al and Bi_2O_3 has a higher reaction rate than the flake Al and Bi_2O_3 system. More investigation is needed to determine if the reaction is due to inadequate mixing of the flake Al and Bi_2O_3 nanopowder.

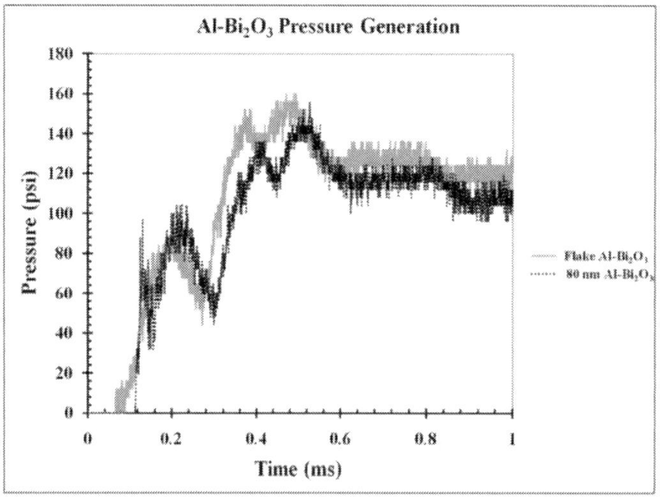

Figure 7: *Dynamic pressure response after ignition of approximately 50 mg of flake Al-Bi_2O_3 and 80 nm Al-Bi_2O_3 MIC materials in a 2.77 cc chamber.*

These reacting systems were also tested using a recoil force setup. The results are presented in Table 1.

Table 1: *Recoil force measurements of flake Al-Bi_2O_3 and 80 nm Al-Bi_2O_3.*

	Flake Al-Bi_2O_3 MIC			80 nm Al-Bi_2O_3 MIC		
Exp	Weight of MIC material (g)	Recoil Force (N)	N/mg	Weight of MIC material (g)	Recoil Force (N)	N/mg
1	0.0068	68	10.00	0.0066	106	16.06
2	0.0088	100	11.36	0.0076	92	12.11
3	0.0053	51	9.62	0.0099	161	16.26
4	0.0088	91	10.34	0.0111	136	12.25
5	0.0056	45	8.04	0.0065	86	13.23
		Average	9.87		Average	13.98

In this case the 80 nm Al-Bi_2O_3 system has shown a higher level of recoil force than the flake Al-Bi_2O_3 system. The recoil force of the aluminum iron oxide system was not measured at this time as it does not generate any appreciable gas.

Conclusions

Results have shown that it is possible to generate similar combustion characteristics for MIC material using a fuel with a different particle morphology but similar specific surface area. ESD testing has shown that for 80 nm Al-Bi_2O_3 MIC in powder form has an ignition threshold of 0.2 mJ, whereas, the flake Al-Bi_2O_3 MIC materials shows an ignition threshold between 0.389 – 0.45 mJ (all measurement conducted on loose powder). Therefore by changing the particle morphology of the fuel in MIC material similar combustion characteristics can be achieved with material that is less sensitive to ESD. More investigation into the mixing of the fuel and oxidizer especially those containing aluminum nanoflakes will be conducted. The effect of flake aluminum concentration of the MIC material's combustion characteristics needs to be more thoroughly investigated.

Acknowledgements

The authors of this paper acknowledge the financial support from the Army Research Development Engineering Center in Picatinny Arsenal, NJ (Contract No. WI5QKN-06-D-0006).

References

1. Bulian, C.J., T.T. Kerr, and J.A. Puszynski. *Ignition Studies of Aluminum and Metal Oxide Nanopowders*. in *Proceedings of the 31st International Pyrotechnics Seminar*. 2004. Ft. Collins, CO.
2. Sanders, V.E., et al., *Reaction Propagation of Four Nanoscale Energetic Composites (Al/MoO_3, Al/WO_3, Al/CuO, and Bi_2O_3)*. Journal of Propulsion and Power, 2007. **23**(4): p. 707-714.
3. Puszynski, J.A., C.J. Bulian, and J.J. Swiatkiewicz, *Processing and Ignition Characteristics of Aluminum-Bismuth Trioxide Nanothermite System*. Journal of Propulsion and Power, 2007. **23**(4): p. 698-706.
4. Bulian, C.J., J.A. Puszynski, and J.J. Swiatkiwicz. *The Effect of Processing Parameters on Ignition Characteristics and ESD Sensitivity of Nanothermites*. in *35th Internation Pyrotechnics Seminar*. 2008. Fort Collins, CO.
5. Umbrajkar, S.M., et al., *Aluminum-Riich Al-MoO_3 Nanocomposite Powders Prepared by Arrested Reactive Milling*. Journal of Propulsion and Power, 2008. **24**(2): p. 192-198.
6. Umbrajkar, S., et al., *Arrested Reactive Milling Synthesis and Characterization of Sodium-Nitrate Based Reactive Composites*. Propellants, Explosives, Pyrotechnics, 2007. **32**(1): p. 32-41.

Aluminum/Nickel (Al/Ni) Heterogeneous Nanostructures: Synthesis, Characterization, and Nano-Heater Applications

Qingzhou Cui, Chemical Engineering, University of Massachusetts Lowell, Lowell, MA

Jacqueline Pelealuw, Mechanical Engineering, University of Massachusetts Lowell, Lowell, MA

Julie Chen, Mechanical Engineering, University of Massachusetts Lowell, Lowell, MA

Zhiyong Gu, Chemical Engineering, University of Massachusetts Lowell, Lowell, MA

Teichi Ando, Mechanical and Industrial Engineering, Northeastern University, Boston, MA

Peter Y. Wong, Mechanical Engineering, Tufts University, Medford, MA

Claus G. Rebholz, Mechanical and Manufacturing Engineering, University of Cyprus, Nicosia, Cyprus

Charalabos C. Doumanidis, Mechanical and Manufacturing Engineering, University of Cyprus, Nicosia, Cyprus

Nanoscale heating sources ("nano-heaters") have received great attention in recent years as enabling tools for nanofabrication and nanomanufacturing. In this presentation, we show various nano-heater structures based on heterogeneous Al/Ni nanostructures. Several methods including chemical reduction, electrochemical deposition, ultrasonic powder consolidation, e-beam evaporation, and sputtering are used to fabricate Al/Ni nanostructures including Al-Ni nanowires, Al-Ni core-shell nanoparticles, nanopowder composite films, and multilayered thin films. The characterization of these nanostructures was performed by field emission scanning electron microscopy (FE-SEM), transmission electron microscopy (TEM), energy dispersive X-ray analysis (EDX), X-ray diffraction (XRD), and differential scanning calorimetry (DSC).

A potential application for such Al/Ni nanostructures relies on the alloy forming reaction between aluminum and nickel, which has been shown to release intense heat from its large reaction enthalpies (-37.85 to -71.65 kJ/mol). Thus the nanoscale Al/Ni heterogeneous structures can be used to construct nano-heater structures or devices. Upon ignition, Ni-Al alloys are formed and one-time localized heat is generated from the exothermic reaction process. The advantages for such type of heating sources include complete enthalpies release, versatile ignition methods (electrical, heat, IR, plasmonic induction, microwave, etc.), conductive reaction product, and easy integration with other systems such as lab-on-a-chip. These advantages could lead to many useful applications such as soldering, adhesive droplet, polymer memory devices, lab on a chip, and hyperthermia for cancer treatment.

See more of this Session: Nano-Energetic Materials
See more of this Group/Topical: Particle Technology Forum

Consolidation of Reactive Nanocomposite Powders

Demitrios Stamatis, Chemical Engineering, New Jersey Institute of Technology, Newark, NJ

Xiaoying Zhu, Chemical Engineering, New Jersey Institute of Technology, Newark, NJ

Mirko Schoenitz, Chemical Engineering, New Jersey Institute of Technology, Newark, NJ

Edward L. Dreizin, Chemical Engineering, New Jersey Institute of Technology, Newark, NJ

There is interest in replacing chemically inert structural weapons components with reactive structures capable of highly exothermic reactions. Consolidated reactive materials are also desired for other applications, including reactive fragments, high density additives to explosives, insensitive pyrotechnic components, etc. Unlike nano-energetic compositions based on mixed nanopowders, reactive nanocomposite powders prepared by Arrested Reactive Milling (ARM) can be readily consolidated to achieve combined characteristics of high reactivity, low porosity, and structural strength. Different consolidation methods can be applied, and as a first step, a simple uniaxial pressing of nanocomposite powders is used in this project. A set of reactive nanocomposite powders with several Al-based thermite compositions prepared by ARM was used to prepare pellet-like consolidated samples. For various mechanical tests, both cylindrical and rectangular pellets were prepared with varied dimensions and varied degrees of compaction. Pellet compaction densities exceeding 90% of theoretical maximum density, were achieved. Despite the presence of Al and oxidizers, including MoO_3, CuO, Bi_2O_3 and others, mixed on the nanoscale in different samples, no reaction was observed to be triggered by the powder compaction at pressures reaching 500 MPa. An experimental technique has been developed to study the thermal ignition initiation of the consolidated samples as a function of their physical and mechanical properties. The experimental technique will be used to develop a theoretical model to describe the ignition behavior of the consolidated materials

See more of this Session: Nano-Energetic Materials
See more of this Group/Topical: Particle Technology Forum

AUTHOR INDEX

AUTHOR INDEX

AUTHOR INDEX

AUTHOR INDEX

AUTHOR INDEX